PLANT DISEASES AND THEIR MANAGEMENT

A Sustainable Approach

PLANT DISEASES AND THEIR MANAGEMENT

A Sustainable Approach

Edited by

Deepa Srivastava, PhD

Rajarshi Kumar Gaur, PhD

Ajay K. Tiwari, PhD

AAP | APPLE ACADEMIC PRESS

First edition published 2025

Apple Academic Press Inc.
1265 Goldenrod Circle, NE,
Palm Bay, FL 32905 USA

760 Laurentian Drive, Unit 19,
Burlington, ON L7N 0A4, CANADA

CRC Press
2385 NW Executive Center Drive,
Suite 320, Boca Raton FL 33431

4 Park Square, Milton Park,
Abingdon, Oxon, OX14 4RN UK

Apple Academic Press exclusively co-publishes with CRC Press, an imprint of Taylor & Francis Group, LLC

Library and Archives Canada Cataloguing in Publication

CIP data on file with Canada Library and Archives

Library of Congress Cataloging-in-Publication Data

CIP data on file with US Library of Congress

ISBN: 978-1-77491-608-7 (hbk)
ISBN: 978-1-77491-609-4 (pbk)
ISBN: 978-1-03272-285-6 (ebk)

About the Editors

Deepa Srivastava, PhD
Assistant Professor, Department of Botany, D. D. U. Gorakhpur University, Gorakhpur, Uttar Pradesh, India

Deepa Srivastava, PhD, is currently working as Assistant Professor in Department of Botany, D. D. U. Gorakhpur University, Gorakhpur, Uttar Pradesh, India, where she was formerly a postdoctoral fellow. With excellent academic achievements and many awards, she possesses extensive knowledge in Botany, especially in plant viruses and medicinal plants. Dr. Srivastava has published over 20 papers in national and international journals and has presented over 50 research papers at international and national conferences. She has written more than five books and six book chapters with publications of international repute. She is a fellow of the International Science Congress Association and Indian Botanical Society. She is also a member of the Botanical Society of America and a life member of eight other academic bodies. She did her PhD in Botany (Plant Virology) at the Department of Botany, DDU Gorakhpur University on the topic "Studies on viral diseases on some medicinal plants with special reference to *Catharanthus roseus*" under the supervision of Prof. (Mrs.) K. Shukla in plant virology.

Rajarshi Kumar Gaur, PhD
Professor, Department of Biotechnology, Deen Dayal Upadhyaya Gorakhpur University, Gorakhpur, Uttar Pradesh, India

Rajarshi Kumar Gaur, PhD, is currently working as Professor in the Department of Biotechnology, Deen Dayal Upadhyaya Gorakhpur University, Gorakhpur, Uttar Pradesh, India. He is handling many national and international grants and collaborative projects on plant viruses and disease management. He received a MASHAV fellowship of the Israeli government for his postdoctoral studies and joined The Volcani Centre, Israel, and Ben-Gurion University, Negev, Israel. In 2007, he received a visiting scientist fellowship from the Swedish Institute Fellowship to work at Umeå University, Sweden. He also received a postdoctoral fellowship from the International Centre for Genetic Engineering and Biotechnology, Italy, in 2008. He has made significant contributions on sugarcane viruses and published 130 national and international papers, authored 20 edited books, and presented about 50 papers in national and international conferences. He is a fellow of many associations, including Linnean Society, Royal Society of Biology, Society of Plant Research, Society of Applied Biology, and International Society of Biotechnology. He has received many awards, including the Prof. B.M. Johri Memorial Award; Society of Plant Research (SPR); Excellent Teaching Award from Astha Foundation, Meerut; UGC-Research Teacher Award, Young Scientist Award-2012 in Biotechnology from the Society of Plant Research (SPR), Meerut; and a Scientific & Applied Research Centre Gold Medal Award-2011 for outstanding contribution in the field of biotechnology. He has visited several laboratories of USA, Canada, New Zealand, UK, Thailand, Sweden, and Italy. His PhD was on molecular characterization of sugarcane viruses, viz., mosaic, streak mosaic, and yellow luteovirus.

Ajay K. Tiwari, PhD
Scientific Officer (Seed), UP Council of Sugarcane Research, Shahjahanpur, UP, India

Ajay K. Tiwari, PhD, is a Senior Scientific Officer and Officer-in-charge at UPCSR-Sugarcane Research and Seed Multiplication Center, Gola, Kheri, UP, India. He has published 95 research and review articles in national and international journals. He has also published 15 book chapters in edited books and has authored ten edited books. He has submitted more than 250 nucleotide sequences of plant pathogens to the The GenBank sequence database. He is Managing Editor of *Sugar Tech Journal* (Springer) and Chief Editor of *Agrica Journal*. Dr. Tiwari is a regular reviewer and member of the editorial boards for many international journals. He is a member of the British Society of Plant Pathology, Indian Phytopathological Society, Sugarcane Technologists Association of India, International Society of Sugarcane Technologists, Society of Sugarcane Research and Promotion, Prof H. S. Srivastava Foundation, and Society of Plant Research. He has been awarded a young researcher award in Italy 2011 and a young scientist award by DST-SERB and was nominated for the Narshiman Award by the Indian Phytopathological Society. He was also awarded a young scientist award by the Chief Minister of the State Government of UP for his outstanding contribution in plant pathology. Dr. Tiwari is also the recipient of many international travel awards given by DST, DBT, and CSIR from India; PATHOLUX from Luxembourg; and International Organization for Migration (IOM) from Brazil. He has visited many countries, including China, Italy, Germany, Vietnam, and Thailand for conferences and workshops. He has been involved in research on molecular characterization and management of agricultural plant pathogens for the last 15 years.

Contents

Contributors

Umme Abiha
Indian Institute of Technology, IDRP- Smart Healthcare, Karwar, Jodhpur, Rajasthan, India
Applied Plant Biotechnology Research Lab, Centre for Plant and Environment Biotechnology, Amity Institute of Biotechnology, Amity University, Noida, Uttar Pradesh, India

Emmanuel Asamoah Adjei
CSIR-Crops Research Institute, Kumasi, Ghana

Joseph Adomako
CSIR-Crops Research Institute, Kumasi, Ghana

Richard Yaw Agyare
CSIR-Savannah Agricultural Research Institute, Nyankpala, Ghana

Aqeel Ahmed
Atta-ur-Rahman School of Applied Biosciences (ASAB), National University of Sciences and Technology (NUST), Islamabad, Pakistan

Muhammad Ali
Atta-ur-Rahman School of Applied Biosciences (ASAB), National University of Sciences and Technology (NUST), Islamabad, Pakistan

Alka
Department of Environmental Sciences, Maharshi Dayanand University Rohtak, Haryana

Blankson Amoabeng
CSIR-Crops Research Institute, Kumasi, Ghana

Francis Osei Amoako Andoh
CSIR-Crops Research Institute, Kumasi, Ghana

Pooja Bhardwaj
ICMR-Regional Medical Research Centre, Gorakhpur, India

Muhammad Faraz Bhatti
Atta-Ur-Rahman School of Applied Biosciences (ASAB), National University of Sciences and Technology (NUST), Islamabad, Pakistan

K. M. G. Chanchala
Sugarcane Research Institute, Sri Lanka

Vanita Chandel
Amity Institute of Virology and Immunology, Amity University, Uttar Pradesh, Noida, Uttar Pradesh, India

Monika Chauhan
Division of Forensic Science, School of Basic and Applied Sciences, Galgotias University, Greater Noida, Uttar Pradesh, India

Nguyen Ngoc Bao Chau
Faculty of Biotechnology, Ho Chi Minh City Open University, Vietnam

Diwakar Chauhan
Division of Chemistry, School of Basic and Applied Sciences, Galgotias University, Greater Noida, Uttar Pradesh, India

Joseph Adjebeng-Danquah
CSIR-Savannah Agricultural Research Institute, Nyankpala, Ghana

Hirawati Deval
ICMR-Regional Medical Research Centre, Gorakhpur, Uttar Pradesh, India

Geeta Dhania
Department of Environmental Sciences, Maharshi Dayanand University Rohtak, Haryana

Karuna Dhiman
College of Horticulture and Forestry, Neri, Hamirpur, Himachal Pradesh, India

Tehsin Fatma
Atta-ur-Rahman School of Applied Biosciences (ASAB), National University of Sciences and Technology (NUST), Islamabad, Pakistan

Sidra Fatima
Atta-Ur-Rahman School of Applied Biosciences (ASAB), National University of Sciences and Technology (NUST), Islamabad, Pakistan

Lovlish Gupta
Division of Forensic Science, School of Basic and Applied Sciences, Galgotias University, Greater Noida, Uttar Pradesh, India

Akash Jangid
School of Agriculture, Lovely Professional University, Phagwara, Punjab, India

Monia Jemni
Regional Research Center in Oasis Agriculture, Degache, Tunisia

Rajni Kant
ICMR-Regional Medical Research Centre, Gorakhpur

Malavika Ram Amanthra Keloth
School of Agriculture, Lovely Professional University, Phagwara, Punjab, India

Ravindra Kumar
ICAR-Indian Agricultural Research Institute, Regional Station, Karnal, Haryana, India

Neha Kaushal
Dr YS Parmar University of Horticulture and Forestry, Nauni, Himachal Pradesh, India

Haris Ahmed Khan
Atta-Ur-Rahman School of Applied Biosciences (ASAB), National University of Sciences and Technology (NUST), Islamabad, Pakistan

Ajay Kumar
Division of Life Science, School of Basic and Applied Sciences, Galgotias University, Greater Noida, Uttar Pradesh, India

Sachin Kumar
Dr BR Ambedkar NIT Jhalandhar, Punjab, India

Sihem BEN Maachia
Regional Research Center in Oasis Agriculture, Degache, Tunisia

Niloufar Mahmoudi
Department of AgroBiotechnology, Peoples' Friendship University of Russia (RUDN University), Moscow, Russia

Amber Malik
Atta-Ur-Rahman School of Applied Biosciences (ASAB), National University of Sciences and Technology (NUST), Islamabad, Pakistan

Raghavendra Reddy Manda
Wageningen University and Research, Wageningen, The Netherlands

Leander Dede Melomey
WACCI-University of Ghana, Legon
West Africa Centre for Crop Improvement (WACCI)-University of Ghana, Ghana

Shweta Meshram
School of Agriculture, Lovely Professional University, Phagwara, Punjab, India

Tulika Mishra
Assistant Professor, Department of Botany, D.D.U Gorakhpur University

Ahmed Namsi
Regional Research Center in Oasis Agriculture, Degache, Tunisia

Yousef Naserzadeh
Department of AgroBiotechnology, Peoples' Friendship University of Russia (RUDN University), Moscow, Russia

Faustina Okyere
CSIR-Crops Research Institute, Kumasi, Ghana

Jacinta Adoma Opoku
CSIR-Crops Research Institute, Kumasi, Ghana

Allen Oppong
CSIR-Crops Research Institute, Kumasi, Ghana

Michael Kwabena Osei
CSIR-Crops Research Institute, Kumasi, Ghana

Isaac Osei-Bonsu
CSIR-Crops Research Institute, Kumasi, Ghana

Shail Pande
Principal DAV Degree College, Gorakhpur, India

Ruth Prempeh
CSIR-Crops Research Institute, Kumasi, Ghana

Meenakshi Rana
School of Agriculture, Lovely Professional University, Phagwara, Punjab, India

Neha Rauniyar
Department of Botany, Deen Dayal Upadhyay Gorakhpur University, Gorakhpur, Uttar Pradesh, India

Narayan Rishi
Amity Institute of Virology and Immunology, Amity University Uttar Pradesh, Noida, Uttar Pradesh, India

Habiba Saeed
Atta-Ur-Rahman School of Applied Biosciences (ASAB), National University of Sciences and Technology (NUST), Islamabad, Pakistan

Adarsh Sharma
School of Agriculture, Lovely Professional University, Phagwara, Punjab, India

Ankita Sharma
Dr YS Parmar University of Horticulture and Forestry, Nauni, Himachal Pradesh, India

Pallavi Sharma
Department of Botany, D.D.U Gorakhpur University, Gorakhpur, Uttar Pradesh, India

Poonam Pandurang Shete
School of Agriculture, Lovely Professional University, Phagwara, Punjab, India

Shardulya Shukla
Medicinal and Aromatic Plant Division, DIHAR-DRDO, Chandigarh, India

Susmita Shukla
Applied Plant Biotechnology Research Lab, Centre for Plant and Environment Biotechnology Amity Institute of Biotechnology, Amity University, Noida, Uttar Pradesh, India

Vinit Pratap Singh
College of Agriculture, Azamgarh Campus, Acharya Narendra Dev University of Agriculture and Technology, Faizabad, Uttar Pradesh, India

Neha Kumari Sonkar
Research Scholar, Department of Botany, D.D.U Gorakhpur University, Uttar Pradesh, India

Deepa Srivastava
Department of botany, Deen Dayal Upadhyay Gorakhpur University, Gorakhpur, Uttar Pradesh, India

Neha Srivastava
ICMR-Regional Medical Research Centre, Gorakhpur, Uttar Pradesh, India

Seweta Srivastava
School of Agriculture, Lovely Professional University, Phagwara, Punjab, India

Rashmi Suman
Amity Institute of Virology and Immunology, Amity University Uttar Pradesh, Noida, Uttar Pradesh, India

A.K. Tiwari
Sugarcane Research and Seed Multiplication Center, Gola, Kheri, Uttar Pradesh, India

Nawale Rutuja Yadaorao
School of Agriculture, Lovely Professional University, Phagwara, Punjab, India

Ankita Yadav
Department of Environmental Sciences, Maharshi Dayanand University, Rohtak, Haryana

Zeeshan Zafar
Atta-Ur-Rahman School of Applied Biosciences (ASAB), National University of Sciences and Technology (NUST), Islamabad, Pakistan

Herbaud Phanuel Fortunat Zohoungbogbo
World Vegetable Center, West and Central Africa (WCA)- Coastal and Humid Regions IITA-Benin Campus, Tri Postal, Cotonou, Benin

Abbreviations

AAO	ascorbate oxidase
ACLSV	apple chlorotic leaf spot virus
ACMV	African cassava mosaic virus
AFB1	aflatoxin B1
AFLP	amplified fragment length polymorphism
AGCaV	apple green crinkle-associated virus
AI	autoinducers
ALV	apricot latent virus
AMF	arbuscular mycorrhiza fungi
AMV	alfalfa mosaic virus
APO	Asian Productivity Organization
APX	ascorbate peroxidase
ARC	Agricultural Research Council
ASAL	*Allium sativum* leaf lectin
ASGV	apple stem grooving virus
ASPV	apple stem pitting virus
BCA	microbial biocontrol agents
BCMV	bean common mosaic virus
BGA	blue-green algae
BGWL	Bermudagrass white leaf
BIOPM	bio-intensification of orchard for pest management
BLRV	bean leaf roll virus
BMV	Brome mosaic bromovirus
BPYV	beet pseudo yellows virus
BYDV	barley yellow dwarf virus
BYMV	bean yellow mosaic virus
CAA	carboxylic acid amides
CABYV	cucurbit aphid-borne yellows virus
cAMP	cyclic adenosine monophosphate
CaMV	cauliflower mosaic virus
CAT	catalase

cDNA	complementary DNA
CGMMV	cucumber green mottle mosaic tobamovirus
CHV1	cryphonectria hypovirus 1
CIPHET	Central Institute of Post-Harvest Engineering and Technology
CLRV	cherry leaf roll virus
CMA	cornmeal agar
CMLV	cherry mottle leaf virus
CMV-P1	cucumber mosaic virus Isolate P1
CNRMV	cherry necrotic rusty mottle virus
CNT	carbon nanotubes
CoX	cytochrome oxidase gene
CRLV	cherry rasp leaf virus
CsPP2	*Cucumis sativus* Phloem protein 2
CTAB	cetyl trimethyl ammonium bromide
CTLV	cherry twisted leaf-associated virus
CVA	cherry virus A
CVYV	cucumber vein yellowing virus
CYSDV	cucumber yellow stunting disorder virus
DAPG	2,4-diacetylphloroglucinol
DGGE	denaturing gradient gel electrophoresis
DMI	demethylation inhibitors
DPPH	2,2-diphenyl 1-2 picrylhydrazyl
DSB	double-stranded break
ECM	extracellular matrix
EDN	ethanedinitrile
EIL	economic injury level
ELCV	enation leaf curl virus
ELISA	enzyme-linked immunosorbent assay
ENP	engineered nanoparticle
EO	essential oils
ESFY	European stone fruit yellows
ETI	triggered immunity
EU	European Union
FCM	flow cytometry
FDI	foreign direct investment
FHB	*Fusarium* head blight

FISH	fluorescence in situ hybridization
FLU	fludioxonil
FPO	farmers 'producers' organizations
GAP	good agricultural practices
GGR	geranyl geranyl reductase
GHI	global hunger index
GO	graphene oxide
GO-AgNP	graphene oxide silver nanoparticle
GPS	global positioning system
GPX	guaiacol peroxidase
GRAS	generally recognized as safe
GST	goods and services tax
GWAS	genome-wide association analyses
HDR	homology direct repair
HIGS	host-induced gene silencing
HMA	heteroduplex mobility assays
HPIM	host–phytopathogen interaction mechanism
IASRI	Indian Agriculture Statistics Research Institute
IBA	indole-3-butyric acid
IBPGR	International Board for Plant Genetic Resource
ICLCV	*Ipomoea* crinkle leaf curl virus
IDM	integrated disease management
IDP	immunodominant membrane proteins
IES	insect-exclusion screening
IFOAM	International Federation of Organic Agriculture movements
IMZ	imazalil
IoT	Internet of Things
IPC	integrated pest control
IPM	integrated pest management
ISEM	immunosorbent electron microscopy
ISR	induction of systemic resistance
IYVV	*Ipomoea* yellow vein virus
JA	jasmonic acid
LAMP	loop-mediated isothermal amplification
LChV-1	little cherry virus 1
LMV	lettuce mosaic virus
LOX	lipoxygenase

LYC	lycopene cyclase
MAS	marker-assisted selection
MBI	melanin biosynthesis inhibitors
MLCV	mungbean leaf curl virus
MLO	mycoplasma-like organisms
MR	moderately resistant
MRL	maximum residue levels
MWCNT	multiwalled carbon nanotubes
MYMIV	mungbean yellow mosaic India virus
MYMV	mungbean yellow mosaic virus
NAA	naphthalene acetic acid
NC	non-circulative
NFSM	National Food Security Mission
NGS	next-generation sequencing
NHB	National Horticulture Board
NHD	National Horticulture Database
NHEJ	non-homologous end-joining
NHM	National Horticulture Mission
NP	nanoparticles
NUST	National University of Sciences and Technology
OF	organic farming
OLCD	okra leaf curl disease
PAM	protospacer adjacent motif
PAMP	pathogen-associated molecular patterns
PBCVd	pear blister canker viroid
PCR	conventional polymerase chain reaction
PDM	plant disease management
PDS	phytoene desaturase
PDV	prune dwarf virus
PepLCV	pepper yellow leaf curl virus
PepMoV	pepper mottle virus
PepYVMV	pepper yellow vein mali virus
PFGE	pulsed-field gene electrophoresis
PGP	plant growth promotion
PGPR	plant growth promoting rhizobacteria
PHL	post-harvest losses
PLRV	potato leafroll virus

PMR4	powdery mildew resistance 4
PNRSV	*Prunus* necrotic ring spot virus
PPSMV	Pigeon pea sterility mosaic virus
PPV	plum pox potyvirus
PRMV	peach rosette mosaic virus
PRR	pattern recognition receptors
PSbMV	pea seed-borne mosaic virus
PSM	pea seed-borne mosaic
PSY	phytoene synthase
PTGS	post-transcriptional gene silencing
PTI	PAMPS-triggered immunity
PVY	potato virus Y
PYLCV	pepper yellow leaf curl virus
PYV	pepper yellows Virus
qPCR	quantitative PCR
RAPD	random amplified polymorphic DNA
RBDV	raspberry bushy dwarf virus
RCA	rolling circle amplification
RFLP	restriction fragment length polymorphism
rGO	reduced graphene oxide
RhM	*Rhynchosia* mosaic
RKVY	Rashtriya Krishi Vikas Yojana
RNAi	RNA interference
ROS	reactive oxygen species
RPA	recombinase polymerase amplification
RT	reverse transcription
RT-PCR	real time-polymerase chain reaction
SA	salicylic acid
SC	sodium carbonate
SCSV	subterranean clover stunt virus
SEM	scanning electron microscopy
SHG	self-help groups
SMD	sterility mosaic disease
SMV	soybean mosaic virus
SNP	silver nanoparticles
SOD	superoxide dismutase
SP	semi-persistent vectors

SPCFV	sweet potato chlorotic fleck virus
SPCSV	management of sweet potato curl Spain virus
SPFMV	sweet potato feathery mottle virus
SPLCCaV	sweet potato leaf curl canary virus
SPLCLaV	sweet potato leaf curl Lanzarote virus
SPLCSV	sweet potato leaf curl Spain virus
SPLCV	sweet potato leaf curl virus
SPLSV	sweet potato leaf speckling virus
SPLV	sweet potato latent virus
SPMMV	sweet potato mild mottle virus
SPMSV	sweet potato mild speckling virus
SPRSV	sweet potato ringspot virus
SPVCV	sweet potato vein clearing virus
SPYDV	Sweet potato yellow dwarf virus
SqVYV	squash vein yellowing virus
SSCP	single-strand conformation polymorphisms
ssDRV	*Sclerotinia sclerotiorum* debilitation-associated RNA virus
SSR	simple sequence repeat
SWCNT	single-walled carbon nanotubes
SWL	sugarcane white leaf
TEM	transmission electron microscopy
TEV	tobacco etch virus
TMV	tobacco mosaic virus
ToCSV	tomato curly stunt begomovirus
ToCV	tomato chlorosis crinivirus
ToRSV	tomato ringspot virus
ToTV	tomato torrado virus
t-RFLP	terminal restriction fragment length polymorphism
TuYV	turnip yellows virus
TYLCV	tomato yellow leaf curl virus
UNEP	U.N. Environment Program
USDA	US Department of Agriculture
UV	short-wave ultraviolet
WACCI	West Africa Center for Crop Improvement
WLD	white leaf disease
WLDP	white leaf disease phytoplasma
ZDS	zeta-carotene desaturase

Preface

Since the dawn of agriculture, generations of farmers have developed techniques for fending against the many ailments that affect crops. Following the early 19th century identification of the root causes of plant diseases, our expanding knowledge of the interactions between pathogens and hosts has allowed us to create a wide range of controls for individual plant diseases. In order to address persistent disease issues, a systematic approach to disease management is necessary, as is interdisciplinary research. In the current era, a focus on regulated pesticides has considerably boosted a number of modern options for disease management in organic farming.

Generations of farmers have created strategies for fending off the various illnesses that harm crops ever since the beginning of agriculture. Our growing understanding of the interactions between pathogens and hosts, which began with the identification of the underlying causes of plant diseases in the early 19th century, has enabled us to develop a variety of controls for specific plant diseases. Interdisciplinary research and a systematic approach to illness management are required to address the problems associated with persistent diseases. The current era has seen a significant increase in the number of contemporary choices for disease management in organic farming due to a focus on controlled insecticides. Alternative disease management methods like biological control are more environmental friendly and lessen the amount of time that people spend in touch with dangerous chemicals and their leftovers. Numerous biocontrol agents, such as fungi and bacteria, have been discovered, but their effective use and further development are still needed. For sustainable agriculture, a deeper comprehension of the complex interactions between the pathogen, plants, and environment is necessary. Beyond field evaluation, a new era in plant disease management has begun with the investigation of microbial communities using culture-independent molecular approaches, such as sequencing technology and genomics data.

This book, *Plant Diseases and Their Management: A Sustainable Approach*, is specifically designed for plant pathology students and scientists and researchers because it contains a wide range of themes about biological, cultural, chemical, and genome engineering controls for plant diseases brought on by viruses, bacteria, phytoplasma, and fungi. Natural materials, organic disease control, and new-generation fungicides can all be utilized to thwart or stop plant pathogen activity.

The goal of this book is to control plant diseases in a sustainable manner while utilizing microbial diversity without harming the environment or endangering human health. Academicians, agricultural-based industries, as well as PhD candidates are anticipated beneficiaries of the book.

CHAPTER 1

Integrated Management Strategies for Shelf-life and Postharvest Diseases in Vegetable Cultivation

MICHAEL KWABENA OSEI[1], JOSEPH ADJEBENG-DANQUAH[2], ISAAC OSEI-BONSU[1], FRANCIS OSEI AMOAKO ANDOH[1], BLANKSON AMOABENG[1], FAUSTINA OKYERE[1], LEANDER DEDE MELOMEY[3], and RICHARD YAW AGYARE[2]

[1]*CSIR-Crops Research Institute, Kumasi, Ghana*

[2]*CSIR-Savannah Agricultural Research Institute, Nyankpala, CSIR-Savannah Agricultural Research Institute, Nyankpala, Ghana*

[3]*WACCI-University of Ghana, Legon, Ghana*

ABSTRACT

Vegetables contain essential mineral elements and form a very important component of the diets of humans for optimum growth and development when consumed in the right state. However, vegetables are highly perishable and susceptible to various pre- and postharvest diseases that reduce their shelf-life and quality from the farm gate to the table. The use of appropriate strategies to prevent these preharvest diseases and postharvest challenges is necessary to reduce losses and render the eventual produce wholesome for human consumption. This chapter highlights vegetable production, utilization, and storage practices in the tropics. Causes of poor shelf-life such as genetic factors, pre- and postharvest handling practices, postharvest diseases, physiological disorders, and microbial agents responsible for spoilage of vegetables are identified. Furthermore, some strategies used for improving the shelf-life of vegetables ranging from genetic improvement,

Plant Diseases and Their Management: A Sustainable Approach.
Deepa Srivastava, Rajarshi Kumar Gaur, and Ajay K. Tiwari (Eds.)

conventional and modern techniques, use of chemical and biological agents, and physical methods for shelf-life extension are elaborated. The chapter then outlines traditional strategies for disease management including preharvest activities, harvesting operations, transportation, packaging and storage conditions, manipulation approaches. Also captured are innovative preharvest techniques for disease management and postharvest treatments such as low temperature treatment, heat treatment, chemical treatment, and use of microbial agents for disease management for extended shelf-life. Finally, the chapter suggests an integrated approach to postharvest disease management which encompasses the use of multiple strategies such as resistant varieties, good agricultural practices, proper pre- and postharvest handling strategies in a harmonious manner that minimizes disease incidences and improves shelf-life in vegetable cultivation.

1.1 INTRODUCTION

Diets sufficient in vegetables provide important nutrients including vitamins, minerals, and dietary fiber (Hailu and Derbew, 2015). However, daily consumption of vegetables globally falls below the WHO's recommended levels though health benefits associated with their consumption are well established (Rekhy and McConchie, 2014). Deficiency in micro-nutrients, often described as the "hidden hunger," is the cause of numerous health problems, high mortality, and low economic productivity in many developing countries where vegetables are often not consumed in sufficient quantities (Grubben et al., 2014). The need to provide fresh vegetables in sufficient quantities with acceptable qualities for the rising global population is imperative (Elik et al., 2019).

The Food and Agricultural Organization (FAO) of the United Nations data put global vegetable production at 1.07 billion tons per annum (FAO, 2016). Regrettably, nearly half of vegetables produced across the world do not reach the dining table (Siegel, 2019). Between 30 and 40% of vegetables produced in the developed countries (Singh et al., 2014) and 50% or more in developing countries (Lastochkina et al., 2019; Singh et al., 2014) go waste. In developing countries, a huge percentage of vegetables is produced by small-scale minimum-input peasant farmers who do not pay attention to the use of improved crop varieties and pesticides to manage pests and diseases (Hailu and Derbew, 2015). In addition, these farmers lack the requisite knowledge and facilities in keeping harvested vegetables under optimum conditions for longer shelf life.

Appropriate postharvest handling of food commodities including vegetables is more important than intensive farming in securing global food security because postharvest losses signify not only waste of food but scarce resources such as water, agro inputs, capital, and human effort (Hailu and Derbew, 2015). Postharvest losses encompass the measurable quantitative and qualitative damage of any food commodity along the postharvest chain (Porat et al., 2018). Often qualitative attributes of vegetables such as edibility, nutritional value, caloric value, and consumer acceptability are more difficult to assess than quantitative losses (Barrett et al., 2010). This underscores the importance of preserving the nutritional value of vegetables through appropriate postharvest handling. Yet, appearance and texture of fresh vegetables remain the most essential attributes that affect consumer acceptance (Mahajan et al., 2018). Postharvest losses also lead to increased unnecessary emission of greenhouse gases in the atmosphere (Moult et al., 2018). Furthermore, efficient postharvest management may also be an effective way of reducing total land area needed for producing food commodities and thus conserving agro ecology (Tscharntke et al., 2012). Minimizing postharvest losses in vegetables enhances food availability at affordable cost to consumers and ensures global food security (Porat et al., 2018).

In the developing world, greater percentage of postharvest losses of vegetables occur at the farm-gate due to lack of ready market for the commodities and unreliable transportation systems to convey goods to market centers. This situation is further compounded by poor road networks and storage facilities. In the developed countries, however, greater percentage of postharvest losses occur on retail shelves (Porat et al., 2018). Vegetables contain high amounts of water and once they are harvested, are prone to evaporation, shriveling, and deterioration if not well handled (Aidoo et al., 2014; Mahajan et al., 2018). The high moisture content promotes attack by microorganisms such as bacteria and fungi (Singh et al., 2014). Moreover, agricultural products are biologically active and thus, biochemical processes including transpiration, respiration, and ripening that hasten quality degradation begin just after harvesting (Adom, 2018). Therefore, the quality of postharvest life of these living plant parts depends on the rate at which they use up stored food reserves and how quick they lose water (Etana et al., 2019). Deterioration of the commodity occurs as soon as the reserved food and water are exhausted (Etana et al., 2019).

Postharvest losses in vegetables are caused mainly by weight loss, fungal and bacterial infections, physiological disorders, and pest infestations. Appropriate postharvest techniques such as temperature control; the most important tool in extending the shelf life of fresh vegetables, the use of low oxygen, and high carbon dioxide are all employed to avoid excessive use of chemicals in postharvest handling of vegetables (Yahia et al., 2017). There are, however, preharvest factors including genetic, environmental, and agronomic or cultural practices that affect postharvest quality of vegetables (Koukounaras et al., 2020). Selecting crop varieties with inherent ability to tolerate postharvest stress is an important step in dealing with postharvest losses. The quality and state of the product at harvest would also determine how long it can be stored for. Preharvest temperature, water, and light intensity have been shown to influence both nutritional quality and postharvest shelf life of vegetables (Weston and Barth, 1997). Agronomic practices such as soil fertility amendment, irrigation, pests and diseases management on the field affect postharvest conditions of vegetables. For instance, poor pest and diseases management of vegetable crop will result in damaged tissues on harvested commodity which will serve as entry point to pathogens that cause deterioration. While soil amendment with the right amount of the various nutrients is desired for vegetables to provide the potential nutrients, excessive amounts of soil nutrients, for example nitrogen, have been found to be associated with storage disorders and poor quality of harvested commodity (Weston and Barth, 1997). The maturity stage at which vegetables are harvested affects shelf life and it is influenced by the intended handling of the commodity. Commodity intended for cold storage or canning may be harvested at a different stage of maturity. Postharvest operations such as sorting, washing, transportation, packaging, storage, and marketing must be well handled to prevent unwarranted infection of pathogens in order to increase shelf life of vegetables. In the developing world, harvested commodities may poorly be packed into vehicles leading to avoidable mechanical injuries and exposing commodities to pathogens. Thus, commodities might deteriorate in transit. For longer postharvest shelf life of vegetable commodities, it is imperative that all the appropriate techniques and activities in vegetable production chain are strictly followed for enhanced global food security. In this chapter, we point out some modern techniques and activities at both the preharvest and postharvest stages in the vegetable production value chain that led to poor postharvest shelf-life. We then propose and discuss the various solutions that may be applicable in reducing postharvest losses in the vegetable production value chain in the tropics.

1.2 VEGETABLE PRODUCTION, UTILIZATION, AND STORAGE

1.2.1 VEGETABLE PRODUCTION IN THE TROPICS

The importance of vegetable production in the tropics cannot be overemphasized. It serves as a major source of employment for several people including women and the youth (Bhatta and Doppler, 2016). Dias (2011) stated that vegetable production is labor intensive and thus has the potential to reduce unemployment among the youth in many countries in the tropics. A lot more people are engaged in one activity or the other along the vegetable value chain from production to marketing. Vegetable production has high rate of returns per unit land area as compared to the main staple crops such as cereals and roots and tubers (Warid, 2018) and can contribute to the economies of many tropical countries. Also, most tropical vegetables have short growth cycles which enable growers to plant crops in succession during the season increasing the rate of returns per unit land area. Traditionally, cultivation of indigenous tropical vegetable crops such as tomato, pepper, amaranth, jute mallow, eggplant, pumpkin etc. was by subsistence farmers who sell the surplus produce to earn extra income (Veenhuizen and Danso, 2007). However, the introduction of exotic vegetable crops such as cabbage, carrot, lettuce, cauliflower, radish, green beans etc. in urban and peri-urban communities has shifted vegetable cultivation from subsistence farming to commercial production in these communities (Gurung et al., 2016). Also, some indigenous vegetable crops have gained high export potential resulting in their commercial production. Similarly, commercialization of vegetable production has increased due to a shift in the food systems from local cuisines to continental cuisines particularly in the major cities. Lastly, rural urban migration has also contributed to the increasing trend in commercial vegetable production in and around major cities. Some of these migrants rely on vegetable production as their main source of income and livelihood (Bryld, 2003).

Vegetable production in the tropics is seasonal and mainly cultivated in the open field on small land holdings under rain-fed conditions (FAO, 2011). However, a substantial quantity of vegetables is produced in the dry season at irrigation sites and along rivers and other water bodies. The production is characterized by low use of external inputs such as inorganic fertilizers, insecticides, and fungicides. The rate of use of a pesticide depends on crop requirement, availability, and cost. Some indigenous vegetable crops have low pest problems and do not require lots of protection with pesticides. However, there are some vegetable crops which have high pest problems and require adequate protection in order to obtain optimum yields. The climate of the tropics (high rainfall, relative humidity, and temperature) creates conducive micro climates favoring disease expression and insect pest damage making vegetable production challenging (Warid, 2018). Growing vegetable crops in the tropics usually lacks the use of modern techniques such as precision agriculture and greenhouse production facilities with the ability to modify the micro climate to support crop growth and development. However, some national agricultural research stations (NARS) and universities have greenhouses for training and research purposes.

Vegetable production in the tropics is dwindling as a result of the presence of numerous pests and diseases which affect crops throughout the growth cycle (FAO, 2017). Several abiotic stresses such as heavy rainfall, drought, heat, and poor soil fertility also contribute significantly to the declining rate of vegetable production in the tropics (Warid, 2018). As such, yields of most vegetable crops under the current production practices are very low. Other factors facing the sector include inadequate supply and high cost of improved vegetable seed and high cost of irrigation facilities (Warid, 2018). Inadequate technical know-how on modern techniques in vegetable cultivation has also contributed to the low productivity in the tropics. Vegetable production is very intensive and requires a lot of labor force throughout the season. Additionally, the small land sizes cultivated make it impractical to mechanize many field operations and as such production relies mainly on human-power. The inability to perform certain tasks at critical periods affects the growth and development of the crops. The labor requirement sometimes becomes nonavailable at peak labor periods such as harvesting which contributes to the high postharvest losses recorded in the tropics. Aside the challenges enumerated above, seasonal glut during the main cropping season coupled with frequent price fluctuations has affected the returns on investment in vegetable production in the tropics (FAO, 2017). The unavailability of well-equipped transportation systems to convey vegetable produce to the market centers has contributed to the high postharvest losses of fresh vegetables from the countryside (Paudel and Adhikari, 2018).

In peri-urban and urban areas, vegetable production is also characterized by small land holdings (Bhatta and Doppler, 2010), very intensive, and all year-round production. During the major season, production is dependent on rainfall whereas in the dry season, the main source of water for irrigation may be from rivers, dugouts, wells, and drainage water. Some of these sources of irrigation water may be unsafe or contaminated water which makes the produce unwholesome for consumption. The source of water or safety of water used for production has negatively affected the consumption of vegetables in most urban and peri-urban communities. In other words, most consumers are unsure of the safety of vegetables produced from certain communities because of the source of water used for production. There is also a growing concern on how insecticides and fungicides are used in vegetable cultivation. These problems can be addressed by the provision of clean water, training on safe use of pesticides, and other good agricultural practices. This will improve vegetable production and consumption in the tropics and eventually improve the livelihoods of the farmers and the health of the populace (Asongwe et al., 2014).

Marketing of indigenous tropical vegetable produce mostly happens at the local markets in the major producing areas. Some of the traders purchase the produce to sell in the nearby market centers while others aggregate the produce in large quantities and transport to the distant city centers. Marketing of vegetable produce in the tropics is not without limitations. The limitations include poor road network, exploitation by aggregators (middlemen), lack of cold storage and processing facilities, and high perishability of some of the produce (Darkey et al., 2014). As indicated above, most of the vegetables produced in the communities are aggregated by large-scale traders and transported to the larger marker centers. These traders often pay much less for the produce at farm gate and retail at higher prices in the cities. Due to the high perishability of most of the vegetables, the farmers have no option than to accept the lower prices quoted by the aggregators (Shrestha et al., 2014). Many countries in the tropics lack well-equipped transportation systems to convey vegetable produce to the market centers and this has contributed to the high postharvest losses of fresh vegetables from the countryside (Paudel and Adhikari, 2018). Lack of infrastructure and processing facilities are major contributing factors to the high postharvest losses annually (FAO, 2011).

Production of indigenous tropical vegetables is mostly to feed the local markets where quality standards are not properly enforced. However, some vegetable crops have high export potential and thus farmers produce them for the export market. They, however, have to produce to meet the standards and specification of the export market. Failure to meet the requirements results in rejection of the consignment and sometimes ban of vegetable produce from a particular country. Also, failure to meet the quantities expected to be supplied on regular basis can have serious contractual issues. This is common under rain-fed production system where cultivation is under the influence of climatic factors. This makes it difficult for farmers to go into contract farming to supply specific quantities of vegetable produce to the export market. Production is limited to the raining season which affects the continuous supply of vegetables all year round.

Addressing the challenges associated with production and marketing will transform the vegetable sector to a profitable business venture that will attract the youth into the vegetable sector. This can be done by a thorough analysis of the challenges along the vegetable value chain with the effective participation of all key actors. Governments must invest in irrigation systems and precooling facilities to ensure all year-round production and extension of the shelf life of the produce. In addition to the precooling facilities, well-ventilated and temperature-controlled transport systems should be subsidized by government to aid in transporting highly perishable vegetable produce to distant markets. Also, government policies on agriculture should create an enabling environment for production and marketing of vegetable crops. Such policies should include easy access to technical support, infrastructure development, and well-organized marketing systems. With an enabling policy environment, farmers can be encouraged to expand their production capacities and even produce for the export market. Such large-scale farmers should be assisted to modernize their production in order to meet the standards of the export market. This can be a major source of foreign exchange for most tropical countries. Small-scale vegetable farmers should also be supported and trained to improve the quality of their produce in order to attract premium prices.

1.2.2 UTILIZATION OF VEGETABLES

Vegetables form a major component of the food systems in the tropics providing a variety of diets for the household. They are not only a rich source of vitamins and minerals but they are a good source of dietary fiber which improves digestion in humans (Sachdeva et al., 2013). They are actually classified as protective foods which help boost the immune system. They can be eaten raw or with minimum cooking. Tropical vegetables are mostly consumed in the form of stew and soup and sometimes as a side dish in the form of salad. These vegetable crops range from leafy vegetables such as amaranth, roselle, jute mallow, nightshade to fruit vegetables such as tomato, okra, pepper, African eggplant, pumpkin, bitter gourd and to bulbs such as onion. Most of these vegetables are utilized in the fresh state even though some tolerate drying. Pepper and okra can be dried, stored, and utilized later. Although several vegetable crops are indigenous to tropical countries, consumption of vegetable crops in most of these countries is below the minimum daily requirement of 300 g (125 g leafy, 100 g roots and tubers, and 75 g others) (Hall et al., 2009). The low consumption of vegetables may have contributed to the widespread of diseases in this part of the world. The low rate of consumption among majority of the populace could be attributable to lack of information on the nutritive value of vegetables. It is therefore imperative to increase the production and consumption of vegetable crops to maintain a healthy population. Awareness on the nutritive value and health benefits of consumption of vegetables should be created to encourage people to incorporate vegetables in their diets. Even though there is also growing concern of the quality and safety of vegetables produced in urban communities because of the quality of irrigation water and indiscriminate use of pesticides, efforts should be made to address these problems and encourage consumers to patronize more vegetable produce.

1.2.3 HANDLING, TRANSPORTATION, AND STORAGE OF VEGETABLES IN THE TROPICS

Handling of vegetable crops encompasses determining the right stage to harvest the produce which is dependent on the use of the produce. For most tropical fruit vegetables, harvesting can be done when the fruits are physiologically or horticulturally mature. Physiological maturity is when a plant or plant part attains full development. However, horticultural maturity refers to the stage when the plant possesses all the attributes for utilization by the final consumer. Depending on the market, some vegetable crops are harvested at the green mature (for long distance markets) and red ripe for the local markets. With regards to vegetable crops sold in long distance markets, harvesting the produce at the green mature stage may reduce postharvest losses as compared to the red ripe stage. Harvesting and postharvest handling of the produce is essential in preserving the quality of the harvested produce. The harvested produce is mostly precooled or cured (particularly in onion) under shade before they are transported to the designated markets. Sorting is carried out to classify the produce into marketable and nonmarketable produce. Vegetable produces with bruises, insect pest damage, and physiological disorders are sorted out. The produce is then graded according to size and color for different markets. The smaller fruits are mostly sold in the local markets while the bigger sized fruits are sold in the larger markets. Transportation of vegetable produce to the city centers is commonly done using big articulator trucks. The harvested produce is normally packaged in wooden crates or boxes or in sacks and loaded in the trucks. The packaging is such that ventilation is reduced which increases deterioration of the produce in transit. The poor nature of most of the roads in most of these parts of the world increases the cost of transportation and contributes significantly to the high postharvest losses recorded annually.

Handling of fresh vegetables in storage is challenging due to the rate at which deterioration occurs as a result of high temperature and relative humidity. This results in a lot of wastage during peak harvest period in most tropical countries. This is so because in most tropical countries, vegetable produce are stored under room temperature. Generally, cold storage facilities either in transit or at the market centers are not common in these areas. However, some supermarkets and malls have cold storage facilities to keep their vegetables fresh with extended shelf-life. In order to reduce the high postharvest losses associated with vegetable cultivation, processing facilities should be built to process the fresh vegetables into other usable forms. This will add value to the produce and reduce wastage at peak harvest.

1.3 CAUSES OF POOR SHELF LIFE IN VEGETABLES

1.3.1 GENETIC FACTORS

Genetic causes of poor shelf life are factors associated with individual genotypes, varieties of crops, or even different species that causes them to have a shorter or longer shelf life compared to others. For example, in tomatoes, varieties that have a soft or thin skin tend to have lower shelf life compared to those with slightly tougher skin. However, too tough a skin may also affect palatability. Therefore, a good balance is needed in the ideal variety. Considerable genetic variation within different vegetable crops allows for crop improvement. Certain genetic variants of crops may have or lack certain genes or set of genes that give them the ability to either withstand attack by pathogens, or make them more susceptible to attack and subsequent damage. Such genes may be responsible for creating a physical barrier such as a tough skin in tomatoes or could be involved in producing certain chemicals such as ethylene which induces quick ripening and subsequent deterioration in many vegetables. Even for some vegetables that do not produce ethylene to any appreciable level, exposure to an external source of ethylene gas leads to rapid deterioration. Different varieties of these vegetables may also lack or have certain genes or set of genes that make them more or less tolerant to ethylene. Breeders can take advantage of knowledge about shelf life regulating genes to breed for enhanced shelf life through traditional or biotechnological approaches. In tomato, repression of a specific gene/transcription factor, SIFR (fruit shelf-life regulator), resulted in downregulation of a number of genes including cell wall regulation associated genes which ultimately resulted in enhanced shelf life (Zhang et al 2018).

1.3.2 PRE- AND POSTHARVEST PRACTICES OR HANDLING

Vegetables are widely used as an excellent source of micronutrients and phytochemicals. Habitual inclusion of fruits and vegetables in the diet may prevent or reduce the risk of several chronic diseases. However, they are perishable products that contain living tissues. The quality retention and prevention of postharvest loss during handling, storage, and marketing is critical; therefore, serious attention is needed to prolong the shelf life of these vegetables and maintain quality. It is estimated that more than 20–22% of the total production of fruits and vegetables are lost due to spoilage at various postharvest handling stages. Both pre- and postharvest practices have strong influence on shelf-life and overall quality of vegetables. Pre- and postharvest treatments are therefore used to minimize the loss of fresh produce as well as to maintain the quality, thereby increasing the shelf life.

Preharvest production practices are critical in developing optimal quality of vegetable crops, which depends on the interaction of genetic, agronomic, and environmental factors. Since quality of vegetables cannot be improved after harvest, identifying the best combinations of these preharvest factors is a key strategy to maximize vegetable quality to meet consumer needs. The major preharvest factors that lead to losses include: poor production conditions such as planting with poor seeds, unfertile soils, drought stress, unknown mature indices, pests and diseases, and poor harvesting techniques. Preharvest practices include cultural and farm management practices such as irrigation, fertilizer application, pest and disease control, starting right from planting (or field preparation) up till the harvest date. Studies have reported effects on shelf life and nutrients value of vegetables are influenced by soil nutrient, cultural practices, and water supply during production (Jiang and Pearce, 2005). Crop production under stressful conditions with poor soil nutrients, water, and salt stress significantly lowers the postharvest life and decreases the nutrient levels of leafy vegetables (Jiang and Pearce, 2005; Sahou et al., 2014). Use of cattle and goat manure has been reported to increase crude protein and leaf biomass yields in some vegetables (Seeiso and Materechera, 2014). Using excessive fertilizer, especially nitrogen (N), aside the negative effect on the environment has been shown to reduce certain nutrients (e.g., vitamin C, potassium, and sucrose) as well as dry matter content of leafy vegetables (Sorensen, 1999). Even, the type or source of fertilizers has effects on certain vegetables. For instance, Abu (2021) reported that single-based fertilizers enhanced shelf-life of lettuce, carrot, and sweet pepper by more than a week compared to compound fertilizers (NPK 15:15:15). Depending on the vegetable or variety, the method of fertilizer application can also affect shelf life. Foliar N application in broccoli makes heads more vulnerable to rots induced by the bacteria, *Erwinia,* after harvest (Bachman and Earles, 2000). Preharvest factors significantly affect the quality and quantity of leafy vegetables. Studies have reported effects on shelf life and nutrients value of vegetables as influenced by soil nutrient, cultural practices, and water supply during production (Jiang and Pearce, 2005). Crop production under stressful conditions with poor soil nutrients, water, and salt stress significantly lowers the postharvest life and decreases the nutrient levels of leafy vegetables (Jiang and Pearce, 2005; Sahou et al., 2014). Use of cattle and goat manure has been reported to increase crude protein and leaf biomass yields in some vegetables (Seeiso and Materechera, 2014). The other production or preharvest factor that influences shelf-life and quality of vegetables is pest and disease management. If pests are not controlled effectively, their attack can leave scars or holes of varying sizes on fruits, leaves, or roots, making them susceptible to attack by microbial spoilage agents and postharvest diseases, culminating in reduced shelf-life. Similarly, when diseases are not managed effectively, pathogenic particles such as spores of fungi may cleave to the vegetables and induce deterioration when the plant tissues become soft due to aging or ripening. Some postharvest diseases actually initiate infection before harvest and develop rapidly after harvest. Excessive- or under irrigation may also influence vegetable quality and shelf life, especially when it happens during or prior to harvesting. High soil moisture levels prior to preharvest reduce brix as well as shelf life in tomatoes (Sibomana et al., 2015). On the contrary, reduced soil moisture levels increase brix and subsequently shelf life but can also decrease yield. A fine balance is therefore needed to achieve optimum fruit quality, shelf life, and yield. Furthermore, irrigation with municipal waste water and other contaminated sources of water, aside the public health risk posed, can also negatively affect shelf life.

Postharvest practices may include practices during harvest as well as those occurring after harvest, throughout the distribution chain until the vegetable is consumed. Poor postharvest handling leads to mechanical damages, such as leaf tearing, crushing, and other physical injuries. The tissue damages then lead to physiological deterioration due to oxidation of phenolic substances (Bachman and Earles, 2000). Injuries stimulate ethylene production in vegetables leading to yellowing and sudden leaf senescence. Additionally, mechanical damage creates openings for microbial infections leading to decay and also increases water loss from the damaged tissue (Bachman and Earles, 2000). Some of these practices are far beyond the control of the farmer. Therefore, only practices within the purview of the farmer will be described here. Examples of postharvest practices that directly or indirectly affect shelf-life include the following:

1.3.2.1 METHOD OF HARVESTING

Harvesting with the wrong tools can result in rips and tears on the skins of vegetables which could facilitate entry by pathogenic organisms, thereby decreasing shelf life. Additionally, rough handling of vegetables during and after harvesting can cause mechanical injuries and lead to reduced shelf life through the previously mentioned channels.

1.3.2.2 STORAGE CONDITIONS

Different vegetables require different storage conditions and temperature. Storing vegetables in cold temperatures slows down metabolism and increases shelf life. However, different vegetables have different tolerance levels to cold temperatures. In particular, temperate vegetables tolerate colder storage temperatures better than tropical vegetables. Therefore, knowing the cold temperature tolerance threshold is essential in prolonging shelf life. Chilling injuries occur when the storage temperature is too cold or the produce is stored for too long in cold temperatures. Chilling injury is a physiological disorder that reduces shelf life of vegetables. Directly exposing certain vegetables to the sun after they have been harvested results in scalding and rapid moisture loss which reduces shelf life. In general, vegetables should be harvested during the cool part of the day such as early morning or late afternoon and produce should be kept in the shade during and after harvest. Anything contrary can lead to poor shelf life of the vegetables. For some vegetables, rapid loss of water through transpiration following harvest reduces shelf life. For such vegetables, it was found that storage in high humidity environment delayed senescence more than storage at cold temperatures (Lownds et al., 1994). Such high humidity conditions may be induced by storing these vegetables in plastic materials such as polyethene bags. The combined effect of high humidity and low storage temperature can extend shelf life to a certain degree (Mier et al., 1995). Storing leafy vegetables together with ethylene-producing fruits or vegetables results in rapid breakdown of chlorophyll causing the leaves to turn yellow. This also applies to fruit vegetables such as cucumbers which are harvested green. Exposure to ethylene gas causes physiological disorders such as russet spotting in lettuce (Bhat, 2009).

1.3.2.3 PACKAGING AND TRANSPORTATION

Using the wrong containers for keeping or transporting vegetables and over-or-under packing harvested produce can also cause mechanical injuries such as compression (crushes), bruises, and abrasions. This ultimately speeds up deterioration and reduces shelf life through attack by pathogenic and microbial spoilage agents.

1.3.3 POSTHARVEST DISEASES AND MICROBIAL SPOILAGE AGENTS

Postharvest diseases are caused by pathogens or microbial spoilage agents and serve to speed up deterioration of vegetables. The causative microbes are mostly bacteria and fungi, although in some cases (e.g., some root and cole crops), viral pathogens present prior to harvest may accelerate infection following harvest (Coates and Johnson, 1997).

Postharvest diseases of vegetables generally fall in two (2) categories based on the mode of initiation and perpetuation of infection:

i. Latent or quiescent infections
ii. "Active" infections

In the first category, infections mostly occur prior to harvest and the pathogen enters a period of inactivity (quiescence) following the initial infection until conditions induced by physiological changes in host tissue due to senescence or ripening become conducive for infection to continue. A typical example is gray mold induced by *Botrytis cinerea.*

In the second group ("Active" infections), the pathogen usually enters through superficial lesions created by pest or mechanical damage during or following harvest. Microbial spoilage agents may also access vegetable tissue through lesions induced by physiological disorders. Infections do not exhibit any clear lag or dormancy phase and thus can be considered as continuously "active" right after initiation. In most cases, the symptoms develop rapidly after harvest as the physiological status of the crop changes. Examples include soft rot (caused by *Erwinia carotovora*) and blue and green mold (by *Penicillium* spp.).

Postharvest diseases, aside the loss caused to produce and consequently income, also pose a public health risk since some of the pathogens produce toxins which can be harmful to humans and livestock. The real threat lies in the use of infected produce in processed food since most consumers will avoid diseased produce.

More examples of common postharvest diseases have been shown in Table 1.1. Bacterial soft rots are widespread in many vegetables since the relatively high pH of most vegetables (compared to low pH of most fruits) is conducive for their growth.

TABLE 1.1 Common Postharvest Diseases and Susceptible Vegetables.

Disease	Causative agent	Susceptible vegetables
Grey mold	*Botrytis cinerea, Botryotinia fuckeliana*	Carrot, cole crops/brassicas, cucurbits, tomato, eggplant, pepper
Black mold rot	*Aspergillus niger*	Onion
Fusarium rot	*Fusarium* spp.	Cucurbits, tomato, eggplant, pepper
Alternaria rot	*Alternaria* spp.	Cole crops, cucurbits, tomato, eggplant, pepper
Cladosporium rot	*Cladosporium* spp.	Tomato, eggplant, pepper
Charcoal rot	*Macrophomina phaseolina*	Cucurbits
Fusarium basal rot	*Fusarium oxysporum f.sp. cepae*	Onion
Phytophthora tot	*Phytophthora porri*	Cole crops
Rhizopus rot	*Rhizopus* spp.	Carrot, cucurbits, tomato, eggplant, pepper
Sclerotium rot	*Sclerotium rolfsii, Arthelia rolfsii*	Carrot, tomato, eggplant, pepper
Watery soft rot	*Sclerotinia* spp.	Carrot, cole crops, tomato, eggplant, pepper
Cottony leak	*Pythium* spp.	Cucurbits, tomato, eggplant, pepper
Chalara	*Chalara thielavioides*	Carrot
Smudge	*Colletotrichum circinans*	Onion
Thielaviopsis rot	*Thielaviopsis basicola*	Carrot
Bacterial soft rot	*Various* spp. *of Bacillus, Erwinia, Lactobacillus, and Pseudomonas, Xanthomonas campestris*	Most vegetables

1.3.4 PHYSIOLOGICAL DISORDERS OF VEGETABLES

Physiological disorders are abnormalities found on crops that have no known pathogenic or pest causative agents. They are often associated with environmental conditions, cultural practices, and nutrient status. These disorders may affect the whole plant or certain parts of the plant including but not limited to leaves, fruits, and roots. Physiological disorders tend to exhibit in vegetables either pre- or postharvest and have effects on both shelf life and marketability/salability of produce.

Physiological disorders can affect shelf life in two (2) main ways. By creating channels for attack by pathogens leading to increased susceptibility to postharvest diseases, some of these physiological disorders may not affect nutritional quality per se but create conditions for easy attack by pathogens leading to further deterioration of the vegetable and reduced shelf life. A typical example is cracking or splitting of cabbage heads or tomato fruit caused by rapid uptake of water when heavy rains follow prolonged dry spells in the field. The crack or split, which may not necessarily affect the taste or nutrient content of the vegetable, could, however, provide an avenue for pathogens to enter, thereby facilitating rapid development of postharvest diseases and consequently reducing shelf-life.

The other way in which physiological disorders influence shelf life is by reducing wholesomeness or attractiveness of produce, thereby rendering them unsalable or with reduced market value. Here also, some of the blemishes such as those seen in russet spotting of lettuce triggered by contact with ethylene gas or temperatures above 3°C, may not necessarily make the vegetable unsafe to eat but the lost visual appeal diminishes salability.

Some common examples of physiological disorders that have effect on shelf life are described in Table 1.2.

TABLE 1.2 Common Physiological Disorders that Affect Shelf Life of Vegetables.

Disorder	Description/causes	Susceptible vegetables
Splitting or cracking	A crack or split of variable size and orientation (i.e. radial or concentric) appears in the edible part of the vegetable at maturity. This happens due to disproportionate expansion between internal and external tissues occurring when plants receive copious amounts of water, usually through rainfall, after a long period of water stress.	Cabbage, tomato, carrot, pepper

TABLE 1.2 *(Continued)*

Disorder	Description/causes	Susceptible vegetables
Blossom end rot	Manifests *in situ* as a necrotic, sometimes sunken, brown to black patch or region at the blossom end of the fruit caused by lack of calcium in the soil or inadequate transport of calcium to growing parts due to rapid growth induced by high nitrogen supply.	Tomato, pepper, eggplant
Gold fleck/ spot/speck	Small golden or yellowish spots appearing on the shoulders of, and giving fruits a dull look. Induced by excessive supply of calcium and phosphate fertilizers leading to accumulation of calcium oxalate crystals under the skin of fruits.	Tomato
Sunscald	Yellow, white, or light-colored patch on the regions of fruits directly exposed to sunlight. It is induced by abrupt exposure of fruits to strong uninterrupted sun rays.	Tomato, pepper
Russet spotting	Characterized by many small, tan to brown spots appearing on the mid ribs of leaves and increasing in severity with storage. Induced by exposure to ethylene gas or temperatures above 3°C. In minimally processed lettuce, high O_2 atmosphere was also found to precipitate this disorder through increased concentration of ethylene in the head of iceberg lettuce (López-Gálvez et al., 2015)	Lettuce
Tipburn	Breakdown of leaf margin tissues of young maturing leaves, presenting as few brown spots on or close to the leaf margin or browning of the entire edge of the leaf. This is caused by temporal calcium deficiency in some growing tissues (particularly young leaves) due to disproportionate distribution of calcium and water in the plant.	Lettuce, brassicas
Watery/ translucent/ leathery scales	Symptoms include thick dark scales between outer dry and inner fleshy scales or glassy, with watery texture and appearing as translucent scales further inward. These develop with storage and are due to imbalances in gas exchange of bulbs, especially with regards to CO_2 concentration.	Onion
Chilling injury	Symptoms vary depending on the vegetable but include the following or combinations thereof: surface lesions/ pitting, early loss of firmness, discoloration of inner tissues, and water-soaked tissues. It is common mostly in tropical (and some temperate) vegetables when they are stored at low (but above freezing) temperatures (>0°C).	Tomato, bell pepper, cucumber, eggplant, okra, lettuce, asparagus, snap bean, potato

1.4 STRATEGIES OF IMPROVING VEGETABLE SHELF LIFE

1.4.1 GENETIC IMPROVEMENT OF VEGETABLES

Harvested vegetables severed from mother plants continually respire to generate energy since the constitutive cells need to be kept alive. These cellular oxidative reactions are enzymatic so the speed of the processes is genetically determined. The cells are persistently exposed to oxidative stress manifested in the production of reactive oxygen species (ROS) resulting from the oxidation of lipids which cause cellular death and eventual spoilage (Das and Roychoudhury, 2014). Plants use antioxidant mechanisms that include enzymes and nonenzymatic antioxidant biomolecules (Gill and Tuteja, 2010). The antioxidative enzymes including superoxide dismutase (SOD), catalase (CAT), ascorbate peroxidase (APX), and guaiacol peroxidase (GPX), and nonenzymatic antioxidants like L-ascorbate, carotenoids, tocopherols, and phenolic compounds, ordinarily serving as stored nutrients, are used to neutralize the oxidative destructive actions of ROS. These cellular activities ultimately result in the reduction of nutritional and health promotional value of vegetables. Plants with high levels of antioxidants, whether constitutive or induced, and increased activity of antioxidant enzymes usually have higher resistance to oxidative damage (Kusvuran et al., 2016). Genotypes of vegetable species (eggplant, cucumbers, melons, squash, beans, okra) that are tolerant to several abiotic stress factors, whether as whole plants or callus cultures, have antioxidant enzymes—SOD, CAT, APX, POD, and GR with relatively high activity (Kusvuran et al., 2016). Germplasm of vegetables varies widely for antioxidant enzyme activity and metabolites content. The biosynthetic pathways of antioxidant biomolecules and the key regulatory enzymes and the corresponding encoding genes have been unraveled—L-ascorbate (Wheeler, Jones, and Smirnoff, 1998), carotenoid (Cunningham and Gantt, 1998), tocopherols (Fritsche et al., 2017; Threlfall and Whistance, 1967), and phenolics (Croft, 1998; Tohge et al., 2017). Such genetic variability within a species is a valuable tool for screening and breeding for vegetables with extended shelf-life. Integrated omics platforms are applicable and useful for the identification of vegetable varieties with such superior attributes to direct the selection of parents during breeding.

Wheeler, Jones, and Smirnoff (1998) identified the L-ascorbate biosynthesis pathway and proposed the involvement of GDP-D-mannose, GDP-L-galactose, L-galactose and L-galactono-1, 4-lactone, and *in vitro* synthesized L-ascorbate from GDP-D-mannose through the other intermediates. There is enormous variability in L-ascorbate levels within species, genotypes, tissues, and developmental stages in higher plants (Mellidou and Kanellis, 2017). In three Chinese non-heading cabbage (*Brassica campestris* ssp. chinensis Makino) varieties, differential expression levels of *D-mannose-6-phosphate isomerase* (*PMI*), *GDP-mannose-3′, 5′-epimerase* (*GME*), *GDP-L-galactose phosphorylase* (*GGP*), and *ascorbate oxidase* (*AAO*) have been identified to be responsible for varying amounts of L-ascorbate accumulation (Ren et al., 2013). Apparently, high expression levels of *PMI*, *GME*, and *GGP* with low *AAO* level contribute to high L-ascorbate accumulation. In radish root, the accumulation of L-ascorbate is mainly affected by biosynthesis and ascribed to the L-galactose route, while *GDP-D-mannose pyrophosphorylase* (*GMP*) plays a key regulatory role in the accumulation during the taproot development (Xu et al., 2013).

A review on the perspective of carotenoid metabolism and regulation in horticultural plants is provided by Hermanns et al. (2020), Stanley and Yuan (2019), and Yuan et al. (2015). Distinct allelic variants of key genes for carotenoid biosynthesis including *phytoene synthase* (*PSY*), *phytoene desaturase (PDS)*, *zeta-carotene desaturase (ZDS*), and *lycopene cyclase*-(*LYC*) for some leafy and nonleafy vegetables have been identified and polymorphic SNPs and InDel markers developed for application in vegetable breeding (Clotault et al., 2008; J. Li et al., 2017; Ma et al., 2017). Besides, other regulatory factors that influence accumulation and stability have also been established. The mutant *orange* (*or*) gene encoding synthesis of the orange proteins which are associated with initiating chromoplast development and division for massive carotenoid sequestration and accumulation, first identified to be responsible for massive accumulation of β-carotene in the mutant orange curd cauliflower (L. Li et al., 2001; Lu et al., 2006), was also later found in some carotenoid accumulating vegetables like potato, carrots, sweetpotato, and squash (Ellison et al., 2018; Gemenet et al., 2020; Lopez et al., 2008; Xanthopoulou et al., 2019). The *or* gene is also known to increase plant resistance to environmental stress (Kim et al., 2018).

Strategies for biofortifying crops with tocopherols have recently been summarized in Mène-Saffrané and Pellaud (2017). Comparative gene expression analyses identified five key regulatory genes: *1-deoxy-D-xylulose-5-P-reductoisomerase* (*DXR*) and *geranyl geranyl reductase*—(*GGDR*) from the MEP; Shikimate (*arogenate dehydrogenase*—*TyrA* and *tyrosine aminotransferase*—*TAT*), and tocopherol-core pathways (*γ-tocopherol methyl transferase 3–TMT3*) pathways to be involved in the differential biosynthesis and accumulation of tocopherols in two soybean varieties (Vinutha et al., 2017).

Phenolic compounds are secondary metabolites found most abundantly in plants. The synthesis of these compounds is promoted by stress via the Shikimate pathway (Francenia Santos-Sánchez et al., 2019). Among others, the primary function of phenolic compounds is to perform the role of antioxidants to protect cells from oxidative damage by neutralizing ROS that are generated because of stress in postharvest vegetables (Meitha et al., 2020). But the pathway is feedback inhibited, the regulatory mechanism remains elusive, making it difficult to use genetic resources for enhanced production of phenolic compounds in vegetables (Yokoyama et al., 2021). Other stored secondary metabolites that are solutes help in maintaining moisture to sustain firmness and freshness by exerting osmotic pressure to prevent particularly wilting in leafy vegetables. Generally, vegetable varieties with thicker outer flesh may better withstand moisture loss and as well resist pathogen attack during storage, so vegetable breeding should be purposefully the development of such lines to promote water retention for shelf-life improvement.

Targeted breeding for colored vegetables for enhanced nutrition may bring about a general knock-on benefit of improving shelf-life if secondary metabolites conferring the color are antioxidants. Additionally, breeding for vegetable varieties that are pest and disease resistant or tolerant to specific pathogens usually reduces microbial and disease-related spoilage thereby extending shelf-life. Specific metabolic pathways with key genes encoding metabolic enzymes responsible for the biosynthesis and accumulation of shelf-life-related metabolites are known; thus, introgression of the relevant genes should be the set objectives for vegetable breeding.

Oxidative enzyme, lipoxygenase, uses polyunsaturated fatty acids (linoleic, linolenic, and arachidonic acids) as substrates in plant tissues to produce aldehydes, ketones, alcohols, and oxoacids through highly oxidizing ROS intermediates (Tressl et al., 1981). These ROS mainly generated in the mitochondria, chloroplasts, and peroxisomes may further oxidize lipids and other macromolecules like DNA, RNA, enzymes, and proteins, eventually leading to membrane damage and cell death and reduced shelf-life (Patel et al., 2018). Accordingly, vegetable varieties with reduced lipoxygenase activity are more desirable to promote shelf-life extension.

While water loss through transpiration is a surface phenomenon, it is an important factor that can be regulated through breeding. Vegetable varieties with tissues having thick waxy cuticle covering that limit surface permeability to water vapor, reduced number, and size of stomata and with smaller surface area-to-volume ratio have reduced transpiration rates (Ben-Yehoshua and Rodov, 2002; van Meeteren and Aliniaeifard, 2016; Nunes and Emond,

2007). Abscisic acid induces closing of stomata to reduce transpiration (Munemasa et al., 2015), so the identification of vegetable varieties with the production of enhanced levels of the hormone may help reduce wilting to maintain longer shelf-life. Breeding of vegetables can purposefully target these traits to prevent excessive water loss by transpiration. Multigene pyramiding can be done by tapping diverse genetic resources using marker-assisted selection techniques and genetic engineering to enhance the levels of antioxidant enzymes and metabolites as well as other relevant traits in vegetables for shelf-life improvement.

1.4.2 AGENTS TO CONTROL SPOILAGE MICROORGANISMS ON VEGETABLES

Vegetables possess high nutritional value as they are rich in vitamins, dietary fibers, amino acids, and minerals (Gastol et al., 2011). The consumption of vegetables is associated with protection against gastro-intestinal cancers (Lunet et al., 2005), cardiovascular diseases, and type 2 diabetes among others (Clifton et al., 2014). Its consumption also helps address micronutrient deficiencies. Nonetheless, after harvest vegetables are highly perishable as a result of water and nutrient loss initiated by physiological deterioration, biochemical changes, and microbial degradation, with a resultant economic loss (Ashiq, 2015; Berg et al., 1986; Praeger et al., 2016). Vegetable spoilage is mainly caused by microorganisms (Hammond et al., 2015).

Postharvest microbial control is a key and essential operation responsible for controlling food-borne diseases, maintaining the quality attributes of vegetables and reducing food losses during storage. Due to the losses and damage incurred from the micro-organisms, strategies geared toward decontamination as a means to reduce microbial actions and packaging conditions that hinder their growth have gained much attention in recent times (Paramithiotis et al., 2017). Ssemanda et al. (2017) demonstrated that activities such as washing or trimming and peeling significantly reduced microbial load. The most common technology for the decontamination of vegetables has been the use of sanitizers (Olanya et al., 2015).

1.4.3 CHEMICAL AGENTS

Several chemical disinfectants and preservatives have been employed over the years in an attempt to reduce microbes in vegetables (Mostafidia et al.,

2020) and the effectiveness of these methods is dependent on concentration, time of treatment, temperature, pH, and the sensitivity of target organisms (Rajwar et al., 2016). The use of chlorine to disinfect fresh products is widely known, though its use presents some demerits. Standard chlorine concentration (50–200 ppm) has proven to be ineffective in the reduction of microbial load in fresh products. The long-term exposure to chlorine irritates the skin and the respiratory tract of workers, while affecting the environment. More so, trihalomethane, a carcinogen, can be formed after chlorine contacts an organic matter (Oliveira et al., 2015).

With hydrogen peroxide, though it exhibited a biocidal effect against *Cryptosporidium parvum* (Kniel et al., 2003), studies deemed it unfit as an optimal disinfectant. Some applications also saw deposition of hydrogen peroxide residue, unsafe for fresh produce (Soliva-Fortuny and Martin-Belloso, 2003). Peracetic acids and weak organic acids (lactic acid, tartaric acid) are effective on low acidic foods with no/minimal effect at high pH. However, they have great potential when combined with other methods such as bacteriocins and physical methods. Lippman et al. (2020) demonstrated this saw reduction in *Salmonella sp.* count when they combined ultraviolet light and peracetic acid as an alternative to chlorine washing for lettuce decontamination. Ozone is a powerful oxidizer that is active in the control of bacteria, molds, and viruses. Its mechanism of action is by oxidizing the membrane of bacteria and destroying them (Carletti et al., 2013). One significant advantage of this method is that it quickly decomposes into harmless molecular oxygen (Papachristodoulou et al., 2018). Its short lifespan in the environment makes it safer than other chemicals. Ozone is more effective at inhibiting *sporogenic bacillus* with food origin, than hydrogen peroxide (Carletti et al., 2013). The use of ozone in vegetables like lettuce, tomatoes (Aguayo et al., 2006), potatoes (Ansorena et al., 2014), and bell peppers (Horvitz and Cantalejo, 2012) has been explored. It reduced mold and bacteria in onions during storage with no effect on chemical composition and sensory feel. Ozone also altered the color of some fruits and vegetables such as peach and carrot, reducing the amount of ascorbic acid in broccoli, and increased phenols and flavonoids in banana. It is however a corrosive gas with a higher instability, which, when released in higher concentrations into the environment, could lead to human death (Carletti et al., 2013).

Chlorine dioxide (ClO_2) is an antiseptic that is currently used to control the growth of microorganisms and has been considered a disinfectant for vegetables (Sun et al., 2019). It is a strong oxidizing agent with a 2.5? higher oxidation capacity than hypochlorite (Wu and Rioux, 2010). Equated to hypochlorite, ClO_2 has a better antimicrobial activity at neutral pH. This gaseous form of chlorine is more effective at inactivating pathogenic cells attached to inaccessible plant parts due to the high diffusion and penetration rate (Sun et al., 2017). It is very effective with short treatment times even at low concentrations in a large pH range, without formation of carcinogens (Sun et al., 2017). Furthermore, it maintains the color, weight, firmness, and sensory properties of vegetables and fruits. Nonetheless, its use might affect the nutritional quality since it can react with phenolics (Sun et al., 2019). On the other hand, mangos treated with ClO_2 retained the green and flesh-color after storage for 21 days. Furthermore, ClO_2 delayed reddening by suppressing the expression of certain genes associated with chlorophyll breakdown and carotenoid synthesis in harvested green peppers (Wei et al., 2019). This can help meet the market and consumers' needs compared to other treatments.

1.4.4 BIOLOGICAL AGENTS

Bacteriocins, ribosomally synthesized antimicrobial peptides, are known to destroy as well as inhibit the growth of other bacteria and are considered to be safe natural bio-preservatives (Barbosa et al., 2018; Martínez et al., 2019). The use of these biological agents in the washing of vegetables for microbial control has been studied in recent years (Ramos et al., 2020; Siroli et al., 2017). The effectiveness of these bio-preservatives is dependent on the chemical and physical properties of the food, and on the storage conditions (Molinos et al., 2005). Conversely, the components of the food can limit the adsorption of bacteriocin to the surface of the vegetable and inactivate them. Again, the natural microbiota on vegetable surface can hinder the bacteriocins (Allende et al., 2007). Consequently, the potential use of bacteriocins ought to be tested in the appropriate food system (Ramos et al., 2020).

Yeast biological control is another promising option for controlling postharvest diseases of fruits and vegetables (Sugiprihatini and Wiyono, 2012). Studies on yeast antagonist action in fruits and vegetables are employed for the control of microbial contaminations at the postharvest phase in cherry tomato and chili among several fruits (Ferraz et al., 2019). Although lots of reports have been made on this biocontrol method, very few have been implemented on a commercial level. Determinants such as plant (vegetable), the type of antagonist, the pathogen, and resident micro flora seem to influence the effectiveness of the yeast antagonist method (Spadaro and Droby,

2016). All fruits and vegetables have thin natural wax which keeps water intact. When these produces are rubbed, wrapped, or washed, the wax is removed, with a consequential increase in amount of respiration (Mostafidi et al., 2020). Edible coatings are therefore used to protect the surfaces of food in order to extend shelf life and reduce microbial spoilage. These materials can also serve as a medium to carry browning agents, colors, flavors, nutrients, and antimicrobial compounds that can increase the life span of the product (Kore et al., 2017). They have been shown to reduce the number of cold-blooded bacteria, coliforms, yeasts, and molds (Putnik et al., 2017). Edible coatings used include pectin, chitosan, alginate, propolis, mango, and cassava among others (Mostafidi et al., 2020). Thus, edible films can be taken into consideration for the biocontrol of foodborne pathogens possibly by immersing food into the extract to preserve the quality, control microbial spoilage, and extend shelf life.

1.4.5 PHYSICAL METHODS

Ultrasonic processing is also employed in controlling microbial spoilage. Ultrasound (US) is a green technology that uses energy transmitted as waves. High intensity ultrasound, alone or in combination with other chemical or organic disinfectants (lactic, acetic, malic, paracetic acids), significantly reduced the bacterial load in tomatoes, lettuce, bell peppers, carrots, and *alfafa* leaves (Pinheiro et al., 2015; Gani et al., 2016; Jiang et al., 2020). It can also preserve color, delay ethylene production as well as most quality attributes of fruits and vegetables. Contrariwise, because of the high waves, this technique might affect other enzymes, nutritional content, and quality properties (color, firmness) of the vegetables (Jiang et al., 2020).

The influence of short wave ultraviolet (UV) rays on vegetables varies depending on the product (species and variety, complete and intact, or vegetables with minimal processing), plant surface, initial microbial load, and the method of application (time, duration, and treatment dose). Low doses of UV tend to improve the quality and nutritional health of fruits and vegetables. These rays have the ability to disinfect and delay the growth of microbial products in fresh produce without affecting the quality (Kim et al., 2013; Mostafidi et al., 2020). Prolonged treatment might have a negative impact on the quality of the vegetable.

Food Irradiation is an economical technology known for reducing post-harvest lesions, improving the health of food and inactivating pathogens and food-borne parasites (Mridha et al., 2017). It also maintains the nutritional components of fresh food and food products. Food irradiation is legally recognized in many countries as a safe and active means to improve on the safety of foods including vegetables. It has been deemed as one of the safe substitutes to disinfecting all types of fruits, while delaying ripening in order to increase shelf life (Mridha et al., 2017).

Modified atmosphere packaging (MAP) is used for controlling microbial decay of freshly cut fruits and vegetables. This method is usually accompanied by refrigeration and cold treatment. In the absence of atmospheric gases inside the pack due to the type of film selected, there is reduction in the amount of respiration of the product which delays the biochemical processes that lead to deterioration. Packaging films often used in MAP are selective with regards to the extent they can be permeated, by atmospheric gases such as O_2, CO_2, and N2. Here, the effect of CO_2 to act as an antimicrobial agent is significantly dependent on the temperature, type of microorganism, growth phase, water activity, and chemical composition of the product. Thus, bacterial load inside the packet can be reduced by manipulating the concentration of gases inside the package. Limitations on the use of MAP in inhibiting microbial growth include temperature, storage time, the volume concentration of modified atmospheric gases, packaging dimensions, and product mass (Putnik et al., 2017).

Other physical methods that are employed as microbial control agents include cold plasma and pulsed light.

Several forms of agents are available for controlling microbial activity, as a means to curb food spoilage in vegetables, thus reducing food loss. These agents are categorized as chemical, biological, and physical agents. Each of these microbial agents has its limitations as well as advantages. The application of these methods will depend on the properties of the vegetable and the consequential effect of the agent on the quality and sensory attributes.

1.5 TRADITIONAL/CONVENTIONAL STRATEGIES FOR POSTHARVEST/CONVENTIONAL STRATEGIES FOR POSTHARVEST DISEASES MANAGEMENT

An effective strategy for postharvest disease management must begin when the plants are growing in the field and continued until the point of consumption (ShiJiang, 2006). This is important because pathogens that cause postharvest decay are mostly acquired before harvest and during harvesting. The strategies for managing these decay pathogens and associated diseases will be grouped into preharvest and postharvest management strategies.

1.5.1 PREHARVEST MANAGEMENT STRATEGIES

1.5.1.1 FARM SANITATION

Good agricultural practices (GAP) are one of the fundamental strategies to manage postharvest diseases of tomato. The type of irrigation used in tomato production plays a significant role in managing postharvest diseases of tomato. Drip, furrow, or seep irrigations are preferred to overhead irrigation since the former irrigation methods do not wet the leaves of the tomato plants. Moisture promotes the spread, growth, and survival of decay pathogens. Farm operations such as trellising of tomato and harvesting should be carried out when moisture in the canopy is dried. Wet plants are more prone to mechanical injury which creates wounds. These wounds become the primary entry point for decay pathogens that cause postharvest diseases of tomato (Mahovic, n.d.).

1.5.1.2 FUNGICIDES AND INSECTICIDES

Fungicides are extensively used to control postharvest diseases of tomato (Barkai-Golan, 2021). Fungicides are applied many times in the growing season to prevent or control pathogens when necessary. Protectant fungicides are applied regularly during the growing season while systematic fungicides are applied to inactivate infections already established and to prevent new infection. Fungicides are also applied after harvest vegetables to arrest pathogen growth and inhibit decay by providing persistent protection (Feliziani et al., 2016). Control of chewing insects that create wounds in fruits is very important during the growing cycle and these have been largely controlled using insecticides or biological control methods.

1.5.1.3 HYGIENE AND HARVESTING

Pathogens can be transferred to plants from contaminated equipment or workers. Frequent hand washing and disinfection of equipment are highly encouraged. The stage of harvesting of vegetables contributes significantly to the postharvest quality of the fruit. Fully ripened vegetables are susceptible to mechanical injuries during harvesting (Toivonen, 2007) and therefore care must be taken to avoid injuries. Harvesting at lower temperature slows down the development of decay pathogens. Temperature directly affects ripening and pathogen growth.

1.5.1.4 PACKAGING

Packaging materials should allow for some aeration to avoid building up of heat due to respiration. Plastic bins are preferred for handling tomato than wooden crates because of the ease of sanitizing the plastic bins. Quaternary ammonia compounds are effective for sanitizing equipment but not approved for direct contact with food (Mahovic, n.d.).

1.5.1.5 STORAGE

Effective control of the postharvest environment is crucial to delay senescence. Treatment for delay ripening directly delays postharvest disease development. Relative humidity of the storage environment also affects the development of postharvest disease. Reducing the humidity can reduce disease levels; however, the level of humidity is a trade-off between minimizing postharvest diseases and water loss (Coates et al., 1997).

1.5.1.6 PRECOOLING AND DISINFECTION

Vegetables harvested at high temperature must be precooled to remove excess field heat. Precooling minimizes metabolic activity, respiration rate, and ethylene production (Shahi et al., 2012). Precooling can be achieved by blowing ambient air over the produce (Mishra and Gamage, 2007). A cheap but effective precooling method is by dipping vegetables in cold water (hydrocooling) mixed with disinfectants. Thiabendazole and sodium hypochlorite are examples of disinfectants that can be used to control postharvest decay. Disinfection of fresh vegetables is essential for postharvest decay due to the rapid mortality of pathogens upon contact. However, they do not deposit antimicrobial residue in treated vegetables (Mahovic, n.d.; Feliziani et al., 2016).

1.5.1.7 REFRIGERATION

Refrigeration is an effective storage method for preserving the quality of vegetables when temperature and relative humidity can be regulated. Some vegetables such as tomato can suffer from chilling injury, uneven ripening, and fungi infection. The chilling injury and postharvest diseases can be avoided with an optimum temperature and relative humidity (Castro et al., 2005; Paull, 1999).

1.5.1.8 POSTHARVEST HEAT TREATMENT

Chilling injury caused by unregulated temperature in the refrigerator can be reduced using postharvest heat treatment (Lurie and Klein, 1992). Heat treatment has adopted to quarantine insects, pests, and pathogens and increase product shelf life since ancient times. Heat treatment has been considered as the most effective physical technique to manage fungi diseases which is the most common in the storage of harvest produce (Qadri et al., 2020). This treatment method involves the use of hot air, vapor heat, and heated water as a source of heat (Mishra and Gamage, 2007).

1.6 EMERGING TECHNOLOGIES FOR THE MANAGEMENT OF POSTHARVEST DISEASES

1.6.1 INNOVATIVE PREHARVEST TECHNIQUE FOR POSTHARVEST DISEASE MANAGEMENT

Vegetables constitute very important components of human diet supplying vitamins and essential minerals for growth and development. However, several preharvest and postharvest pathogens affect vegetables causing various levels of economic damage. Some common pathogens which affect vegetables include fungi (Eukaryotes) and bacteria (Prokaryotes) which reduce the quality and shelf-life of the produce and therefore the market value of these crops (Sharma et al., 2009; Barkai-Golan, 2005). These pathogens cause diseases through their inoculum which are left dormant on their surfaces before harvest until conditions are conducive for their development (Mohajar et al., 2015). Several control measures are used to manage these disease-causing organisms before and after harvest ranging from farm sanitation to timely harvesting of the produce as well as chemical control of the pathogens and organisms whose activities predispose the vegetables to these pathogens. Besides these, there is an emerging method of modern technologies aimed at controlling these pathogens on vegetables prior to harvesting and in storage. Some of these include:

1.6.2 PREHARVEST TREATMENTS

1.6.2.1 CHEMICAL TREATMENT

Postharvest use of pesticides to manage pathogens in fruit vegetables is often not allowed in several countries due to health concerns. However, for commodities with longer shelf-life in storage, some selected chemicals which are not phytotoxic and permitted by the local food additive laws are used (Singh and Sharma, 2018). Organic compounds such as sodium O-phenylphenate, dichloran, Sec-butylamine, etc are effective in preventing decay in fungal species such as *Penicillium* sp and *Rhizopus* sp. These chemicals are sometimes applied to the produce before harvesting in order to prevent the pathogens from fully establishing themselves before harvesting of the produce (Hosseinnejad and Jafari, 2016; Singh and Sharma, 2018). Preharvest spraying of fruit vegetables with copper-based compounds has been found to inhibit spores formation on the fruits thereby preventing brown rot. The use of Chitosan, a partially deacetylated derivative of chitin, is used to inhibit growth of bacteria and fungi on fruit vegetables (Hosseinnejad and Jafari, 2016). This compound is highly stable with a neutral pH and is capable of inhibiting microbial activities of several food-borne pathogens (Alishahi and Aïder, 2012).

1.6.2.2 SANITIZATION

During harvesting, and postharvest handling, surfaces of vegetables may develop wounds which may serve as entry points for these pathogens. Sanitization involves ridding the vegetables (especially fruit vegetables) of any pathogens that may cause decay in the produce. The source of inoculum of the pathogen needs to be removed to avoid the proliferation of the spores or fruiting bodies which may help spread the pathogens. The surfaces of fruits or vegetables are often washed with water, steam, or chemicals to eliminate these pathogens (Singh and Sharma, 2018; El-Ramady et al., 2015). This prevents the creation of conducive environment for the pathogens to cause any economic damage (Barkai-Golan, 2015).

1.6.2.3 BIOLOGICAL CONTROL

This method involves development of crop varieties that are resistant to most plant diseases. The approach is to introgress disease into already farmer preferred cultivars of the crop (Wilson and Wisniewski, 1989). Since ripening in some fruit vegetables softens the skin and predisposes them to attack by pathogens, developing crop varieties with extended shelf-life can be an effective way of managing these decay pathogens. Osei et al. (2020) reported the possibility of improving the shelf-life of tomato through the introgression of genes from a ripening mutant tomato, Alc-LA 3134. The success of this approach will ensure that such a variety when used in combination with other disease management strategies would minimize postharvest diseases in vegetables. Another biological control strategy is to elicit resistance responses of the produce. During processes such as wound healing, lignin-like substances are produced by the plant cells that can serve as protection against decay pathogens (Baudoin and Eckert, 1985). Chemicals such as ethylene can be sprayed into the surfaces of the vegetables to induce enzyme activities which produce phenolic compounds and phytoalexins that can help combat postharvest diseases in vegetables. The use of antagonistic microorganisms such as *Candida* sp and *Pichia guilliemondi* induces phytoalexins production and in some cases chitinase, causing the disposition of papillae along the host cell walls (Wisniewski and Wilson, 1992). This results in induced resistance of the host to the decay pathogens. Natural plant products such as essential oils have antimicrobial properties and can be extracted and sprayed on the harvestable portions to prevent the buildup of these decay pathogens prior to harvest (Kanmani and Rhim, 2014; Sharma and Pongener, 2010). Some of these extracts contain high levels of phenolic compounds which can be added to edible coatings and applied to the harvested produce to prevent the development of diseases and extend the shelf-life (Silva-Weiss et al., 2014).

1.6.3 POSTHARVEST TREATMENTS

1.6.3.1 PHYSICAL TREATMENT OF THE PRODUCE

Harvested vegetables such as fruits and root vegetables can be treated with gamma radiation which destroys any pathogen that may be on the surface and can even penetrate the underlying skin. Though sometimes the doses lethal to these pathogens can affect the quality of the vegetables especially the fruits, matured fruits rarely suffer adverse damage when subjected to gamma radiation (Johnson et al., 1990). Irradiation combined with other technologies and induced disease resistance have been found to be very efficient in combating postharvest diseases and thereby maintaining the quality of fruits under extended storage (Jeong and Jeong, 2017).

1.6.3.2 LOW-TEMPERATURE TREATMENT

Low temperatures slow down the ageing of the produce and the growth of pathogens thereby inhibiting their activities (Barkai-Golan, 2005). Ripening of fruit vegetables is a specialized form of the maturity process that reduces the plant's defensive mechanisms and also softening of the physiological barriers to pathogen entry (Prusky, 1996). Ability to manage temperature helps to slow down the ageing process and minimize postharvest physiological deterioration, a condition that predisposes vegetables to decaying pathogens. Temperatures below 0?C have been found to be effective against rhizopus and other decay fungi. Singh and Sharma (2018) reported that every 10?C increase in temperature within the range of 0?C to 30?C increases the metabolic activities; therefore reducing the temperature can slow down this process.

1.6.3.3 HEAT TREATMENT OF PRODUCE

Most postharvest disease pathogens cannot penetrate the surface of vegetables unless there are wounds or cuts to facilitate their entry. Surface treatment of such vegetables can kill most of such pathogens. Harvested vegetables especially fruits are subjected to high temperatures often around 50–55?C to control most pathogens. However, this temperature range can also damage the produce. Li et al. (2015) reported that temperatures above 65?C can lead to cell damage which will eventually affect the quality of the fruit. Heat treatment of tomato fruit at about 37–42C prior to cold storage can slow down the ripening process and also increases their pathogenic resistance during storage (Arah et al., 2016). Heat treatment is done in several ways such as hot water rinsing, vapor heat, hot dry air, etc (Itoh, 2003; Fallik et al., 2001; Barkai-Golan and Phillips, 1991). Most vegetables are very sensitive to high temperatures so factors to consider in heat treatment should include heat sensitivity of the pathogen and the fruit, the location of the targeted organism whether on the fruit or inside the fruit (Singh and Sharma, 2018).

1.6.3.4 POSTHARVEST CHEMICAL TREATMENT OF PRODUCE

One other way of managing postharvest diseases is the treatment of wounded surfaces of these produce with chemicals to prevent the decay organisms from proliferating. Traditionally, indigenous people have used woodash to treat cut surfaces of agricultural produce to prevent growth of decay organisms such as *Rhizopus* sp. Chemicals such as biphenyl, thiobendazol, and carbendazim have been used to control postharvest decay in fruits and vegetables. Pathogens such as Pythium sp, Rhizopus sp, Alternallia solani, and Fusarium have been successfully controlled with aluminum-containing salts (Kolaei et al., 2013; Lorenz, 1988). The use of growth regulators such as 2,4-D, 2,4,5-T, Napthtalane Acetic Acid, gibberellic acids, etc has also been successfully used in controlling fungal diseases in tomato, bringal, and potatoes caused by *Rhizopus stolonifera* (Thakur et al., 1974). Waxing using food-grade waxes has been used to treat fruits and vegetables to seal tiny injuries and scratches that result from harvesting and postharvest operations. Apart from inhibiting the activities of microorganisms, the wax also helps to prevent water loss and improve the shelf-life of the produce (El-Ramady et al., 2015).

1.6.3.5 THE USE OF MICROBIAL ANTAGONISTS

Naturally occurring microbial agents which are antagonistic to postharvest pathogens have been isolated and used to prevent postharvest diseases in vegetables. Singh and Sharma (2018) reported that when fruits and vegetables are washed, they tend to decay faster than those what were left unwashed. Studies have indicated that artificial antagonists can be used to control postharvest diseases in fruits and vegetables and have actually been found to be more effective than other biocontrol methods (Wang et al., 2008; Yang et al., 2006; Xi and Tian, 2005; Saligkaris et al., 2002). Sometimes, these microbial antagonists are used in combination with chitosan-based coatings which come in the form of essential oils, nanoparticles, and acids used to maintain the storage quality and prolong the shelf life vegetables (Xing et al., 2016; Jianglian and Shaoying, 2013).

1.6.3.6 INTEGRATED STRATEGIES OF POSTHARVEST DISEASES MANAGEMENT OF VEGETABLES

This approach encompasses the use of two or more of the different postharvest disease management methods in an integrated manner. It ranges from the use of host plant resistance, good agricultural practices in vegetables cultivation, timely harvesting, proper handling during harvesting and postharvest operations and good curing processes that prevent decay organisms from causing economic damage to the produce (Karabulut et al., 2005). The possibility of integrating the different control strategies depends on how compatible they are. For instance, Mostafari et al. (2011) employed a combination of biological control and irradiation and found that combination of gamma radiation and microbial antagonists such as *Penicilium fluorescens* was effective in the control of *P. expansum* than either method alone. Irradiation of carrots with UV-C (wavelength below 280 nm) was also found to induce resistance to postharvest decay (Mercer et al., 1993). A sound knowledge of biology and ecology of the pathogen, its sources of inoculum, host range (including alternate host), factors that influence growth and development, as well as conditions that favor their activities in the host plants are very important in selecting the appropriate control strategies.

1.7 CONCLUSION

Vegetables are nutritious crops that form an essential component of the human diet. They can be consumed fresh or used in preparation of various delicious meals. To ensure that the optimum nutritional benefits are derived from vegetables, they have to consumed in the right state. Unfortunately, vegetables are highly perishable crops and as such their quality and wholesomeness deteriorates with time. However, there are techniques and technologies that can be used to enhance their shelf life. Enhancing the shelf life benefits multiple stakeholders including consumers and producers (farmers) as well as many people along the vegetable distribution chain.

To better enhance shelf life of vegetables, there is the need to understand the causative factors for poor shelf life. These include genetic factors whereby some varieties or types of vegetables have a gene or set of genes that enable them to have prolonged shelf life compared to others. Aside these, pre- and postharvest practices can also influence the shelf life and quality of vegetables. Preharvest practices include fertilizer application, irrigation, as well as pest and diseases management. On the other hand, certain postharvest practices have the potential to increase or decrease storage life and quality of vegetables. These pre- and postharvest practices induce changes in shelf life of vegetables by eliciting physiological disorders and rapid deterioration through postharvest diseases. Most postharvest diseases of vegetables are caused by fungal and bacterial pathogens present during harvesting or are acquired through handling after harvest or in storage.

Consequently, any approaches aimed at prolonging shelf life should either slow down entry by pathogens, reduce the rate of normal physiological changes that occur in vegetables following harvest or rid the surface of produce of the propagules of disease-causing microorganisms. Such feats may be achieved by the use of specific external chemical, biological, and physical agents or modifying the ambient micro-environment of the produce at the level of individual vegetables as in the case of modified atmosphere packaging (MAP) or varying the concentration of different gases, particularly CO_2, in the storage environment of the whole produce. Another approach is through modification of the host resistance to attack by pathogenic and microbial vegetable spoilage organisms through development of new varieties that have shelf life enhancing genes or lack receptors for certain gases such as ethylene.

While some traditional techniques for prolonging shelf life of vegetables work effectively, limitations or costs associated with their use necessitate development of new technologies or enhancement upon existing ones. Consequently, emerging technologies have been and are continuously being developed to further prolong the shelf life and maintain the quality of vegetables. Thus, it is safe to conclude that the area of improving shelf life of vegetables is under active research with the development of better technologies and integrated approaches.

KEYWORDS

- **disease management**
- **microbial agents**
- **preharvest diseases**
- **postharvest diseases**

REFERENCES

Abu, M. Effect of Compound and Single-based Fertilizers on Shelf Life of Exotic/Minor Vegetable Crops Cultivated for the Fresh Market. *Hortic. Int. J.* **2021,** *5* (2), 90–95.

Adom, D. The Role of Good Post-harvest Management Practices in the Development of the Agriculture Sector in Ghana. In *Trends and Prospects in Post-harvest Management of Horticultural Crops*; 2018, pp 169–188.

Adhikari S. Country Paper: Nepal (2). In *APO. 2006. Postharvest Management of Fruit and Vegetables in the Asiapacific Region*; Asian Productivity Organization (APO) and FAO, 2006; pp 200–208.

Aguayo, E.; Escalona, V. H.; Artés, F. Effect of Cyclic Exposure to Ozone Gas on Physicochemical, Sensorial and Microbial Quality of Whole and Sliced Tomatoes. *Postharvest Biol. Technol.* **2006,** *39* (2), 169–177.

Aidoo, R.; Danfoku, R. A.; Mensah, J. O. Determinants of Postharvest Losses in Tomato Production in the Offinso North District of Ghana. *J. Dev. Agric. Econ.* **2014,** *6* (8), 338–344.

Allende, A.; Martínez, B.; Selma, V.; Gil, M. I.; Suárez, J. E.; Rodríguez, A. Growth and Bacteriocin Production by Lactic Acid Bacteria in Vegetable broth and their Effectiveness at Reducing Listeria Monocytogenes In Vitro and in Fresh-Cut Lettuce. *Food Microbiol.* **2007,** *24,* 759–766.

Alishahi, A.; Aïder, M. Applications of Chitosan in the Seafood Industry and Aquaculture: A Review. *Food Bioproc. Tech.* **2012,** *5* (3): 817–830.

Ansorena, M. R.; Moreira, M. R.; Roura, S. I. Combined Effect of Ultrasound, Mild Heat Shock and Citric Acid to Retain Greenness, Nutritional and Microbiological Quality of Minimally Processed Broccoli (Brassica Oleracea L.): An Optimization Study. *Postharvest Biol. Technol.;* **2014,** *94,* 1–13.

Arah, I. K.; Ahorbo, G. K.; Anku, E. K.; Kumah, E. K.; Amaglo, H. Postharvest Handling Practices and Treatment Methods for Tomato Handlers in Developing Countries: A Mini Review. *Adv. Agric.* **2016,** *2016,* 1–8.

Arah, I. K.; Amaglo, H.; Kumah, E. K.; Ofori, H. Preharvest and Postharvest Factors Affecting the Quality and Shelf Life of Harvested Tomatoes: A Mini Review. *Int. J. Agro.* **2015,** *2015.* https://doi.org/10.1155/2015/478041

Ashiq, S. Natural Occurrence of Mycotoxins in Food and Feed: Pakistan Perspective. *Compr. Rev. Food Sci. Food Saf.* **2015,** *14* (2), 159–175.

Lippman, B.; Yao, S.; Huang, R.; Chen, H. Evaluation of the Combined Treatment of Ultraviolet Light and Peracetic Acid as an Alternative to Chlorine Washing for Lettuce Decontamination. *Int. J. Food Microbiol.* **2020,** *323,* 108590.

Bachman, J.; Earles, R. Postharvest Handling of Fruits and Vegetables. Horticulture Technical Note, 2000; pp 1–19.

Barbosa, M. S.; Jurkiewicz, C.; Landgraf, M.; Todorov, S. D.; Franco, B. D. G. M. Effect of Proteins, Glucose and NaCl on Growth, Biosynthesis and Functionality of Bacteriocins of Lactobacillus Sakei subsp. sakei 2a in Foods during Storage at 4°C: Tests in Food Models. *LWT* **2018,** *95,* 167–171.

Barkai-Golan, R. *Postharvest Diseases of Fruits and Vegetables*; A Division of Reed Elsevier India, Pvt. Limited: New Delhi, 2005.

Barrett, D. M.; Beaulieu, J. C.; Shewfelt, R. Color, Flavor, Texture, and Nutritional Quality of Fresh-Cut Fruits and Vegetables: Desirable Levels, Instrumental and Sensory Measurement, and the Effects of Processing. *Crit. Rev. Food Sci. Nutr.* **2010,** *50* (5), 369–389.

Baudoin, A. B. A. M.; Eckert, J. W. Development of Resistance Against *Geotrichum candidum* in Lemon Peel Injuries. *Phytopathologyphytopathology* **1985,** *74,* 174–179.

Bhat, K. L. *Physiological Disorders of Vegetable Crops*; Daya Publishing House: India, 2009.

Ben-Yehoshua, S.; Rodov, V. Transpiration and Water Stress. In *Postharvest Physiology and Pathology of Vegetables*; 2002; pp. 111–159.

Berg, J. D.; Roberts, P. V.; Matin, A. Effect of Chlorine Dioxide on Selected Membrane Functions of Escherichia-coli. *J. Appl. Bacteriol.* **1986,** *60* (3), 213–220.

Carletti, L.; Botondi, R.; Moscetti, R.; Stella, E.; Monarca, D.; Cecchini, M.; et al. Use of Ozone in Sanitation and Storage of Fresh Fruits and Vegetables. *J. Food Agric. Environ.* **2013,** *11* (3 and 4), 585–589.

Clifton, P. M.; Petersen, K. S.; Blanch, N.; Keogh, J. B. How do Fruit and Vegetables Prevent Heart Disease and Type 2 Diabetes? *Curr. Opin. Lipidol.* **2014,** *25* (2), 155–156.

Clotault, J.; Peltier, D.; Berruyer, R.; Thomas, M.; Briard, M.; Geoffriau, E. Expression of Carotenoid Biosynthesis Genes During Carrot Root Development. *J. Exp. Bot.* **2008,** *59* (13), 3563–3573.

Coates, L.; Johnson, G. Postharvest Diseases of Fruit and Stored Vegetables. In *Plant Pathogens and Plant Diseases*; Brown, J. F., Ogle, H. J., Ed.; University of New England Printery: NSW: Armidale, 1997; pp 533–548.

Croft, K. D. The Chemistry and Biological Effects of Flavonoids and Phenolic Acids. *Ann. N. Y. Acad. Sci.* **1998,** *854,* 435–442.

Cunningham, F. X.; Gantt, E. Genes and Enzymes of Carotenoid Biosynthesis in Plants. *Annu. Rev. Plant Physiol. Plant mol. Biol.* **1998,** *49* (1), 557–583.

Das, K.; Roychoudhury, A. Reactive Oxygen Species (ROS) and Response of Antioxidants as ROS-Scavengers during Environmental Stress in Plants. *Front. Environ. Sci.* **2014,** *2* (DEC), 1–13.

Dinah, K. African Leafy Vegetables Pre-harvest and Post-harvest Constrains and. *Researchgate* **2019,** *2018.*

Ellison, S. L.; Luby, C. H.; Corak, K. E.; Coe, K. M.; Senalik, D.; Iorizzo, M.; Goldman, I. L.; et al. Carotenoid Presence is Associated with the or Gene in Domesticated Carrot. *Genetics* **2018,** *210* (4), 1497–1508.

El-Ramady H. R.; Domokos-Szabolcsy E.; Abdalla N. A.; Taha H. S.; Fári M. Postharvest Management of Fruits and Vegetables Storage. *Sustain. Agric. Rev.* **2015,** *15,* 66–132.

Elik, A.; Yanik, D. K.; Istanbullu, Y.; Guzelsoy, N. A.; Yavuz, A.; Gogus, F. Strategies To Reduce Post-harvest Losses For Fruits and Vegetables. *Strategies,* **2019,** *5* (3), 29–39.

Etana, M. B.; Fufa, B. O.; Aga, M. C. A Detailed Review on Common Causes of Postharvest Loss and Quality Deterioration of Fruits and Vegetables in Ethiopia. *J. Biol. Agric. Healthcare* **2019,** *9* (7), 48–52.

Ferraz, P.; Cássio , F.; Lucas, C. Potential of Yeasts as Biocontrol Agents of the Phytopathogen Causing Cacao Witches' Broom Disease: is Microbial Warfare a Solution? *Front Microbiol.* **2019,** *10,* 1766.

Food and Agriculture Organization of the United Nations (FAO) [Online]. 2016. Available at: http://www.fao.org/faostat/en/#data

Francenia Santos-Sánchez, N.; Salas-Coronado, R.; Hernández-Carlos, B.; Villanueva-Cañongo, C. Shikimic Acid Pathway in Biosynthesis of Phenolic Compounds. In *Plant Physiological Aspects of Phenolic Compounds*; 2019. pp 1–15.

Fritsche, S.; Wang, X.; Jung, C. Recent Advances in Our Understanding of Tocopherol Biosynthesis in Plants: An Overview of Key Genes, Functions, and Breeding of Vitamin E Improved Crops. *Antioxidants* **2017,** *6* (99), 1–18.

Gastol, M.; Domagala-Swiatkiewicz, I.; Krosniak, M. Organic Versus Conventional—A Comparative Study on Quality and Nutritional Value of Fruit and Vegetable Juices. *Biol. Agric. Hortic.* **2011,** *27* (3–4), 310–319.

Gemenet, D. C.; Da Silva Pereira, G.; De Boeck, B.; Wood, J. C.; Mollinari, M.; Olukolu, B. A.; Diaz, F.; et al. Quantitative Trait Loci and Differential Gene Expression Analyses Reveal The Genetic Basis For Negatively Associated B-Carotene and Starch Content In

Hexaploid Sweetpotato [Ipomoea Batatas (L.) Lam.]. *Theor. Appl. Genet.*, **2020,** *133* (1), 23–36. https://doi.org/10.1007/s00122–019–03437-7

Gil, M. I.; Selma, M. V.; López-Gálvez, F.; Allende, A.; Fresh-cut Product Sanitation and Wash Water Disinfection: Problems and Solutions. *Int. J. Food Microbiol.* **2009,** *134,* 37–45.

Gill, S. S.; Tuteja, N. Reactive Oxygen Species and Antioxidant Machinery in Abiotic Stress Tolerance in Crop Plants. *Plant Physiol. Biochem.* **2010,** *48* (12), 909–930. doi: 10.1016/j.plaphy.2010.08.016

Gogo, E.; Opiyo, A.; Ulrichs, C.; Huyskens-Keil, S. Postharvest Treatments of African Leafy Vegetables for Food Security in Kenya: A Review. *African J. Agric. Sci.* **2016.**

Grubben, G.; Klaver, W.; Nono-Womdim, R.; Everaarts, A.; Fondio, L.; Nugteren, J. A.; Corrado, M. Vegetables to Combat the Hidden Hunger in Africa. *Chron. Hortic.* **2014,** *54* (1), 24–32.

Hailu, G.; Derbew, B. Extent, Causes and Reduction Strategies of Postharvest Losses of Fresh Fruits and Vegetables—A Review. *J. Biol. Agric. Healthcare* **2015,** *5* (5), 49–64.

Hammond, S. T.; Brown, J. H.; Burger, J. R.; Flanagan, T. R.; Fristoe, T. S.; Mercado-Silva, N.; et al. Food Spoilage, Storage, and Transport: Implications for a Sustainable Future. *Bioscience* **2015,** *65* (8), 758–768.

Hartz, T. K. Vegetable Production Best Management Practices to Minimize Nutrient Loss. *Horttechnology* **2006,** *16* (3), 398–403. doi: 10.21273/horttech.16.3.0398

Hermanns, A. S.; Zhou, X.; Xu, Q.; Tadmor, Y.; Li, L. Carotenoid Pigment Accumulation in Horticultural Plants. *Hortic. Plant J.* **2020,** *6* (6), 343–360.

Horvitz, S.; Cantalejo, M. J. Effects of Ozone and Chlorine Postharvest Treatments on Quality of Fresh-Cut Red Bell Peppers. *Int. J. Food Sci. Technol.* **2012,** *47* (9), 1935–1943.

Hosseinnejad, M.; Jafari, S. M. Evaluation of Different Factors Affecting Antimicrobial Properties of Chitosan. *Int. J. Biol. Macromol.* **2016,** *85,* 467–475.

Jiang, T.; Pearce, D. Shelf-Life Extension of Leafy Vegetables: Evaluating the Impacts. *Impact Assessment Series Report* **2005,** *32,* 62.

Jeong, M.-A.; Jeong, R.-D. Applications of Ionizing Radiation for the Control of Postharvest Diseases in Fresh Produce: Recent Advances. *Plant Pathol.* **2017,** *67,* 18–29. doi: 10.1111/ppa.12739.

Jianglian, D.; Shaoying, Z. Application of Chitosan Based Coating in Fruit and Vegetable Preservation: A Review. *J. Food Process. Technol.* **2013,** *4* (5), 227–230. doi: 10.4172/2157-7110.1000227

Johnson, G.; Boag, T. S.; Cook, A. W.; Izard, M.; Panitz, M.; Sangchote, S. Interaction of Postharvest Disease Control Treatments and Gamma Irradiation on Mangoes. *Ann. Appl. Biol.* **1990,** *116,* 245–257.

JICA. *Vegetable Farming Techniques Manual*, 2016; pp 1–138.

Kanmani, P.; Rhim, J. W. Antimicrobial and Physical-Mechanical Properties of Agar-based Films Incorporated with Grapefruit Seed Extract. *Carbohydr. Polym.* **2014,** *102,* 708–716.

Karabulut, O. G.; Arslan, U.; Kadir, I.; Gul, K. Integrated Control of Postharvest Diseases of Sweet Cherry with Yeast Antagonist and Sodium Bicarbonate Applications within a Hydrocooler. *Postharvest Biol. Technol.* **2005,** *37,* 135–141.

Kolaei, E. A.; Cenatus C.; Tweddell, R. J.; Avis, T. J. Antifungal Activity of Aluminium-Containing Salts against the Development of Carrot Cavity Spot and Potato Dry Rot. *Ann. Appl. Biol.* **2013,** *163* (2): 311–317.

Kim, H. S.; Ji, C. Y.; Lee, C. J.; Kim, S. E.; Park, S. C.; Kwak, S. S. Orange: A Target Gene for Regulating Carotenoid Homeostasis and Increasing Plant Tolerance to Environmental Stress in Marginal Lands. *J. Exp. Bot.* **2018,** *69* (14), 1–8.

Kim, Y. H.; Jeong, S. G.; Back, K. H.; Park, K. H.; Chung, M. S.; Kang, D. H. Effect of Various Conditions on Inactivation of Escherichia Coli O157: H7, Salmonella Typhimurium, and Listeria Monocytogenes in Fresh-cut Lettuce Using Ultraviolet Radiation. *Int. J. Food Microbiol.* **2013,** *166* (3), 349–355.

Kniel, K. E.; Sumner, S. S.; Lindsay, D. S.; Hackney, C. R.; Pierson, M. D.; Zajac, A. M.; et al. Effect of Organic Acids and Hydrogen Peroxide on Cryptosporidium Parvum Viability in Fruit Juices. *J. Food Prot.* **2003,** *66* (9), 1650–1657.

Kore, V. T.; Tawade, S. S.; Kabir, J. Application of Edible Coatings on Fruits and Vegetables. *Imp. J. Interdiscip. Res.* **2017,** *3* (1), 591–603.

Koukounaras, A.; Bantis, F.; Karatolos, N.; Melissas, C.; Vezyroglou, A. Influence of Pre-harvest Factors on Postharvest Quality of Fresh-Cut and Baby Leafy Vegetables. *Agronomy* **2020,** *10* (2), 172.

Kusvuran, S.; Kiran, S.; Ellialtioglu, S. S. Antioxidant Enzyme Activities and Abiotic Stress Tolerance Relationship in Vegetable Crops. In *Abiotic and Biotic Stress in Plants—Recent Advances and Future Perspectives*; 2016, pp 481–506.

Lastochkina, O.; Seifikalhor, M.; Aliniaeifard, S.; Baymiev, A.; Pusenkova, L.; Garipova, S.; . . . Maksimov, I. Bacillus spp.: Efficient Biotic Strategy to Control Postharvest Diseases of Fruits and Vegetables. *Plants* **2019,** *8* (4), 97.

Li, J.; Zhang, Y.; Ding, Q.; Li, H.; Liu, L.; Wang, F.; Gao, J. Transcriptome Analysis of Orange Head Chinese Cabbage (Brassica Rapa L. ssp. Pekinensis) and Molecular Marker Development. *Int. J. Genomics*, **2017,** *2017*, 1.

Li, L.; Paolillo, D. J.; Parthasarathy, M. V.; Dimuzio, E. M.; Garvin, D. F. A Novel Gene Mutation That Confers Abnormal Patterns of ?-Carotene Accumulation in Cauliflower (Brassica Oleracea var. Botrytis). *Plant J.* **2001,** *26* (1), 59–67.

Lopez, A. B.; Van Eck, J.; Conlin, B. J.; Paolillo, D. J.; O'Neill, J.; Li, L. Effect of the Cauliflower or Transgene on Carotenoid Accumulation and Chromoplast Formation in Transgenic Potato Tubers. *J. Exp. Bot.* **2008,** *59* (2), 213–223.

Lownds, N. K.; Banaras, M.; Bosland, P. W. Postharvest Water Loss and Storage Quality of Nine Pepper (Capsicum) Cultivars. *Hortsciencehortscience* **1994,** *29,* 191–193.

López-Gálvez, F.; Ragaert, P.; Haque, M. A.; Eriksson, M.; Van Labeke, M. C.; Devlieghere, F. High Oxygen Atmospheres can Induce Russet Spotting Development in Minimally Processed Iceberg Lettuce. *Postharvest Biol. Technol.* **2015,** *100,* 168–175. https:/doi.org/10.1016/j.postharvbio.2014.10.001

Lorenz, G. Dicarboximide Fungicides: History of Resistance Development and Monitoring Methods. In *Fungicide Resistance in North America*; Delp, C. J., Ed.; APS Press: St. Paul, 1988; pp 45–51.

Lu, S.; Van Eck, J.; Zhou, X.; Lopez, A. B.; Halloran, D. M. O.; Cosman, K. M.; Conlin, B. J.; et al. The Cauliflower or Gene Encodes a DnaJ Cysteine-rich Domain-containing Protein that Mediates High Levels of B-Carotene Accumulation. *Plant Cell* **2006,** *18* (December), 3594–3605.

Lunet, N.; Lacerda-Vieira, A.; Barros, H. Fruit and Vegetables Consumption and Gastric Cancer: A Systematic Review and Meta-analysis of Cohort Studies. *Nutr. Cancer Int. J.* **2005,** *53* (1), 1–10.

Ma, J.; Xu, Z.; Tan, G.; Wang, F.; Xiong, A. Distinct Transcription Profile of Genes Involved in Carotenoid Biosynthesis Among Six Different Color Carrot (Daucus Carota L.) Cultivars. *Acta Biochim. Biophy. Sin.* **2017,** *49* (9), 817–826.

Mahajan, B. C.; Tandon, R.; Kapoor, S.; Sidhu, M. K. Natural Coatings for Shelf-Life Enhancement and Quality Maintenance of Fresh Fruits and Vegetables—A Review. *J. Postharvest Technol.* **2018,** *6* (1), 12–26.

Martínez, B.; García, P.; Rodríguez, A.; Swapping the Roles of Bacteriocins and Bacteriophages in Food Biotechnology. *Curr. Opin. Biotechnol.* **2019,** *56,* 1–6.

Meir, S.; Rosenberger, Y.; Aharon, Z.; Grienber, S.; Fallik, E. Improvement of the Postharvest Keeping Quality and Colour Development of Bell Pepper by Packaging with Polyethylene Bags at a Reduced Temperature. *Postharvest Biol. Technol.;* **1995,** *5,* 303–309.

Meireles, A.; Giaouris, E.; Simões, M.; Alternative Disinfection Methods to Chlorine for Use in the Fresh-Cut Industry. *Food Res. Int.* **2016,** *82,* 71–85.

Mellidou, I.; Kanellis, A. K. Genetic Control of Ascorbic Acid Biosynthesis and Recycling in Horticultural Crops. *Front. Chem.* 2017.

Mène-Saffrané, L.; Pellaud, S. Current Strategies for Vitamin E Biofortification of Crops. *Curr. Opin. Biotechnol.* 2017.

Mercier, J.; Arul, J.; Ponnanapalan, R.; Boulet, M. Induction of 6-Methoxymellein and Resistance to Storage Pathogens in Carrot Slices By UV-C. *J. Phytopathol.* **1993,** *137,* 44–54.

Mohajer, S.; Taha, R. T.; Yaacob, J. S.; Kumari, A. Post Harvest Management of the Pests and Some Important Diseases of the Fruits. In *Recent Trends in Post Harvest Technology and Management*; Manglam Publications: India, 2015; Vol. 1, pp 1–12. https://umexpert.um.edu.my/file/publication/00009939_108037.pdf

Molinos, A. C.; Abriouel, H.; Ben Omar, N.; Valdivia, E.; Lopez, R. L.; Maqueda, M.; Canamero, M. M.; Galvez, A. Effect of Immersion Solutions Containing Enterocin AS-48 on Listeria Monocytogenes in Vegetable Foods. *Appl. Environ. Microbiol.* **2005,** *71,* 7781–7787.

Moult, J.; Allan, S.; Hewitt, C.; Berners-Lee, M. Greenhouse Gas Emissions of Food Waste Disposal Options for UK Retailers. *Food Policy* **2018,** *77,* 50–58.

Mridha, F.; Huque, R.; Khatun, A.; Islam, M.; Hossain, A.; Hossain, M. A.; et al. Effects of Gamma Irradiation on Antioxidant Markers, Microbial Population and Sensory Attributes of Strawberry (Fragaria × Ananassa Duch.) cv. Festival. *Plant Cell Biotechnol. Mol. Biol.* **2017,** *18* (5 and 6), 208–218.

Muchoki, C. N.; Imungi, J. K.; Lamuka, P. O. Changes in Beta-carotene, Ascorbic Acid and Sensory Properties in Fermented, Solar-dried and Stored Cowpea Leaf Vegetables. *African J. Food Agric.* **2007**.

Munemasa, S.; Hauser, F.; Park, J.; Waadt, R.; Brandt, B.; Schroeder, J. I. Mechanisms of Abscisic Acid-Mediated Control of Stomatal Aperture. *Curr. Opin. Plant Biol.* **2015,** doi:10.1016/j.pbi.2015.10.010

Narayanasamy, P. Integrated Systems for the Management of Postharvest Diseases. In *Postharvest Pathogens and Disease Management*; John Wiley & Sons, Inc., 2006; Vol. 12, pp 537–554. https://doi.org/10.1002/0471751987.ch12

Number, S. *J. Agric. Sci. Technol. A* **2013,** *3* (1), 1–6.

Nunes, C. N.; Emond, J.-P. In *Relationship between Weight Loss and Visual Quality of Fruits and Vegetables*, Proceeedings of Florida State Horticultural Society, 2007; Vol. 120; pp 235–245. Retrieved from http://www.envirotainer.

Olanya, O. M.; Annous, B. A.; Taylor, J. Effects of Pseudomonas Chlororaphis and Gaseous Chlorine Dioxide on the Survival of Salmonella Enterica on Tomatoes. *Int. J. Food Sci. Technol.* **2015,** *50* (5), 1102–1108.

Oliveira, M.; Abadias, M.; Usall, J.; Torres, R.; Teixidó, N.; Viñas, I. Application of Modified Atmosphere Packaging as a Safety Approach to Fresh-Cut Fruits and Vegetables—A Review. *Trends Food Sci. Technol.* **2015,** *46* (1), 13–26.

Osei, M. K.; Danquah, A.; Danquah, E.; Blay, E.; Adu-Dapaah, H. Gene Action of Shelf-Life and Other Fruit Quality Traits in a Cross Between a Regular Cultivar and *Alc* Mutant of Tomato. *Agric. Food Sci. J. Ghana* **2020,** *13,* 1225–1237.

Patel, R.; Rinker, L.; Peng, J.; Chilian, W. M. Reactive Oxygen Species: The Good and the Bad. *Reactive Oxygen Species (ROS) In Living Cells*; 2018.

Papachristodoulou, M.; Koukounaras, A.; Siomos, A. S.; Liakou, A.; Gerasopoulos, D. The Effects of Ozonated Water on the Microbial Counts and the Shelf Life Attributes of Fresh-Cut Spinach. *J. Food Process. Preserv.* **2018,** *42* (1).

Paramithiotis, S.; Drosinos, E. H.; Skandamis, P. Microbial Ecology of Fruits and Fruit-Based Products. In *Quantitative Microbiology in Food Processing—Modeling the Microbial Ecology*; De Souza Sant'Ana, A., Ed.; John Wiley & Sons, Ltd: New York, 2017; pp 358–381.

Prange, R. K. Pre-harvest, Harvest and Post-harvest Strategies for Organic Production of Fruits and Vegetables. *Acta Hortic,* **2012,** *933* (March 2012), 43–50. https://doi.org/10.17660/actahortic.2012.933.3

Praeger, U.; Herppich, W. B.; Hassenberg, K. Aqueous Chlorine Dioxide Treatment of Horticultural Produce: Effects on Microbial Safety and Produce Quality—A Review. *Crit. Rev. Food Sci. Nutr.* **2016,** 1–16.

Pobiega, K.; Kraśniewska, K.; Gniewosz, M. Application of Propolis in Antimicrobial and Antioxidative Protection of Food Quality—A Review. *Trends Food Sci. Technol.* **2019,** *83,* 53–62.

Porat, R.; Lichter, A.; Terry, L. A.; Harker, R.; Buzby, J. Postharvest Losses of Fruit and Vegetables During Retail and in Consumers' Homes: Quantifications, Causes, and Means of Prevention. *Postharvest Biol. Technol.;* **2018,** *139,* 135–149.

Prusky, D. Pathogen Quiescence in Postharvest Diseases. *Annu. Rev. Phytopathol.* **1996,** *34,* 413–434.

Putnik, P.; Roohinejad, S.; Greiner, R.; Granato, D.; Bekhit, A. E. D. A.; Kovačević, D. B. Prediction and Modeling of Microbial Growth in Minimally 847 Processed Fresh-Cut Apples Packaged in a Modified Atmosphere: A Review. *Food Control* **2017,** *80,* 411–419.

Rajwar, A.; Srivastava, P.; Sahgal, M. Microbiology of Fresh Produce: Route of Contamination, Detection Methods, and Remedy. *Crit. Rev. Food Sci. Nutr.* **2016,** *56* (14), 2383–2390.

Rekhy, R.; Mcconchie, R. Promoting Consumption of Fruit and Vegetables for Better Health. Have Campaigns Delivered on The Goals? *Appetite* **2014,** *79,* 113–123.

Ren, J.; Chen, Z.; Duan, W.; Song, X.; Liu, T.; Wang, J.; Hou, X.; et al. Comparison of Ascorbic Acid Biosynthesis in Different Tissues of Three Non-heading Chinese Cabbage Cultivars. *Plant Physiol. Biochem.* **2013,** *73,* 229–236.doi:10.1016/j.plaphy.2013.10.005

Sforcin, J. M.; Bankova, V. Propolis: is there a Potential for the Development of New Drugs? *J. Ethnopharmacol.* **2011,** *133,* 253–260.

Sharma, R. R.; Pongener, A. Natural Products for Postharvest Disease Control of Horticultural Produce: A Review. *Stewart Postharvest Rev.* **2010,** *4,* 1–9.

Sharma, R. R.; Singh, D.; Singh, R. Biological Control of Postharvest Diseases of Fruits and Vegetables by Microbial Antagonists: A Review. *Biol. Control.* **2009,** *2009* (50), 205–221.

Silva-Weiss, A.; Bifani, V.; Ihl, M.; Sobral, P. J. A., Gomez-Guillen, M. C. Polyphenol-Rich Extract from Murta Leaves on Rheological Properties of Film-Forming Solutions based on Different Hydrocolloid Blends. *J. Food Eng.* **2014,** *140,* 28–38.

Singh, D.; Sharma, R. R. Postharvest Diseases of Fruits and Vegetables and Their Management. In *Postharvest Disinfection of Fruits and Vegetables*; Siddiqui, M. W. Ed.; Academic Press, 2018; pp 1–52.

Siegel, K. R. Insufficient Consumption of Fruits and Vegetables among Individuals 15 Years and Older in 28 Low-And Middle-Income Countries: What can be Done? *J. Nutr.* **2019,** *149* (7), 1105–1106.

Singh, V.; Hedayetullah, M.; Zaman, P.; Meher, J. Postharvest Technology of Fruits and Vegetables: An Overview. *J. Postharvest Technol.* **2014,** *2* (2), 124–135.

Siroli, L.; Patrignani, F.; Serrazanetti, D. I.; Vernocchi, P.; Delchierico, F.; Russo, A.; Torriani, S.; Putignani, L.; Gardini, F.; Lanciotti, R. Effect of Thyme Essential Oil and Lactococcus Lactis CBM21 on the Microbiota Composition and Quality of Minimally Processed Lamb's Lettuce. *Food Microbiol.* **2017,** 68.

Soliva-Fortuny, R. C.; Martin-Belloso, O. New Advances in Extending the Shelflife of Fresh-Cut Fruits: A Review. *Trends Food. Sci. Technol.* **2003,** *14* (9), 341–353.

Sorensen, J. Nitrogen Effects on Vegetable Crop Production and Chemical Composition. *Acta Hortic.* **1999,** *506,* 41–50. doi: 10.17660/actahortic.1999.506.4

Spadaro, D.; Droby, S. Development of Biocontrol Products for Postharvest Diseases of Fruit: The Importance of Elucidating the Mechanisms of Action of Yeast Antagonists. *Trends Food Sci. Technol.* **2016,** *47,* 39–49.

Stanley, L.; Yuan, Y. W. Transcriptional Regulation of Carotenoid Biosynthesis in Plants: So many Regulators, so Little Consensus. *Front. Plant Sci.* **2019,** *10.*

Sugiprihatini, D. W. I.; Wiyono, S. Selection of Yeasts Antagonists as Biocontrol Agent of Mango Fruit Rot Caused by Botryodiplodia Theobromae. *Microbiol. Indones.* **2012,** *5* (4).

Sun, X.; Bai, J.; Ference, C.; Wang, Z.; Narciso, J.; Zhou, K. Antimicrobial Activity of Controlled-Release Chlorine Dioxide Gas on Fresh Blueberries. *J. Food Prot.* **2014,** *77,* 1127–1132.

Thakur, D. P.; Chennulu, V. V.; Kanwar, Z. S.; Kherata, R. B. S. Growth Regulators in Control of Postharvest Fungal Disease of Fruits and Vegetables. *Indian Phytopathol.* **1974,** *27,* 532–536.

Threlfall, D. R.; Whistance, G. R. Biosynthesis of Tocopherols and Biogenetically Related Compounds. *Angew. Chem.* **1967,** *6* (11), 369–396.

Tohge, T.; De Souza, L. P.; Fernie, A. R. Current Understanding of the Pathways of Flavonoid Biosynthesis in Model and Crop Plants. *J. Exp. Bot.* **2017,** *68* (15), 4013–4028.

Tressl, R.; Bahri, D.; Engel, K.-H. Lipid Oxidation in Fruits and Vegetables. In *Quality of Selected Fruits and Vegetables of North America*; 1981; pp 213–232.

Tscharntke, T.; Clough, Y.; Wanger, T. C.; Jackson, L.; Motzke, I.; Perfecto, I.; ?? Whitbread, A. Global Food Security, Biodiversity Conservation and the Future of Agricultural Intensification. *Biol. Conserv.* **2012,** *151* (1), 53–59.

Van Meeteren, U.; Aliniaeifard, S. Stomata and Postharvest Physiology. In *Postharvest Ripening Physiology of Crops*; Pareek, S., Ed.; CRC Press: New York, USA, 2016; pp 157–216.

Vinutha, T.; Bansal, N.; Kumari, K.; Prashat, R. G.; Sreevathsa, R.; Krishnan, V.; Kumari, S.; et al. Comparative Analysis of Tocopherol Biosynthesis Genes and its Transcriptional Regulation in Soybean Seeds. *J. Agric. Food Chem.* **2017,** *65* (50), 11054–11064.

Wang, Y.; Li, X.; Bi, Y.; Ge, Y.; Li, Y.; Xie, F. Postharvest ASM or Harpin Treatment Induce Resistance of Muskmelons against *Trichothecium Roseum*. *Agric. Sci. China* **2008,** *7* (2), 217–223.

Weston, L.; Barth, M. Preharvest Factors Affecting Postharvest Quality of Vegetables. *Hortscience* **1997,** *32* (5), 812–816.

Wheeler, G. L.; Jones, M. A.; Smirnoff, N. The Biosynthetic Pathway of Vitamin C in Higher Plants. *Nature* **1998,** *393* (6683), 365–369.

Wilson, C. L.; Wisniewski, M. E. Biological Control of Postharvest Diseases of Fruits and Vegetables: An Emerging Technology. *Annu. Rev. Phytopathol.* **1989,** *27,* 425–441.

Wisniewski, M. E.; Wilson, C. L. Biological Control of Postharvest Diseases of Fruits and Vegetables: Recent Advances. *Hortic. Sci.* **1992,** *27,* 94–98.

Wu, V. C. H.; Rioux, A. A Simple Instrument-Free Gaseous Chlorine Dioxide Method for Microbial Decontamination of Potatoes During Storage. *Food Microbiol.* **2010,** *27,* 179–184.

Xanthopoulou, A.; Montero-Pau, J.; Mellidou, I.; Kissoudis, C.; Blanca, J.; Picó, B.; Tsaballa, A.; et al. Whole-Genome Resequencing of Cucurbita Pepo Morphotypes to Discover Genomic Variants Associated with Morphology and Horticulturally Valuable Traits. *Hortic. Res.* **2019,** *6* (1), 1–17.

Xi, L.; Tian, S. P. Control of Postharvest Diseases of Tomato Fruit by Combining Antagonistic Yeast with Sodium Bicarbonate. *Sci. Agric. Sin.* **2005,** *38* (5), 950–955.

Sun, X.; Zhou, B.; Luo, Y.; Ference, C.; Baldwin, E.; Harrison, K.; Bai, J. Effect of Controlled-Release Chlorine Dioxide on the Quality and Safety of Cherry/Grape Tomatoes. *Food Control* **2017,** *82,* 26–30.

Sun, X.; Baldwin, E.; Bai, J. Applications of Gaseous Chlorine Dioxide on Postharvest Handling and Storage of Fruits and Vegetables—A Review. *Food Control* **2019,** *95*, 18–26.

Xing, Y.; Xu, Q.; Li, X.; Chen, C.; Ma, L.; Li, S.; Che, Z.; Lin, H. Chitosan-based Coating with Antimicrobial Agents: Preparation, Property, Mechanism, and Application Effectiveness on Fruits and Vegetables. *Int. J. Polym. Sci.* **2016,** *2016,* 1–24. doi: 10.1155/2016/4851730

Xu, Y.; Zhu, X.; Chen, Y.; Gong, Y.; Liu, L. Expression Profiling of Genes Involved in Ascorbate Biosynthesis Andrecycling During Fleshy Root Development in Radish. *Plant Physiol. Biochem.* **2013,** *70,* 269–277.

Yahia, E. M.; et al. Influence of Postharvest Technologies and Handling Practices on Phytochemicals in Fruits and Vegetables. In *Fruit and Vegetable Phytochemicals: Chemistry and Human Health*; 2nd ed., 2017; pp 609–662.

Yang, D. M.; Bi, Y.; Chen, X. R.; Ge, Y. H.; Zhao, J. Biological Control of Postharvest Diseases with Bacillus Subtilis (B1 Strain) on Muskmelons (*Cucumis melo* L. cv. Yindi). *Acta Hortic.* **2006,** *712* (2):735–739.

Yokoyama, R.; De Oliveira, M. V. V.; Kleven, B.; Maeda, H. A. The Entry Reaction of the Plant Shikimate Pathway is Subjected to Highly Complex Metabolite-Mediated Regulation. *Plant Cell* **2021,** *33* (3), 671–696.

Yuan, H.; Zhang, J.; Nageswaran, D.; Li, L. Carotenoid Metabolism and Regulation in Horticultural Crops. *Hortic. Res.* **2015,** *2* (2), 1–11.

Zhang, B. D.; Huang, C. X.; Zhang, L. Y.; Wang, J.; Huang, X. Q.; Zhao, Y.; Liu, Y.; Li, C. C. Application of Chlorine Dioxide Microcapsule Sustained-Release Antibacterial Films for Preservation of Mangos. *J. Food Sci. Technol.* **2019,** *56,* 1095–1103.

Zhang, L.; Zhu, M.; Ren, L.; Li, A.; Chen, G.; Hu, Z. The SLFSR Gene Controls Fruit Shelf-life in Tomato. *J. Exp. Bot.* **2018,** *69* (12), 2897–2909. doi:10.1093/jxb/ery116

CHAPTER 2

Management of Fruits to Increase Shelf Life

YOUSEF NASERZADEH and NILOUFAR MAHMOUDI

Department of AgroBiotechnology, Peoples' Friendship University of Russia (RUDN University), Moscow, Russia

ABSTRACT

The lengthy process of harvesting fruit before it reaches the consumer, which includes various stages of transportation, storage, and storage, etc. causes the fruits to be constantly exposed to quality loss and spoilage. As a result, according to annual statistics, more than 35% of fruit leaves the sale cycle and goes to the garbage disposal center. The high humidity of freshly harvested horticultural products has increased the rotting of crops resulting in decreased warehousing life. Therefore, nowadays increasing the warehousing life of fruits and vegetables to reduce waste is one of the most important challenges in the Agricultural Product sector. Multiple solutions to increase the shelf life of these products during warehousing are such as chemical and atmospheric heat treatments are done. Nowadays, the use of antimicrobial food coatings is expanding due to the increasing demand of consumers for the use of fresh products with minimal chemical additives and on the other hand, the desired quality and with the feature of maintaining environmental health. In different parts of this research, different methods of management and storage conditions of fruits; the effect of temperature and humidity; chitosan coating and edible nanocoating; supervising conditions of storing of sensitive fruits that quickly become soft (e.g., strawberries); new methods such as the use of natural preservatives, edible coating, packing with modified atmosphere, gamma radiation and the use of high hydrostatic pressure are explained.

Plant Diseases and Their Management: A Sustainable Approach.
Deepa Srivastava, Rajarshi Kumar Gaur, and Ajay K. Tiwari (Eds.)

2.1 INTRODUCTION

Fruits and vegetables are commonly consumed as a good source of micronutrients and phytochemicals. Consuming fruits and vegetables on a regular basis may help to prevent or reduce the risk of several chronic diseases (Naserzadeh and Mahmoudi, 2018; Rodríguez-Martínez et al., 2020; Christensen et al., 2021; Ncama et al., 2018). However, because they are perishable products containing living tissues, quality retention and postharvest, loss prevention during handling, storage, and retailing are critical. More than 20–22 percent of total fruit production is estimated to be lost due to spoilage at various postharvest handling stages (Sridhar et al., 2021).

Postharvest treatments are used to reduce fresh produce loss while also maintaining quality and thus increasing shelf life (Taormina, 2021; Ijaz et al., 2020; Chen et al., 2021). It is classified into three categories: chemical, physical, and gaseous treatments. To retard browning reactions, inhibit ethylene biosynthesis, reduce respiration rate and water loss, and reduce the incidence of postharvest diseases, chemical treatments such as hydrogen peroxide, chlorine-based solutions, peroxyacetic acid, organic acids, nitric oxide, and sulfur dioxide are used (Pirozzi et al., 2020).

Furthermore, chemical treatments can reduce texture deterioration and microbial growth in fresh produce (Douglas et al., 2020). The main physical treatments used to prevent postharvest loss of fruits and vegetables are heat treatments, edible coating, and irradiation. Heat treatments are used to control insect infestations, diseases, and to avoid chilling injuries (de Oliveira Filho et al., 2021). Edible coatings act as a moisture barrier while also preserving color, texture, and natural aroma (Ijaz et al., 2020; Bratovcic, 2021). Irradiation is used to prevent tuber, bulb, and root sprouting (Chen et al., 2021). Ozonation, 1-methyl propene, control atmospheric packaging, and modified atmospheric packaging are examples of gaseous treatments. He also aids in the preservation of cell wall integrity, peel color, the prevention of senescence, the reduction of decay, the slowing of the respiration rate, and the prevention of deterioration (Şen and Kılıç, 2021; Yousuf et al., 2021; Olivas and Barbosa-Cánovas, 2005).

2.2 INTRODUCING NEW PACKAGING IN THE FOOD INDUSTRY

The purpose of food packaging at the same time is to increase the storage time and protection of food completely from the risk of internal and external spoilage and oxidative factors, as well as better and easier transportation of food (Han et al., 2004; Tripathi et al., 2004). If the material of packaging raw materials is selected from a type that food is visible, its quality can be better understood and the purchase of such products will be important for the consumer (Meredith et al., 2014; Church, 1994). In this research, we will introduce 5 new methods of food packaging industry:

1. Modified Atmospheric Packaging (MAP): this method is widely used in industry today and is to enter the atmosphere in a food package with a special combination of carbon dioxide, nitrogen, and oxygen.
2. Nanotechnology: this technology can be effective in cases such as increasing resistance to penetration in coatings, increasing wall properties (mechanical, thermal, chemical and microbial), increasing heat resistance, spreading active antimicrobials and antifungal surfaces.
3. Intelligent sensors: these sensors can inform or warn the producer, seller, or consumer about some characteristics related to the food that is packaged or the characteristics of the environment in which the food is stored.
4. Packaging in smart containers for ready meals: by making new containers that are designed to extend the shelf life and the possibility of tracking food products, the possibility of quality problems or spoilage of food is reduced.
5. Inhibitors in the manufacture of utensils in order to preserve the taste and odor of volatile foods: Absorbents can absorb unwanted gaseous molecules that cause an unpleasant odor inside the container space and the use of these inhibitors in the construction of the container, it reduces the absorption, return, and emission of gas inside the container space, thus maintaining the quality of the food.

Today, developed countries use nano-packaging methods, smart sensors, and smart containers to package prepared foods (Shamaila et al., 1992). The new approach of the packaging industry focuses on ensuring the health and quality of food (controlling microbial growth, delaying oxidation, improving supervision, preserving the taste and smell of food) along with ease of use and packaging stability.

2.3 CHARACTERISTICS OF FILMS, FOOD COATINGS, AND THEIR ROLE IN INCREASING SHELF LIFE AND FOOD MANAGEMENT

Food films and coatings are placed as thin layers of food on the surface of food or between their compounds and are considered as one of the basic ways to control physiological, microbial, and physicochemical changes in food (Han et al., 2004). Their main structure is based on natural polymers with special properties. Their function is to create a barrier against the transfer of materials (water, gas, fat). Preservation and transfer of food components and additives (dyes, flavorings, etc.) prevent the growth of microorganisms in food surface as well as their mechanical protection. Other potential benefits of edible films include easier handling, improved quality, longer shelf life, biodegradability, ease of access, and low cost of packaging materials (Al-Tayyar et al., 2020). On the other hand, synthetic polymer packaging materials have replaced the reduction of residues, recyclable sources, and food coatings. All these factors are the basis of studies conducted by various researchers in the field of protective, mechanical, and sensory properties of various films and food coatings in the last three decades (Ncama et al., 2018; Kumar et al., 2021). Oral films and coatings are made from biopolymers. The biopolymers that make up the film include proteins, polysaccharides (carbohydrates and gums), and lipids. Plasticizers and other additives, along with the biopolymers that make up the film, are used to change the physical and functional properties of the films.

2.4 HISTORY OF THE USE OF FILMS AND FOOD COATINGS

Wax coatings were used on citrus fruits in the 12th and 13th centuries, but were commercially used only on apples and pears from 1930 onward (Gheorghita Puscaselu et al., 2020). Lipid coatings from previous years prevent shrinkage on meat and cheese (Suhag et al., 2020). Collagen films are used as sausage coatings and some methylcellulose films are used as bags for dry food. Wax coatings on fruits and vegetables, ornamental coatings on candies, and sugar coatings on nuts are the most common commercial applications of edible coatings (Dos Santos et al., 2020). In the pharmaceutical industry, sugar coatings on tablets and films and gelatin are used for soft capsules. The use of cellulose compounds (such as carboxymethylcellulose, dehydroxypropyl cellulose, and methylcellulose) as a coating for fruits, vegetables, meat, fish, nuts, bakery, confectionery, and other agricultural products is increasing.

2.5 REASONS FOR USING FOOD FILMS AND COATINGS

Every food item suffers from some kind of problem of material transfer. Moisture transfer, oxygen permeation, loss or absorption of aroma or oil are examples of these problems. Using films and food coatings can partially improve these problems and improve the stability, texture, taste, and aroma of food (Dehghani et al., 2018). Also, using controlled amounts of antioxidants and antimicrobials can further improve nutrition (Ganiari et al., 2017). Many foods break down or disintegrate during the manufacturing, transportation, packaging, or use processes. The more product, the better appearance, and the easier usage are the potential benefits of edible films and coatings. Finally, consumers recognize the quality of food in its brightness, color, and sensory properties and these are the traits that are affected by food films and coatings.

2.5.1 THE ADVANTAGES OF EDIBLE FILMS OVER OTHER SYNTHETIC POLYMER PACKAGING MATERIALS

1. Consumption of films with the packaged product without discarding any residue.
2. If films are used, environmental pollution is reduced, because they are easier to convert into recycled materials.
3. Films increase the organoleptic properties of packaging materials. In this way, various compounds such as flavorings, colors, and sweeteners are transferred with them.
4. Films can complement the nutritional value of food.
5. Films can be used to pack small amounts of food, such as peas, beans, and strawberries.
6. Inside heterogeneous foods, films bond different layers between compounds. They can also be designed to prevent the removal of moisture and the migration of soluble substances in foods such as pizza, cookies, and candy.
7 These films can be used as transmitters of antimicrobials and antioxidants.
8 Films can be easily used to produce micro-capsules of food flavorings and fermenting agents.

2.6 INGREDIENTS FOR FOOD FILMS AND COATINGS

The main constituents of the biopolymer film are proteins, fats, and polysaccharides. These materials can be used alone or in combination (Ribeiro et al., 2021). The physical and chemical properties of biopolymers have a great impact on the properties of the films made from them. The ingredients of the film can be hydrophobic or hydrophilic.

2.7 THE FILM BASED ON POLYSACCHARIDE

Polysaccharides used in the preparation of films and edible coatings include starch, nonstarchy carbohydrates, gums, and fibers. The chain of polysaccharides is simpler than that of proteins, but the spatial arrangement of polysaccharide structures is more complex and unpredictable and therefore has a higher molecular weight than proteins (Menossi et al., 2021). Most carbohydrates are neutral while some gums are negatively charged. Polysaccharide films, due to their hydrophilic nature, are permeable to water vapor but have good resistance to oxygen, so they can reverse the oxidation of fats.

2.8 PROTEIN-BASED FILM

Proteins are commonly used as film constituents. They are macromolecules with molecular structures and special amino acid chains. The second, third, and fourth structures of proteins can be easily modified by thermal denaturation, pressure, radiation, mechanical treatment, acids, alkalis, metal ions, salts, chemical hydrolysis, enzymatic treatment, and chemical crosslinking (Chen et al., 2019). The most distinctive feature of proteins in comparison with other constituents of denaturing film is the spatial arrangement of electrostatic charges and their dual nature of friendliness (Jin et al., 2020). Many factors affect the spatial arrangement of proteins, including charge density and hydrophobic–hydrophobic balance, and by controlling these factors, the physical and mechanical properties of the prepared films can be determined. Protein films are permeable to moisture but have good resistance to oxygen (Schmid and Müller, 2019).

2.9 THE FILM BASED ON LIPID COMPOUNDS

Other constituents of films are lipids and resins, but they are not polymers and biopolymers are not the right term for them. However, they are degradable and sticky (Pisoschi et al., 2018). Most oral lipids and resins are solid but soft at room temperature and have a specific phase transition temperature. Due to their hydrophobic nature, films or coatings made of lipid materials have a very high resistance to water vapor. They also have little surface energy (Fabra et al., 2018). Lipids can be combined with other film constituents such as proteins or polysaccharides as emulsion particles or multilayer coatings to increase water permeability. Resins, which include clear lacquer, poplar gum, are used as important coating compounds to make products shiny, and as a polishing and edible coating in pharmaceuticals, confectionery, fruits, and vegetables.

2.10 PLASTICIZER

Protein and polysaccharide films are often hard and brittle due to the interaction between the polymer bonds. The addition of a plasticizer increases the interactions in the polymer structure and reduces the glass transition temperature and improves the flexibility of the film, increases the elongation of the film, and reduces the modulus elasticity. Unfortunately, plasticizers reduce the ability of films to block the transfer of moisture, oxygen, aroma, and oil (Janjarasskul et al., 2020). Plasticizers increase the molecular motions of polymer molecules in such a way that they can reduce the ratio of the crystalline region to the bisexual region and thus reduce the glass transition temperature (Rangaraj et al., 2021). Therefore, the addition of plasticizers is not only effective on the elastic modulus and other mechanical properties, but also the resistance of films and coatings to the penetration of water vapor and gases. Most plasticizers are very hydrophilic (Xu et al., 2018). Thus, they can adsorb water molecules and form a large hydro-dynamic hydro–plasticizer complex. In the case of protein and polysaccharide edible films, plasticizers cause degradation of hydrogen bonds within and between molecules, increasing the distance between polymer molecules and decreasing the ratio of the crystalline region to amorphous. Water is a very good plasticizer but it dehydrates easily at low relative humidity, so adding hydrophilic chemical plasticizers can reduce water loss by dehumidifying, reducing, and increasing the amount of composite water, and maintaining high water activity.

2.10.1 TYPES OF PLASTICIZERS

1. Materials that can form high hydrogen bonds. Polymers react by destroying polymer–polymer bonds, creating a greater distance between the polymer branches.

2. Materials that can react with large amounts of water, as a result, the moisture content is higher and a greater hydrodynamic radius is obtained.

The performance of plasticizers is a function of factors such as the size and shape of the plasticizer molecules, the number of hydrogen atoms and their spatial distance within the structure of the plasticizer, and the capacity to bind to water. In addition to the hydrogen bonding, the driving forces between molecules of similar charge or between polar or nonpolar polymers can increase the distance between the polymers, leading to plasticizing on the structures of the charged polymer film. Therefore, compared to unloaded polymer films such as starch, the flexibility of charged polymer films can be changed by pH, and the addition of salt at a similar level affects water activity more.

2.11 ADDITIVES

Edible films and coatings can carry a variety of active ingredients such as emulsifiers, antioxidants, antimicrobials, flavorings, and dyes, thus enhancing food quality and safety. Emulsifiers are surfactants of hydrophobic and hydrophilic nature that can reduce the surface tension of the water–lipid or water–air contact surface (de Carvalho and Conte-Junior, 2021). They are essential for the formation of protein–lipid or polysaccharide–lipid composite films. They also change the surface energy to control the adhesion and wetting capacity of the film surface. Although many biopolymers have certain levels of emulsifier capacity, emulsifiers should also be used in film-forming formulations to produce emulsion-lipid films. In the case of protein films, some of the film-forming proteins have sufficient emulsifying capacity due to their amphiphilic structure (Senturk Parreidt et al., 2018). Antioxidants and antimicrobials can be added to the film-forming formulation to obtain active packaging, which protects food products from oxidation and microbial spoilage, thereby improving food quality and safety.

2.12 INCREASE THE SHELF LIFE OF HORTICULTURAL PRODUCTS USING BIODEGRADABLE COATINGS

Various factors affect fruit shelf life, including harvesting, pre- and post-harvest processes, packaging, transportation, temperature and humidity control, and most importantly, marketing. To have a good fruit life, it must be effectively protected from spoilage, and the drop in moisture must be slowed down in a way that preserves the appearance of fresh fruit (Sothornvit and Krochta, 2000). Fruits lose some of their flavors during long-term storage, even if they control spoilage and moisture loss (Aulin et al., 2010). However, by considering the metabolism of the fruit and the use of different storage methods, chemical reactions that lead to the change of the unpleasant flavor in the fruit can be delayed during long-term storage or their speed can be reduced. Packaging is also an important factor in product protection (Jost et al., 2014). Because synthetic packaging materials are environmentally friendly, degradable edible packaging and coatings have been proposed for replacement or concomitant use with synthetic packaging. Therefore, these packages not only increase the shelf life of horticultural products but also greatly eliminate the problem of environmental pollution.

Biodegradable coatings made from natural materials form a thin layer on the surface of fruits or vegetables. The use of these coatings in increasing the shelf life of horticultural products has grown in the last two decades (Chen, 1995). The external appearance of the fruit is an important feature in the quality of the fruit. This feature is the first thing that the buyer pays attention to. Although there is not always a direct relationship between the external situation and the food quality of the product, the appearance of the product has a great impact on market acceptance. Different types of biodegradable oral waxes and coatings are used to improve the appearance of the product (color and shine).

2.12.1 BENEFITS OF WAXING AND COATING THE PRODUCT

2.12.1.1 IMPROVE PRODUCT APPEARANCE

Waxed fruits and vegetables are more radiant (Baldwin et al., 1995). These products also retain their fresh color and appearance for a longer time.

This issue is very important from the perspective of marketing and consumer judgment. Waxing and coating also improve the interior color of some products and prevent discoloration.

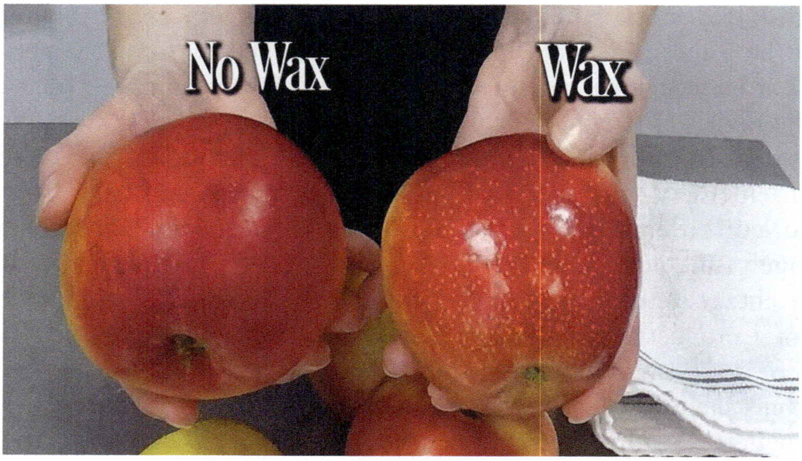

FIGURE 2.1 A picture of an apple covered with wax.
Source: Still from: How To Clean Wax Off Apples (0:55)

2.12.1.2 LESS WEIGHT LOSS

A cuticle layer that prevents moisture loss naturally covers all fruits and vegetables. At the same time, some water vapor can escape from the pores, cuticle, and clefts below the cuticle. During the waxing process, a thin, sticky film of the coating material is applied to the surface of the fruit. This coating closes the cuticle holes, which greatly reduces the amount of water vapor released. Loss of moisture from fruits and vegetables leads to wilting and shrinkage of the product (Baldwin et al., 1995). This has a negative effect on the appearance of the product and causes the product to lose its fresh and desirable state. Stress related to dehydration leads to dryness and as a result the skin withers and looks plastic (Douglas et al., 2020). Wrinkling of the skin has an adverse effect on gas exchange and metabolic changes because the enzymatic activity and ion flow change due to a decrease in the cell's internal pressure. In general, a 5–6 percent drop in water is acceptable before the fruit loses its marketability. This drop in humidity will reduce the quality of the product and make it impossible to market (Ijaz et al., 2020). The longer the product is expected to last, the greater the importance of waxing in reducing weight loss. Applying a coat can reduce weight loss by 30–40%. Loss of moisture in the fresh product has led to undesirable tissue changes that have a negative impact on the edible quality of the product. Tissue stiffness is reduced and the tissue loses its strength or in some cases becomes dry and perforated.

2.12.1.3 REDUCE ECONOMIC LOSSES

Water is the main component of fresh fruits and vegetables and usually makes up 80–90% of the weight of the fresh product, and when the fresh product is harvested, it begins to lose moisture through the process of perspiration (water evaporation) and respiration (Senturk Parreidt et al., 2018). This has led to weight loss, which is economically undesirable. Manufacturers usually sell their products by weight and have lower economic returns as weight decreases.

2.12.1.4 REDUCE POSTHARVEST CORRUPTION

Waxing prevents the entry of fungal and bacterial pathogens into the product. Waxing creates a hydrophobic surface that is not suitable for the growth of pathogens (Menossi et al., 2021). It can also be added to fungicide wax.

2.12.1.5 LONGER SHELF LIFE AFTER HARVEST

Fruits and vegetables are living organisms that continue to breathe after harvest. The wax creates a modified atmosphere inside the product in which the amount of oxygen decreases and the amount of carbon dioxide increases (Escamilla-García et al., 2018). This reduces the rate of respiration and increases the life of the product. Increased product life after harvest the product has a shelf life in the market. Less susceptible to frosts Tropical fruits and vegetables are sensitive to frostbite, which is a physiological injury that occurs at low temperatures. The amount of frost depends on the temperature and the period to which the product is exposed. This problem usually occurs at temperatures between 1 and 12 degrees Celsius depending on the product.

2.13 DISADVANTAGES OF COVERAGE

2.13.1 CREATE AN ARTIFICIAL LOOK

Waxing changes the product and makes it artificial. In some cases, consumers prefer the product to have its natural color and appearance and prefer it to a glossy coated product.

2.13.2 COST INCREASE

Coating operations increase the cost of the product and its preparation time. The added cost is related to the labor cost and equipment required to apply the wax or coating plus the cost of the wax itself.

2.13.3 SUPERFICIAL BURNS

When the wax or coating is used improperly, it can cause burns on the surface of the product (Escamilla-García et al., 2018). This usually happens when the product is immersed in molten waxes.

2.13.4 SHRINKAGE OF THE COATING (PLASTERING)

If the wax is exposed to high temperatures or humidity, it will wrinkle on the surface of fruits or vegetables. This usually happens when the waxed product is transferred from cold storage to higher temperatures (Pirozzi et al., 2020). When the temperature difference between the product and the ambient temperature is large, the product sweats. Water droplets cause some waxes, such as slag, to dissolve and white residues to appear opaque to the product (Fig. 2.1). High concentrations of wax exacerbate this problem. Bleaching also occurs when large amounts of moisture are present during coating.

FIGURE 2.2 Damage to the wax layer in apple fruit as a result of perspiration.

Source: Still from: Apples and Wax - What You Should Know (2:26)

2.13.5 CREATES AN UNPLEASANT TASTE

All fruits and vegetables are living organisms that need to absorb oxygen and release carbon dioxide into their natural metabolism. Waxing reduces gas exchange through the skin and thus reduces the amount of internal oxygen and increases carbon dioxide. Low levels of oxygen not only stimulate anaerobic respiration and the production of compounds with an unpleasant odor and taste, but also cause skin disorders that negatively affect the appearance of the product (Suhag et al., 2020). The applied wax layer should not be too thick. Various coatings are used in the food industry. The most common are starch, methylcellulose, pectin, chitosan, and paraffin waxes, carnauba, slag, and polyethylene. Each of these materials has unique properties in terms of strength, gas exchange, and physical properties. The amount of coverage used for each product is very small. For example, a gallon of carnauba or liquid slag using a roller brush can be about 7500 tomatoes and about 97,500 cucumbers or peppers. Polyethylene wax is the cheapest wax to cover a wide range of fruits and vegetables.

2.14 FOOD COATINGS

These coatings are thin layers of material that block the transfer of moisture-oxygen and solutes in food and can be eaten by the consumer. There is a major reduction in the quality and quantity of fresh fruits between harvest and consumption. With a better understanding of the respiration process of fresh fruits, there are several techniques to increase shelf life. Controlled atmospheric storage is used to preserve fruits by reducing their quality changes and reducing quantity during storage (Han et al., 2004). Edible coatings on fresh fruits can be used as an alternative method of storing the modified atmosphere by making qualitative changes and reducing the quantity by modifying and controlling the internal atmosphere of the fruits.

2.14.1 HISTORY

The first oral wax coating increased shelf life compared to uncoated fruits. Hot Molten Paraffin Waxes (1930): For coatings of fresh fruits such as pears, the results showed that coating peaches with wax emulsions reduced vapor transfer. Water reduced respiration rate and increased durability.

2.14.2 CHITIN AND CHITOSAN

Transparent film was used as an oral coating on fruits. In 1982, a mixture of sucrose fatty acid esters (as emulsifiers), sodium carboxymethylcellulose, and meno and diglycerides was found to be effective in preserving fruitless, odorless, and nontoxic fruits (Radi et al., 2017). Zein coatings on tomato surfaces delayed color and reduced weight and maintained strength during storage.

2.14.3 TYPES OF FOOD COATINGS

A. Polysaccharide coatings
B. Protein coatings
C. Lipid coatings
D. Composite coatings

2.14.3.1 POLYSACCHARIDE COATINGS

1. Alginate
2. Pectin
3. Carrageenan
4. Starch
5. Dextrins
6. Cellulose derivatives
7. Microbial gums

2.14.3.1.1 Alginate

Alginate is extracted from brown algae. The concentration of polyvalent cations (especially calcium), the rate of addition and its contact time, pH, and the presence of other compounds (such as hydrocolloids, fillers, etc.) are effective in the formation of alginate film. Alginate coatings are applied to beef, pork, poultry, fish, and whole lamb carcasses in a two-step procedure.

2.14.3.1.2 Pectin

Pectin is a structural polysaccharide and its constituent units are type D galacturonic acid with varying degrees of methyl ester (Sridhar et al., 2021).

The method of production, properties, and application or consumption is the same as alginate. The films are odorless, shiny, crisp, and slightly salty. Since the vapor permeability of water at 25°C and the relative humidity of 50% is relatively high, the use of a lipid protective coating can increase the performance of these films.

2.14.3.1.3 Carrageenan

Carrageenan is a red algae extract and is one of the sulfated polysaccharides. As soon as the hot aqueous solution of the polymer has cooled, a gel is formed (Ganiari et al., 2017). The formation of the gel is probably due to the conversion of the double helix structure into a three-dimensional polymer network. Carrageenan gel coating delays the reduction of food moisture.

2.14.3.1.4 Starch

Ethering a portion of starch containing high levels of amylose with propylene oxide gives derivatives that are more soluble in water that are flexible, transparent, antiadhesive, and sealable. They are hydrated at high relative humidity (more than 78%). As a result, resistance to oxygen transfer decreases.

2.14.3.1.5 Dextrins

Compounds resulting from the hydrolysis of starch have been proposed as a protective coating. Although they are hydrophilic, they show limited resistance to moisture transfer.

2.14.3.1.6 Cellulose Derivatives

Cellulose is insoluble in water due to its structural form. Alkali is used to swell the cellulose structure in water. It is then combined with chloroacetic acid, methyl chloride or propylene oxide, carboxymethylcellulose, methyl cellulose, and hydroxypropyl cellulose to form a film. Meanwhile, hydroxypropyl cellulose polymers, which are in fact a thermoplastic resin, are molded in the molten state and used to produce a film (Mohebbi et al., 2012).

Cellulose derivative coatings are relatively resistant to water penetration and are not affected by oils and fats and most nonpolar organic solvents.

2.14.3.1.7 Microbial Gums

From this group of compounds, we can name dextran, scleroglucan, cardalan, and plulan. The dextran coating prevents the reduction of moisture in many foods such as shrimp, meats, dried fruits, cheese, butter, and frozen foods. Scleroglucan also has potential applications as a coating. Plulan produces powerful, resilient films (Ncama et al., 2018). Films made from water-insoluble and alkali-soluble cardan films are three times stronger than amylose films and one-third as cellulose films. The main drawback is their high price and the fact that some laws, such as the laws of the European Economic Community, do not allow their use.

2.14.3.2 PROTEIN COATINGS

1. Zein
2. Gluten
3. Casein
4. Whey proteins
5. Gelatin

2.14.3.2.1 Zein

It is the protein part of corn that is soluble in alcohol. Zein films are water resistant. Zein is used in the formulation of oral coatings for tablets, confectionery, and fruit slices. The film is produced by spraying its solution on the surface of the product or immersing the product in it. It contains an authorized softener such as glycerin, antioxidant, and dye, which is formed on the surface of the product after the solvent evaporates.

2.14.3.2.2 Gluten

Gluten is a water-insoluble protein of wheat flour. Disulfide bonds are thought to play an important role in the formation of gluten film. Disulfide bonds between two polypeptide rings or extracellular bonds of the same polypeptide proteins increase the strength of the film. Due to the high intermolecular forces, it will be crisp. Therefore, the addition of softener is necessary to improve the mechanical properties of the film. Films made from gluten are pure and clear, but commercial gluten produces a matte film due to the gelatinization of the available starch (Pirozzi et al., 2020). Gliadin films are clear and elastic, but gluten-free films are brittle and inelastic. Gluten films are a good barrier against oxygen and carbon dioxide. The biggest obstacle to exploiting these films is their high water permeability (Rais and Sheoran, 2015). Gluten is used to coat peanuts and to encapsulate dyes and flavorings. Microencapsulation of unsaturated fatty acids by spray-dried gliadin powder reduces oxidation and produces the least amount of peroxide compared to other samples.

2.14.3.2.3 Casein

Caseins are a specific group of milk proteins with specific characteristics, such as ester phosphate bonds, high levels of proline, low cysteine content, and low solubility at pH4–5. Casein is also used to make an oral film. The resulting films are sticky, dark, soluble in alkali, and resistant to water penetration.

2.14.3.2.4 WHEY PROTEINS

The good properties of milk proteins make them a good raw material for the preparation of edible film. Milk proteins have, firstly, good nutritional value and, secondly, numerous functional properties, which are very important for making films.

2.13.3.2.5 GELATIN

Gelatin is a collagen-derived protein compound that contains 18 types of amino acids. Gelatin produces reversible gels due to heat. The gel forms when the aqueous suspension solutions of the polypeptide are cooled. Gelatin films are strong and transparent and form after the gel dries.

2.14.3.3 LIPID COATINGS

1. Acetyl glycerides
2. Waxes
3. Surface activating compounds

2.14.3.3.1 Lipid Films

The use of fat coatings on food products has a long history, especially on confectionery and fresh produce (fruit coatings). The types of lipid compounds used as protective coatings include acetylated monoglycerides, natural waxes, and surfactants. Because these compounds have low polarity, the main function of these coatings is generally to prevent moisture transfer.

2.14.3.3.2 Acetyl Glycerides

Mono-stearate glycerol is reacted with anhydrous acetic acid and acetylated. Acetyl glycerides are present in the solid phase in the form of alpha polymorphs, which makes these compounds flexible (Senturk Parreidt et al., 2018). They are impermeable to oxygen and their resistance to moisture transfer depends largely on the water vapor pressure gradient throughout the film (Sothornvit and Krochta, 2000). These compounds are used for pieces of meat and poultry, cheese, raisins, nuts, nuts and candy, and chocolate chips to prevent dehydration during storage.

2.14.3.3.3 Waxes

Most waxes are very resistant to moisture transfer compared to other lipid and non-lipid films. If pectinate is applied as a film on dried films, it prevents moisture from escaping (Sridhar et al., 2021). The property of high resistance to water penetration is related to the molecular structure of these compounds.

2.14.3.3.4 Surface Activating Compounds

These compounds reduce surface water activity and moisture loss by evaporation; in other words, they inhibit evaporation. Microbial growth is delayed due to reduced water activity by the use of surfactants, for example, by inhibiting the emergence of potato tubers because of the accumulation of carbon dioxide in the tissue, chlorophyll synthesis is prevented.

2.14.3.3.5 Compound Coatings

Composite films can contain a mixture of polysaccharide–protein and lipid films. In order to overcome the negative effect of one of the mentioned compounds or to benefit from the practical features of each of them, composite or multi-combination films are used (Kumar et al., 2021). Composite films, like the other films mentioned, have the property of preventing the passage of moisture-gas and dissolved materials. For example, the use of sodium alginate coating containing dextrin and carboxymethylcellulose solution to protect meat, poultry, and raw fish from discoloration and water loss during refrigeration has been suggested.

2.15 PROBLEMS WITH FOOD COATINGS

Even though some food coatings have been used successfully, modifying the indoor atmosphere by consuming food coatings can cause unpleasant odors related to increasing the concentration of carbon dioxide and decreasing the concentration of oxygen (Meredith et al., 2014). The results of a study showed that waxing apples and pears hinders normal ripening speed and if enough wax is used, respiration is greatly hindered and alcoholic flavors are developed by anaerobic fermentation.

Apples coated with sucrose fatty acid ester did not reduce harmful changes in firmness, yellowing, and weight loss. Tomatoes coated with 2.6 mm of zein film produced alcoholic and bad taste aromas that were associated with low oxygen content and high carbon dioxide content (Han et al., 2004). Excessive increase in film thickness and coating causes harmful effects due to reduction of oxygen concentration below a desirable level and increase of carbon dioxide concentration as a result of anaerobic fermentation. Each fruit and variety needs a different gas composition. Therefore, depending on the needs of each material, a suitable coating should be selected.

2.16 THE EFFECT OF COLD STORAGE ON NUTRITIONAL VALUE AND TASTE OF FOOD

Type of climate, soil, amount of fertilizer, method of cultivation, season and method of harvest, ripening rate and transport conditions are important factors in storing these products. For better and longer storage while preserving its nutritional value and sensory properties, the products should be transported to the cold storage at a very short time after harvest, and the product should be completely healthy, clean, and tidy. When transferred to the refrigerator, it should not have surface moisture, but it should be free of mud, dirt, and dust (products that are to be stored in the refrigerator for some time should be wrapped in a porous wrapper). Fruits and vegetables are harvested instantly, like a piece of ice that is exposed to adverse environmental conditions. It is gradually melting, while losing the quality of the food and the food itself, which it is (Han et al., 2004).

They are microbial swabs. For example, sweet corn in 24 h 1.4% of your blood sugar is lost due to normal room pain. Therefore, all products must be processed within 24 h at their storage temperature. In the cold, in the case of perishable or environmentally sensitive materials such as berries, this time is reduced to 2–3 h (El-Ramady et al., 2015). On the other hand, many other processes are routinely performed before packaging products and storing them in packaged or layered form in the cold. Processes are cleaning, sorting, grading, etc.

For these reasons, most large fruit and vegetable production centers today use the preliminary cooling process to prevent undesirable changes immediately after harvest, and the product until it reaches the centers. Keep or supply, keep fresh and healthy, and most importantly, prevent the pressures on the refrigerators to remove the initial heat of the incoming products. Storage at temperatures close to 0°C delays the ripening and aging of the crop by reducing cell respiration, ethylene production, water loss, and biological activities in general (Fabra et al., 2018). Therefore, cold storage does not reduce the nutritional value, as a destructive process, and even preserves more of its nutritional value and taste properties than normal environmental conditions (Ganiari et al., 2017). One of the destructive changes that should always be prevented is frostbite. Frostbite basically means that the storage temperature of the product is too low so that the respiratory enzymes in the fruit or vegetable stop working. These disorders lead to the accumulation of toxins due to shortness of breath and unwanted biochemical cycles and taste, texture, and color.

It causes undesirable changes to the product. It is obvious that the optimal storage temperature of different fruits and vegetables is different from each other, so frostbite also occurs for each in different conditions. Tropical fruits, such as bananas, are more sensitive to cold, with temperatures below 13–13°C. They suffer from frostbite, but for example, pears can easily withstand temperatures slightly below zero (Gheorghita Puscaselu et al., 2020). Symptoms of frostbite include: sunken surface, drying and subsequent plating, burnt spots (orange) or red spots (lemon), browning of the product, formation of undesirable fluffy texture (peach), and loss synthesis power of carotene pigments (potato). The table below lists the various effects on different products.

TABLE 2.1 Symptoms of Frostbite in Some Fruits Below Safe Temperature.

The product	Safe temperature	Symptoms of frostbite between 0°C and safe temperature
Apple	2.2–3.3°C	Browning of internal tissues
Avocado	4.4–7.2°C	Brownish fleshy texture
Banana	11.7–13.3°C	Skin opacity—browning of internal tissues
Grapefruit	10°C	Dipping—superficial burns
Lemon	11.1–12.7°C	Depression—Production of red spots on the outer surface
Mango	10–12.7°C	Burnt gray spots on the skin
Olive	7.2°C	Depression—brown spots on the skin
Orange	7.2°C	Depression—brown spots on the skin
Pineapple	7.2–10°C	Produces a dark green color on the outer skin

In addition to excessive temperature drop, other factors such as increased concentration, CO_2, reduced oxygen, long shelf life, and reduced relative humidity also contribute to frostbite (Mohebbi et al., 2012). As we know, a group of changes led to the arrival of the product, which is the appropriate changes in texture, nutritional value, and color. Changes may also occur, including the breakdown of large starches or large sugar molecules into smaller sugars with greater sweetness, reduced concentrations of edible acids, protein balance, the appearance of latent pigments, and softening of tissues with the gradual decomposition of pectin.

2.17 INVESTIGATION OF THE EFFECT OF SILVER NANOPARTICLES ON THE SHELF LIFE OF CITRUS SINENSIS

Primary small, bitter, and kernel oranges, which are due to genetic engineering as well as superior varieties and sufficient fertilizer, are sweeter and more fragrant in arid and desert areas (Han et al., 2004; Deseta et al., 2021). Citrus growers are currently using unconventional methods to control the growth of storage fungi, such as benomyl and tecto 60 fungicide. Benomyl fungicide is very dangerous for the health of consumers due to its penetration into the fruit and its use has been declared unauthorized for this case (Radi et al., 2017; Mohebbi et al., 2012; Guerreiro et al., 2015). It is noteworthy that today world is trying to minimize the use of chemical treatments to reduce waste and increase the shelf life of citrus fruits after harvest (Baldwin et al., 1995). For this purpose, different methods have been studied, which silver has more advantages than other antimicrobial agents and is toxic to countless species of bacteria, fungi, algae, and some viruses. Following the first published report of the antimicrobial properties of colloidal silver, silver nanoparticles were found to be a potent agent against many pathogens (Rais and Sheoran, 2015; El-Ramady et al., 2015). One way for consumers to have more access to citrus products is to avoid wasted time until consumption, which is ignored. Citrus fruits cannot be stored in normal conditions for a long time, and if they can be stored, a lot of facilities, technical and industrial equipment are needed (Anjum Malik et al., 2016). Silver nitrate is one of the most common antibacterial agents that is not used today due to its risks to human health and the environment (Ncama et al., 2018). But in order to evaluate the results in the fruit's shelf life, it has to be used and it certainly penetrates into the fruit tissue. In order to use it practically in fruit shelf life, the best way to reduce its harmful effects would be to cover this nanoparticle with harmless materials. In recent years, the use of silver nanoparticles has become popular due to its antimicrobial properties (Olivas and Barbosa-Cánovas, 2005). The presence of silver nanoparticles is more effective than other silver compounds in controlling and killing disease-causing microorganisms due to its high surface area, which leads to better contact with bacteria (Church, 1994). Decay and weight loss due to transpiration are important and limiting factors for the preservation of orange fruit. Green and blue molds infect fruits through wounds that are formed during harvest and transportation on fruits and if effective control is not done, they will cause waste after harvesting citrus fruits before reaching the consumer (Pirozzi et al., 2020). The antimicrobial effects of different types of nanomaterials such as copper, zinc, titanium, magnesium, gold, and silver have been studied. But silver nanoparticles have shown the greatest antimicrobial effect because it has a good antimicrobial effect against bacteria, viruses, and other eukaryotic microorganisms (Bratovcic, 2021; Church, 1994). Silver is one of the metals that has been suggested as an effective antibacterial agent that can kill more than 650 types of microorganisms such as bacteria and viruses. By increasing the surface to volume ratio of silver at the nanoscale, its antibacterial properties are improved (Meredith et al., 2014).

Silver accumulates in alfalfa (Taormina, 2021), the presence of copper and silver ions in the environment through the production of reactive oxygen species and by increasing the expression of genes involved in the biosynthesis of flavonoids, and increasing the content of flavonoids and anthocyanin reduces environmental stress. Silver nitrate is one of the most common antibacterial agents that is not used today due to its risks to human health and the environment (de Oliveira Filho et al., 2021).

In recent years, the use of silver nanoparticles has become popular due to its antimicrobial properties (Tripathi et al., 2004). The presence of silver nanoparticles during storage is an effective factor in controlling and killing disease-causing microorganisms. Silver nanoparticles are more efficient than other silver compounds due to their high surface area, which leads to better contact with bacteria (Church, 1994). In the experiment, Martinez Romero et al. (2006) by combining storage with a modified atmosphere and coating with aloe Vera gel were able to significantly preserve the quality of the cherry fruit compared to the control and increase its shelf life (Olivas and Barbosa-Cánovas, 2005). A study by Hagen Meyer and Shaw (1991) and Aslam et al. (2020) showed that fruit coating treatments may in some cases cause an alcoholic taste and odor and thus a reduction in food quality (Yousuf et al., 2021). Mikael et al. (2007) in their research on tomatoes showed that nano-technology silicone coating protects tomatoes from oxygen and moisture, and coated tomatoes have a longer shelf life (Han et al., 2004). However, no detailed study has been conducted in the world to investigate the effect of silver nanoparticles on the shelf life of oranges. In general, according to the results of research, nanoparticles have a significant effect on fruit diameter and length, acidity, solids, vitamin C, electrolyte leakage, pulp weight, fruit flesh weight, and percentage of rot. Nanoparticles also have a significant effect on the quality of oranges during storage (Pisoschi et al., 2018).

2.18 THE EFFECT OF POSTHARVEST TREATMENTS ON THE QUALITY AND QUANTITY OF FRESH VEGETABLES READY FOR CONSUMPTION DURING THE STORAGE PERIOD

The use of vegetables in human nutrition, their taste characteristics, their usefulness in maintaining health, the short postharvest life and microbial contamination of vegetables, it is become necessary to research on to increase the shelf life and reduce the microbial load. The present chapter describe the shelf life increase and reduce the microbial load of two high-consumption edible vegetables such as Mentha and Lepidium sativum using salicylic acid (0, 0.5, and 1 mM), calcium chloride (0, 1, and 2%), and vegetable essential oil (0, 500, and 1000 ppm) (Pisoschi et al., 2018). The experiment was performed as a factorial experiment in a randomized complete block design with seven treatments and two maintenance periods in three replications.

The results showed that with increasing the storage period in both vegetables from three days to six days, the rate of spoilage, electrolyte leakage, proline, and relative humidity content increased and the amount of chlorophyll decreased (Radi et al., 2017). Among the treatments used, salicylic acid and calcium chloride increased the shelf life of both vegetables, so that these substances reduced the apparent index of spoilage and electrolyte leakage, and increased the relative moisture content and chlorophyll content. Concentrations of 0.5 mM salicylic acid and 1% calcium chloride were more effective.

According to the results, only the use of plant essential oil significantly reduced the microbial load in vegetables and other treatments were not effective on this feature. In general, salicylic acid and calcium chloride can be recommended to increase the shelf life of vegetables during storage and plant essential oil to reduce the microbial load of the studied vegetables (Rais and Sheoran, 2015). Fresh fruits and vegetables are essential components of the human diet and there is considerable evidence of their health and nutritional benefits. Consumers expect fresh and processed vegetables and fruits, in addition to high nutritional and organoleptic quality, to have a high shelf life.

Quality fresh fruits and vegetables are a combination of characteristics including appearance, texture, taste, nutritional value, and safety (Rodríguez-Martínez et al., 2020). Quality of ready-to-eat fruits and vegetables depends on preharvest agricultural operations (genotype, climatic conditions, cultivation operations), ripening (maturity and ripening), postharvest factors (physical damage during harvesting and handling), temperature, relative humidity management, complementary processes, and also depends on the product, complementary processes, changing the product atmosphere and harvesting method. A proper quality assurance program should consider all of these factors that affect the quality of healthy fruits and vegetables and their freshly cut products. Low-processed vegetables refer to ready-to-eat products (Schmid and Müller, 2019) or freshly sliced products. In fact, vegetables and fruits are raw that are washed or peeled and sliced for the consumer before packaging (Şen and Kılıç, 2021). These under-processed products also refer to cases where the processes performed on the product are not like the process of canning or freezing, but in any case they increase the value of the product before distribution.

Production operations for freshly harvested crops, especially pruning and peeling, increase respiration and ethylene production, intensify water loss (Senturk Parreidt et al., 2018), stimulate the accumulation of secondary metabolites, cell breakdown, and enzymatic browning. Tissue damage may also cause microbial growth and proliferation and eventually product spoilage. Microbial growth in tissues is affected by the chemical composition of the tissue (Sothornvit and Krochta, 2000). The technology of maintaining the safety of the freshly cut product must control pathogenic microorganisms and spoilage, and at the same time prevent severe processes and reduce tissue growth and viability. These methods are based on the use of a combination of hurdles. Bacterial contamination with Listeria monocytogenes, Salmonella, and Escherichia coli can occur in fresh vegetables at various stages of growth, during harvest, postharvest processes, or during transplantation.

The prevalence of diseases caused by fruits and vegetables has increased significantly in recent years. Adding disinfectants to the rinse water can significantly reduce the cell population of plankton bacteria and thus reduce the risk of secondary contamination (Sridhar et al., 2021). Such a reduction improves the safety and durability of the product, as a result, workshops and packaging plants for fresh vegetables can be used to increase the shelf life of the product by disinfection and disinfection, and consequently make it easier for consumers to consume fresh vegetables. The rate of reduction of pathogenic microorganisms during the washing and disinfection process is not sufficient to ensure the quality of food (Gheorghita Puscaselu et al., 2020). The microbial, sensory, and nutritional shelf life of processed vegetables and fruits should be at least 4–7 days (Guerreiro et al., 2015).

Reduction of yeasts and molds in processed and ready-to-eat vegetables is essential to improve the quality of the final product and increase shelf life. Washing vegetables with natural microbial flora water reduces it by approximately 1 logarithmic unit. Therefore, antimicrobial compounds are needed to reduce the microbial load in rinsing water systems. Numerous studies have shown that chlorine used in FDA-approved concentrations is not effective in removing human pathogens and spoilage microorganisms. In addition, it may combine with organic matter in water to form mutagenic compounds. Chlorine substitutes such as chlorine dioxide, acidified sodium chlorite, peroxyacetic acid, and some essential oils have been suggested recently (El-Ramady et al., 2015). No efficacy and is approved by the FDA as a disinfectant for vegetables and fruits. Persidine is a 15% food grade disinfectant based on the most effective and strongest biocidal compound in the world, Prastic Acid. It has an official license up to 500 ppm of oral residue on surfaces in contact with food and a high level disinfectant effective on all microorganisms. Unlike chlorine, which has a toxic residue, it is not harmful, has an edible residue, and maintains the quality, nutritional value and vitamins of vegetables, and rinsing is eliminated in the final stage.

Its antimicrobial action is mainly related to the production of reactive oxygen species that cause DNA damage, cell membrane rupture, and enzymatic blockage. The total number of bacteria present in fresh vegetables when entering the plant depends on the type of bacteria and the existing conditions and usually varies from 102 to 107 per gram (Menossi et al., 2021). Consumers' desire to use natural products free of any process and chemical and unsafe preservatives has led manufacturers to look for safer processes to increase the shelf life of cut products. Nonvolatile compounds such as plant essential oils due to studies have shown that many essential oils have antimicrobial properties, including the essential oils of bay leaf, sweet pepper, coriander, black cumin, cardamom, fennel, parsley, mustard, mint, dill, cloves, oregano, rosemary, thyme, sage and vanilla can be named. The active ingredients in essential oils often have a broad spectrum of antimicrobials against pathogenic microbes of food origin and spoilage bacteria. Of course, many of these essential oils are effective against a specific microbial species. Major volatile compounds have antimicrobial activity and flavoring essential oils, including isoprenoids, organic sulfur, aldehydes, and alcohols. The antimicrobial activity of essential oils is due to their chemical structure and especially the presence of hydrophilic functional groups such as hydroxyl groups of phenolic compounds or lipophils of some essential oils.

Due to the effect of chemical preservatives used in food products on human health, essential oils have been proposed as safe compounds to increase the shelf life of freshly cut fruits and vegetables. The process of disinfection and washing process causes physiological effects including ethylene production, increased respiration, membrane loss, water loss, susceptibility to microbial spoilage, chlorophyll loss, pigment formation, reduced acidity, increased sweetness, volatile flavors, tissue softening, enzymatic browning leads to lipolysis and fat oxidation. Salicylic acid and methyl salicylate are natural plant materials and play an important role in growth, plant defense responses, stimulation of acquired systemic resistance, and transmission of symptoms (Han et al., 2004). Salicylic acid increases the activity of the enzyme phenylalanine ammonia lysis (Menossi et al., 2021), which is a key enzyme in the metabolism of Phenylpropanoid, and the formation of trans kinamic acid by demineralization of the phenylalanine catalytic aromatic amino acid. With increasing the activity of this enzyme, the synthesis and accumulation of phenolic compounds increase and finally phenolic compounds with their antioxidant properties increase tissue resistance to biotic and abiotic stresses.

Salicylic acid and methyl salicylate are plant-marking molecules that play a key role in plant growth and defense responses (Menossi et al., 2021). Salicylic acid is a simple phenolic compound that is involved in regulating a number of plant growth and developmental processes, including stomatal opening and closing, seed germination, ion uptake, sex differentiation, and stimulation of disease resistance. It also interferes with the biosynthesis and action of ethylene in plants. The use of salicylic acid has been reported to stimulate the accumulation of phenylproponides, including coumarin metabolites. However, the relationship between salicylic acid, PAL, and phenolic compounds in the biosynthesis of materials is unknown (Olivas and Barbosa-Cánovas, 2005).

The inhibitory properties of salicylic acid on the activity of ACC oxidase have also been proven in apple fruit discs. Salicylic acid also inhibits the activity of the enzyme lipoxygenase (LOX) in kiwifruit discs, which in turn reduces free radical production and ethylene biosynthesis. Application of salicylic acid and jasmonic acid reduces frostbite (Ribeiro et al., 2021). In foliar application of salicylic acid on some characteristics after harvest of tomato fruit, the amount of firmness and thickness of the skin has increased, in apricots, the amount of acidity and antioxidant activity has been observed higher than in control fruits. Salicylic acid treatment had a significant effect on preventing the increase of phenylalanine ammonialysis and sugar content in increasing the shelf life of pomegranate. This is effective in controlling postharvest lesions and shelf life of apricot, pomegranate, and tomato fruits. The use of natural calcium sources as preservatives along with its enrichment effect and nutritional value can have significant benefits in industry as well as for consumers.

Calcium compounds such as lactate, chloride, phosphate, and propionate and calcium gluconate are widely used to increase the shelf life and firmness of these products. Calcium nitrate and carbonate are further (Rodríguez-Martínez et al., 2020) added to foods to increase their nutritional value. Choosing the right source of calcium depends on many factors, such as bioavailability, solubility, and the effects of using it on flavor and interfering with food components. Calcium chloride is one of the compounds that has been extensively researched and is used in the processing industry of whole or sliced fruits and vegetables.

2.19 MODIFIED ATMOSPHERE PACKAGING (MAP)

Basically, a food packaging is expected to keep its quality as long as possible. Although most foods under the production process reach relatively stable chemical and microbial conditions that can be maintained for a long time despite proper packaging, some foods such as fresh fruits and vegetables undergo thermal processing conditions (Rodríguez-Martínez et al., 2020). Or they do not tolerate cold and because of their biological structure, they perform respiration long after it is arranged. Although the respiration and chemical reactions inside the plant slow down a lot after it is sorted out, it does not stop, which means that the chemical activity inside the plant is taking place and making changes in the plant.

FIGURE 2.3 Modified atmosphere packaging.

Image source: Onsite Gas System. https://www.onsitegas.com/modified-atmosphere-packaging/

On the other hand, the manufacturer's goal is to supply these products fresh and unprocessed. However, in the case of some of these products, such as oranges, apples, or bananas, they can be picked before they are fully ripe, or in the case of cucumbers and citrus, waxy substances can be used to cover the surface of the fruit and create a barrier against plant respiration (Meredith et al., 2014). It slows down the chemical reactions inside the plant by restricting the entry of oxygen, but in many cases, such as tomatoes or lettuce and celery, etc., which have a very short life or a sensitive and different texture and shape, or in the case of fish and seafood, there are no suitable methods (Pirozzi et al., 2020). Therefore, considering the purpose of offering a new product, making changes and processing on the plant itself is not the solution to this problem; as a result, the creators of the food packaging industry came up with the idea of changing the environment in a way that would slow down the plant's respiration rate. Although this was initially done in food warehouses, a similar solution eventually led to the idea of MAP (modified atmosphere packaging) or atmospheric packaging. In the 1930s, when fruits and vegetables were shipped, CO_2 gas was used to increase the shelf life of these products. The United Kingdom was the first country to use a modified atmospheric system to increase the shelf life of pork and fish, but this method gradually expanded to the size of countries and today it is mostly used for fresh fruits and vegetables. According to researchers, the release of oxygen from the environment alone is not enough to preserve food. In this method of food preservation, two points are very important; one is the temperature and the other is the MAP process itself. Reducing the temperature and using this method reduces the speed of chemical reactions (Taormina, 2021). Modified atmosphere packaging means replacing the air in the package with a mixture of different gases; typically, a mixture of carbon dioxide, nitrogen, and oxygen.

In modified atmosphere packaging, the empty space inside the package is first vacuumed and then the desired gas composition is replaced by this space under pressure, and thus the enzymatic and microbial activities of the product are controlled. MAP packaging is seemingly a very simple method; emptying the air inside the package and replacing it with a mixture of gases; but in practice it also has many technical complexities. Determining the ratio of gases and keeping this atmosphere inside and around the food is not easy and in some cases impossible.

As mentioned earlier, fresh fruits and vegetables are living and breathing organisms, but the intensity of this respiration and, consequently, the rate of chemical reactions are not the same in all of these fruits and vegetables; for example, in the case of mushrooms, the respiration rate is high, and therefore this product is very perishable and has a short life, but not in the case of onions and potatoes.

Therefore, the temperature factor is very important in the shelf life of products, especially products that have MAP packaging, because the rate of respiration, oxygen consumption, and CO_2 production increase with increasing temperature and decrease with decreasing it, so it is clear that decreasing temperature increases this shelf life. Various combinations of gases are often used to modify the atmosphere inside the package. The ratio of 80% CO_2 and 20% O_2 is a common combination. Of course, this packaging technology is also used for seafood, which is also very perishable. In packaging of products such as fish, the goal is to reduce the respiration and activity of microorganisms that cause food spoilage. The packaging of these products usually uses 30–40% N2, 20% O_2, and 30–40% CO_2 and is stored at 0–2°C.

CO_2 gas is the most used in MAP packaging due to its desirable properties (Radi et al., 2017). This gas better preserves vitamins in meat products, lowers the pH of the environment, and also acts as an antimicrobial and reduces their activity. Also, by absorbing water and hydrating the cell wall, it delays the drying of the product and increases its freshness. In general, the properties of MAP packaging include:

1. Increases the shelf life of fresh products.
2. Maintains desirable quality properties.
3. The appearance of the packages is due to the possibility of using clear, attractive, and customer-friendly films.

4. The packaging is odorless and easy to transport and label.
5. Preservatives are not used in food to increase the shelf life of the product. Of course, CO_2 gas itself is a preservative and is slightly absorbed by the surface layer of the food.

And we can say about the disadvantages of this method

1. The cost of this packaging method is relatively high and it is about twice as expensive as the vacuum method.
2. The devices and its raw materials are expensive.
3. The shape and size of the packages are more voluminous than the vacuum method and therefore increase the cost of storage and transportation.
4. A mistake in the composition of the gas can cause the product to be crushed or leakage in the container, which can be controlled by correcting the composition of the gas and using a layer of absorbent paper at the bottom of the container.
5. Although the ratio of gases is determined at the time of mixing, it is not always constant during storage and may change.

Modified atmosphere packaging should not be confused with CAP controlled atmosphere packaging! In controlled atmosphere packaging, not only the composition of the gases and their ratio are determined, but also the methods try to always control the ratio of these gases.

KEYWORDS

- **edible coating**
- **fruit management**
- **increased shelf life**
- **edible nanocoating**
- **MAP packing**

REFERENCES

Al-Tayyar, N. A.; Youssef, A. M.; Al-Hindi, R. R. Edible Coatings and Antimicrobial Nanoemulsions for Enhancing Shelf Life and Reducing Foodborne Pathogens of Fruits and Vegetables: A Review. *Sustain. Mater. Technol.* **2020,** e00215.

Anjum Malik, A.; Ahmed, N.; Babita, C. H.; Gupta, P. Plant Extracts in Post-harvest Disease Management of Fruits and Vegetables—A Review. *J. Food Process. Technol.* **2016,** *7* (592), 2.

Aslam, R.; Alam, M. S.; Saeed, P. A. Sanitization Potential of Ozone and Its Role in Postharvest Quality Management of Fruits and Vegetables. *Food Eng. Rev.* **2020,** *12* (1), 48–67.

Aulin, C.; Gällstedt, M.; Lindström, T. Oxygen and Oil Barrier Properties of Microfibrillated Cellulose Films and Coatings. *Cellulose* **2010,** *17* (3), 559–574.

Baldwin, E. A.; Nisperos-Carriedo, M. O.; Baker, R. A. Use of Edible Coatings to Preserve Quality of Lightly (and slightly) Processed Products. *Crit. Rev. Food Sci. Nutr.* **1995,** *35* (6), 509–524.

Bratovcic, A. Physical–Chemical, Mechanical and Antimicrobial Properties of Bio-Nanocomposite Films and Edible Coatings. *Int. J. Res. Appl. Sci. Biotechnol.* **2021,** *8* (5), 151–161.

Chen, H. Functional Properties and Applications of Edible Films Made of Milk Proteins. *J. Dairy Sci.* **1995,** *78* (11), 2563–2583.

Chen, H.; Wang, J.; Cheng, Y.; Wang, C.; Liu, H.; Bian, H.; et al. Application of Protein-based Films and Coatings for FOOD PACKAGING, A Review. *Polymers* **2019,** *11* (12), 2039.

Chen, Y.; Fanourakis, D.; Tsaniklidis, G.; Aliniaeifard, S.; Yang, Q.; Li, T. Low UVA Intensity During Cultivation Improves the Lettuce Shelf-life, an Effect that is Not Sustained at Higher Intensity. *Postharvest Biol. Technol.* **2021,** *172,* 111376.

Christensen, F. M. M.; Solheim-Bojer, C.; Dukovska-Popovska, I.; Steger-Jensen, K. Developing New Forecasting Accuracy Measure Considering Product's Shelf Life: Effect on Availability and Waste. *J. Cleaner Prod.* **2021,** *288,* 125594.

Church, N. Developments in Modified-atmosphere Packaging and Related Technologies. *Trends Food Sci. Technol.* **1994,** *5* (11), 345–352.

de Carvalho, A. P. A.; Conte-Junior, C. A. Food-derived Biopolymer Kefiran Composites, Nanocomposites and Nanofibers, Emerging Alternatives to Food Packaging and Potentials in Nanomedicine. *Trends Food Sci. Technol.* 2021.

de Oliveira Filho, J. G.; Miranda, M.; Ferreira, M. D.; Plotto, A. Nanoemulsions as Edible Coatings, A Potential Strategy for Fresh Fruits and Vegetables Preservation. *Foods* **2021,** *10* (10), 2438.

Dehghani, S.; Hosseini, S. V.; Regenstein, J. M. Edible Films and Coatings in Seafood Preservation. A Review. *Food Chem.* **2018,** *240,* 505–513.

Deseta, M. L.; Sponton, O. E.; Erben, M.; Osella, C. A.; Frisón, L. N.; Fenoglio, C.; et al. Nanocomplexes based on Egg White Protein Nanoparticles and Bioactive Compounds as Antifungal Edible Coatings to Extend Bread Shelf Life. *Food Res. Int.* **2021,** *148,* 110597.

Dos Santos, C. A.; Ingle, A. P.; Rai, M. The emerging role of metallic nanoparticles in food. *Appl. Microbiol. Biotechnol.* **2020,** *104* (6), 2373–83.

Douglas, G. L.; Zwart, S. R.; Smith, S. M. Space Food for Thought: Challenges and Considerations for Food and Nutrition on Exploration Missions. *J. Nutr.* **2020,** *150* (9), 2242–2244.

El-Ramady, H. R.; Domokos-Szabolcsy, É.; Abdalla, N. A.; Taha, H. S.; Fári, M. Postharvest Management of Fruits and Vegetables Storage. In *Sustainable Agriculture Reviews*; Springer, 2015; pp 65–152.

Escamilla-García, M.; Rodríguez-Hernández, M. J.; Hernández-Hernández, H. M.; Delgado-Sánchez, L. F.; García-Almendárez, B. E.; Amaro-Reyes, A.; et al. Effect of an Edible Coating based on Chitosan and Oxidized Starch on Shelf Life of Carica papaya L.; and Its Physicochemical and Antimicrobial Properties. *Coatings* **2018,** *8* (9), 318.

Fabra, M. J.; Falcó, I.; Randazzo, W.; Sánchez, G.; López-Rubio A. Antiviral and Antioxidant Properties of Active Alginate Edible Films Containing Phenolic Extracts. *Food Hydrocoll.* **2018,** *81,* 96–103.

Ganiari, S.; Choulitoudi, E.; Oreopoulou, V. Edible and Active Films and Coatings as Carriers of Natural Antioxidants for Lipid Food. *Trends Food Sci. Technol.* **2017,** *68,* 70–82.

Gheorghita Puscaselu, R.; Lobiuc, A.; Dimian, M.; Covasa, M. Alginate: From Food Industry to Biomedical Applications and Management of Metabolic Disorders. *Polymers* **2020,** *12* (10), 2417.

Guerreiro, A. C.; Gago, C. M.; Faleiro, M. L.; Miguel, M. G.; Antunes, M. D. The use of Polysaccharide-based Edible Coatings Enriched with Essential Oils to Improve Shelf-life of Strawberries. *Postharvest Biol. Technol.* **2015,** *110,* 51–60.

Han, C.; Zhao, Y.; Leonard, S.; Traber, M. Edible Coatings to Improve Storability and Enhance Nutritional Value of Fresh and Frozen Strawberries (Fragaria × ananassa) and Raspberries (Rubus ideaus). *Postharvest Biol. Technol.* **2004,** *33* (1), 67–78.

Ijaz, M.; Zafar, M.; Afsheen, S.; Iqbal, T. A review on Ag-nanostructures for Enhancement in Shelf Time of Fruits. *J. Inorg. Organomet. Polym. Mater.* **2020,** *30* (5), 1475–1482.

Janjarasskul, T.; Tananuwong, K.; Phupoksakul, T.; Thaiphanit, S. Fast Dissolving, Hermetically Sealable, Edible Whey Protein Isolate-based Films for Instant Food and/or Dry Ingredient Pouches. *LWT* **2020,** *134,* 110102.

Jin, S.; Li, K.; Gao, Q.; Zhang, W.; Chen, H.; Li, J. Development of Conductive Protein-Based Film Reinforced by Cellulose Nanofibril Template-directed Hyperbranched Copolymer. *Carbohydr. Polym.* **2020,** *237,* 116141.

Jost, V.; Kobsik, K.; Schmid, M.; Noller, K. Influence of Plasticiser on the Barrier, Mechanical and Grease Resistance Properties of Alginate Cast Films. *Carbohydr. Polym.* **2014,** *110,* 309–319.

Kumar, R.; Ghoshal, G.; Goyal, M. Biodegradable Composite Films/Coatings of modified Corn Starch/Gelatin for Shelf Life Improvement of Cucumber. *J. Food Sci. Technol.* **2021,** *58* (4), 1227–1237.

Menossi, M.; Cisneros, M.; Alvarez, V. A.; Casalongué, C. Current and Emerging Biodegradable Mulch Films based on Polysaccharide Bio-composites. A Review. *Agron. Sustainable Dev.* **2021,** *41* (4), 1–27.

Meredith, H.; Valdramidis, V.; Rotabakk, B.; Sivertsvik, M.; McDowell, D.; Bolton, D. Effect of Different Modified Atmospheric Packaging (MAP) Gaseous Combinations on Campylobacter and the Shelf-life of Chilled Poultry Fillets. *Food Microbiol.* **2014,** *44,* 196–203.

Mohebbi, M.; Ansarifar, E.; Hasanpour, N.; Amiryousefi, M. R. Suitability of Aloe Vera and Gum Tragacanth as Edible Coatings for Extending the Shelf Life of Button Mushroom. *Food Bioprocess Technol.* **2012,** *5* (8), 3193–3202.

Naserzadeh, Y.; Mahmoudi, N. Chelating Effect of Black Tea Extract Compared to Citric Acid in the Process of the Oxidation of Sunflower, Canola, Olive, and Tallow Oils. *Int. J. Agric. Biosyst. Eng.* **2018,** *12* (5), 5.

Ncama, K.; Magwaza, L. S.; Mditshwa, A.; Tesfay, S. Z. Plant-based Edible Coatings for Managing Postharvest Quality of Fresh Horticultural Produce: A Review. *Food Packag. Shelf Life* **2018,** *16,* 157–167.

Olivas, G.; Barbosa-Cánovas, G. Edible Coatings for Fresh-cut Fruits. *Crit. Rev. Food Sci. Nutr.* **2005,** *45* (7 and 8), 657–670.

Pirozzi, A.; Pataro, G.; Donsì, F.; Ferrari, G. Edible Coating and Pulsed Light to Increase the Shelf Life of Food Products. *Food Eng. Rev.* **2020,** 1–26.

Pisoschi, A. M.; Pop, A.; Cimpeanu, C.; Turcuş, V.; Predoi, G.; Iordache, F. Nanoencapsulation Techniques for Compounds and Products with Antioxidant and Antimicrobial Activity—A Critical View. *Eur. J. Med. Chem.* **2018,** *157,* 1326–1345.

Radi, M.; Firouzi, E.; Akhavan, H.; Amiri, S. Effect of Gelatin-based Edible Coatings Incorporated with Aloe Vera and Black and Green Tea Extracts on the Shelf life of Fresh-cut Oranges. *J. Food Qual.* **2017,** 2017.

Rais, M.; Sheoran, A. Scope of Supply Chain Management in Fruits and Vegetables in India. *J. Food Process. Technol.* **2015,** *6* (3).

Rangaraj, V. M.; Rambabu, K.; Banat, F.; Mittal, V. Natural Antioxidants-based Edible Active Food Packaging: An Overview of Current Advancements. *Food Biosci.* **2021,** 101251.

Ribeiro, A. M.; Estevinho, B. N.; Rocha, F. Preparation and Incorporation of Functional Ingredients in Edible Films and Coatings. *Food Bioprocess Technol.* **2021,** *14* (2), 209–231.

Rodríguez-Martínez, V.; Velázquez, G.; Welti-Chanes, J.; Torres, J. A. Applications of Probabilistic Engineering in Food Moisture Management to Meet Product Quality, Safety, and Shelf-Life Requirements. In *Water Activity in Foods: Fundamentals and Applications*; 2020, pp 501–520.

Schmid, M.; Müller, K. Whey Protein-based Packaging Films and Coatings. In *Whey Proteins*; Elsevier, 2019; pp 407–437.

Şen, D. B.; Kılıç, B. Effects of Edible Coatings Containing Acai Powder and Matcha Extracts on Shelf Life and Quality Parameters of Cooked Meatballs. *Meat. Sci.* **2021,** *179,* 108547.

Senturk Parreidt, T.; Müller, K.; Schmid, M. Alginate-based Edible Films and Coatings for Food Packaging Applications. *Foods* **2018,** *7* (10), 170.

Shamaila, M.; Powrie, W.; Skura, B. Sensory Evaluation of Strawberry Fruit Stored under Modified Atmosphere Packaging (MAP) by Quantitative Descriptive Analysis. *J. Food Sci.* **1992,** *57* (5), 1168–1184.

Sothornvit, R.; Krochta, D. J. Plasticizer Effect on Oxygen Permeability of β-Lactoglobulin Films. *J. Agric. Food Chem.* **2000,** *48* (12), 6298–6302.

Sridhar, A.; Ponnuchamy, M.; Kumar, P. S.; Kapoor, A. Food Preservation Techniques and Nanotechnology for Increased Shelf Life of Fruits, Vegetables, Beverages and Spices: A Review. *Environ. Chem. Lett.* **2021,** *19* (2), 1715–1735.

Suhag, R.; Kumar, N.; Petkoska, A. T.; Upadhyay A. Film Formation and Deposition Methods of Edible Coating on Food Products: A Review. *Food Res Int.* **2020,** *136,* 109582.

Taormina, P. J. Purposes and Principles of Shelf Life Determination. In *Food Safety and Quality-Based Shelf Life of Perishable Foods*; Springer, 2021; pp 1–26.

Tripathi, P.; Dubey, N.; Banerji, R.; Chansouria, J. Evaluation of Some Essential Oils as Botanical Fungitoxicants in Management of Post-harvest Rotting of Citrus Fruits. *World J. Microbiol. Biotechnol.* **2004,** *20* (3), 317–321.

Xu, Q.; Cao, W.; Xu, L.; Liu, Y.; Zhang, H.; Yin, T.; et al. Mechanical Property Stability of Soy Protein Isolate Films Plasticized by a Biological Glycerol-based Polyester and Application in the Preservation of Fresh-cut Apples. *J. Food Process. Preserv.* **2018,** *42* (12), e13829.

Yousuf, B.; Wu, S.; Siddiqui, M. W. Incorporating Essential Oils or Compounds Derived Thereof into Edible Coatings: Effect on Quality and Shelf Life of Fresh/fresh-cut Produce. *Trends Food Sci. Technol.* 2021.

CHAPTER 3

Postharvest Management and Shelf Life of Fresh Produce

ALKA, ANKITA YADAV, and GEETA DHANIA

Department of Environmental Sciences, Maharshi Dayanand University Rohtak, Haryana, India

ABSTRACT

Fruits and vegetables are highly putrescible in certain environmental conditions through time. Certain environmental factors like temperature, humidity, and some extremes reduced the shelf life of fruits and vegetables, due to which their quality deteriorates and a large amount of food loss occurs. As growers and shippers have to face rejection of fewer shipments, there is a great need to prolong the shelf life of fresh fruits and vegetables. Once these are harvested, they become metabolically active and their ripening/senescence processes start and then it becomes necessary to control externally in order to extend the postharvest quality. Postharvest management of fresh produce tends to slower down the physiological process and reduces the threat of growth of microbes and prevents them from being contaminated. Apart from raw techniques of managing the storage time, these postharvest damages can be controlled by various physical, chemical, and gaseous treatments.

3.1 INTRODUCTION

Fresh produce includes the fruits and vegetables that have not been processed in any way. The ovary of flowering plant produces a seed-like structure named as fruit, while the root, stems, leaves, and shoot produce the edible part named as vegetable. Fruits and vegetables are rich source of vitamins, minerals and are full of dietary fibers. Quality and quantity of

Plant Diseases and Their Management: A Sustainable Approach.
Deepa Srivastava, Rajarshi Kumar Gaur, and Ajay K. Tiwari (Eds.)

vegetable and fruits plays a vital role in the healthy diet of an individual. All the micro and macro nutrients which are beneficial for health are not present in individual fruits or vegetables. So, everyone should consume a variety of fruits and vegetables each day. A balanced diet of fruits and vegetables helps in controlling anger and stress, reduces threat of cardiac strokes, prevents certain types of mutagenic diseases like cancer, lowers down the peril of eye and gastro-intestinal issues, and also maintains sugar level which helps in maintaining the appetite of an individual. Weight can be loosened by consuming nonstarchy and nonfatty vegetables and fruits like pears, apple, and green-leafy vegetables (Bertoia et al., 2015). Their low glycemic loads maintain blood sugar spikes which leads to increasing hunger. The American Institute for Cancer Research and World Cancer Research Fund presented a report, which revealed that vegetables having no starch content like lettuce and green leafy vegetables such as cabbage, broccoli, garlic, onion, and fruits like apple defend us from various types of mutagenic diseases like cancer including mutagenic tumors of mouth, voice box, stomach throat oesophagus whereas mostly fruits prevent from respiratory tumors like lung cancer (Wiseman, 2008).

Fruits and vegetables need optimal postharvest techniques in order to increase the shelf life and to maintain the storage strength. The whole system of garden fresh fruits and vegetables commerce is speedily growing to one with an improved emphasis on enhancing the quality and reducing the expenses by rationalizing their dissemination. This tendency is driven by the increasing demand of consumers for handiness in food making and consuming, together with easy packaging of product and form, preservation of qualitative parameters, and perennial accessibility around the whole year, all together are essential part in the one-upmanship of retailers (Mupondwa, 2009).

3.2 SHELF LIFE OF FRESH PRODUCE

It is the time period for which the product maintains the quality and remains in a form that is easily accepted by the customers. Quality of fresh-picked and fresh-cut fruits and vegetables is an amalgamation of various aspects, characteristics, or properties that regulates the worth of produces for the consumer. Texture, appearance, color, texture, aroma or flavor, and dietary or nutritional values are the parameters that help in deciding the quality of fresh produces. Each quality parameter has its own importance and the relative importance of these parameters largely depends on the type of product and eating type that either it is eaten directly in fresh form or after a slightly processing. At the time of purchasing fresh produces, buyers judge the quality of fresh and freshly cut fruits and vegetables on the basis of certain factors like appearance and freshness, i.e., the "best before" date. Later, the consequent buying depends upon the satisfaction of the buyer in the name of texture and flavor excellence while eating the particular produce. Upon making comparison to other food products, the quality of entire fruit and vegetable is dependent on the particular variety of plant, preharvesting practises, harvesting methods, climatic conditions, ripening stage at harvesting time and postharvesting conditions and practices, making shelf life prediction a tough process. The quality of fruits and vegetables, and accordingly the quality of freshly cut items, is also affected by handling procedures, conditions, and stage. Methodology of preparation (proper washing and treatment, sharpness of cutting tools, cutting techniques, size and surface area of cut pieces, and removal of moisture from surface) and following handling conditions (cooling rate, packaging, sanitation conditions, proper management of optimum conditions of temperature, and relative humidity at the time of their distribution) are other factors that influence the quality of freshly cut fruits and vegetables.

3.2.1 CAUSES OF REDUCED SHELF LIFE

In order to increase the shelf life, we must know what harms the fresh produce.

The following are the primary elements that can influence the shelf life of fresh

- Ethylene poisoning
- Changes in temperature and extremes
- Moisture or humidity
- Microbial development
- Harm to the body

3.2.2 DIFFERENT METHODS OF SHELF LIFE EXTENSION

Methods to extend shelf life can vary for different types of fruits and vegetables, but there are some solutions for most of the commodities.

Different ways of extending shelf life are given below:

3.2.2.1 PACKAGING IMPROVEMENTS AND MODIFIED ATMOSPHERE PACKAGING

Extending the shelf life of fruits and vegetables is as simple as improving packing techniques. Packaging that improves shelf life can reduce the pace of respiration, ripening, and microbiological development. A longer shelf life provides for more seasonality, less food loss, and a better possibility of keeping freshness. The gases that surround an item can be controlled with modified atmosphere packaging. Fruit's shelf life could be extended, for example, by reducing the oxygen that surrounds it. Controlling ethylene exposure can be as simple as using pallet covers, carton containers, and sheets.

3.2.2.2 IMPROVING TREATMENT AND HANDLING PROCEDURES

After the time a fruit or vegetable is harvested, you have to be sure that your handling methods and treatment are safe to them and are not harming them in any way. Use of contaminated waters for cleaning and washing of fresh produce and use of contaminated boxes for their storage processes are some of the examples of poor treatment methods. Also, you have to make sure that fresh produce is not subjected to excessive forces that will help you to ensure that they are not physically damaged. They are so delicate that even minor things or minor force or friction can harm shelf life. You have to note that it is important to keep the standard operating procedures to ensure proper handling of fruits and vegetables.

3.2.2.3 FINDING TEMPERATURE WEAKNESSES IN COLD CHAIN

Keeping a low temperature helps in limiting the decaying process and slowing down the microbial growth. You do not have to compromise with any part of the cold chain between harvest and the customer if this happened; it can harm the products.

One study showed that a delay of four hours between harvest and precooling of strawberries increased water loss by almost 50% and it also affects the appearance in a negative manner upon arrival at the distribution center. There are a few useful methods to record the temperature throughout the cold chain. This data can also be used to enable shelf life prediction.

3.2.2.4 MONITORING HUMIDITY TO IDENTIFY IMPROVEMENT AREAS

In case of fresh produce, humidity monitoring is as important as the temperature tracking. These two are often compiled together, as well. Having high humidity helps to reduce the moisture loss in fruits and vegetables. Refrigeration can cause humidity loss, so by managing proper humidity levels, you can effectively reduce shrinking of fresh produces. IoT devices track both temperature and humidity of products from the agricultural production to the customer to identify areas for improvement.

3.2.2.5 INDIVIDUAL ITEM COVERINGS

Plant-based natural coverings on fruits and vegetables have created a noise in present marketing systems. One company named Apeel has even raised a fund of few hundred million dollars for its solution. The degradation process of produce was initiated by losing water content and through the oxidation process in the presence of oxygen. Item coverings of Apeel work by keeping the moisture content inside and oxygen outside. Apeel is a member of Friends of Champions 12.3 alongside One-Third, as both are now dedicated to helping halve the food loss and waste by the year 2030. Some fresh produce stores have implemented the necessary use of plastic coverings on fresh produce individually to extend the shelf life. It is best to avoid plastic wherever possible. However, using plastic coverings can be better for the environment than letting food go to waste. If this shocks you, you should learn more about the effects of food waste.

3.2.2.6 SMART STICKERS

Some studies show how certain compounds can reduce the effects of ethylene or limit microbial growth to extend shelf life. One company named StixFresh claims that by just placing a sticker on the fruit, the shelf life of fresh fruit can be extended up to 14 days. While there is plenty of validation in the supply chain left to perform, it is certainly an interesting idea that requires little effort.

3.2.2.7 ABSORBENT FOOD TRAYS AND PADS

There is a lot of invention around pads that can go in the containers of freshly cut produce. These pads are able to absorb excessive amount of juices and help in slowing down the respiration rate of product. These are already in common use in the industries, but it is worthy to keep up with the increasing capabilities of absorbent pads.

3.2.2.8 SWITCHING VARIETIES

Humans have always wanted to create new types that taste better and make more money since the dawn of agriculture. Because we now send our product all over the world as our planet becomes more dispersed, developing kinds of produce with longer shelf life is a priority. The US Department of Agriculture (USDA), for example, just released a strawberry with a longer shelf life. After a week in cold storage, just 29% of the strawberries decayed, compared to 93% of Camarosa strawberries. If you are open to trying new varieties, it is a good idea to keep an eye on what is available and weigh your options.

3.2.2.9 NATURAL ANTIMICROBIALS

Herbs, spices, or plant extracts can be used as an antimicrobial that certainly can help to prevent spoiling of fresh produce and can extend their shelf life. To meet the customer's demand of safer ingredients, it becomes important to seek attention toward the natural alternatives in place of harmful chemicals. A good profit can be earned at the selling point by attracting even more customers.

3.3 POSTHARVEST LOSS OF FRESH PRODUCE

Postharvest loss of fresh is also a major concern in this review. It addresses postharvest food loss concepts, the relevance of perishable crops (fruits and vegetables), the causes of food loss, and the impact of environmental factors on food loss. Also, some of the difficulties were attempted to be solved by the study's findings. In many nations, postharvest loss is a source of worry for food and nutrition security and malnutrition. Fruit and vegetable loss may reach up to 50% (Gustavsson et al., 2011); therefore minimizing fruit and vegetable loss and waste is one of the most crucial matters for supporting the growing world population properly in the future (Ray and Ravi, 2005). Vegetable and fruit demand is rising in lockstep with population increase. According to the figures from the United Nations Food and Agriculture Organization, 1000 million tonnes of fresh veggies were produced worldwide in 2016 across fifty seven million hectares of land. For 60 million hectares of land, the value was 860 million tonnes of fresh fruit (Elik et al., 2019), nearly half of all agricultural goods are lost before they reach the point of consumption. Furthermore, in emerging and developing countries, this percentage is considerably greater (Gustavsson et al., 2011).

The term "post-harvest loss" relates directly to quantifiable, food loss in the postharvest process. From the moment of harvest, through agricultural processing, selling, and meal preparation, to the consumer's final choice to eat or reject the food, this system is made up of interrelated activities (De Lucia and Assennato, 1994).

Any harm to the amount and quality of product from the time of harvest till utilization is termed as postharvest losses. In wealthy nations, postharvest losses in fresh produce are estimated to range from 2% to 30%, whereas in underdeveloped countries, losses range from 30% to 55% (FAO, 2011; Buzby et al., 2014). This massive food loss places a major financial, societal, and ecological burden on society, motivating to improve present methods and understand the opportunities to decrease waste. Postharvest losses may occur in both, quality and quantity of fresh commodities.

Qualitative Losses:

According to Kader (2005), qualitative losses may include loss in edibility, nutrient content, caloric value, and consumer acceptability which may vary considerably among nations and cultures (Kader, 2005).

Quantitative Losses:

According to Kader et al. (2012), over 1 billion tonnes of food generated each year is wasted. In developed countries, much food is thrown while still fit for human eating (more than 50% of losses occur at the retail stage), but in underdeveloped countries, food losses occur early in the food supply chain at the postharvest and processing stages (Kader et al., 2012), as stated in Table 3.1.

TABLE 3.1 The Loss of Original Output in Various Areas Throughout the Supply Chain.

Area	While production (%)	During handling and storage (%)	During packaging (%)	While distributing (%)	Consumption stage (%)	Overall (%)
Russia and Europe region	20	5	2	10	19	56
Oceania and North America region	20	4	2	12	28	66
Developed Asia region	10	8	2	8	15	43

3.3.1 *WHY SHOULD POSTHARVEST LOSSES BE REDUCED?*

Reducing these postharvest losses is crucial to feeding the global people in the coming years (Gustavsson et al., 2011). One of the most pressing global issues today is maintaining food supply for a rising world population while also guaranteeing protracted sustainable growth. In order to feed the world's seven billion people by 2050, the Organization (FAO) estimates that food output must increase by about 60–70% (Nellemann et al.).

Fresh produce losses cannot be prevented, but they may be avoided by taking measures like storing it at a low temperature, controlling humidity while storage, properly packing and transporting it, and so on (Ahmad and Siddiqui, 2015). It is critical to understand where losses occur in the food supply chain so that probable causes may be identified and optimal postharvest practises can be improved (Gustavsson et al., 2011; Porat et al., 2018). High levels of unemployment, food insecurity, and disease have all been linked to a lack of postharvest management technology and processes, such as control of temperature to sustain the supply chain, processing and packaging, and packing.

Fruits and vegetables are actively growing, the heat output, humidity, the surrounding air during and after cultivation, and the nature and level of infection by microbes or pests, all affect their value in the market. In storage, they lose moisture, succumb to decay due to pests and rodents, lose energy and nutrients (Desai and Pathak, 1970; Majumdar and Pathak, 1991), and experience physical losses, along with a loss in quality arising from metabolic disorders, greening (vegetables), root growth, budding, build-up, and seed germination. In addition, postharvest losses are caused by mycotoxins, which are produced by plant parts through fungal attack. So to minimize postharvest losses, it is crucial to adopt suitable postharvest handling, packing, transportation, and storage methods (Elik et al., 2019).

3.3.2 WHAT CAUSES POSTHARVEST LOSSES IN FRESH PRODUCE?

There are internal and external factors contributing to postharvest loss.

3.3.3 HOW DO THESE FACTORS CAUSE/ARE RESPONSIBLE FOR POSTHARVEST LOSSES?

- As a general note, inappropriate harvesting procedures can also lead to losses: rough handling, late picking, inadequacy, or bad design.
- The primary cause of the loss is the high cost of precooling and the lack of knowledge on the cost-benefits of precooling technologies.

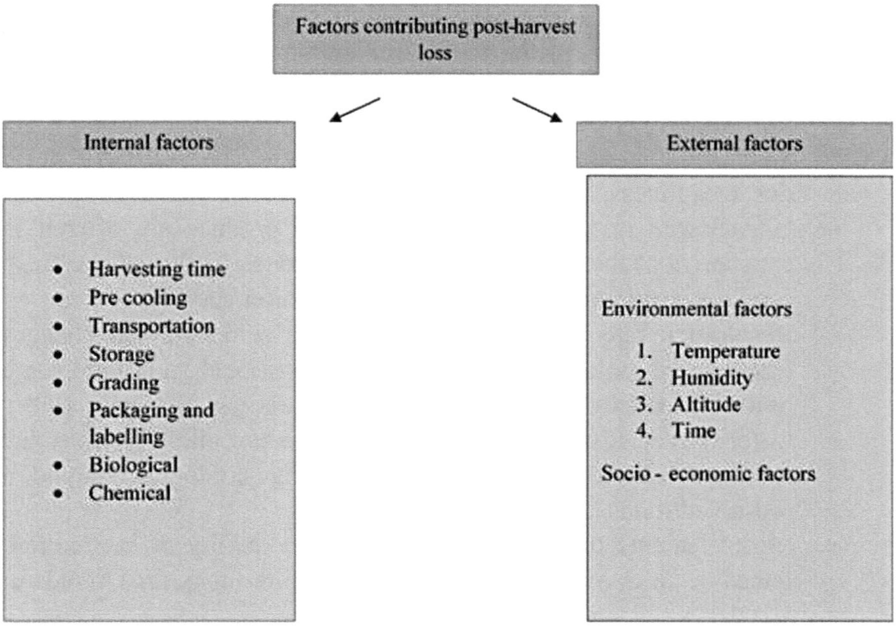

FIGURE 3.1 Factors contributing the postharvest loss.

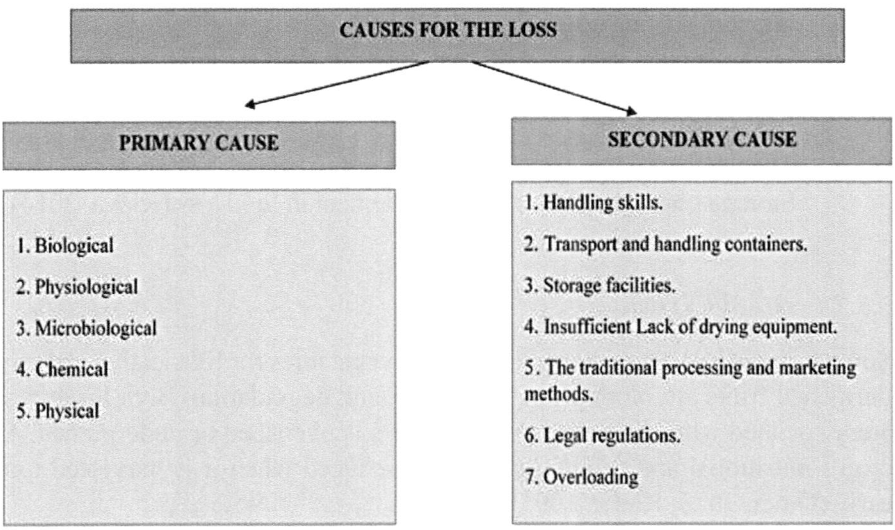

FIGURE 3.2 Causes for the postharvest loss.

- Lack of infrastructure (roads, bridges, etc.) and refrigeration are the primary obstacles in the transportation stage of the supply chain. Plants cannot be transported efficiently in most poor nations because the roads are inadequate (Kader, 2002).
- To ensure long-term storage, it is necessary to have sufficient facilities, cleanliness, and monitoring.
- Unpackaged or knotted bundles of fresh produce are offered to consumers at the grocery store level. Because of this, if the fresh food is not sold promptly, it has a much shorter shelf life.
- Respiration rate, ethylene generation, and compositional changes (related with color, texture, flavor, and nutritional value) are some of the biological reasons of degradation (Kitinoja and Gorny, 1999).
- Color, flavor, texture, and nutritional value are all affected by the spontaneous reaction of many of the chemical components found in preserved foods (Atanda et al., 2011).
- Most elements that damage or reduce the quality of horticulture goods occur at a quicker pace as temperature increases (Atanda et al., 2011),
- Because fresh horticulture goods have a high relative moisture content, they must be kept in a way that minimizes relative moisture loss and wilting (Atanda et al., 2011).
- High-altitude storage will consequently tend to extend storage life and reduce food losses, provided that it is kept out of direct sunlight (FAO, 1983).
- The longer the food is held, the higher the danger of damage and loss. Storage time is therefore a crucial element in food loss (Kiaya, 2014).

3.3.3.1 HARVESTING

Some farmers harvest premature fruits and vegetables for financial purposes; unripened fruits are more prone to damage and degradation. Both instances are associated with some disorder; may be it is overriped or underpinned. A crop's nutritional and economic worth is reduced when it is harvested too early (Ömer, 2018; Kader, 1995).

3.3.3.2 STORAGE AND TRANSPORTATION

Perishable food must be transported in a cold chain to guarantee that it reaches the customer in good condition. As per the International

Refrigeration Institute (IIR), about 20 percent of the total of fresh produce in underdeveloped nations rotted as a result of not using the cooler (IIR, 2009). Organizing and storing items allows for better time management, allowing for more relaxed marketing and consuming activities. The major source of postharvest losses in poor nations is a lack of appropriate storage facilities (FAO, 2013).

Note: Cold chain in place that ensures a product's temperature remains constant from manufacturing to consumption starting with pre-cooling, cold storage, cold shipping, and chilled display cases, an effective cold chain management system is essential. Manufacturing to consumption.

Perishable goods cannot be effectively preserved in underdeveloped nations due to a lack of sufficient transportation, bad roads, and ineffective logistics management. Furthermore, in many nations, loading and unloading activities are performed by inexperienced and uneducated personnel who, in general, do not handle items with care. Azabaaolu (2018) claims that this causes mechanical harm to agricultural items (Ömer, 2018).

3.3.3.3 INADEQUATE PACKING

Packaging materials that allow moisture to escape are not used. Fresh markets and wholesale marketplaces are where farmers sell their produce. Fresh produce is offered in unpackaged form or in bundles at the retail level. If fresh food is not sold immediately, this sort of market handling drastically lowers its shelf life (Kiaya, 2014).

Sad to say, in many impoverished nations or portions of countries, extremely low quality materials are utilized for packing purposes. The level of cheapness may be noticed when some sensitive fruits and vegetables are packaged in very low quality bags, causing harm to them. According to several surveys, 46% of fresh produce is wrapped in fabric bundles, 30% is placed in open baskets, and 8% is packed without any packaging either (Kitinoja and AlHassan, 2010).

3.3.3.4 CUSTOMER WASTE

Fruit and vegetables lose a considerable amount of their value during the eating stage; according to a study (Food et al., 2011), fruit and vegetables account for more than 30% of household waste, while another study (WRAP, 2008) puts the figure at approximately 40%. When it comes to shopping, one of the first things that comes to mind is

- Over-purchasing.
- Inadequate planning
- Inadequate management of household storage, among other factors (Porat et al., 2018).

3.3.3.5 BIOLOGICAL

A bird's feathers and hairs of animals can contaminate the food, making it unsuitable for consumption by humans later, resulting in weight loss and reduced nutritional value for humans.

3.3.4 CHEMICAL

The spontaneous response of many of the chemical components in preserved foods affects color, flavor, texture, and nutritional value. It is possible for foods to lose their flavor or become discolored or softened during storage because of enzyme-activated reactions.

3.3.5 THE INFLUENCE OF THE ENVIRONMENT ON FOOD LOSS

The amount and quality of a harvest are influenced by climatic variables such as air, moisture, precipitation, and heat (Grolleaud, 2002). U.N. environment programme (UNEP) is engaged in improving the health and well-being of both people and the environment.

When food loss is reduced and malnourished countries are fed, the environmental damage caused by farming methods is also reduced. As a result, UNEP promotes food conservation because of its favorable environmental impacts.

3.3.6 WHAT CAN BE DONE TO DECREASE POSTHARVEST LOSSES AND WASTE?

The most frequent method of preserving fruit quality and decreasing postharvest losses is to delay ripening and senescence. When fresh produce is harvested, it uses food stocks for oxygen and emits carbon dioxide and heat from the carbohydrates that were stored in the product before it was

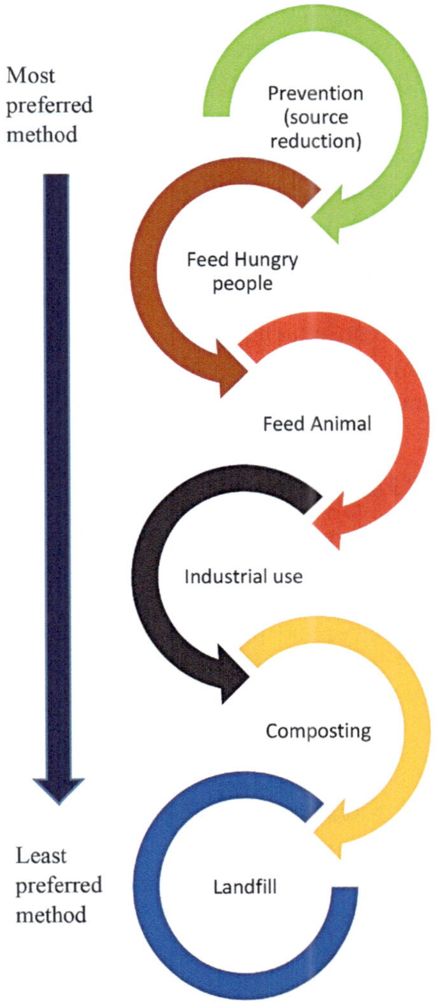

FIGURE 3.3 Steps to be considered while minimizing the loss waste.

harvested. A postharvest therapy that blocks respiration will help preserve product quality by slowing deterioration processes and increasing shelf life. The significance of the plant hormone ethylene in the ripening process is well known. Ethylene is helpful since it slows down the ripening process.

It is becoming more common to use a chemical-free method of extending the shelf life of fruit, such as controlled atmosphere (Paulauskienė et al., 2020), or dynamic controlled atmosphere (Koríćanac et al., 2020; Dällenbach et al., 2020) for apples and kiwifruit.

Ethylene deficiency may significantly slow down the maturation process of climacteric fruits, as well as reducing physiologic problems like apple burn (Dällenbach et al., 2020; Jung and Choi, 2020). Growth regulator therapies such as salicylic acid and the polyamine utrescine have significant potential to expand postharvest treatments and, as a consequence, increase the storage period of non-climacteric fruits such as mandarins (Ennab et al., 2020). Retarding microbial activity and decreasing losses can be accomplished with the use of chemical treatments, which are extremely effective. Because of the potentially hazardous by-products and residues, this technique may have significant health consequences for both humans and the environment. The public is becoming increasingly concerned about the usage of these substances (Shimshoni et al., 2019).

3.4 HARVESTING FACTORS

3.4.1 TIME OF HARVESTING

An ideal maturity date for fruits and vegetables must be determined in order for them to be harvested properly and efficiently. When to harvest fresh fruits and vegetables has been studied extensively (dos Santos Neto et al., 2017). Examples of ripeness criteria for certain fruits and vegetables include: fruit opening and hardness, juice content, textural qualities, softness, color, and nutritional composition such as glucose, starch, fat, protein content, solubility, and total acidity (Mencarelli et al., 2005). As an example, strawberries collected at the 2/3 stage of its life cycle might be stored at room temperature for up to 3 days. One-third of the developed fruits had a longer shelf life at room temperature, but the quality decreased (Rahman et al., 2016). In order to determine the level of maturity of grapes, a variety of criteria, such as amount of sugar, alkalinity, and acidic profiles, are often employed (Mencarelli et al., 2005). Tomato age, morphology, shade and shape, and seeds falling when fruit is sliced, and turning of the color from green to pink are some maturity indicators (Okiror et al., 2017). Harvesting in the early morning is preferable since harvesters are more efficient at that time (Kiaya, 2014).

3.4.2 TEMPERATURE AFTER HARVESTING

In order to reduce postharvest losses and extend shelf life, it is crucial to control the temperature after harvesting. There are a variety of ways for chilling fresh fruit after harvest, including hydro cooling, condensation, and room cooling (Elansari, 2009). Optimal storage temperatures and humidity levels of some vegetables and fruits are given in Table 3.2.

TABLE 3.2 Some Vegetables and Fruits' Optimal Storage Temperatures and Humidity Levels Are Listed Here (Sibanda and Workneh, 2020).

Vegetable/fruits	Optimum temperature (°C)	Optimum relative humidity (%)
Potatoes	5–16	90–95
Onions	1–2	65–70
Tomatoes	13–15	80–85
Mango	13	85–95
Banana	13–14	90–95
Guava	5–10	90

3.4.3 METHOD OF HARVESTING

Unfortunately, horticultural produce can be permanently harmed by poor harvesting techniques if they are not corrected. In order to prevent damage during harvest, it is important to standardize harvesting procedures for all fruits and vegetables. Harvesting methods include manual and mechanized harvesting, as well as a combination of the two. To avoid harm and waste, harvesters should be careful when cutting, chopping, plucking, and holding or extracting the fruit or vegetable from the plant. Using only your hands or sharp objects, you may harvest the entire crop, or just a portion of it. Sharp and clean materials are required (Food and Agriculture Organization of the United Nations, 1989).

3.4.4 HANDLING

As a harvesting container you may use numerous materials such as plastic, wooden, or fiber; the pickers should be able to easily handle them. Citrus fruits and avocados are examples of crops that are gathered in bags. Harvesting crops such as potatoes and onions are often collected in sacks. Plastic containers that can be cleaned and sterilized between are also good. Smooth edged buckets are ideal for collecting tomatoes. Harvesting containers should be selected based on the type of crop.

3.4.5 TRANSPORTATION AND STORAGE

Multiple stages are required to get a product from a farm to a consumer and this can be quite expensive. Distance, freshness, and the value affect how fresh fruit and vegetables are transported (Harris, 1988). Improved long-distance transportation networks are needed for the retention of quality. A proper temperature management system and air circulation system are the most essential ways to assure quality preservation of perishables (Vigneault et al., 2009; Hertog et al., 2014). For perishable products, maintaining a cold chain is the best way to preserve their quality and extend their shelf life (Negi and Anand, 2014). Other elements that impact the quality of product include: starting quality, ambient humidity and moisture loss, gases concentration, damage and physical discomfort, and transport circumstances (road conditions, time of day, etc.) (Vigneault et al., 2009).

Whatever method is used, the principles of transport are the same:

- Careful Lifting
- Rapid Transport To Preserve Quality And Freshness
- Proper Maintenance For Physical Damage;
- Maintenance Of Correct Temperature Throughout The Transport;
- Good Ventilation
- Maintenance of Proper Humidity Levels Or Minimum Water Loss
- Providing Shade from Heat and Rain (Paltrinieri, 2016).

To avoid nutritional deficiencies, food storage is necessary. The approach, on the other hand, is entirely scientific. There are a few basic concepts that should be understood before storing fruits and vegetables (Khan et al., 2017). Different types of storage structures exist across the world. Food must be kept cool, and the product must be of good quality when it is first placed in storage (Kiaya, 2014). Cold storage is the most common method for lowering metabolic activity.

3.4.6 IMPROVING THE PACKAGING PROCESS IN ORDER TO REDUCE POSTHARVEST LOSSES

Fresh fruit and vegetables have the ability to last for a long time (shelf life) even after they are harvested. However, maintaining their shelf life requires certain atmospheric conditions because they cannot manage themselves under normal conditions (for example, they lose moisture and become dry) (Verma, 2000). Each type of fresh produce has unique packaging requirements. Some require protection against germs, while others require protection against physical harm. Similarly many fruits and vegetables require temperature and humidity control (Ramaswamy, 2014; Ahvenainen, 2003; Ahmad and Siddiqui, 2016).

3.4.7 PROCESSING

In addition to stabilizing and diversifying the food supply, processing also offers job possibilities. As a result, green vegetables have a lower perishability rate. As well as being more stable and digestible, processed food allows for a more varied diet, allowing customers access to a larger variety of items, vitamins, and minerals. This includes drying, salting, fermenting (fermentation), and pickling (pickling). Quality characteristics such as shape, structure, and taste are not maintained as a result of limited processing treatment, and product degradation might occur quickly (Huxsoll et al., 1989).

TABLE 3.3 Postharvest Food Waste Minimization (Elik et al., 2019).

Problems	Alternatives to consider
Cuts and wounds (overfilling and transport injury)	Use a container with a flat surface and do not overfill it, maximizing individual coating and padding
Loss of water or dryness and growth of microbes	Utilizing packaging strategies like modified atmosphere packaging and a good ray exposure
Poor aeration	Incorporation of breathable packaging materials
Consumer's over purchasing	Consumer packaging should be reduced

3.5 HOW TO PREVENT FRUIT AND VEGETABLE POSTHARVEST DISEASES?

Three distinct techniques are used to manage postharvest infections in fruits and vegetables. It focuses on preventing infections, eradicating those that are already present, and preventing the pathogens; the major goal is to preserve them disease-free until they are sold or eaten (Barkai-Golan, 2005; Sharma and Alam, 1998). For the prevention of illness, physical, chemical, and biological techniques can be utilized.

3.5.1 PHYSICAL TREATMENT

Heat is the most often used physical therapy. Traditionally, hot water dips, hot water rinsing and brushing, vapor, hot air, and curing were all methods of applying it (Fallik, 2004). Low-temperature storage or high-temperature treatments, as well as magnetic fields and radiation (such as sound and ultrasound), can be utilized as physical treatments (Eckert and Sommer, 1967).

TABLE 3.4 Gamma Rays are Being Used to Treat Postharvest Illness in Fruits (Beraha et al., 1957).

Fresh produce	Dosage required	Limit of tolerability
Apple	Approx. 150	Up to 150
Apricot	Approx. 200	Up to 100
Orange	Approx. 200	Up to 200
Strawberry	Approx. 200	Up to 200
Tomato	Approx. 300	Up to 150

3.5.2 TREATMENTS USING CHEMICALS

Infection and marketing techniques play a role in the therapy. It is important for citrus, which has a rather lengthy postharvest life, to avoid initial infection and sporulation in order to prevent contamination of neighboring fruits. Berry postharvest life is brief, and treatment focuses on keeping gray mold from spreading to other strawberries after harvest. So the treatment must be in line with how it will be sold (Singh and Sharma, 2018).

3.5.3 TREATMENT USING BIOLOGICAL ADDITIVES

Plant extracts have several advantages, including inexpensive manufacturing costs, minimal toxicity, and high biodegradability (Maswada and Abdallah, 2013). Furthermore, the majority of the extracts are high in polyphenols, which might increase the antioxidant characteristics of edible coatings (Silva-Weiss et al., 2014). The use of edible coatings to preserve food product quality is a low-cost, ecologically friendly method with numerous advantages, including biodegradability and the ability to provide a selectively permeable barrier against gases and water vapor, decreasing microbial assault (Dhall, 2013).

3.6 POSTHARVEST TREATMENTS

To maintain the fresh looking quality of fresh commodities having high nutritional values, many postharvest approaches like physical treatments, chemical treatments, and gaseous treatments have been utilized. These postharvest treatments are typically bounded to the suitable management of temperature in stored condition. These postharvest treatments help in managing the quality and reduce the wastage of fresh commodities.

3.6.1 PHYSICAL TREATMENTS

3.6.1.1 HEAT TREATMENT

For harvested fresh fruits and vegetables, heat treatment works as an alternative to the chemical treatments. Heat of saturated hot water vapors, hot dry air, hot water dip, and hot water rinse while brushing are some of the heating treatments for fresh commodities. Heat treatments can be provided either in a short time duration of 1 h or for a longer time duration, i.e., up to 4 days. The above given heat treatments have been provided to preserve the color of lettuce, asparagus, celery, broccoli, kiwi fruits, and green beans, to secure the hardness of tomatoes, potatoes, strawberries, and carrots, to avoid the expansion of overripening flavors in cantaloupe and other species of melons, and to improve the durability of peaches, grapes, bean sprouts, plums, and others. Hot water works as a better energy vector as compared to air and has provided comparable declines in decaying processes by fungi. In case of grapefruit, blue molds initiated by *Penicillium sp.* have been controlled by the process of hot water dip, by dipping them for 2 min in hot water at 50°C (Schirra et al., 2000). The structure and composition of epicuticular waxes may also be affected by hot water treatments. After heating process, wrapping of wounds and cracks presented and antifungal substances formed in the wax are the possible ways or modes of working actions (Tahir et al., 2009). At present, commercial scale applications or uses of heat treatments are restricted. *Heat* treatment offers a better alternative approach toward usage of fungicide; hot water dip treatment has been used in Germany to store the organic apples in order to prevent them from fungi infection. Immediate application of heating treatments to fruits after removing them from a long time usage of controlled conditions for their storage and after a few days of their storage in cold conditions can provide a new alternative for subsequently enhancing their storage life timings (Maxin et al., 2012). Although acceptance rates of heat treatment technology are less entertained by fruits and vegetables cultivators due to high costs of energy used and also due to the requirement of extra labor in the peak hours of working during the harvest time period, some of the general benefits of heat treatments are given below:

- These make changes in the physiology of fresh fruits and vegetables. For example, it reduces the injuries caused in chill conditions and helps in delaying their ripening processes by inactivating the degradative enzymes due to heat.
- It kills contamination done by critical insects.
- By making control on the arrival of fungi it prevents them from fungal decay (Maxin et al., 2012).

3.6.1.2 EDIBLE COATING

Edible coatings are the thin layers uniformly applied on the outer surface of fresh produces in order to improve the waxy nature of cuticle. This coating acts as an alternative for natural barriers on the surfaces of fresh produces from where the cuticle layer has been detached because of certain reasons. Edible coatings are made up of hydrophilic or hydrocolloid groups (materials composed of polysaccharide and protein) or hydrophobic groups (waxes composed of lipids), or a combination of both the groups with the intention of making improvement in the functions of the coating (Gol et al., 2013). Some of the edible coatings like mineral oils, aloe vera, chitosan, proteins, polyvinyl acetate, and cellulose-based materials are known to add some needed characteristics toward fresh commodities such as taste, no residual odor, act as a better barrier, and have good antimicrobial activities (Dhall, 2013). During last some years, a large number of researches and innovations have been focused on the synthesis of edible coatings by using both synthetic and natural sources in a manner to save fresh commodities from various kinds of pathological and physiological damages, some of them are mentioned in Table 3.5.

3.6.1.3 IRRADIATION

In irradiation treatment, fresh produces like fruits and vegetables are exposed to radiant energy emitted from electronic beams (used high energy electrons) and γ rays; these radiations penetrate inside the fresh produces and break down the molecular bonds present within them, including the damage to DNA of living organisms. Also ionizing radiation from caesium-137 and cobalt-60, or electron beams generated from machines are used as a source of irradiation to extend the shelf life of fresh commodities (Farkas, 2014). Low doses of irradiation of 0.3–0.7 kGy on mango delayed ripening and extended shelf life by 3–4 days (Mahto and Das, 2013). According to Pandey

TABLE 3.5 Different Types of Coating Materials and Their Purposes.

Coating material	Purpose of coatings
Based on wax of candelilla	Antioxidant and antimicrobial properties, and maintains quality
Arabic gum, soya bean gum, jojoba wax, and glycerol	Whole quality
Aloe vera gel and Shellac	Maintaining quality
Carboxymethyl cellulose and soy proteins	Antioxidant properties and water barrier
Zein and chitosan	Antioxidant properties and water barrier
Coconut, sunflower oil and beeswax	Antimicrobial and antioxidant properties, and maintaining quality
Carboxymethyl cellulose, alginate, and pectin base	Antioxidant, water barrier
Methyl cellulose and chitosan	Antimicrobial, antioxidant, and acts as O_2, CO_2, water barrier
Carboxymethyl cellulose and soy proteins	Antioxidant properties and acts as water barrier
Pectin	Whole quality
Aloe vera gel	Whole quality
Chitosan, agar and acetic acid (collectively)	Antimicrobial properties and O_2, CO_2 barrier
Rice bran oil and whey protein	Water barrier and maintains quality
Chitosan sugar	Maintains quality
Sucrose and polyester	Antioxidant properties and acts as a water barrier
Gellan and alginate	Acts as O_2, CO_2, water barrier

et al. (2013), a 1 kGy irradiation treatment was shown to be the only dose effective in extending the shelf life of litchi fruit while maintaining other quality features. Irradiation has been approved for the treatment of potato and onion sprouting, as well as strawberry deterioration. While much of the focus of irradiation on fresh fruits and vegetables has been on prolonging shelf life and minimizing decay, it has long been recognized that irradiation is efficient at killing, sterilizing, or preventing the development of a wide variety of insect pests on perishable fresh fruits and vegetables. Contrary to popular belief, subjecting food to irradiation does not render it radioactive. The irradiation technique causes only minor chemical changes in food and has no effect on its nutritional value. Irradiated food is safe and healthful, according to extensive study and testing (Ferrier, 2010).

3.6.2 CHEMICAL TREATMENTS

3.6.2.1 ANTIMICROBIAL AND ANTI-BROWNING AGENTS

The rising incidence of foodborne illness outbreaks has increased regulatory agencies', producers', and consumers' concerns about the microbiological safety of fresh fruits and vegetables over the last decade. Cucumber, sprouts, celery, lettuce, watercress, and leeks have all been linked to outbreaks (Warriner et al., 2009; Caleb et al., 2013). Antimicrobial and anti-browning compounds, which are divided into chemical and natural/bio-based agents, offer the option of maintaining safety. Peroxyacetic acid, electrolyzed water, chlorinated solutions, hydrogen peroxide, and organic acids are some examples of chemical-based agents. Because of its powerful oxidizing characteristics and low cost, a chlorine-based solution such as NaClO has been one of the most regularly used disinfectants for fresh produce (Artés et al., 2009). However, its antibacterial activity is reliant on chlorine levels, and high amounts may produce taste and aroma problems in treated products. Furthermore, chlorine-based chemicals have been shown to be ineffective in reducing microbial load on fresh vegetables (Baskaran et al., 2013).

Peroxyacetic acid is a powerful oxidizer that produces no hazardous by-products (Carrasco and Urrestarazu, 2010). On lettuce, strawberries, cantaloupe, and apples, peroxyacetic acid was found to be efficient in controlling E. coli O157:H7 and *L. monocytogenes* (Rodgers et al., 2004). When lettuce was treated with peroxyacetic acid, there was a 5 log reduction in *Enterobacter sakazakii* (Kim et al., 2006). According to Landfeld *et al.* (2010), decontamination treatment with peroxyacetic acid reduced the initial load of aerobic mesophilic bacteria by roughly 4 log units and yeasts and molds by 3.5 log units, with no additional microbial growth seen during storage.

Because of its potential to create other cytotoxic oxidizing species, such as hydroxyl radicals, hydrogen peroxide has antibacterial, sporicidal, and inhibitory properties. Melons, oranges, apples, prunes, tomatoes, whole grapes, and fresh-cut food can all be treated with hydrogen peroxide to improve shelf life and lower natural and pathogenic bacteria populations. Hydrogen peroxide treatment, on the other hand, necessitates a long period of administration and can harm some product. It is also generally considered safe for several food uses, but it has not been approved as an antibacterial agent (Cengiz and CertelM, 2013). In contrast to tap water washing, Lopez-Galvez et al. (2013) discovered that the newly created hydrogen peroxide-based sanitizers caused a considerable increase in the respiration rate and electrolyte leakage of fresh-cut iceberg lettuce.

Bae et al. (2011) showed that organic acids (acetic, lactic, and malic acids) mixed with modified atmosphere packaging had inhibitory effects on foodborne pathogens such as *E. coli* O157:H7, *S. Typhimurium*, and *L. monocytogenes* on cabbage. Internalization of germs and inaccessible places within fresh produce, such as the calyx, are some of the factors that limit the efficacy of antibacterial and anti-browning chemicals. These restrictions underscore the need for new antibacterial and anti-browning agent application techniques.

3.6.2.2 NITRIC OXIDE

Nitric oxide is a highly reactive free radical gas that serves as a multifunctional signaling molecule in a variety of plant physiological processes, including fruit ripening, fresh fruits, and vegetables senescence (Wendehenne et al., 2004). Endogenous nitric oxide concentrations in fresh fruits and vegetables decrease with maturation and senescence, allowing for exogenous administration to counteract the effects. Optimal nitric oxide levels delay the climacteric phase of many tropical fruits, extend postharvest storage life by inhibiting ripening and senescence, inhibit ethylene biosynthesis, lower ethylene output, and, as a result, delay fruit ripening (Singh et al., 2013). Nitric oxide also inhibits ethylene production and regulates gene expression through cross-communication with other phytohormones. Apple, Chinese bayberry, banana, strawberry, kiwifruit, tomato, mango, pear, peach, plum, Chinese winter jujube fruit, papaya, and loquat have all been observed to benefit from the application of nitric oxide (Manjunatha et al., 2012).

3.6.2.3 SULFUR DIOXIDE

Sulfur dioxide is extensively employed on table grapes to prevent deterioration during storage, either through initial field fumigation followed by weekly storage room fumigation or through delayed release from in-package pads containing sodium metabisulfite (Palou et al., 2010). Other fruits, such as litchi, fig, banana, lemon, or apple, have also been examined for postharvest degradation with sulfur dioxide technology (Sivakumar et al., 2010). According to Cantn et al. (2012), sulfur dioxide fumigation followed by controlled atmosphere storage (3 percent oxygen + 6 or 12 percent carbon dioxide) for fresh blueberries is a potential postharvest technique for reducing deterioration, extending market life, and maintaining high nutritional value. Rivera et al. (2013) showed that sulfur dioxide is an effective and viable solution for minimizing the danger of blueberry gray mold deterioration during storage, and that it could be used for export. The sulfur dioxide concentration required to limit fungal growth can cause injuries in grape fruits and stems, and sulfite residues can be harmful to some people, as well as maintaining the texture of some fruit species such as pomaces, inadequate de-sulphiting, and incomplete re-coloring of red fruits. Nonetheless, because of the benefits of universal antibacterial action and cost-effectiveness, sulfur dioxide therapy is widely used.

3.6.3 GASEOUS TREATMENTS

3.6.3.1 OZONE

Ozone can replace standard sanitizing agents, according to recent research and commercial uses. Ozone is a very corrosive, naturally occurring gas with powerful, extremely reactive oxidizing characteristics. Ozone has 1.5 times the oxidizing potential of chlorine and 3000 times the potential of hypochlorous acid, according to reports. Antimicrobial contact times are typically four to five times shorter than chlorine contact times. At practical and safe concentrations, ozone damages bacterial cell walls faster than chlorine and is more efficient than chlorine against thick-walled spores of plant diseases and animal parasites (Huyskens-Keil et al., 2011). In comparison to untreated fruit, Ali (2014) found that fruit exposed to 2.5 ppm ozone had higher lycopene content, total soluble solids, β-carotene content, antioxidant activity and ascorbic acid content as well as less weight loss at day 10. The sensory qualities of ozone-treated papaya were similarly superior in sweetness and overall acceptability, indicating that ozone can be used as a nonthermal and safe food preservation method for fresh fruits and vegetables. Ozone can be used to sterilize process water, cold storage, and washing systems. Apples, potatoes, cherries, plums, carrots, peaches, garlic, onions, kiwi, and table grapes have all been used for commercial purposes (Suslow, 2004). Ozone, on the other hand, has a difficult time penetrating natural openings or wounds. More research is needed to determine the potential and limitations of using ozone for postharvest treatment of fresh fruits and vegetables to improve their quality and safety.

3.6.3.2 ETHYLENE

On horticultural fresh commodities, ethylene gas that is originated from within and the ethylene produced from outside have both good and bad effects when treated. Externally applied ethylene gas has a number of advantages, including inducing ripening process, improving color of fruits, enhancing the quality in various crops, for example, in mangoes, bananas, avocados, tomato, kiwifruit, tomatoes, persimmon, mangoes, and removing green coloring from the skin of citrus fruits (Wills et al., 2007; Singh and Singh, 2012). Shorter storage life, discoloration or browning, softening of fruits and russet spots appearance in lettuce, yellowing of leafy green vegetables and cucumbers, and amplified vulnerability of fresh fruits and vegetables to decay, all these have been recorded as negative impacts of exposure to ethylene in postharvest phases of horticulture produces (Saltveit, 1999; Wills, 2005). Therefore, ethylene management of ethylene gas plays a very crucial role in maintaining quality and safe postharvest life of fresh produces, both climacteric and non-climacteric. By controlling composition of atmospheric gases and temperature, we can extend the durability of produces. The most commonly used commercial strategies for maintaining horticultural commodities involve their storage at low temperatures, hindering biosynthesis and action of ethylene gas, and minimizing exposure of fresh produces to ethylene during ripening, harvesting, storing, and transport (Watkins, 2002). Newly designed ethylene measurement equipment will allow critical quantities to be detected during storage and transit (Janssen et al., 2014). For the storage of fresh horticulture goods, Martínez-Romero et al. (2009) designed a carbon and heat hybridized scrubber for ethylene. A cartridge heater was securely connected to the activated carbon and 1% palladium (Pd) in the apparatus. Heat pulses cause ethylene oxidation to rise and activated carbon to self-regenerate.

3.6.4 EMERGING TECHNOLOGIES

Decontaminating fresh fruits and vegetables with plasma is a relatively new technology. Plasma is made up of ionized gas molecules that have been separated by the application of energy. They can generate high or low temperatures, which are referred to as hot or cold plasma, depending on the manner of particle activation and the excitation energy (Niemira, 2012). Cold plasma can be created at atmospheric pressure by converting argon gas to plasma at a radio frequency of 27 MHz (Baier et al., 2013) or by an electric discharge between two electrodes separated by dielectric barriers (Pankaj et al., 2013). The erosion of microbial spore surface atom by atom through adsorption of reactive free radicals ("etching"), direct destruction of DNA via UV irradiation, and volatilization of compounds from the spore surface by UV photons through intrinsic photo-desorption have all been proposed as mechanisms for microbial spore inactivation in plasma environments. According to Fernández et al. (2013), roughly 15 minutes of cold gas plasma treatment was required to achieve 2.72, 1.76, and 0.94 log reductions in viable cells of *S. enteric* sv. *Typhimurium* lettuce, strawberry surfaces, and potato tissue, respectively, under ideal operating conditions. Baier et al. (2013) found that plasma treatment at 20 W for 1 min successfully inactivated *E. coli* by 4 log-cycles on fresh corn salad leaves. More research, however, is needed to fully comprehend the function of microbial cell structure, physiology, and stress resistance mechanisms in plasma resistance. Furthermore, the impact of plasma treatment on food enzymes, fresh fruits, and vegetables postharvest quality features warrants further investigation. Further research is needed on gas safety, consumer perception, and the translation of laboratory scale to big commercial scale.

3.7 CONCLUSION

A large number of physical, chemical, and gaseous treatments and technologies exist to reduce the postharvest losses of fresh fruits and management. After that a large amount of food waste occurs due to enormous factors including changes in the atmospheric conditions, harvesting methods, bad handlings, and many more. Hence, growers, dealers, and shippers have to face rejection from customers and gain less profit. Although the existing technologies are useful in reducing the postharvest losses and help in extending the shelf life of fresh produces, but they can reach up to the margin. Hence, more research based on postharvest technology should be continued to improve the efficiency of the existing treatments as well as safety issues should be viewed. However, more detailed studies are required to assess the effect of novel treatments on fresh produce.

KEYWORDS

- **fruits and vegetables**
- **shelf life**
- **postharvest management**
- **contamination**
- **ripening**
- **senescence**
- **microbial growth**

REFERENCES

Ahmad, M. S.; Siddiqui, M. W. *Factors Affecting Postharvest Quality of Fresh Fruits Postharvest Quality Assurance of Fruits*; Springer, 2015; pp 7–32.

Ahmad, M. S.; Siddiqui, M. W. Postharvest Quality Assurance of Fruits. *Chapter* **2016,** *2,* 7–12.

Ahvenainen, R. *Novel Food Packaging Techniques*; Elsevier, 2003.

Ali, A.; Ong, M. K.; Forney, C. F. Effect of Ozone Pre-conditioning on Quality and Antioxidant Capacity of Papaya Fruit during Ambient Storage. *Food Chem.* **2014,** *142,* 19–26. doi:10.1016/j.foodchem.2013.07.039

Artés, F.; Gómez, P.; Aguayo, E.; Escalona, V.; Artés-Hernández, F. Sustainable Sanitation Techniques for Keeping Quality and Safety of Fresh-cut. *Postharvest Biol. Technol.* **2009,** *51,* 287–296. doi:10.1016/j.postharvbio.2008.10.003

Atanda, S. A.; Pessu, P. O.; Agoda, S.; Isong, I. U.; Ikotun. The Concepts and Problems of Post-harvest Food Losses in Perishable Crops. *Afr. J. Food Sci.* **2011,** *5* (11), 603–6013

Bae, Y. M.; Choi, N. Y.; Heu, S.; Kang, D. H.; Lee, S. Y. Inhibitory Effects of Organic Acids Combined with Modified Atmosphere Packaging on Foodborne Pathogens on Cabbage. *J. Korean Soc. Appl. Biol. Chem.* **2011,** *54,* 993–997. doi:10.1007/BF03253191

Baier, M.; Foerster, J.; Schnabel, U.; Knorr, D.; Ehlbeck, J.; Herppich, W. B.; Schlüter, O. Direct Non-thermal Plasma Treatment for the Sanitation of Fresh Corn Salad Leaves: Evaluation of Physical and Physiological Effects and Antimicrobial Efficacy. *Postharvest Biol. Technol.* **2013,** *84,* 81–87. doi:10.1016/j.postharvbio.2013.03.022

Barkai-Golan, R. *Postharvest Diseases of Fruits and Vegetables*. A Division of Reed Elsevier India, Pvt. Limited: New Delhi; 2005.

Baskaran, S. A.; Upadhyay, A.; Kollanoor-Johny, A.; Upadhyaya, I.; Mooyottu, S.; Amalaradjou, M. A. R.; Schreiber, D.; Venkitanarayanan, K. Efficacy of Plant-derived Antimicrobials as Antimicrobial Wash Treatments for Reducing Enterohemorrhagic Escherichia Coli O157:H7 on Apples. *J. Food Sci.* **2013,** *78,* M1399–M1404. doi:10.1111/1750-3841.12174

Beraha, L.; Ramsey, G. B.; Smith, M. A.; Wright, W. R. Gamma Radiation for Possible Control of Postharvest Diseases of Apples, Strawberries, Grapes, and Peaches. In *Phytopathology*; 1957; Vol. 47 (1); pp 4–4. 3340 Pilot Knob Road, St Paul, MN 55121 USA: Amer. Phytopathological. Soc.

Bertoia, M. L.; Mukamal, K. J.; Cahill, L. E.; Hou, T.; Ludwig, D. S.; Mozaffarian, D.; Willett, W. C.; Hu, F. B.; Rimm, E. B. Changes in Intake of Fruits and Vegetables and Weight Change in United States Men and Women Followed for up to 24 Years: Analysis From Three Prospective Cohort Studies. *PLoS Med.* **2015,** *12* (9), e1001878.

Buzby, J. C.; Wells, H. F.; Hyman, J. *The Estimated Amount, Value, and Calories of Postharvest Food Losses at the Retail and Consumer Levels in the United States; EIB-121*; U.S. Department of Agriculture, Economic Research Service: Washington, DC, USA, 2014.

Caleb, O. J.; Mahajan, P. V.; Al-Said, F. A.; Opara, U. L. Modified Atmosphere Packaging Technology of Fresh and Fresh-cut Produce and the Microbial Consequences: A Review. *Food Bioprocess Technol.* **2013,** *6,* 303–329. doi:10.1007/s11947-012-0932-4

Cantín, C. M.; Minasa, I. S.; Goulas, V.; Jiménez, M.; Manganaris, G. A.; Michailides, T. J.; Crisosto, C. H. Sulfur Dioxide Fumigation Alone or in Combination with CO2-enriched Atmosphere Extends the Market Life of Highbush Blueberry Fruit. *Postharvest Biol. Technol.* **2012,** *67,* 84–91. doi:10.1016/j.postharvbio.2011.12.006

Carrasco, G.; Urrestarazu, M. Green Chemistry in Protected Horticulture: The use of Peroxyacetic Acid as a Sustainable Strategy. *Int. J. Mol. Sci.* **2010,** *11,* 1999–2009. doi:10.3390/ijms11051999

Cengiz, M. F.; Certel, M. Effects of Chlorine, Hydrogen Peroxide, and Ozone on the Reduction of Mancozeb Residues on Tomatoes. *Turk. J. Agric. For.* **2013,** *38,* 1–6. doi:10.3906/tar-1307-14

Dällenbach, Laura J.; Thomas, Eppler; Simone, Bühlmann-Schütz; Markus, Kellerhals; Andreas, Bühlmann. Pre- and Postharvest Factors Control the Disease Incidence of Superficial Scald in the New Fire Blight Tolerant Apple Variety "Ladina". *Agronomy* **2020,** *10* (4), 464. doi: 10.3390/agronomy10040464

De Lucia, M.; Assennato, D. *Agricultural Engineering in Development: Postharvest Operations and Management of Food Grains*. FAO Agricultural Services Bulletin No. 93. Rome: FAO; 1994.

Desai, B. G.; Pathak, V. N. Phosphorus and Ascorbic Acid Contents of Mango Fruits Infected by Diplodia Natalensis Pole Evans. *Phytopathol. Z.* **1970,** *69* (1), 36–37.

Dhall, R. K. Advances in Edible Coatings for Fresh Fruits and Vegetables: A Review. *Crit. Rev. Food Sci. Nutr.* **2013,** *53,* 435–450.

Dhall, R. K. Advances in Edible Coatings for Fresh Fruits and Vegetables: A Review. *Crit. Rev. Food Sci. Nutr.* **2013,** *53,* 435–450. doi:10.1080/10408398.2010.541568

dos Santos Neto, J. P.; de Assis, M. W. D.; Casagrande, I. P.; Júnior, L. C. C.; de Almeida Teixeira, G. H. Determination of 'Palmer'mango Maturity Indices Using Portable Near Infrared (vis-nir) Spectrometer. *Postharvest Biol. Technol.* **2017,** *130,* 75–80.

Eckert, J. W.; Sommer, N. F. Control of Diseases of Fruits and Vegetables by Postharvest Treatment. *Annu. Rev. Phytopathol.* **1967,** *5,* 391–432.

Elansari, A. M. Design Aspects in the Precooling Process of Fresh Produce. In *Quality Retention During Postharvest Handling Chain*; Sivakumar, D., Ed., 2009; pp 49–57.

Elik, A.; Yanik, D. K.; Istanbullu, Y.; Guzelsoy, N. A.; Yavuz, A.; Gogus F. Strategies to Reduce Post-Harvest Losses for Fruits and Vegetables. *Strategies* **2019,** 5 (3), 29–39.

Ennab, Hassan A.; Mervat, A. El-Shemy; Shamel, M. Alam-Eldein. Salicylic Acid and Putrescine to Reduce Post-Harvest Storage Problems and Maintain Quality of Murcott Mandarin Fruit. *Agronomy* **2020,** *10* (1), 115. doi: 10.3390/agronomy10010115

Fallik, E. Prestorage Hot Water Treatments (Immersion: Rinsing and Brushing). *Postharvest Biol. Technol.* **2004,** *32,* 125–134.

FAO. Food Loss Prevention in Perishable Crops L [Online]. 1983; http://www.fao.org/docrep/s8620e/s8620e06.htm

FAO. *Global Food Losses and Food Waste–Extent, Causes and Prevention*; FAO: Rome, Italy, 2011.

Farkas, J. Food Technologies: Food Irradiation. *Encycl. Food Safety* **2014,** *3,* 178–186.

Fernández, A.; Noriega, E.; Thompson, A. Inactivation of Salmonella Enterica Serovar Typhimurium on Fresh Produce by Cold Atmospheric Gas Plasma Technology. *Food Microbiol.* **2013,** *33,* 24–29. doi:10.1016/j.fm.2012.08.007

Ferrier, P. Irradiation as a Quarantine Treatment. *Food Policy* **2010,** *35,* 548–555. doi: 10.1016/j.foodpol.2010.06.001

Food and Agriculture Organization of the United Nations (FAO). Food Losses and Waste in Turkey, Country Report, 48 s [Online]. 2013. Available at: http://www.fao.org/3/a-au824e.pdf

Food and Agriculture Organization of the United Nations (FAO). Global Food Losses and Food Waste – Extent, Causes and Prevention [Online]; 2011. Available at: http://www.fao.org/docrep/014/mb060e/mb060e.pdf.

Food and Agriculture Organization of the United Nations (FAO). Prevention of Post-harvest Food Losses: Fruits, Vegetables, and Root Crops. A Training Manual [Online]; 1989. Available at: http://www.fao.org/docrep/T0073E/T0073E03.htm#5.%20Harvesting%20and%20field%20handling

Gol, N. B.; Patel, P. R.; Rao, T. V. R. Improvement of Quality and Shelf Life of Strawberries with Edible Coatings Enriched with Chitosan. *Postharvest Biol. Technol.* **2013,** *85,* 185–195. doi:10.1016/j.postharvbio.2013.06.008

Grolleaud, M. Post-harvest Losses: Discoverying the Full Story. In *Overview of the Phenomenon of Losses During the Post-harvest System*; Rome, Italy: FAO, Agro Industries and Post-Harvest Management Service (AGSI); 2002.

Gustavsson, J.; Cederberg, C.; Sonesson, U.; Otterdijk, R.; Meybeck, A. *Global Food Losses and Food Waste, Food and Agriculture Organization of the United Nation (FAO)*. Swedish Institute for Food and Biotechnology: Düsseldorf, Interpack, 2011.

Gustavsson, J.; Cederberg, C.; Sonesson, U.; Otterdijk, R.; Meybeck, A. *Global Food Losses and Food Waste, Food and Agriculture Organization of the United Nation (FAO)*. Swedish Institute for Food and Biotechnology, Düsseldorf, Interpack; 2011.

Harris, S. R. *Production is Only Half the Battle. A Training Manual in Fresh Produce Marketing for Eastern Caribbean*; 1988.

Hertog, M. L.; Uysal, I.; McCarthy, U.; Verlinden, B. M.; Nicolaï, B. M. Shelf Life Modelling for First-expired-first-out Warehouse Management. *Philos. Trans. R. Soc. Lond. A: Math., Phys. Eng. Sci.* **2014,** *372* (2017), 20130306.

Huxsoll, C. C.; Bolin, H. L.; King Jr, A. D. Physicochemical Changes and Treatments for Lightly Processed Fruits and Vegetables. In *Quality Factors of Fruits and Vegetables, 3. 3*; Jen, Ed.; American Chemical Society Series. Series No. 405, Washington, DC, 1989; pp 203–215.

Huyskens-Keil, S.; Hassenberg, K.; Herpich, W. B. Impact of Postharvest UV-C and Ozone Treatment on Textural Properties of White Asparagus (Asparagus officinalis L.). *J. Appl. Bot. Food Qual.* **2011,** *84,* 229–234.

International Refrigeration Institute (IIR). The Role of Refrigeration in Worldwide Nutrition. 5th Informatory Note on Refrigeration and Food [Online]. Available at: http://www.iifiir.org/userfiles/file/publications/notes/NoteFood_05_EN.pdf

Janssen, S.; Schmitt, K.; Blanke, M.; Bauersfeld, M. L.; Wöllenstein, J.; Lang, W. Ethylene Detection in Fruit Supply Chains. *Phil. Trans. R. Soc. A* **2014,** *372,* 20130311. doi:10.1098/rsta. 2013.0311

Jung, S.-K.; Choi, H.-S. Browning of Early and Late-Harvested 'Empire' Apples Affected by Cold Storage and 1-MCP. *Agronomy* **2020,** *10* (7), 1050. doi:10.3390/agronomy10071050

Kader, A. A. In *Increasing Food Availability by Reducing Postharvest Losses of Fresh Produce*, Proceedings of the 5th Int. Postharvest Symp., Mencarelli F., Tonutti P., Eds. Acta Hort. 682, ISHS 2005. USA, 2005; pp 2169–2176.

Kader, A. A. Maturity, Ripening, and Quality Relationships of Fruit-vegetables. In *Strategies for Market Oriented Greenhouse Production 434*; 1995 Mar 11, pp 249–256.

Kader, A. A. *Postharvest Technology of Horticultural Crops*; University of California Agriculture and Natural Resources; 2002.

Kader, A. A.; Kitinoja, L.; Hussein, A. M.; Abdin, O.; Jabarin, A.; Sidahmed, A. E. *Role of Agroindustry in Reducing Food Losses in the Middle East and North Africa Region*; 2012.

Khan, F. A.; Bhat, S. A.; Narayan, S. *Storage Methods for Fruits and Vegetables. Sher-e-Kashmir University of Agricultural Sciences and Technology of Kashmir*; Shalimar; 2017.

Kiaya, V. Post-harvest Losses and Strategies to Reduce Them. Technical Paper on Postharvest Losses, Action Contre la Faim (ACF). 2014, 25.

Kim, H.; Ryu, J.; Beuchat, L. Survival of Enterobacter Sakazakii on Fresh Produce as Affected by Temperature, and Effectiveness of Sanitizers for its Elimination. *Int. J. Food Microbiol.* **2006,** *111,* 134–143. doi:10.1016/j.ijfoodmicro.2006.05.021

Kitinoja, L.; AlHassan, H. Identification of Appropriate Postharvest Technologies for Small Scale Horticultural Farmers and Marketers in Sub-saharan Africa and South Asia—Part 1. In *Postharvest Losses and Quality Assessments. Paper Presented at the XXVIII International Horticultural Congress on Science and Horticulture for People (IHC2010)*; International Symposium on 934; 2010.

Kitinoja, L.; AlHassan, H. Identification of Appropriate Postharvest Technologies for Small Scale Horticultural Farmers and Marketers in Sub-saharan Africa and South Asia—part 1. In *Postharvest Losses and Quality Assessments. Paper Presented at the XXVIII International Horticultural Congress on Science and Horticulture for People (IHC2010)*; International Symposium on 934; 2010.

Kitinoja, L.; Gorny, J. R. Postharvest Technology for Small-scale Produce Marketers: Economic Opportunities, Quality and Food Safety. In *Postharvest Technology for Small-scale Produce Marketers: Economic Opportunities, Quality and Food Safety;* 1999, (21).

Korićanac, A.; Miletić, N.; Popović, B.; Mitrović, O.; Lukić, M.; Pešaković, M.; Tomić, J. The Effect of ULO and NA Storage on Changes in the Quality of Apple Fruit (Malus domestica Borkh.) During Shelf Life. *Agronomy* **2020,** *10* (1), 25. doi:10.3390/agronomy10010025

Landfeld, A.; Erban, V.; Kováˇríková, E.; Houška, M.; Kýhos, K.; Pruchová, J.; Novotná, P. Decontamination of Cut Carrot by Persteril R Agent based on the Action of Peroxyacetic Acid. *Czech J. Food Sci.* **2010,** *28,* 564–571.

Lopez-Galvez, F.; Ragaert, P.; Palermo, L. A.; Eriksson, M.; Devlieghere, F. Effect of New Sanitizing Formulations on Quality of Fresh-cut Iceberg Lettuce. *Postharvest Biol. Technol.* **2013,** *85,* 102–108. doi:10.1016/j.postharvbio.2013.05.005

Mahto, R.; Das, M. Effect of Gamma Irradiation on the Physico-chemical and Visual Properties of Mango (Mangifera indica L.), cv. Dushehri and Fazli Stored at 20°C. *Postharvest Biol. Technol.* **2013,** *86,* 447–455. doi:10.1016/j.postharvbio.2013.07.018

Majumdar, V. L.; Pathak, V. N. Effect of Hot Water Treatment on Post-harvest Diseases of Guava (Psidium guajava L.) Fruits. *Acta Bot. Indica* 1991.

Manjunatha, G.; Gupta, K. J.; Lokesh, V.; Mur, L. A. J.; Neelwarne, B. Nitric Oxide Counters Ethylene Effects on Ripening Fruits. *Plant Signal. Behav.* **2012,** *7,* 476–483. doi:10.4161/psb.19523

Martínez-Romero, D.; Guillén, F.; Castillo, S.; Zapata, P. J.; Serrano, M.; Valero, D. Development of a Carbon-heat Hybrid Ethylene Scrubber for Fresh Horticultural Produce Storage Purposes. *Postharvest Biol. Technol.* **2009,** *51,* 200–205. doi:10.1016/j.postharvbio.2008.07.013

Maswada, H. F.; Abdallah, S. A. In Vitro Activity of Three Geophytic Plant Extracts Against Three Post-harvest Pathogenic Fungi. *Pak. J. Biol. Sci.* **2013,** *16* (23), 1698–1705.

Maxin, P.; Weber, R. W. S.; Pedersen, H.; Williams, M. Control of a Wide Range of Storage Rots in Naturally Infected Apples by Hot-water Dipping and Rinsing. *Postharvest Biol. Technol.* **2012,** *70,* 25–31. doi:10.1016/j.postharvbio.2012.04.001

Maxin, P.; Weber, R. W. S.; Pedersen, H.; Williams, M. Control of a Wide Range of Storage Rots in Naturally Infected Apples by Hot-water Dipping and Rinsing. *Postharvest Biol. Technol.* **2012,** *70,* 25–31. doi:10.1016/j.postharvbio.2012.04.001

Mencarelli, F.; Salcini, M. C.; Bellincontro, A. Consumer Risk in Storage and Shipping of Raw Fruit and Vegetables. In *Improving Saf. Fresh Fruit Vegetables*; Woodhead Publishing: 2005; pp 556–598.

Mupondwa, E. K. Economic Benefits of Controlled Atmosphere Storage and Modified Atmosphere Packaging. In *Modified and Controlled Atmospheres for Storage, Transportation, and Packaging of Horticultural Commodities*; Yahia, E. M., Ed.; CRC Press: Boca Raton, FL, 2009; pp 527±552.

Negi, S.; Anand, N. Supply Chain Efficiency: An Insight From Fruits and Vegetables Sector in India. *J. Operat. Supply Chain Manag.* **2014,** *7* (2), 154–167.

Nellemann, C.; MacDevette, M.; Manders, T.; Eickhout, B.; Svihus, B.; Prins, A. G.; Kaltenborn, B. P. Eds. *The Environmental Food Crisis-The Environment's Role in Averting Future Food Crisis*: UNEP: Nairobi.

Niemira, B. A. Cold Plasma Decontamination of Foods. *Annu. Rev. Food Sci. Technol.* **2012,** *3,* 125–142. doi:10.1146/annurev-food-022811-101132

Okiror, P.; Lejju, J. B.; Bahati, J.; Rugunda, G. K.; Sebuuwufu, C. I. Maturity Indices for Tomato (Solanum lycopersicum L.), cv. Ghalia 281 in Central Uganda. *Afr. J. Agric. Res.* **2017,** *12* (14), 1196–1203.

Ömer, Azabağaoğlu M. Investigating Fresh Fruit and Vegetables Losses at Contemporary Food Retailers. *Sosyal Bilimler Araştırma Dergisi.* **2018,** *7* (4), 55–62.

Palou, L.; Serrano, M.; Martinez-Romero, D.; Valero, D. New Approaches for Postharvest Quality Retention of Table Grapes. *Fresh Produce* **2010,** *4,* 103–110.

Paltrinieri, G. *Handling Fruits, Vegetables and Root Crops*; Food and Agriculture Organization of United Nations; 2016.

Pandey, N.; Joshi, S. K.; Singh, C. P.; Kumar, S.; Rajput, S.; Khandal, R. K. Enhancing Shelf Life of Litchi (Litchi Chinensis) Fruit Through Integrated Approach of Surface Coating and Gamma Irradiation. *Radiat. Phys. Chem.* **2013,** *85,* 197–203. doi:10.1016/j.radphyschem.2012.11.003

Pankaj, S. K.; Misra, N. N.; Cullen, P. Kinetics of Tomato Peroxidase Inactivation by Atmosphere Pressure Cold Plasma Based on Dielectric Discharge. *Innov. Food Sci. Emerg. Technol.* **2013,** *19,* 153–157. doi:10.1016/j.ifset.2013.03.001

Paulauskienė, A.; Tarasevičienė, Ž.; Žebrauskienė, A.; Pranckietienė, I. Effect of Controlled Atmosphere Storage Conditions on the Chemical Composition of Super Hardy Kiwifruit. *Agronomy* **2020,** *10* (6), 822. doi:10.3390/agronomy10060822

Porat, R.; Lichter, A.; Terry, L. A.; Harker, R.; Buzby, J. Postharvest Losses of Fruit and Vegetables During Retail and in Consumers' Homes: Quantifications, Causes, and Means of Prevention. *Postharvest Biol. Technol.* 2018

Porat, R.; Lichter, A.; Terry, L. A.; Harker, R.; Buzby, J. Postharvest Losses of Fruit and Vegetables During Retail and in Consumers' Homes: Quantifications, Causes, and Means of Prevention. *Postharvest Biol. Technol.* **2018,** *139,* 135–49.

Rahman, M. M.; Moniruzzaman, M.; Ahmad, M. R.; Sarker, B.; Alam, M. K. Maturity Stages Affect the Postharvest Quality and Shelf-Life of Fruits of Strawberry Genotypes Growing in Subtropical Regions. *J. Saudi Soc. Agric. Sci.* **2016,** *15* (1), 28–37.

Ramaswamy, H. S. *Post-harvest Technologies of Fruits & Vegetables*; DEStech Publications, Inc.; 2014.

Ray, R.; Ravi, V. Post Harvest Spoilage of Sweetpotato in Tropics and Control Measures. *Crit. Rev. Food Sci. Nutr.* **2005,** *45* (7–8), 623–644.
Rivera, S. A.; Zoffoli, J. P.; Latorre, B. A. Determination of Optimal Sulfur Dioxide Time and Concentration Product for Postharvest Control of Gray Mold of Blueberry Fruit. *Postharvest Biol. Technol.* **2013,** *83,* 40–46. doi:10.1016/j.postharvbio.2013.03.007
Rodgers, S.; Cash, J.; Siddiq, M.; Ryser, E. A Comparison of Different Chemical Sanitizers for Inactivating Escherichia Coli O157:H7 and Listeria Monocytogenes in Solution and on Apples, Lettuce, Strawberries, and Cantaloupe. *J. Food Prot.* **2004,** *67,* 721–731.
Saltveit, M. E. Effect of Ethylene on Quality of Fresh Fruits and Vegetables. *Postharvest Biol. Technol.* **1999,** *15,* 279–292. doi:10.1016/S0925-5214(98)00091-X
Schirra, M.; D'Hallewin, G.; Ben-Yehoshua, S.; Fallik, E. Host–pathogen Interactions Modulated by Heat Treatment. *Postharvest Biol. Technol.* **2000,** *21,* 71–85. doi:10.1016/S0925-5214(00) 00166-6
Sharma, N.; Alam, M. M. *Postharvest Diseases of Horticultural Perishables*; International Book Distributing Co.: Lucknow; 1998.
Shimshoni, J. A.; Bommuraj, V.; Chena, Y.; Sperling, R.; Barel, S.; Kaye, Y.; Fallik, E.. Residual Distribution Kinetics of Pesticides in Cherry Tomato Peel, Pulp, and Fruit as a Function of Irrigation Water Salinity, Household Rinsing, and Storage Regimen. *Agronomy* **2019,** *9* (12), 800. doi:10.3390/agronomy9120800
Sibanda, S.; Workneh, T. S. Potential Causes of Postharvest Losses, Low-cost Cooling Technology for Fresh Produce Farmers in Sub-Sahara Africa. *Afr. J. Agric. Res.* **2020,** *16* (5), 553–66.
Silva-Weiss, A.; Bifani, V.; Ihl, M.; Sobral, P. J. A.; Gomez-Guillen, M. C. Polyphenol-Rich Extract From Murta Leaves on Rheological Properties of Film-Forming Solutions Based on Different Hydrocolloid Blends. *J. Food Eng.* **2014,** *140,* 28–38.
Singh, D.; Sharma, R. R. Postharvest Diseases of Fruits and Vegetables and Their Management. *Postharvest Disinfec. Fruits Vegetables* **2018,** 1–52. doi:10.1016/b978-0-12-812698-1
Singh, S. P.; Singh Z. Postharvest Oxidative Behaviour of 1-Methylcyclopropene Treated Japanese Plums (Prunus salicina Lindell) during Storage Under Controlled and Modified Atmospheres. *Postharvest Biol. Technol.* **2012,** *74,* 26–35. doi:10.1016/j.postharvbio.2012.06.012
Singh, Z.; Khan, A. S.; Zhu, S.; Payne, A. D. Nitric Oxide in the Regulation of Fruit Ripening: Challenges and Thrusts. *Stewart Postharvest Rev.* **2013,** *4,* 3.
Sivakumar, D.; Terry, L. A.; Korsten, L. An Overview on Litchi Fruit Quality and Alternative Postharvest Treatments to Replace Sulfur Dioxide Fumigation. *Food Rev. Int.* **2010,** *26,* 162–188. doi:10.1080/87559121003590516
Suslow, T. V. *Ozone Applications for Postharvest Disinfection of Edible Horticultural Crops*; Publication 8133. University of California, Division of Agriculture and Natural Resources: Oakland, CA; 2004.
Tahir, I. I.; Johansson, E.; Olsson, M. E. Improvement of Apple Quality and Storability by a Combination of Heat Treatment and Controlled Atmosphere Storage. *HortScience* **2009,** *44,* 1648–1654.
Verma, L. R. *Postharvest Technology of Fruits and Vegetables: General Concepts and Principles (Vol. 1)*; Indus Publishing; 2000.
Vigneault, C.; Thompson, J.; Wu, S.; Hui, K. C.; LeBlanc, D. I. Transportation of Fresh Horticultural Produce. *Postharvest Technol. Horticult. Crops*, **2009,** *2,* 1–24.

Warriner, K.; Huber, A.; Namvar, A.; Fan, W.; Dunfield, K. Recent Advances in the Microbial Safety of Fresh Fruits and Vegetables. *Adv. Food Nutr. Res.* **2009,** *57,* 155–208.

Watkins, C. B. Ethylene Synthesis, Mode of Action, Consequences and Control. In *Fruit Quality and its Biological Basis*; Knee, M. Ed.; Sheffield Academic Press: Sheffield, UK, 2002; pp 180–222.

Wendehenne, D.; Durner, J.; Klessig, D. F. Nitric Oxide: A New Player in Plant Signalling and Defence Responses. *Curr. Opin. Plant Biol.* **2004,** *7,* 449–455. doi:10.1016/j.pbi.2004.04.002

Wills, R. Minimizing the Harmful Effects of Ethylene on the Quality of Fruit and Vegetables. In *Environmentally Friendly Technologies for Agricultural Produce Quality*; Ben-Yehoshua, S., Ed.; Taylor and Francis: London, UK, 2005; pp 133–148.

Wills, R.; McGlasson, B.; Graham, D.; Joyce, D. *Postharvest: An Introduction to the Physiology and Handling of Fruit, Vegetables and Ornamentals*; Boston, MA: CABI Publishing, 2007.

Wiseman, M. The Second World Cancer Research Fund/American Institute for Cancer Research Expert Report. Food, Nutrition, Physical Activity, and the Prevention of Cancer: A Global Perspective: Nutrition Society and BAPEN Medical Symposium on 'Nutrition Support in Cancer Therapy'. *Proc. Nutr. Soc.* **2008,** *67* (3), 253–256.

WRAP. Helping Consumers Reduce Fruit and Vegetable Waste: Final Report; 2008 [Online]. Available at: http://www.wrap.org.uk/sites/files/wrap/WRAP%20RTL044-001%20Final%20report.pdf.

CHAPTER 4

Soil Microflora: As Promisable Approach for Effective Disease Management

ANKITA SHARMA[1], NEHA KAUSHAL[1], KARUNA DHIMAN[2], SHARDULYA SHUKLA[3], and SACHIN KUMAR[4]

[1]YS Parmar University of Horticulture and Forestry, Nauni, India

[2]College of Horticulture and Forestry, Neri, India

[3]Medicinal and Aromatic Plant Division, DIHAR-DRDO, Chandigarh, India

[4]Dr BR Ambedkar NIT, Jhalandhar, India

ABSTRACT

Threat to global food security is on alarming stage as balance between the rate of food production and consumption is decreasing. In addition to this, exponential increase in human population and decrease in cultivable land has worsened the situation further. Therefore, a great concern has been arisen toward the development of a sustainable agriculture system to fulfil the demand of continuously growing human population. Protection of crops from insects, pests, and effective disease management is one of main aspects which can play an important role in meeting the requirement of food quality and quantity. It has been estimated that approximately 20–40% of world's food loss occurs due to infestation of different pathogens every year. Further, green revolution has resulted in the development of high yielding varieties with reduced resistance to insects and pests resulting in the utilization of more insecticides and pesticides for the efficient disease control. Nowadays, a great concern has been raised toward the harmful impacts of chemical pesticides on human health and environment. Further, develoment of resistance among pathogens against chemical pesticides has posed a

Plant Diseases and Their Management: A Sustainable Approach.
Deepa Srivastava, Rajarshi Kumar Gaur, and Ajay K. Tiwari (Eds.)

major challenge in front of agricultural scientists inspiring them to look for alternative approaches for disease management. Application of soil micro flora is one of such approaches. It is a well-known fact that plant microbe interaction plays an important role in disease development as well as in the control of phytopathogens. Further application of microbiont decreases the perception of excess pesticide application making them more acceptable among consumers. It also leaves positive effects on soil health and environment. These beneficial microbes can be utilized as antagonists against phytopathogens, as biocontrol agents against various pests and insects, for inducing resistance in plants or as a source of nutrition for the development of healthy plants etc. Therefore, application of soil microflora can prove to be an efficient tool for effective disease management. Thus, present chapter will be written with the aim of understanding the various approaches by which beneficial soil microflora can be used in the management of diseases.

4.1 INTRODUCTION

Pathogenic microorganisms of plants, also known as "phytopathogens," are the transmissible biotic (living) agents responsible for diseases causing in plants by disrupting cell metabolism either through enzymes, toxins, growth regulators secretion, or by absorbing chemical sources of energy from the respective host cells for their own benefit. However, a few pathogens can also cause diseases by dwelling and spreading through xylem or phloem, leading to inhibition of ascending or downward movement of water sugars, respectively. Different environmental conditions viz., temperature variations, humidity, mineral, and pollutants occurring at or below a definite threshold value make the plants more vulnerable for the pathogen infection and spreading of diseases (Fig. 4.1) (Agrios, 2006).

4.1.1 *WHAT IS DISEASE?*

An aberrant condition with profound negative effects on structural or functional organization of an organism except for any external abrasion is called "disease." It is also known as any abnormal deviation from the conventional working state of an individual (Farwah et al., 2020). But according to the standard definition given by American Phytopathological Society and British Mycological Society, "Any malfunctioning process, caused by continuous irritation leading to development of adversity producing signs and symptoms, is known as disease." Further, any biotic agents, responsible for the induction of disease in a plant, are referred to as pathogenic agents. Plants can be manifested by diseases at any stage of their life cycle, that is, from seed sowing to storage (Pandey et al., 2016).

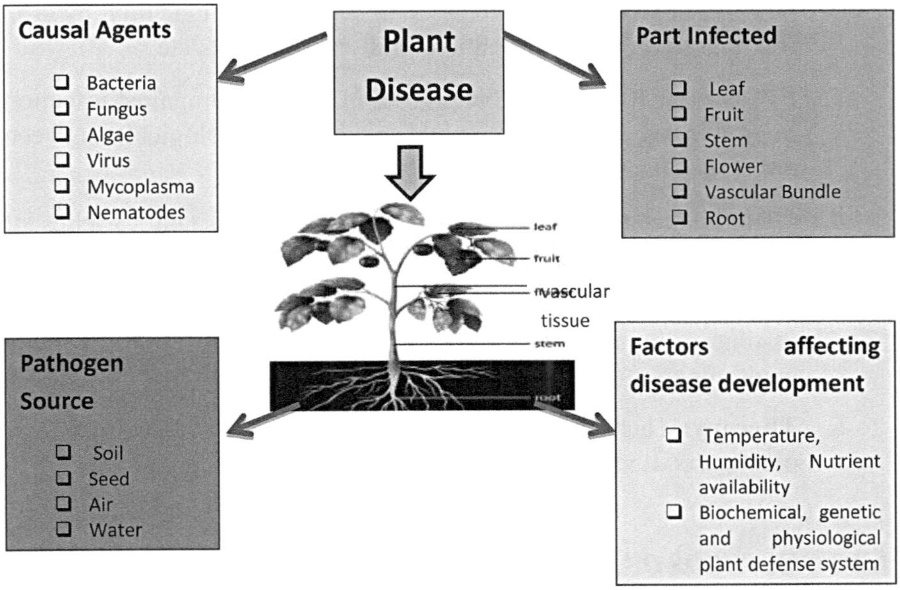

FIGURE 4.1 Plant diseases: type, source of pathogens, and factors affecting disease development.

4.2 PLANT DISEASE MANAGEMENT (PDM)

Disease management practices include various methods, measures, and strategies to control or manage phytopathogens. Plant disease management is carried out with the aim of reducing economic losses caused by various diseases. Conventionally, it has been known as "control," but modern social and environmental values declare "control" as being very absolute and rigid term.

4.2.1 FUNDAMENTAL PRINCIPLES OF PLANT DISEASE MANAGEMENT

The six elementary principles behind effective disease management are exclusion, eradication of pathogens, protection, and development of resistance against pathogens, therapy, along with eschewance of insect vectors and weed hosts (https://extension.unr.edu/publication.aspx.PubID=2647).

1. **Exclusion of pathogen:** It involves the prohibition on introduction of pathogen inoculum in fields. It can be achieved by inspection,

certification, and treatment of seeds, along with implementation of quarantine rules.

2. **Eradication:** It includes removal or reduction of inoculum by crop rotation, sanitation, elimination of alternative hosts etc.
3. **Protection:** It implies protection of healthy plant against infection by providing "barriers" (physical/chemical/biological) between concerned pathogens and host plants.
4. **Genetic resistance:** It includes growing of resistant varieties of crops against susceptible cultivars.
5. **Avoidance:** Avoidance involves prevention of diseases by selection of time or a site with no inoculum or when there is no favorable environment.
6. **Therapy:** Therapy includes treatment of already-infected plants or plant parts. It involves chemotherapy and thermotherapy.

4.2.2 METHODS OF PLANT DISEASE MANAGEMENT

A. Cultural Methods

1. **Host eradication:** Eradication of host is an important step executed to eliminate infected plants from fields. These infected plants act as immediately available source of pathogen inoculum within the crop species. It is a routine exercise in most nurseries, greenhouses, and fields in order to avoid the expansion of different diseases, for example, cucumber mosaic virus (CMV) surpasses winters mainly in perennial plants, such as wild floras. Therefore, the removal of such volunteer plants where pathogen pass unfavorable environmental conditions is enough to eradicate diseases wholly most of the times or to reduce enormously the magnitude of pathogen inoculums, responsible for reinfections in the next season. Further, for some infectious agents two alternative hosts are required to accomplish their complete life cycles, for example, *Puccinia graminis tritici* needs wheat and barberry, pine and currant (*Ribes*) for *Cronartium ribicola*, and cedar and apple are requisite for *Gymnosporangium juniperi-virginianae*. In such cases, the removal of the economically less valuable alternative host plants drastically alters the life process

of the concerned pathogen resulting in the control of the disease (Keel and Defago, 1997; Agrios, 2006).

2. **Crop rotation:** Continuous cultivation of the same crop in the same land piece makes it more vulnerable for the attack of pathogens resulting in great economic losses to crop cultivation. Number of many soil-borne pathogens, infecting either a few plant species or whole families, can be lowered by cropping plant species belonging to non-host species or families for 3 or 4 years. Sufficient disease management through crop rotation is possible for soil invaders pathogens with the ability to survive only on living plants. In few cropping systems the field is left unplanted for the next whole year or to one cropping season after tilling. During this period, accumulated debris and pathogen inoculum are demolished by microorganisms. But some other cropping systems employ herbicides and reduced ecofallow. Many diseases, such as stalk rot of sorghum and corn, attributable to *Fusarium moniliforme*, have been removed dramatically in many systems (Farwah et al., 2020).
3. **Sanitation:** Sanitation of field is another important approach carried out through cultural practices for the management of disease spread. Infectious infuse remained over plants may multiply in the soil or on the plant parts and can develop to spread epidemic in the next growing season. Various measures used for sanitation include burning of infected crop debris and grass seeds. Further, workers should wash their hands before handling some plants like transmission of tobacco mosaic virus can be reduced to infected tomato by mere washing of hands of smoke addicted worker. Frequent disinfection of knives used for propagative stock preparation and for pruning, such as potato tubers/ plum, can prove to be helpful in the reduction of pathogens spread due to such appliances. Splashing the soil off from farm equipment before taking it to another field may also help to prevent the spread of any pathogens further (Singh and Chawla, 2012).
4. **Traps and mulches:** Transmission of several diseases is carried out through various insect vectors like aphid or by peppers. Usage of erect, and polyethylene sheets attracts a considerable number of aphids which stick to polythene. Although it is done basically to control the populations of insects, but indirectly, this

technique also helps in lowering the quantity of virus inoculum reaching to the crop (Mehrotra, 2003).

5. **Application of fertilizers:** A well-nurtured plant has more disease-resistant capacity than a nutrient-deficient/toxic plant. An undernourished plant will be more stressed leading to greater susceptibility to disease attack. However, excessive utilization of fertilizer can also make plants more prone to disease. For example, calcium deficiency promotes wilt incidence of tomato (Agrios, 2006).

B. Physical methods

Different physical approaches used to control various diseases in plants include temperature variations (high or low), implementation of radiation with varying wavelengths, and dry air. Cultivation of crops in greenhouses (glass/plastic) is another approach which provides physical barriers between pathogens and their vectors (https://www.gardeningknowhow.com/garden-how-to/soil-fertilizers/sterilizing-soil.html).

1. **Heat treatment**

Soil disinfection by heat: Heat carried in aerated steam or hot water can be used to sterilize soil in greenhouses, seed beds, and cold frames. Sterilization of soil is carried out either in special vessels (soil sterilizers), where steam is applied under pressure, or on the greenhouse workbenches, in which steam is piped and permitted to disperse through the soil. At 50°C, nematodes, most of oomycetes, and other water-borne molds are destroyed. Further, many phytopathogens including fungi and bacteria, as well as some worms, slugs, and centipedes, are often killed at temperatures varying between 60 and 70°C. At about 80°C, remaining plant pathogenic bacteria, weed propagules, viruses, and insects in plant debris are killed. For example, close to the boiling point, that is, 95 and 100°C, many viruses, like tobacco mosaic virus (TMV), are destroyed. This type of disinfection of soil can also be obtained with the help of electrically produced heat rather than through steam/hot water only.

Soil Solarization: Spreading of polythene sheet during hot summer days on humid soil increases the temperature top 5 cm of soil to 52°C compared to unmulched soil (37°C). This increase in temperature is quite enough to kill soil-borne pathogens including fungi, nematodes, and bacteria near to the soil surface, resulting in the reduction of the potential pathogen inoculum of the various diseases.

Hot-Water Treatment: Different propagules like seeds, bulbs, and nursery stock are soaked in hot water in order to destroy infectious agent present on outside and inside seed coverings, bulb scales. Efficiency of this technique is based on the principle that dormant propagules of the plants organs can resist to higher temperatures in comparison to their corresponding pathogens for a given period of time. For example, seeds are dipped in warm water (52°C) for about 11 min in the prevention of loose smut of wheat caused by *Ustilago tritici,* whereas for the prevention of *Ditylenchus infections,* bulbs are soaked in water for 3 h at 43°C temperature.

Hot-Air Treatment: Treatment of certain storage plant organs with warm air removes the excess moisture from their surface and speeds up the process of wound healing, resulting in the prevention of their infection. For example, keeping sweet potatoes at 28–32°C for a fortnight is useful in fast healing of wounds and avoids the occurrence of *Rhizopus* and soft-rotting bacterial infection.

Radiation therapy: Different types of electromagnetic radiations like UV, X-, and γ rays along with particulate radiations, such as α and β particles, have been investigated to assess their capability to control the occurrence of different diseases in fruits and vegetables after harvesting, for example, γ rays have been utilized to control postharvest diseases of peaches, strawberries, and tomatoes due to their respective fungal pathogens.

Host-plant resistance: Development of resistance in host plant is one of the important approaches to control diseases in vegetable crops. Usage of disease-resistant varieties is one of the effective methods to control diseases as it is environment-friendly and does not require additional cost. For the prevention of some infections viz., wilts, bacterial blights smuts, nematodes, and rusts, genetic resistance is the only solution (Sharma and Rodomiro, 2002).

TABLE 4.1 Genetic-Resistant Varieties of Vegetables.

Crops	Varieties	Diseases
Tomato	Pant Bahar	Verticilium wilt
	HissarAnmol and Hissar Gaurav	Leaf curl virus
	BWR-5 and BWR-1	Bacterial wilt
Brinjal	Pant Rituraj	Bacterial wilt
	Pusa purple cluster	Bacterial wilt
	PusaBhairav	Phomopsis blight
	PusaSamrat	Bacterial wilt, Phomopsis blight & Shoot and fruit borer.
Chilli	PusaJawala	Leaf curl virus
	Punjab Lal	Leaf curl virus
Potato	KufriGirdhari	Late blight

C. **Chemical methods:**

Applications of chemicals are generally carried out in order to stop the infection on the upper surfaces of plant or to eradicate the already-infested pathogens. It includes soil fumigation, disinfection of warehouses, handling equipment, trap and control of numeral insect vectors.

Chemical treatment of soil: A few fungicides, that is, metalaxyl, captan, diazoben, pentachloronitrobenzene (PCNB), and chlorone-bare are added to the soil as dusts, granules, or in the form of liquid drenches to control various soil-borne pathogens like *Pythium,* causal agent of seedling blights, damping-off, rots of crown, and roots, etc.

1. **Fumigation:** Application of chemical fumigants like chloropicrin, metam sodium, methyl bromide, and dazomet is an important approach to control nematodes and other soil-borne pathogens and pests. Contact nematicides, including fensulfothion, aldicarb, carbo-furan, and ethoprop, are successful against nematodes and insects and can be applied either before or after planting (https://www.thedirtdoctors.com/types-of-chemicals-used-in-fumigation/)
2. **Disinfestation of warehouses:** Walls and floors of warehouses should be cleaned with bleach (1 pound in 5 gallons of water), or with the application of some other sanitizing agent. Fumiga-tion of airtight ware houses with relative humidity at 100% along with temperature ranged between 25 and 30°C can be carried out

effectively with chloropicrin (tear gas) at the rate of 1 pound/1000 cubic feet.

3. **Control of insect vectors:** Usage of insecticides to control various insect carriers has been proved to be a fairly successful and recommended approach. For example, oil sprays have been found effective to control various aphid-borne viruses like CMV (*cucumber mosaic virus)* on cucumber (Sharma and Rodomiro, 2002).

4.2.3 NEED OF NEW GREEN REVOLUTION

Green revolution in the 20th century has made remarkable contribution to providing the food security around the globe. This revolution was based on three steps. First, development and monocultures of plants of carefully bred or genetically modified high-yielding varieties of major cereal crops like rice, wheat, and maize. Second, input of high doses of fertilizers, pesticides along with heavy irrigation water to obtain maximum yields. Third, multiple cropping by enhancing the number of crops per plot of land annually. However, this increased productivity with the use of excessive fertilizers and pesticides has been achieved on the cost of environmental degradation and development of pesticide-resistant pathovars. Therefore, a need has been felt for a fresh green revolution for sustainable production of food, fiber, and fuel without causing harm to the environment. This new green revolution needs to be based on the application of biofertilizers and biopesticides without affecting the productivity of crops to a lesser level such as usage of soil microbiome and their products for control of phytopathogens or by rhizospheric engineering (Timmusk et al., 2017). Application of microbial culture inoculation is not a new concept as legume-based rhizosphere inoculation has been reported since the inception of the 20th century (Desbrosses and Stougaard, 2011).

Continuous changing environment has imposed challenge on agricultural scientists to look for new approaches for the development of highly productive abiotic stress-resistant crops in decreasing area under cultivation (Pachauri et al., 2014). Soil microbiota living in the close vicinity to plant plays a significant role in the survival of plant under extreme condition of environment by providing nutrient and protection from pathogenic microorganisms. Soil microbiome interacts with the plants either positively by enhancing its survivability by stress deterioration or negatively by causing diseases. These positive interactions can be utilized for increasing disease-resistant among cultivars.

4.3 TYPES OF DIFFERENT PLANT MICROBES INTERACTION IN SOIL

Microorganisms existing in the soil environment have a prominent place in the functioning of ecology. A vast number of microbes such as bacteria and fungi are present in the soil adhering to the roots of plant. These microorganisms carried out different interactions with each other as well as with plants, that is, microbe–microbe interaction and plant–microbe interaction. The different interactions carried out by these microorganisms may have positive or negative effect (Ding et al., 2019; Wille et al., 2019). The interrelationship among microbial community and plants is very multifaceted. Positive interactions occurring between plants and microorganisms have a beneficial effect on plants through direct or indirect ways. The positive microbial alliance results in plant protection from unfavorable environmental conditions besides enriching plant growth under normal condition. The negative microbial interactions constitute mainly pathogen–plant interactions. The kind of correlation whether it will be beneficial, or antagonistic, is taken into consideration by the sort of microbial species together with the mechanism of action retained by the microorganism (Vacheron et al., 2013).

4.3.1 BENEFICIAL PLANT-MICROBE INTERACTION

Beneficial plant–microbe interactions can be segregated into direct mechanisms and indirect mechanisms. Direct mechanisms include the ability to supply nutrients and modifying plant hormone levels. Indirect mechanisms include diminishing the destructive effect of numeral pathogens in the form of biological control agents, root colonizers, and environmental protectors. Numeral plant growth promoting microbes residing in the close proximity of the plant roots depend on root exudates for their growth survival (Whipps, 1990). Root exudates are released by the roots of living plants and contain a number of compounds such as polysaccharides, proteins, ions, carbon-based compounds, sterols, and many other chemical compounds. Broadly, there are two types of plant–microbe associations which are beneficial for the plants: (1) highly specialized interaction, in which a mutualistic correlation is present; and (2) commensal relationship where microbial interaction depends upon the release of nutrients from plant roots.

4.3.1.1 HIGHLY SPECIALIZED INTERACTIONS

Highly specialized interactions are apparent in soil amid various classes of organisms such as lichens, bacterial legume symbiosis, and mycoorhizae.

Mycorrhizal association results from the interaction of fungus and plant roots. Symbiotic interaction with mycorrhizae occurs approximately with roots of 95% of vascular plants. Mycorrizal association begins when fungal spores present in the soil germinate and a thread-like structure called hyphae emerges from the fungal spore and enters into the epidermal layer of roots of plant and colonizes plant roots extensively. After colonization, fungus spreads out the great network of hyphae throughout the soil to increase the absorptive surface area. These fungal hyphae have the same functions as plant root hairs. As a result of effectively extended root system, fungi have large influence on uptake of water, better mineral nutrition, alteration in soil structure and root architecture, induction of ABA, thereby increasing plant growth and tolerance to stress (Jakobsen et al., 2005; Querejeta et al., 2003). On the other hand, fungi takes up carbohydrate from the plants for nutrition. Mycorrhiza is mainly of two types, that is, endomycorrhizae (fungal growth inside the root) and ectotrophic mycorrhizae (fungal growth outside the root). Ectotrophic mycorrhiza aid in assimilation of minerals, adsorption of nutrients, and also acts as receptacle. Mycorrhizae fungi play a crucial role against numeral types of stresses through improvement in antioxidant system and osmolytes production (Quiroga et al., 2020). Bacterium root nodule system shares a number of relationships between free-living soil *Rhizobium* and host plant's roots. The roots of the leguminous plants release a set of chemical substances or attractants such as flavonoids. In the presence of these flavonoids, specific rhizobial cells start movement in the direction of root hairs leading to the development of nod (nodulation) factors. These factors are present on the surface of bacteria and are responsible for the binding to the lectin proteins available on the surface of the root hairs leading to hair curling around *Rhizobial* cells. These enter the root hair by degrading root cells and extend through the root hair to reach different cortical cells and form nodule by cell division. These nodules are the site for biological nitrogen fixation. Cyanobacteria plant interaction involves *Anabaena* and *Nostoc* species which are common symbionts in lichens, *Azolla, Anthoceros*, and cycadroots.

4.3.1.2 COMMENSAL INTERACTIONS

Commensal interaction is an abiding relationship in which one member gains benefits while the other member neither benefits nor harms. The commensal, that is, the species that gains advantage from the interaction may obtain nutrients and support from the other one. PGPR directly promote nutrient uptake or availability by nitrogen fixation, mineral solubilization, mineralization of organic compounds, and production of hormones. On the other hand, PGPR improve soil fertility by a variety of indirect mechanisms such as antibiosis, siderophore production, HCN production, production of lytic enzymes and expolysaccharides etc. (Egamberdieva and Lugtenberg, 2014). Sandhya et al., (2009) reported the ability of PGPR to produce exo-polysaccharides beneficial for plants under water-deficient conditions. Phosphate and potassium solubilizing bacteria play important part in escalating the accessibility of phosphorus and potassium—the key minerals to the plant (Panhwar et al., 2014). Further, PGPRs also aid in the protection of plant from harmful impacts of plant pathogens. The plant protection could be attained either by nutrient limitation to the pathogen or by lysis of cell wall of pathogens. The availability of certain nutrients could be decreased by the secretion of siderophores which limits the availability of iron to pathogens (Bhattacharyya and Jha, 2012). The PGPR have also been found to provide plant protection from diseases through lysis of cell walls of the pathogens by different enzymes such as chitinases, cellulases, glucanases, proteases, and lipases (Beneduzi et al., 2012). Ahmad et al. (2015) observed enhancement in essential nutrient uptake and accumulation of antioxidants and osmolytes after the treatment of Indian mustard with *Trichoderma harzianum*.

4.3.2 ANTAGONISTIC PLANT–MICROBE INTERACTION

The antagonistic plant–microbe interactions are inimical to plants as the attacking microbes may be saprophytic and kill the living cells in the colonizing plants. Many microorganisms are deleterious to plants. The relationship of plants and phytopathogenic microorganisms is diverse and these phytopathogenic microorganisms have the ability to colonize plant effectively. The destructive effect of these microorganisms may be a consequence of production of certain compounds which are detrimental for the growth of the target plant or due to the supererogation of specific growth regulators. For example, bacterial cyanide has an inhibitory effect on plants (Agbodjato et al., 2015). Sometimes, amount of concentration of these substances determines their role whether it is positive or negative. Horner et al. (2012) observed the production of inhibitors of various enzymes including proteases, glucanases, and cellulases by *P. oligandrum* which further results in mycoparasitism. Bacterial infections are mainly caused by two groups of bacteria, that is, proteobacteria and actinobacteria. Most important features of bacterial invaders are their virulence factors determining disease severity caused by them. They include secretion of extracellular enzymes, phytohormones, toxins, and type I to VI secretion system, which varies with species to species (Frederick et al., 2001). Diseases resulting due to the invasion of *Phytophthora* include soybean root rot, sudden oak death, cocoa black pod, and dieback (Davidson, et al., 2005).

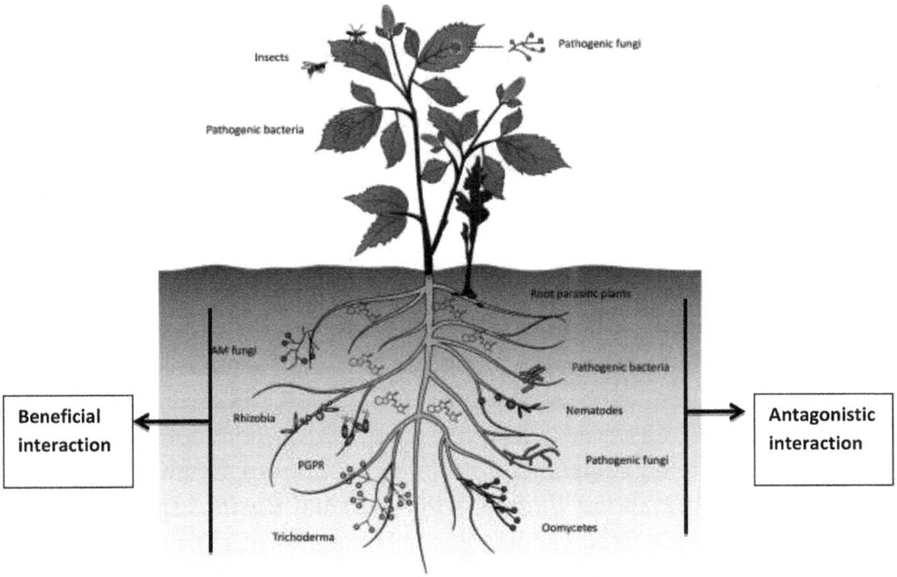

FIGURE 4.2 Different types of plant–microbe interaction in soil (Lopez-Rayaz et al., 2017).
Source: Reprinted with permission from Lope-Rayad et al., 2017

4.4 PLANT GROWTH PROMOTING RHIZOBACTERIA (PGPRS) AND DISEASE MANAGEMENT

Plant growth promoting rhizobacteria (PGPRs) are the microbe which are present in the rhizospheric area of plant and promote plant growth. PGPRs not only promote nutrient assimilation and plant growth but also are identified as biocontrol agents as it suppresses a wide variety of disease-causing organisms. These PGPRs are antagonistic to pathogens and can be successfully incorporated into disease management system in agriculture field. It presents a highly efficient, economical, and ecofriendly alternative approach in place of pesticides (Emmert and Handelsman, 1999). These organisms can adjust to the rhizospheric condition and could colonize in the host roots aggressively which is the unique feature of using PGPRs as biocontrol agents (Dunne et al., 1997). Thus to reduce chemical input in agriculture soil microorganisms are the best possible alternative (Compant et al., 2005). Although PGPRs are studied for their nutrient assimilation and plant growth promoting role worldwide, many commercial formulations based on PGPRS were used by organic growers but the role of PGPRs in reducing plant diseases has attracted much attention. Till now full potential of these bacteria has not been achieved but work done is quite good and offers organic growers some of their first effective control of serious plant diseases.

How PGPRs elicitate plants' defense is gaining attention nowadays (Sarbadhikary and Mandal, 2017). Rhizospheric population of *Fusarium oxysporum* was declined using root dip application of *Aspergillus niger, Pseudomonas fluorescenes, Bacillus subtilis*, and *Penicillum digitatum* (Khan and Khan, 2002). A number of pathogens like *Bemisia tabaci, Fusarium, Botrytis cinerea, Colletotrichum orbiculare, Rhizoctonia solani, Macrophomina phaseolina, Ralstonia solanacearum*, and *Pectobacterium carotovorum* were inhibited using systemic resistance inducers PGPR genera such *as Alcaligenes, Chryseobacterium, Paenibacillus*, and *Pseudomonas* in potato, tomato, and Chinese cabbage (Murthy et al., 2014; Abdallah et al., 2016). Han et al. (2006) observed a suppressive effect on common scab disease in pot assay using an antagonistic strain of *Bacillus* sp. Senthilkumar et al. (2008) found that *Paenibacillus* sp. (HKA-15) and *Bacillus* sp. (HKA-121) could be used as potential candidates for biocontrol of charcoal rot fungus as well as for soybean plant growth promotion. A plant growth-promoting bacteria *Bacillus megaterium* L8 is reported to control *Pythium aphanidermatum* leading to seeding damping off disease in cucumber (Liang et al., 2011). A pot and field experiment was done to evaluate the effect of *Pseudomonas fluorescence* and *Bacillus subtilis*. The results showed that using these rhizobacteria together also led to a decrease in damping-off disease besides increasing nodulation. In Ethiopia, Kurabachew and Wydra (2013) evaluated the potential of *B. cereus* BC1AW and *P. putida* PP3WT strains for the inhibition of *R. solanacearum* in various tomato genotypes in vitro as well as in vivo conditions and it was found that inoculums of these strains efficiently reduced the occurrence of bacterial wilt King Kong genotype of tomato. 2. Rai and Nabti (2017) observed 65.6% reduction in bacterial wilt and 63% reduction of the pathogen in the soil compared to the control when treated with *Pseudomonas protegens* RS-9. A microbial consortium of *Bacillus thuringiensis* and *Trichoderma viride* showed 30.76% suppression of early blight disease and resulted in achieving a high yield of potato (2.78 t/ha^{-1}) compared to the control under field condition. VBLR10 and VBLR39, bacterial strains were isolated from rhizospheric soil of healthy tomato plants in a wilt-infected field and were studied for antimicrobial property against *Alternaria alternata*, *Penicillium digitatum*, *Helminthosporium oryzae*, and *Fusarium oxysporum* (Sarbadhikary and Manda, 2017). Antagonistic activity of two *Bacillus velezensis* isolates against *Ralstonia solanacearum* and *Fusarium oxysporum* was evaluated by Cao et al. (2018). In another study antagonistic activity of many soil-borne isolates of *Bacillus* sp. and *Pseudomonas* sp. was identified against a number of phytopathogens. It was found that bio-surfactants and siderophores production and bacilysin biosynthesis attributed to their antagonistic activity (Caulier et al., 2018).

Different antagonistic mechanisms used by these bacteria include antibiotics production, secretion of hydrolytic and cell wall lytic enzymes, biosynthesis of siderophore, induction of systemic resistance (ISR), competition for space and substrate (Kim et al., 2003; Ramamoorthy et al., 2001; Nakkeeran et al., 2004).

4.4.1 PRODUCTION OF ANTIBIOTICS

Antibiotics are small molecular weight secondary metabolic compounds produced by microorganisms. These can affect growth and metabolic activities of other microorganisms. A positive correlation is observed between disease suppression and antibiotic production in many biocontrol strains of *Pseudomonas* using genetic analysis study (Vincent et al., 1991). A number of antibiotics have been isolated from various biocontrol bacterial agents. *Bacillus* species which is predominately present in soil produce around 167 different types of antibiotics (Maksimov et al., 2011). Polymyxin, circulin, and colistin major antibiotics of *Bacillus* showed antagonistic property toward *Alternaria solani*, *Botryosphaeria ribis*, *Aspergillus flavus*, *Colletotrichum gloeosporioides*, *Helminthosporium maydis*, and *Fusarium oxysporum* (Maksimov et al., 2011). AFM (antifungal metabolites) like HCN, phenazines, 2,4-diacetylphloroglucinol (DAPG), pyrrolnitrin, viscosinamide, pyoluteorin, and tensin are reported to be produced by various rhizobacteria (Bhattacharyya and Jha, 2012). Antibiotics induce fungistasis, inhibit spore development, cause breakdown of fungal mycelia, or exert fungicidal impacts to suppress different pathogens (Fravel, 1988). DAPG is one of the antibiotics produced by *Pseudomonas* and it was reported that it is responsible for suppressing phytopathogens (Cronin et al., 1997). De Souza et al. (2003) observed a rapid decline in take-all disease in wheat (*Gaeumannomyces graminis* var. tritici) with increasing population of

DAPG antibiotic-producing *Pseudomonas* species. Schouten also reported antagonistic activity of DAPG against pathogenic *F. oxysporum* (Schouten et al., 2004). Furthermore, soil enriched with population of DAPG-producing *Pseudomonads* was found suppressive to *Fusarium* wilt of peas naturally (Landa et al., 2002). Similarly, phenazine antibiotic synthesized by *P. fluorescens* was also investigated for its antagonistic ability (Chin-a-woeng et al., 2000). The basic role of phenazine in disease elimination was confirmed by the results that showed phenazine-deficient mutant's inability to inhibit the same disease (Mazurier et al., 2009). Proof of biocontrol with the help of antibiotics has also been spotted in *Paenibacillus polymyxa* (AB15) which showed inhibitory activity against *Colletotrichum acutatum,* causal agent of anthracnose in pepper (Lamsal et al., 2012). Considerable success has been achieved in the control of many phytopathogens with a few exceptions like pathogens with the ability to develop antibiotic resistance.

4.4.2 NUTRIENTS AND NICHE COMPETITION

Rhizobacteria as one of the dominant bacteria in rhizospheric region competes with other sol microorganisms for the space and nutrients. Competition of these PGPRs with several pathogenic sp. has been observed as a vital strategy for biocontrol, but there is no experimental proof (Kamilova et al., 2005). PGPRs show a negative form of association as they are predominately present in the rhizosphere and take control of whole metabolic activities. PGPRs also have some properties like flagellum presence, chemotaxix, and secretion of root exudates which enhanced their survival in rhizosphere and made them the most competent group of microorganisms (Lugtenberg and Kamilova, 2009). Keeping in view this concept it was claimed that this mechanism could help in biocontrol strain selection (Kamilova, 2008). They also proved this experimentally by applying a mixture of rhizospheric strain onto sterilized seeds and repeating this for two to three cycles to prove that isolated bacteria were as good as or even better biocontrol agents (Mahajan and Shirkot, 2014). A mutant study also confirmed the proposed mechanism (Kamilova et al., 2005; Validov et al., 2007). One of such mechanisms present naturally in PGPRs is synthesis of siderophore. These siderophores have the ability to chelate iron metal ions making it inaccessible to phytopathogens and proliferation and root colonization of the phytopathogens will be inhibited (Chaiharn et al., 2009; Miethke and Marahiel, 2007; Sayyed and Chincholkar, 2009). Iron is a very important nutrient because it is required by microorganisms for the synthesis of ATP, heme, reduction of nucleotide precursors of DNAetc. (Saraf et al., 2011). Thus, siderophore-producing bacteria act as biocontrol agents because it leads to iron deficiency in pathogen (Lemanceau and Albouvette, 1993). The role of PGPRs in the phytopathogens inhibition was reviewed by Sindhu et al. (1997). Some studies show suppression of *Colletotrichum gossypii* by siderophore-synthesizing rhizobacteria (Sindhu et al., 1997) and antagonistic activity against several bacterial and fungal plant pathogens of *Pseudomonas* strain RBT 13, a sidephore producing fluorescent bacteria was reported by Johri et al. (1997). Chaiharn et al. (2009) studied that siderophore-producing isolates suppressed the rice pathogen and Macrophomina *phaseolina*, causalagent of charcoal rot of peanut (Gupta et al., (2002). In another study, siderophore-producing pseudomonad was found to significantly reduce the root rot disease in chickpea (Akhtar and Siddiqui, 2009). But only those siderophore-synthesizing microbionts can prove to be strong biological control agents which have high root-colonizing potential (Chincholkar et al., 2000).

4.4.3 BIOSYNTHESIS OF LYTIC ENZYMES

Lytic enzymes such as chitinase, b 1-3 glucanase, lipase, cellulose, and protease produced by PGPRs can cause lysis of major fungal pathogens and thus can be suggested as an essential type of biocontrol (Markowich and Kononova, 2013). This makes PGPR an effective biocontol agent because such strains can be used for growth as well as for inhibition of plant pathogenic fungi (Garbeva et al., 2011; Ran et al., 2005). These lytic enzymes generally lyse cell wall of phytopathogens because their cell wall is made up of chitin. Chitinase enzyme comes under hydrolytic class which can break glycosidic bond in chitin (Kim et al., 2003). Thus, the degradation of the cell wall of phytopathogen with the help of chitinase can prove to be another potential approach of biocontrol. Isolation of chitin degrading biocatalysts has been communicated from *Streptomyces* (Kim et al., 2003), bacterialisolates from marine water, alkaline water, and rhizosphere soil (Park et al., 2000; Loni et al., 2011; Kishore and Pande, 2007; Idris et al., 2007; Kamil et al., 2007). The chitinase produced by these microbes has been found to show antagonistic and antifungal bioactivity toward numerous disease-causing fungi (Singh et al., 1999; Frankowski et al., 2001; Gohel et al., 2005; Kim et al., 2008; Ramos-Solano et al., 2010). Velusamy and Kim

(2011) reported chitin degrading bioactivity of *Enterobacter* species toward fungal hyphae. A few biocontrol bacterial strains were also found to produce another set of biocatalysts like cellulases, proteases, β-1,3 glucanases, and lipases to hydrolyze the cell wall of many fungal pathogens (Glick, 2012). Palumbo et al. (2005) have carried out an investigation to decipher the role of beta-1, 3-glucanase on the biocontrol abilities of *Lysobacter enzymogenes* strain C3 against *Phytium* sp. causal agent of bipolaris leaf spot caused by the application of lytic enzyme which has also been reported to biocontrol blight (*Phytophthora capsica)* and *Fusarium infection* in pepper (Jung et al., 2005; Hariprasad et al., 2011) and beet by *Pythium ultimum* (Dunne et al., 1997). Chaiharn et al. (2008) demonstrated the inhibitory capability of PGPR by the release of chitinase, proteolytic enzymes, β 1, 3 glucanase, and cellulase in small amount (Cattelan et al., 1999). Chitinases producing parasitic molds and *Trichoderma* species have also been tested against *R. necatrix* (Hoopen and Krauss, 2006; Harman et al., 2004). In another study, defense against anthracnose pathogen in mango mediated by lytic enzymes of *Pseudomonas fluorescens* and other biocontrol organisms was observed (Vivekananthan et al., 2004).

4.4.4 HYDROGEN CYANIDE

It is a volatile pale blue liquid or colorless gas with a mild, almond-like smell. Production of hydrogen cyanide is known as cyanogenesis and it is produced by some bacterial species as secondary metabolites (Schippers et al., 1990). HCN is mainly linked to pseudomonas sp. which can be identified by using different techniques described by Lorck (1948). This toxic compound interferes with the cellular respiration as it is a strong inhibitor of many metalloenzymes, specifically copper containing cytochrome C oxidase (Blumer and Hass, 2000). Thus, it inhibits the synthesis of ATP (adenosine-5′-triphosphate), energy house of cell, leaving the cell energy deficit, resulting in death of cell. Glycine is the main precursor for HCN production. Many rhizobacteria when supplied with glycine in their culture medium are capable of producing HCN. Role of HCN production in the control of root pathogens is not clear till now; however, there are numerous effects of HCN on different plant types. Identification of HCN production by rhizospheric bacteria has been reported by many investigators (Lorck, 1948; Castric, 1975; Askeland and Morrison, 1983). Synthesis of HCN by Gram-negative bacteria like *Pseudomonas fluorescens, P. aeruginosa*, and

Chromobacterium violaceum is tested by finding change in color of indicator paper dipped in picric acid solution in color (from yellow to brownish red) (Askeland and Morrison, 1983). HCN has been reported to inhibit the growth of aerobic root pathogens (Schippers et al., 1990). Voisard et al. (1989) observed that wild-type strain of fluorescent *Pseudomonas* CHAO is a potent HCN-producer with the ability to inhibit black root rot of tobacco whereas to curb canker of tomato. Voisard et al., (1989) reported the use of *P. fuorescens* strain CHA0 to stop the expansion of black root rot of tobacco by *Thielaviopsis basicola* (Laville et al., 1998). HCN production also led to resistance against nematodes such as in one study *P. chlororaphis* strain O6 was found to be efficient to control the root knot nematode on tomato with the synthesis of HCN as a chief inhibitor (Lee et al., 2011). Nandi et al. (2015) also reported that pyrrolnitrin and HCN are involved in the antagonistic activity of *P. chlororaphis* PA23 against *C. elegans*. *Pseudomonas chlororaphis* O6 also exhibit nematicidal activity against *Meloidogyne hapla* by producing hydrogen cyanide (Kang et al., 2018).

4.4.5 INDUCED SYSTEMIC RESISTANCE (ISR)

It is an enhanced defensive capacity of plants which is triggered by chemical or biological inducers and protects plants not only from initial attacks but also against future attacks by pathogen (Van loon et al., 1998). SAR (systemic acquired resistance) and ISR (induced systemic resistance) are the two clearly defined forms of induced resistance. SAR is triggered by plant pathogens and ISR by root colonizing mutualistic microbes like *Pseudomonas simiae* (syn. *Pseudomonas fluorescens*), or *Trichoderma* spp. (Zhang et al., 2009; Pieterse et al., 2012, 2014; Alizadeh et al., 2013; Zamioudis et al., 2014, 2015; Zhao et al., 2014; Martínez-Medina et al., 2017; Verbon et al., 2017). These two forms can be identified with respect to each other based on the nature of elicitor and regulatory pathways involved, but they have same components also (Pieterse et al., 1998, 2002, 2012, 2014; Van Loon et al., 1998; Choudhary et al., 2007). The gene and gene product included in such types of biological control processes have not been investigated thoroughly. Induced systemic resistance (ISR) involves the usage of organic acid and phytohormones like salicylic acid (SA), jasmonic acid (JA), and ethylene for signaling and induction of host plant defense response against numerous varieties of plant pathogens (Handelsman and Stabb, 1996; Niranjan et al., 2005; Beneduzi et al., 2012; Pieterse et al., 2014). It was found that increased level of salicylic acid in systemic tissues also induces SAR which leads to expression of *PATHOGENESIS-RELATED* (*PR*) genes, responsible for redox-regulated protein NON-EXPRESSOR OF *PR* GENES1 (NPR1)-based defence responses in plants (Pieterse et al., 1998, 2002, 2012, 2014; Van Loon et al., 1998; Choudhary et al., 2007). But, jasmonic acid, and ethylene mediate ISR by an SA-independent pathway and it functions without *PR* gene activation (Pieterse et al., 1998, 2002, 2012, 2014; Van Loon et al., 1998; *Choudhary* et al., 2007). But in some cases synthesis and assimilation of SA is requisite for the ISR process (Ryu et al., 2003; Alizadeh et al., 2013). Microbial species and plant species also affect signaling pathways required in the induction of ISR (Ryu et al., 2003; Jankiewicz and Koltonowicz, 2012; Alizadeh et al., 2013). In rhizospheric bacterial strains, ISR response is observed with increment in physical and mechanical firmness of cell wall and also by the adaptation of their physical and biochemical metabolism toward environmental stress (Labuschagne et al., 2010). In a broader scale, the application of PGPR strain in the form of seed coat has been observed to tremendously increase the ISR against *Colletotrichum lagenarium* causal agents of anthracnose in cucumber, bacterial wilt (*Erwinia tracheiphila)*, and *Pseudomonas syringae* (angular leaf spot) (Zehnder et al., 2001). Rice leaves showed induced systemic resistance against *Xanthomonas oryzae* pv. oryzae when induced by *Pseudomonas fluorescens* Pf1 (Vidhyasekaran et al., 2001). Ivaicoli et al. (2003) studied that induced systemic resistance was induced after root inoculation of *Arabidopsis thaliana* with *Pseudomonas fluorescens* CHA0.

4.5 BOTTLENECKS ASSOCIATED WITH LAB TO THE FIELD TRANSFER

Although technology of application of bioinoculants holds an immense potential in improving agricultural productivity, its lab to field transfer faces a number of challenges which need to be addressed. For example, development of PGPR formula depends on in vitro screening of isolates for production of phytohormones, nitrogen fixation, phosphate solubilization etc. However, sometimes, high producing clones are unable to perform in natural condition whereas less producing clones perform better by using alternative mechanism not decipherable under laboratory conditions. Therefore, it is very necessary to select a correct clone able to perform under natural soil condition among other microbiota (Cardinale et al., 2015).

Developing cold storage chain for enhancing shelf life of microbial consortia is another challenge faced for commercial popularity of PGPR-based pesticides and fertilizers. Further, suitable carriers are required for effective colonization and competition against native fauna and flora under natural field conditions. In addition, frequent fumigation of soil poses another for the survivability of bio communities (Dangi et al., 2017). Green revolution leading to the enhanced food production also came with increased usage of pesticides. Combination of bioremediation along with PGPR can prove to be an important approach for increasing crop production further. Additionally, training of farmers and suppliers will aid in achieving the expected outcomes from the PGPRs for efficient disease management (Itelima et al., 2018).

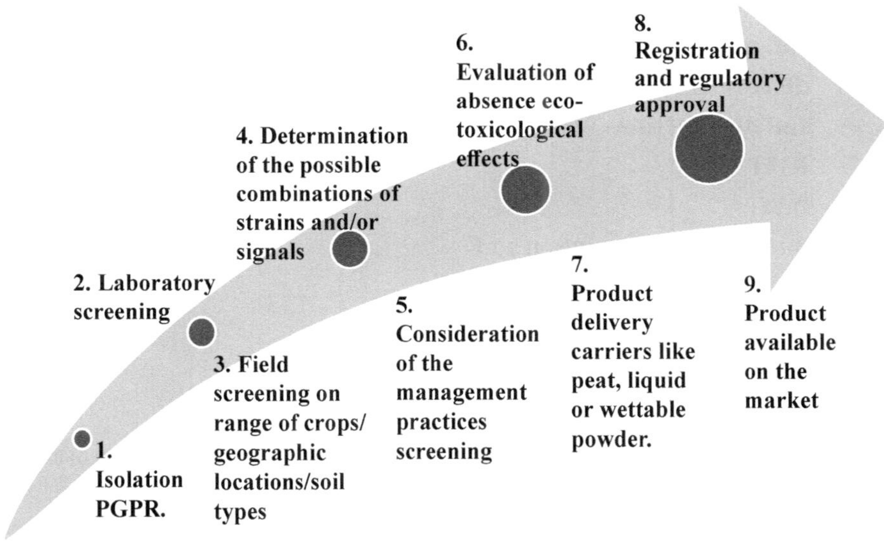

FIGURE 4.3 Steps involved in commercial product formulation of PGPRs.

4.6 CONCLUSION

Around the globe, millions of tons of crops are sacrificed due to the manifestation of diseases every year. A number of physical and chemical methods are employed to control these pathogens such as application of large tons of pesticides. Although these methods have been proved successful in controlling plant diseases, these have given birth to secondary problems including deterioration of quality of land, increased susceptibility of crops toward diseases, and development of more resistant pathovars. Therefore, an urgent need has been felt to develop eco-friendly and sustainable approaches for effective disease management. Usage of soil microbiota is one of such approach.The relationships between soil microflora and plants are as ancient as the evolution of crops on this planet earth and this association has been proved beneficial for global food production. It has been well understood that many members of soil microbiome offer a huge potential for the development of disease-controlling and plant growth-stimulating agents. However, commercial exploitation of this approach suffers some limitations which need to be addressed carefully which can be used for a great deal.

KEYWORDS

- **soil microflora**
- **disease management**
- **soil disinfection**
- **HWT**
- **HAT**
- **microbes interaction in soil**

REFERENCES

Abdallah, R. A. B.; Mokni-Tlili, S.; Nefzi, A.; Jabnoun-Khiareddine, H.; Daami-Remadi, M. Biocontrol of *Fusarium* wilt and Growth Promotion of Tomato Plants Using Endophytic Bacteria Isolated from *Nicotiana glauca* Organs. *Biol. Control.* **2016,** *97*, 80–88.

Abdullahi, S.; Simon, S.; Babychan, M. Effect of Bioagents and Their Consortia in the Management of Early Blight Disease of Potato. *Int. J. Agric. Res. Innov. Technol.* **2016,** *5* (3), 495–498.

Agbodjato, N. A.; Noumavo, P. A.; Baba-Moussa, F., Salami, H. A.; Sina, H. A.; Sezan, H.; Bankole, A; Adjanohoun, L. B.-M. Characterization of Potential Plant Growth Promoting Rhizobacteria Isolated from Maize (*Zea mays* L.) in Central and Northern Benin (West Africa). *Appl. Environ. Soil Sci.* **2015,** 1–9.

Agrios, G. N. *Plant Pathology*, 5th ed.; 2006; pp. 4–72 and 254–286.

Akhtar, M. S.; Siddiqui, Z. A. Use of Plant Growth Promoting Rhizobacteria for the Biocontrol of Root-Rot Disease Complex of Chickpea. *Aust. Plant. Pathol.* **2009,** *38*, 44–50.

Alizadeh, H.; Behboudi, K.; Ahmadzadeh, M.; Javan-Nikkhah, M.; Zamioudis, C.; Pieterse, C. M. J. et al. Induced Systemic Resistance in Cucumber and *Arabidopsis thaliana* by the Combination of *Trichoderma harzianum* Tr6 and *Pseudomonas* sp. Ps14. *Biol. Control.* **2013,** *65*, 14–23.

Askeland, R. A.; Morrison, S. M. Cyanide Production by *Pseudomonas fluorescens* and *Pseudomonas aeruginosa*. *Appl. Environ. Microbiol.* **1983,** *45*, 1802–1807.

Beneduzi, A.; Ambrosini, A.; Passaglia, L. M. P. Plant Growth-Promoting Rhizobacteria (PGPR): Their Potential as Antagonists and Biocontrol Agents. *Genet Mol Bio.* **2012,** *35*, 1044–1051.

Bhattacharya, P. N.; Jha D. K. Plant Growth-Promoting Rhizobacteria (PGPR): Emergence in Agriculture. *World J. Microbiol. Biotechnol.* **2012,** *28*, 1327–1350.

Blumer, C.; Haas, D. Mechanisms, Regulation and Ecological Role of Bacterial Cyanide Biosynthesis. *Arch. Microbiol.* **2000,** *173* (3), 170–177.

Cao, Y.; Pi, H.; Chandrangsu, P.; Li, Y.; Wang, Y.; Zhou, H.; Cai, Y. Antagonism of Two Plant-Growth Promoting *Bacillus velezensis* Isolates Against *Ralstonia solanacearum* and *Fusarium oxysporum*. *Sci. Rep.* **2018,** *8* (1), 4360.

Cardinale, M.; Ratering, S.; Suarez, C.; Zapata Montoya, A. M.; Geissler-Plaum, R.; Schnell, S. Paradox of Plant Growth Promotion Potential of Rhizobacteria and Their Actual Promotion Effect on Growth of Barley (*Hordeum vulgare* L.) Under Salt Stress. *Microbiol. Res.* **2015,** *181*, 22–32.

Castric, P. A. Hydrogen Cyanide, a Secondary Metabolite of *Pseudomonasaeruginosa*. *Can. J. Microbiol.* **1975,** *21*, 613–618.

Cattelan, M. E.; Hartel, P. G.; Uhrmann, J. J. Screening of Plant Growth Promoting Rhizobacteria to Promote Early Soybean Growth. *Soil. Sci. Soc. Am. J.* **1999,** *63*, 1670–1680.

Caulier, S.; Gillis, A.; Colau, G.; Licciardi, F.; Liépin, M.; Desoignies, N.; Bragard, C. Versatile Antagonistic Activities of Soil-Borne *Bacillus* spp. and *Pseudomonas* spp. Against *Phytophthora infestans* and Other Potato Pathogens. *Front. Microbiol.* **2018,** *9*, 143.

Chaiharn, M.; Chunhaleuchanon, S.; Lumyong, S. Screening Siderophore Producing Bacteria as Potential Biological Control Agent for Fungal Rice Pathogens in Thailand. *World J. Microbiol. Biotechnol.* **2009,** *25* (11), 1919–1928.

Chaiharn, M.; Chunhaleuchanon, S.; Kozo, A.; Lumyong, S. Screening of Rhizobacteria for Their Plant Growth Promoting Activities. *Kmitl. Sci. Tech. J.* **2008,** *8* (1), 18–23.

Chin-a-woeng, T. F. C.; Bloemberg, G. V.; Mulders, I. H. M.; Dekkers, L. C.; Lugtenberg, B. J. J. Root Colonization by Phenazine-1-Carboxamide Producing Bacterium *Pseudomonas chlororaphis* PCL1391 Is Essential for Biocontrol of Tomato Foot and Root Rot. *Mol. Plant Microbe Interact.* **2000,** *13* (12), 1340–1345.

Chincholkar, S. B., Chaudhari, B. L., Talegaonkar, S. K., Kothari, R. M. Microbial Iron Chelators: A Tool for Sustainable Agriculture. In: *Biocontrol Potential and Their Exploration in Crop Disease Management*; Upadhayay, R. K., Mukherji, K. G., Chamola, B. P., Eds., Vol. 1; Kluwer Academic Press: New York, 2000; pp 49–70.

Choudhary, D. K.; Prakash, A.; Johri, B. N. Induced Systemic Resistance (ISR) in Plants: Mechanism of Action. *Indian J. Microbiol.* **2007,** *47*, 289–297.

Compant, S.; Duffy, B.; Nowak, J.; Clément, C.; Barka, E. A. Use of Plant Growth-Promoting Bacteria for Biocontrol of Plant Diseases: Principles, Mechanisms of Action, and Future Prospects. *Appl. Environ. Microbiol.* 2005, *71* (9), 4951–4959.

Cronin, D.; Möenne-Loccoz, Y.; Fenton, A.; Dunne, C.; Dowling, D. N.; O'Gara, F. Role of 2, 4-Diacetylphloroglucinol in the Interactions of the Bio-Control Pseudomonad Strain F113 with the Potato Cyst Nematode *Globodera rostochiensis*. *Appl. Environ. Microbiol.* **1997,** *63*, 1357–1361.

Dangi, S.; Tirado-Corbalá, R.; Gerik, J.; Hanson, B. Effect of Long-Term Continuous Fumigation on Soil Microbial Communities. *Agronomy* **2017,** *7*, 37.

Davidson, J. M.; Wickland, A. C.; Patterson, H. A.; Falk, K. R.; Rizzo, D. M.. Transmission of *Phytophthora ramorum* in Mixed-Evergreen Forest in California. *Phytopathology* **2005,** *95*, 587–96.

De Souza, J. T.; Weller, D. M.; Raaijmakers, J. M. Frequency, Diversity and Activity of 2,4- Diacetylphloroglucinol Producing Fluorescent *Pseudomonas* spp. in Dutch Take-all Decline Soils. *Phytopathology* **2003,** *93*, 54–63.

Desbrosses, G. J.; Stougaard, J. Root Nodulation: A Paradigm for How Plant-Microbe Symbiosis Influences Host Developmental Pathways. *Cell Host Microb.* **2011,** *10*, 348–358.

Dileep Kumar, B. S. Crop Improvement and Disease Suppression by a *Bacillus* spp. SR2 from Peanut Rhizosphere. *Indian J. Exp. Biol.* **1996,** *34*, 45–48.

Ding, L. J.; Cui, H. L.; Nie, S. A.; Long, X. E.; Duan, G. L.; Zhu, Y. G. Microbiomes Inhabiting Rice Roots and Rhizosphere. *FEMS Microb. Eco.* **2019,** *95*, 1–10.

Dunne, C.; Crowley, J. J.; Moenne-Loccoz, Y.; Dowling, D. N.; de Bruijn,.F. J.; O'Gara, F. Biological Control of *Pythium ultimum* by *Stenotrophomonas maltophilia* W81 Is Mediated by an Extracellular Proteolytic Activity. *Microbiology* **1997,** *143*, 3921–3931.

Egamberdieva, D.; Lugtenberg, B. Use of Plant Growth-Promoting Rhizobacteria to Alleviate Salinity Stress in Plants. In: *Use of Microbes for the Alleviation of Soil Stresses*, Vol. 1; Springer: New York, 2014; pp 73–96.

Emmert, E. A. B.; Handelsman, J. Biocontrol of Plant Disease: A (Gram-) Positive Perspective. *FEMS Microbiol. Lett.* **1999,** *171*, 1–9.

Farwah, S.; Hussain, K.; Rizvi, S.; Hussain, S. M.; Rashid, M; Maqboollone, S. Introduction to Disease and Integrated Disease Management in Solanaceous Vegetables. *Int. J. Curr. Microbiol. App. Sci.* **2020,** *11*, 2791–2807.

Frankowski, J.; Lorito, M.; Scala, F.; Schmidt, R.; Berg, G.; Bahl, H. Properties of Two Chitinolytic Enzymes of *Serratia plymuthica* HRO-C48. *Arch. Microbiol.* **2001,** *176*, 421–426.

Fravel, D. R. Role of Antibiosis in the Biocontrol of Plant Diseases. *Annu. Rev. Phytopathol.* **1998,** *26*, 75–91.

Frederick, R. D., Ahmad, M.; Majerczak, D. R.; Arroyo-Rodriguez, A. S.; Manulis, S.; Coplin, D. L. Genetic Organization of the *Pantoea stewartii* Sub sp. *stewartii hrpgene* Cluster and Sequence Analysis of the hrpA, hrpC, hrpN, and wtsE operons. *Mol. Plant-Microbe Interact* **2001,** *14*, 1213–1222.

Garbeva, P.; van Veen, J. A.; van Elas, J. D. Assessment of the Diversity and Antagonism Towards *Rhizoctonia solani* AG3, of *Pseudomonas* Species in Soil from Different Agricultural Regimes. *FEMS Microbiol. Ecol.* **2004,** *47*, 51–64.

Girlanda, M.; Perotto, S.; Moenne-Loccoz, Y.; Bergero, R.; Lazarri, A.; Defago, G.; Bonfante, P.; Luppi, A. M. Impact of Biocontrol *Pseudomonas ûuorescens* CHA0 and a Genetically Modiûed Derivative on the Diversity of Culturable Fungi in the Cucumber Rhizosphere. *Appl. Environ. Microbiol.* **2001,** *67*, 1851–1864.

Glick, B. R. Plant Growth-Promoting Bacteria: Mechanisms and Applications. *Scientifica* **2012,** 963401.

Gnanamanickam, S. S. *Biological Control of Crop Diseases*; Dekker: New York, 2002.

Gohel, V.; Singh, A.; Misuria, V.; Phadnis, A.; Chhatpar, H. S. Bioprospecting and Antifungal Potential of Chitinolytic Microorganisms. *Afr. J. Biotech.* **2005,** *5* (2), 54–72.

Gupta, C. P.; Dubey, R. C.; Maheshwari, D. K. Plant Growth Enhancement and Suppression *Macrophomina phaseolina* Causing Charcoal Rot of Peanut by Fluorescent *Pseudomonas*. *Biol. Fertil. Soils* **2002,** *35*, 399–405.

Han, H. S.; Supanjani; Lee, K. D. Effect of Co-inoculation with Phosphate and Potassium Solubilizing Bacteria on Mineral Uptake and Growth of Pepper and Cucumber. *Plant Soil Environ.* **2006,** *52* (3), 130–136.

Handelsman, J.; Stabb, E. V. Biocontrol of Soil Borne Plant Pathogens. *Plant Cell* **1996,** *8*, 1855–1869.

Hariprasad, P.; Divakara, S. T.; Niranjana, S. R. Isolation and Characterization of Chitinolytic Rhizobacteria for the Management of *Fusarium* Wilt in Tomato. *Crop Prot.* **2011,** *30*, 1606–1612.

Harman, G. E.; Howel, C. H.; Viterbo, A.; Chet, I.; Lorito, M. *Trichoderma* Species-Opportunistic, Virulent Plant Symbionts. *Nat. Rev. Microbiol.* **2004,** *2*, 43–56.

Hoopen, G. M.; Krauss, U. Biology and Control of *Rosellinia bunodes*, *Rosellinia necatrix* and *Rosellinia pepo*: A Review. *Crop. Prot.* **2006,** *25*, 89–107.

Horner, N. R.; Grenville-Briggs, L. J. West, P. The Oomycete *Pythium oligandrum* Expresses Putative Effectors During Mycoparasitism of *Phytophthora infestans* and Is Amenable to Transformation. *Fungal Biol.* **2012,** *16*, 24–41.

Iavicoli, A.; Boutet, E.; Buchala, A.; Métraux, J. P. Induced Systemic Resistance in *Arabidopsis thaliana* in Response to Root Inoculation with *Pseudomonas fluorescens* CHA0. *Mol. Plant Microbe Interact.* **2003,** *16*, 851–858.

Idris, H. A.; Labuschagne, N.; Korsten, L. Screening Rhizobacteria for Biological Control of *Fusarium* Root and Crown Rot of Sorghum in Ethiopia. *J. Biocontrol.* **2007,** *40*, 97–106.

Itelima, J. U.; Bang, W. J.; Onyimba, I. A.; Oj, E. A Review: Biofertilizer; A Key Player in Enhancing Soil Fertility and Crop Productivity. *J. Microbiol. Biotechnol. Rep.* **2018,** *2*, 22–28.

Jakobsen, I., Chen, B. D., Munkvold, L., Lundsgaard, T., Zhu, Y. G. Contrasting Phosphate Acquisition of Mycorrhizal Fungi with that of Root Hairs Using the Root Hairless Barley Mutant. *Plant Cell Environ.* **2005,** *28*, 928–938.

Jankiewicz, U.; Koltonowicz, M. The Involvement of *Pseudomonas* Bacteria in Induced Systemic Resistance in Plants (Review). *Appl. Biochem. Microbiol.* **2012,** *48*, 244–249.

Johri, B. N., Rao, C. V. S., Goel, R. Fluorescent Pseudomonads in Plant Disease Management. In: *Biotechnological Approaches in Soil Microorganisms for Sustainable Crop Production*; Dadarwal, K. R., Ed.; Scientific Publisher: Jodhpur, 1997; pp 193–223.

Jung, W. J.; Jin, Y. L.; Kim, K. Y.; Park, R. D.; Kim, T. H. Changes in Pathogenesis-Related Proteins in Pepper Plants with Regard to Biological Control of *Phytopthora* Blight with *Paenibacillus illinoisensis*. *Biocontrol* **2005,** *50*, 165–178.

Kamil, Z.; Rizk, M.; Saleh, M.; Moustafa, S. Isolation and Identification of Rhizosphere Soil Chitinolytic Bacteria and Their Potential in Antifungal Biocontrol. *Global J. Mol. Sci.* **2007,** *2* (2), 57–66.

Kamilova, F.; Lamers, G.; Lugtenberg, B. Biocontrol Strain *Pseudomonas fluorescens* WCS365 Inhibit Germination of *Fusarium oxysporum* Spores in Tomato Root Exudates as Well as Subsequent Formation of New Spores. *Environ. Microbiol.* **2008,** *10*, 2455–2461.

Kamilova, F.; Validov, S.; Azarova, T.; Mulders, I.; Lugtenberg, B. Enrichment for Enhanced Competitive Plant Root Tip Colonizers Selects for a New Class of Biocontrol Bacteria. *Environ. Microbiol.* **2005,** *7*, 1809–1817.

Kang, B. R.; Anderson, A. J.; Kim, Y. C. Hydrogen Cyanide Produced by *Pseudomonas chlororaphis* o6 Exhibits Nematicidal Activity Against *Meloidogyne hapla*. *Plant Pathol. J.* **2018,** *34*, 35–43.

Keel, C. and Defago, G. Interactions Between Beneûcial Soil Bacteria and Root Pathogens: Mechanisms and Ecological Impact. In: *Multitrophic Interactions in Terrestrial Systems*; Gange, A. C., Brown, V. K., Eds.; Blackwell Science Ltd: London, 1997; pp 27–46.

Khan, M. R.; Khan, S. M. Effect of Root-Dip Treatment with Certain Phosphate Solubilization Microorganisms on the Fusarial Wilt of Tomato. *Bioresour. Technol.* **2002,** *85*, 213–215.

Khokhar, M. K.; Gupta, R. Integrated Disease Management. *J. Popular Kheti* **2014,** *2*, 87–91.

Khoury, W. E.; Makkouk, K. Integrated Plant Disease Management in Developing Countries. *J. Plant Pathol.* **2010,** *9*, 35–42.

Kim, K. J.; Yang, Y. J.; Kim, J. G. Purification and Characterization of Chitinase from *Streptomyces* sp. M-20. *J. Biochem. Mol. Biol.* **2003,** *36* (2), 185–189.

Kim, Y. C.; Jung, H.; Kim, K. Y.; Park, S. K. An Effective Biocontrol Bioformulation Against *Phytophthora* Blight of Pepper Using Growth Mixtures of Combined Chitinolytic Bacteria Under Different Field Conditions. *Eur. J. Plant Pathol.* **2008,** *120*, 373–382.

Kishore, G. K.; Pande, S. Chitin-Supplemented Foliar Application of Chitinolytic *Bacillus cereus* Reduces Severity of *Botrytis* Gray Mold Disease in Chickpea Under Controlled Conditions. *Lett. Appl. Microbiol.* **2007,** *44*, 98–10.

Kurabachew, H.; Wydra, K. Characterization of Plant Growth Promoting Rhizobacteria and Their Potential as Bioprotectant Against Tomato Bacterial Wilt Caused by *Ralstonia solanacearum*. *Biol. Cont.* **2013,** *67* (1), 75–83.

Labuschagne, N. T.; Pretorius, A. H.; Idris. Plant Growth Promoting Rhizobacteria as Biocontrol Agents Against Soil-Borne Plant Diseases. In: *Plant Growth and Health Pro-Moting Bacteria, Microbiology Monographs*; Maheshwari, D. K., Ed.; Springer: Verlag Berlin Heldelberg, 2010; pp 211–230.

Lamsal, K.; Kim, S. W.; Kim, Y. S.; Lee, Y. S. Application of Rhizobacteria for Plant Growth Promotion Effect and Biocontrol of Anthracnose Caused by *Colletotrichum acutatum* on Pepper. *Mycobiology* **2012,** *40* (4), 244–251.

Landa, B. B.; Mavrodi, O. V.; Raaijmakers, J. M.; Gardener, B. B. M.; Thomashow, L. S.; Weller, D. M. Differential Ability of Genotypes to Colonize the Roots of Pea Plants. *Appl Environ Microbiol.* **2002,** *68* (7), 3226–3237.

Laville, J.; Blumer, C.; Von Schroetter, C.; Gaia, V.; De fago, G.; Keel, C.; Haas, D. Characterization of the hcnABC Gene Cluster Encoding Hydrogen Cyanide Synthase and Anaerobic Regulation by ANR in the Strictly Aerobic Biocontrol Agent *Pseudomonas fuorescens* CHA0. *J. Bacteriol.* **1998,** *180*, 3187–3196.

Lee, J. H., Ma, K. C., Ko, S. J., Kang, B. R., Kim, I. S., Kim, Y. C. Nematicidal Activity of a Nonpathogenic Biocontrol Bacterium, *Pseudomonas chlororaphis* O6. *Curr. Microbiol.* **2011,** *62*, 746–751.

Lemanceau, P.; Albouvette, A. Suppression of *Fusarium* Wilts by Fluorescent *Pseudomonas*: Mechanism and Applications. *Biocon. Sci. Technol.* **1993,** *3*, 219–234.

Liang, J.; Tao, R.; Hao, Z.; Wang, L.; Zhang, X. Induction of Resistance in Cucumber Against Seedling Damping-off by Plant Growth –Promoting Rhizobacteria (PGPR) *Bacillus megaterium* Strain L8. *Afr. J. Biotechnol.* **2011,** *10* (36), 6920–6927.

Loni, P. P.; Shyam, S.; Bajekal. Alkaline chitinase from *Bacillus firmus* SBPL-05 isolated from alkaline-saline environment of Lonar lake. Indian J. Fundam. Appl. Life. Sci. **2011,** 1, 3, 161–165.

López-Ráez, J. A.; Shirasu, K.; Foo, E. Strigolactones in Plant Interactions with Beneficial and Detrimental Organisms: The Yin and Yang. *Trends Plant Sci.* **2017,** *22*, 527–537.

López-Cruz, M.; Crossa, J.; Bonnett, D.; Dreisigacker, S.; Poland, J.; Jannink, J. L. et al. Increased Prediction Accuracy in Wheat Breeding Trials Using a Marker × Environment Interaction Genomic Selection Model. *Genes Genomes Genet.* **2015,** *5*, 569–582.

Lorck, H. Production of Hydrocyanic Acid by Bacteria. *Physiol. Plant.* **1948,** *11*, 142–146.

Lugtenberg, B.; Kamilova, F. Plant-Growth Promoting Rhizobacteria. *Annu. Rev. Micro.* **2009,** *63* (1), 541–556.

Mahajan, R.; Shirkot, C. K. PGPR: Their Role in Plant Growth and Disease Management. In: *Approaches and Trends in Plant Disease Management*; Sharma, M.; Gupta, S. K., Eds.; Scientific Publishers: Jodhpur, 2014; pp 85–122.

Maksimov, I. V.; Abizgidina, R. R.; Pusenkova, L. I. Plant Growth Promoting Rhizobacteria as Alternative to Chemical Crop Protectors from Pathogens. *Appl. Biochem. Microbiol.* **2011,** *47*, 333–345.

Markowich, N. A.; Kononova, G. L. Lytic Enzymes of *Trichoderma* and Their Role in Plant Defense from Fungal Diseases: A Review. *Appl. Biochem. Microbiol.* **2003,** *39*, 341–351.

Martínez-Medina, A.; Van Wees, S. C. M.; Pieterse, C. M. J. Airborne Signals from *Trichoderma* Fungi Stimulate Iron Uptake Responses in Roots Resulting in Priming of Jasmonic Acid Dependent Defences in Shoots of *Arabidopsis thaliana* and *Solanum lycopersicum*. *Plant Cell Environ.* **2017,** *40*, 2691–2705.

Mazurier, S. Phenazine Antibiotics Produced by Fluorescent Pseudomonads Contribute to Natural Soil Suppressiveness t*o Fusarium* wilt. *ISME J.* **2009,** *3* (8), 977–991.

Mehrotra, R. S. *Plant Pathology*, 2nd ed.; McGraw Hill Publication Company Ltd: New Delhi, 2003; p 771.

Miethke, M.; Marahiel, M. A. Siderophore-Based Iron Acquisition and Pathogen Control. *Microbiol. Mol. Biol. Rev*. **2007,** *71* (3), 413–451.

Murthy, K. N.; Uzma, F.; Chitrashree, C. S. Induction of Systemic Resistance in Tomato Against *Ralstonia solanacearum* by *Pseudomonas fluorescens*. *Am. J. Plant Sci.* **2014,** *5* (12), 1799–1811.

Nakkeeran, S.; Kavitha, K.; Mathiyazhagan, S.; Fernando, W. G. D.; Chandrasekar, G.; Renukadevi, P. Induced Systemic Resistance and Plant Growth Promotion by *Pseudomonas chlororaphis* Strain PA-23 and *Bacillus subtilis* Strain CBE4 Against Rhizome Rot of Turmeric (*Curcuma longa* L.). *Can. J. Plant Pathol.* **2004,** *26*, 417–418.

Nandi, M.; Selin, C.; Brassinga, A. K. C.; Belmonte, M. F.; Fernando, W. D.; Loewen, P. C.; De Kievit, T. R. Pyrrolnitrin and Hydrogen Cyanide Production by *Pseudomonas chlororaphis* Strain PA23 Exhibits Nematicidal and Repellent Activity Against *Caenorhabditis elegans*. *PLoS One* **2015,** *10*, e0123184. DOI: 10.1371/journal.pone.0123184.

Niranjan, R.; Shetty, H. S.; Reddy, M. S. Plant Growth-Promoting Rhizobacteria: Potential Green Alternative for Plant Productivity. In: *PGPR: Biocontrol and Biofertilization*; Siddiqui, Z. A., Ed.; Springer: Dordrecht, 2005; pp 197–216.

Pachauri, R. K.; Allen, M. R.; Barros, V. R.; Broome, J.; Cramer, W.; Christ, R. et al. Switzerland: Intergovernmental Panel on Climate Change. In: *Synthesis Report on contribution of Working Groups I, II and III to the fifth Assessment Report of the Intergovernmental Panel on Climate Change IPCC*; Pachauri, R. K.; Meyer, L. A., Eds.; Intergovernmental Panel on Climate Change: Geneva, 2014.

Palti, J. *Cultural Practices and Infectious Crop Diseases*; Springer-Verlag: Berlin, 1981.

Palumbo, J. D.; Yuen, G. Y.; Jochum, C. C.; Tatum, K.; Kobayashi, D. Y. Mutagenesis of β-1,3-Glucanase Gene in *Lysobacter enzymogenes* C3 Results in Reduced Biological

Control Activity Toward Bipolaris Leafspot of Tall Fescue and Pythium Damping-off of Sugarbeet. *Phytopathology* **2005,** *95*, 701–707.

Pandey, A. K.; Sain, S. K.; Singh, P. A Perspective on Integrated Disease Management in Agriculture. *Res. J. Bio Bull.* **2016,** *2*, 13–29.

Park, S. H.; Lee, J. H.; Lee, H. K. Purification and Characterization of Chitinase from a Marine Bacterium, *Vibrio* sp. 98CJ11027. *J. Microbiol.* **2000,** *38* (4), 224–229.

Pieterse, C. M. J.; Van der Does, D.; Zamioudis, C.; Leon-Reyes, A.; Van Wees, S. C. M. Hormonal modulation of plant immunity. *Annu. Rev. Cell Dev. Biol.* **2012,** *28*, 489–521.

Pieterse, C. M. J.; Van Wees, S. C. M.; Ton, J.; Van Pelt, J. A.; Van Loon, L. C. Signalling in Rhizobacteria-Induced Systemic Resistance in *Arabidopsis thaliana*. *Plant Biol.* **2002,** *4*, 535–544.

Pieterse, C. M. J.; Van Wees, S. C. M.; van Pelt, J. A.; Knoester, M.; Laan, R.; Gerrits, H. et al. A Novel Signaling Pathway Controlling Induced Systemic Resistance in *Arabidopsis*. *Plant Cell* **1998,** *10*, 1571–1580.

Pieterse, C. M. J.; Zamioudis, C.; Berendsen, R. L.; Weller, D. M.; Van Wees, S. C. M.; Bakker, P. A. H. M. Induced Systemic Resistance by Beneficial Microbes. *Annu. Rev. Phytopathol.* **2014,** *52*, 347–375.

Querejeta, J. I., Barea, J. M., Allen, M. F., Caravaca, F., Roldan, A. Differential Response of δ13C and Water Use Efficiency to Arbuscular Mycorrhizal Infection in Two Arid Land. *Woody Plant Species Oecologia* **2003,** *135*, 510–515.

Quiroga, G.; Erice, G.; Aroca, R.; Delgado-Huertas, A.; Ruiz-Lozano, J. M. Elucidating the Possible Involvement of Maize Aquaporins and Arbuscular Mycorrhizal Symbiosis in the Plant Ammonium and Urea Transport Under Drought Stress Conditions. *Plants* **2020,** *9*, 148.

Rai, A.; Nabti, E. Plant Growth-Promoting Bacteria: Importance in Vegetable Production. In: *Microbial Strategies for Vegetable Production*; Zaidi, A., Khan, M. S., Eds.; Springer, 2017; pp 23–48.

Ramamoorthy, V.; Viswanathan, R., Raghuchander, T.; Prakasam, V.; Samiyappan, R. Induction of Systemic Resistance by Plant Growth Promoting Rhizobacteria in Crop Plants Against Pests and Diseases. *Crop Prot.* **2001,** *20*, 1–11.

Ramos Solano, B.; García, J. A. L.; Villaraco, A. G.; Algar, E.; Cristobal, J. G.; Mañero, F. J. G. Siderophore and Chitinase Producing Isolates from the Rhizosphere of *Nicotiana glauca* Graham Growth and Induce Systemic Resistance in *Solanum lycopersicum* L. *Plant Soil.* **2010,** *334*, 189–197.

Ran, L. X.; Liu, C. I.; Wu, G. J.; van Loon, L. C.; Bakker, P. A. H. M. Suppression of bacterial wilt in *Eucalyptus urophylla* by Fluorescent *Pseudomonas* spp. in China. *Biol. Control.* **2005,** *32*, 111–120.

Ryu, C. M.; Hu, C. H.; Reddy, M. S.; Kloepper, J. W. Different Signaling Pathways of Induced Resistance by rhizobacteria in *Arabidopsis thaliana* Against Two Pathovars of *Pseudomonas syringae*. *New Phytol.* **2003,** *160*, 413–420.

Sandhya, V.; Ali, S. K. Z.; Grover, M.; Reddy, G.; Venkateswarlu, B. Alleviation of Drought Stress Effects in Sunflower Seedlings by Exopolysaccharides Producing *Pseudomonas putida* Strain GAP-P45. *Biol. Fertil. Soil.* **2009,** *46*, 17–26.

Saraf, M.; Jha, C. K.; Patel, D. The Role of ACC Deaminase Producing PGPR in Sustainable Agriculture. In: *Plant Growth and Health Promoting Bacteria Microbiology*; Maheshwari, D. K., Ed., Vol. 18; Springer: Steinbuchel A. Germany, 2011; pp 365–386.

Sarbadhikary, S. B.; Mandal, N. C. Field Application of Two Plant Growth Promoting Rhizobacteria with Potent Antifungal Properties. *Rhizosphere* **2017,** *3*, 170–175.

Sayyed, R. Z.; Chincholkar, S. B. Siderophore Producing *A. feacalis* More Biocontrol Potential Vis-a'-vis Chemical Fungicide. *Curr. Microbiol.* **2009,** *58* (1), 47–51.

Schippers, B.; Bakker, A.; Bakker, P.; van Peer, R. Beneficial and Deleterious Effects of HCN-Producing Pseudomonads on Rhizosphere Interactions. *Plant Soil.* **1990,** *129*, 75–83.

Schouten, A.; Berg, G.; Van Den; Edel-hermann, V.; Steinberg, C.; Alabouvette, C.; de Vos, C. H.; Lemanceau, P.; Raaijmakers, J. M. Defense Responses of *Fusarium oxysporum* to 2,4- Diacetylphloroglucinol, a Broad-Spectrum Antibiotic Produced by *Pseudomonas fluorescens*. *Mol. Plant Microbe.* **2004,** *17* (11), 1201–1211.

Senthilkumar, M.; Swarnalakshmi, K.; Govindaswami, V.; Lee, Y. K.; Annapurna, K. 2008. Biocontrol Potential of Soyabean Bacterial Endophytes Against Charcoal Rot Fungus *Rhizoctonia bataicol*a. *Curr. Microbiol.* **2008,** *58* (4), 288–293.

Sharma, H. C.; Rodomiro, O. Host-Plant Resistance to Insects: An Ecofriendly Approach for Pest Management and Environment Conservation. *J. Environ. Biol.* **2002,** *23*, 111–135.

Sindhu, S. S.; Suneja, S.; Dadarwal, K. R. Plant Growth Promoting Rhizobacteria and Their Role in Crop Productivity. In: *Biotechnological Approaches in Soil Microorganisms for Sustainable Crop Production*; Dadarwal, K. R., Ed.; Scientific Publishers: Jodhpur, 1997; pp 149–193.

Singh, P. P.; Shin, Y. C.; Park, C. S.; Chung, Y. R. Biological Control of *Fusarium* Wilt of Cucumber by Chitinolytic Bacteria. *Phytopathology* **1999,** *89*, 92–99.

Singh, V. K.; Chawla, S. *Cultural Practices: An Eco-Friendly Innovative Approach in Plant Disease Management*, 2012.

Timmusk, S.; Behers, L.; Muthoni, J.; Muraya, A.; Aronsson, A. Perspectives and Challenges of Microbial Application for Crop Improvement. *Front. Plant Sci.* **2017,** *9*, 1–10.

Vacheron. J.; Desbrosses, G.; Bouffaud, M.; Touraine, B.; Moenne-Loccoz, Y.; Muller, D.; Legendre, L.; Wisniewski-Dye, F.; Prigent-Combaret, C. Plant Growth-Promoting Rhizobacteria and Root System Functioning. *Front. Plant Sci.* **2013,** *4*, 1–19.

Validov, S. *Biocontrol of Tomato Foot and Root Rot by Pseudomonas Bacteria in Stonewool.* PhD Thesis, Leiden University, 2007. http://hdl.handle.net/1887/12480

Van Loon, L. C.; Bakker, P. A.; Pieterse, C. M. J. Systemic Resistance Induced by Rhizospheric Bacteria. *Annu. Rev. Phytopathol.* **1998,** *36*, 453–483.

Velusamy, P.; Kim, K. Y. Chitinolytic Activity of *Enterobacter* sp. KB3 Antagonistic to *Rhizoctonia solani* and Its Role in the Degradation of Living Fungal Hyphae. *Int. Res. J. Microbiol.* **2011,** *2* (6), 206–214.

Verbon, E. H.; Trapet, P. L.; Stringlis, I. A.; Kruijs, S.; Bakker, P. A. H. M.; Pieterse, C. M. J. Iron and Immunity. *Annu. Rev. Phytopathol.* **2017**, *55*, 355–375.

Vidhyasekaran, P.; Kamala, N.; Ramanathan, A.; Rajappan, K.; Paranidharan, V.; Velazhahan, R. Induction of Systemic Resistance by *Pseudomonas fluorescens* Pf1 Against *Xanthomonas oryzae* pv. Oryzae in Rice Leaves. *Phytoparasitica* **2011,** *29*, 155–166.

Vincent, M. N.; Harrison, L. A.; Brackin, J. M.; Kovacevich, P. A.; Mukerji, P.; Weller, D. M.; Pierson, E. A. Genetic Analysis of the Antifungal Activity of a Soilborne Pseudomonas Aureofaciens Strain. *Appl. Environ. Microbiol.* **1991,** *57*, 2928–2934.

Voisard, C.; Keel, C.; Haas, D.; Defago, G. Cyanide Production by Pseudomonas *fluorescens* Helps Suppress Black Root Rot of Tobacco Under Gnotobiotic Conditions. *EMBO J.* **1989,** *8*, 351–358.

Whipps, J. M. Carbon Economy. In: *The Rhizosphere*; Lynch, J. M., Ed.; Wiley: Chichester, 1990, pp 59–99.

Wille, L.; Messmer, M. M.; Studer, B.; Hohmann, P. Insights to Plant–Microbe Interactions Provide Opportunities to Improve Resistance Breeding Against Root Diseases in Grain Legumes. *Plant, Cell Environ.* **2019,** *42* (1), 20–40.

Zamioudis, C.; Hanson, J.; Pieterse, C. M. J. β-Glucosidase BGLU42 is a MYB72-Dependent Key Regulator of Rhizobacteria-Induced Systemic Resistance and Modulates Iron Deficiency Responses in *Arabidopsis* Roots. *New Phytol.* **2014,** *204*, 368–379.

Zamioudis, C.; Korteland, J.; Van Pelt, J. A.; van Hamersveld, M.; Dombrowski, N.; Bai, Y. et al. Rhizobacterial Volatiles and Photosynthesis-Related Signals Coordinate MYB72 Expression in *Arabidopsis* Roots During Onset of Induced Systemic Resistance and Iron-Deficiency Responses. *Plant J.* **2015,** *84*, 309–322.

Zehnder, G. W.; Murphy, J. F.; Sikora, E. J.; Kloepper, J. W. Application of Rhizobacteria for Induced Resistance. *Eur. J. Plant. Pathol.* **2001,** *107*, 39–50.

Zhang, H.; Sun, Y.; Xie, X.; Kim, M. S.; Dowd, S. E.; Paré, P. W. A Soil Bacterium Regulates Plant Acquisition of Iron via Deficiency Inducible Mechanisms. *Plant J.* **2009,** *58*, 568–577.

Zhao, L.; Wang, F.; Zhang, Y.; Zhang, J. Involvement of *Trichoderma asperellum* Strain T6 in Regulating Iron Acquisition in Plants. *J. Basic Microbiol.* **2014,** *54*, S115–S124.

https://extension.unr.edu/publication.aspx.PubID=2647

https://www.gardeningknowhow.com/garden-how-to/soil-fertilizers/sterilizing-soil.html

https://www.thedirtdoctors.com/types-of-chemicals-used-in-fumigation

CHAPTER 5

Plant Diseases and Management Approaches in Organic Versus Conventional Farming

NEHA SRIVASTAVA, POOJA BHARDWAJ, HIRAWATI DEVAL, and RAJNI KANT

ICMR-Regional Medical Research Centre, Gorakhpur, India

ABSTRACT

Organic farming (OF) has expanded worldwide with a high growth rate (more than 10%). Organic crop production regenerates the health of the soil ecosystem and humans. OF relies on natural cycles and processes, and biodiversity to adapt to the local habitat rejecting the use of chemical fertilizers or pesticides. The occurrence of plant diseases in organic and conventional farming is not new and both require fundamentally different management approaches. In this chapter, a comparative review of plant diseases in organic and conventional farming is done and variations in management approaches are analyzed to provide better options for plant disease management and control in OF practices.

5.1 INTRODUCTION TO ORGANIC FARMING

OF is cost-effective, ecologically, and socially responsible way of farming (Finckh et al., 2015). It provides a supply of safe and healthy food products with less nutrient and energy loss (Bruggen et al., 2015). First, practices in organic crop production promote the health of the ecosystem and secondly no synthetic pesticides and fertilizers used (Finckh et al., 2015). In 2013,

Plant Diseases and Their Management: A Sustainable Approach.
Deepa Srivastava, Rajarshi Kumar Gaur, and Ajay K. Tiwari (Eds.)

the global market of organic products reached $US 72 and increased tremendously worldwide in the past 20 years (Willer et al., 2015). The OF standards are formulated by the International Federation of Organic Agriculture Movements (IFOAM) which laid down the idea that all natural processes in ecosystem are interdependent and in OF management should be achieved through self-regulation of natural processes (Birkhofer et al., 2008). The occurrence of diseases in OF requires fundamentally different management approaches from conventional farming. As a result of conventional farming, epidemic of many plant diseases caused by fungi, bacteria, nematodes and viruses have emerged (Berkelmans et al., 2003; Finckh and Wolfe 2015; Bruggen and Termorshuizen 2003; Bruggen and Semenov 2015) . Root diseases occurred due to loss of soil organic matter and degradation of soil quality. High nitrogen concentrations and imbalance in plant nutrition make plants susceptible to pathogens causing root and foliar diseases (Datnoff et al., 2007). Unlike conventional farming, OF systems generally have higher plant diversity, high water holding capacity, crop rotation, better cation exchange capacity and cover cropping, higher biomass, diverse activity of soil microorganisms, lesser nutrient loss and increased internal cycling (Aparna et al., 2014; Gomiero et al., 2011; Tuck et al., 2014). These differences in OF and conventional farming lead to variability in the occurrence and intensity of plant disease that needs to be reviewed. Further, this aspect also requires the contribution of various disease management approaches that could be useful for farming practitioners and agriculture researchers for greater ecological sustainability. In this chapter, we will discuss the different plant diseases in organic and conventional farming and practices to manage these diseases, concluded with some suggestions for research on plant diseases and their effective management in OF.

5.2 PLANT DISEASES IN ORGANIC VERSUS CONVENTIONAL FARMING

5.2.1 SOIL-BORNE PLANT DISEASES

Root diseases are caused by various soil-borne pathogens and quality of soil and its health strongly affects the root diseases in both organic as well as conventional farming. Common plant pathogens affected crop and its effective management is summarized in Table 5.1.

TABLE 5.1 Soil Borne Plant Diseases and Their Management Strategy in OF System

Plant pathogen	Plant disease	Management strategy in organic farms
Fusarium spp.	Foot rot in wheat barley (cereals)	Management of organic fields according to European Union (EU) guidelines; organic farm fields should be located in a lower lying and wetter area than conventional farm fields
	Fusaria moxysporum wilt in flax and melon	Use of composted manure in organic farm fields
	Fusarium virguliforme death syndrome in soyabean	Transitions in organic productions: pasture, grain rotation, or vegetable crops (with or without amendments)
Pratylenchus spp.	Root disease in corn or sorghum in rotation with soybean and oat	Management of organic fields according to National Organic Program guidelines
	Pratylenchus thornei root knot nematode disease in wheat and oats	Winter cover crops; use of manure in organic farms
Bipolaris sorokiniana	Damping-off in wheat and barley	Management of organic fields as per EU guidelines
Gaeumannomyces graminis	Root rot and take-all in wheat and barley	Management of organic fields as per EU guidelines
Rhizoctonia spp.	Root rot and eyespot in wheat, barley, rye, and triticale	Shallower tillage, lower nitrogen content, no synthetic pesticides in organic fields
	Damping off in *Brassica* spp.	Practicing mixed cropping, management of crop residue, use of botanical pesticides
	Rhizoctonia solani in soyabean	Transition of organic productions from pasture, grain rotation, or vegetable crops (with or without amendments)
	R. solani black scurf in potato	Use of rapeseed as precrop, use of biocontrol agents at organic and conventional farms one on one
	Damping-off in tomato	Winter cover crops, use of composted manures and rotations in organic farms
	Root rot in strawberry	Mulch of straw and no fumigation in organic farms
Heterodera avenae	Root knot nematode disease in wheat and oats	Production of winter cover crops and use of composted manures
Ditylenchus dipsaci	Root knot nematode disease in wheat and oats	Production of winter cover crops, use of composted manures and rotations in organic farms
Plasmodiophora brassicae	Club root disease in *Brassica* spp.	Practicing mixed cropping, management of crop residue, use of botanical pesticides in organic farms

TABLE 5.1 *(Continued)*

Plant pathogen	Plant disease	Management strategy in organic farms
Rhizorhapis suberifaciens	Corky root disease in lettuce	Practicing longer crop rotation and use of compost in organic farms; use of copper fungicides
Ralstonia solanacearum	Brown rot disease in potato	Management of organic farm fields according to EU guidelines
Streptomyces scabies	Scab disease in Potato	Use of rapeseed as precrop, use of biocontrol agents at organic and conventional farms one on one
Helminthosporium solanii	Silver scurf in potato	Use of rapeseed as precrop, use of biocontrol agents at organic and conventional farms one on one
Sclerotium rolfsii	Southern blight disease in tomato	Organic treatments including mulch and various amendments in soil of organic field
Pyrenochaetalyco persici	Corky root in tomato	Organic productions as per California regulations
Phytophthora parasitica	root rot in tomato	Organic productions as per California regulations
Verticillium spp.	In tomato and strawberry	Mulch of straw and no fumigation in organic farms

Several long-term farming experiments show that cereal crops are less susceptible to root diseases in the OF system as compared to the conventional farming system (Hiddink et al., 2005; Letourneau et al., 2006; Bruggen 1995; Bruggenet al., 2003). Due to the biodiversity richness of soil and low nitrogen contents in organically managed soil, the incidence of *Fusarium* infections, that is, *Fusarium graminearum or Fusarium culmorum* are low in organic farms. Non-pathogenic *Fusarium* species are often found dominant over pathogen *Fusarium* species in organic farms (Bruggen et al., 2003). Severity due to *F. culmorum* (foot rot), *F. nivale* (snow mold), *Pseudocercosporella herpotrichoides* (eyespot), and *Rhizoctonia cerealis* (sharp eyespot) are correlated with total nitrogen applied in the field soil (Bruggen et al., 2003). Conventional fields converted into organic fields long term ago have a low incidence of *G. graminis* (take-all) infection while recently converted organic farms have high incidence of *G. mycesgraminis* (take-all) as compared to conventional farms (Bruggen et al., 2003) which may be due to change from monocropping to crop rotation practices.

In comparison to organic farms, conventional farms growing cereal crops and corn were more prone to disease caused by Nematodes. Comparatively higher numbers of predatory omnivorous nematodes are responsible for the reduction in the population of plant parasitic nematodes in organic farms (Brikhofer et al., 2008; Finckh et al., 2015).

The addition of fresh organic matter in organic fields for vegetable production increases the risk of damping-off by the growth of facultative saprophytic pathogens (*Pythium* spp.). With the addition of organic materials, there was initially an increase in damping-off followed by an increase or decrease in disease incidence in an oscillatory fashion (He et al., 2010, 2012). Thus damping-off depends upon planting time since the incorporation of organic materials (Darby et al., 2006) as seedling damping-off of tomatoes is either reduced (*R. solani*) or the same (*Pythium aphanidermatum*) in OF compared to conventional farming (Grunwald et al., 2000; Mansour et al., 2014). However, seedling damping-off of *Brassica* spp. (*R. solani*) found reduced in OF (Shrestha et al., 2014).

Root diseases are less common in older plants of organic farms as compared to conventional farming fields, for example, *P. brassicae* (club root) and *Sclerotinia sclerotiorum* (stalk root) of *Brassica* spp. in Nepal (Shrestha et al., 2014). *S. rolfsii* infection in tomatoes (southern blight) was reduced by a few amendments in soil by mulching and cotton gin trash. This amendment increases the population of natural control agents, that is, *Trichoderma* spp., increases the water holding capacity of soil, enhanced nitrogen content, microbial biomass, and biodiversity (Bulluck and Ristaino 2002; Liu et al., 2007). Addition of composted manure in melons-producing organic farms led to a reduced incidence of Fusarium wilt in melons compared to conventional farms (Yogev et al., 2011). Likewise, incidence of Fusarium wilt in flax found less in organic farms compared to conventional farms which is due to use of organic composts that cause high pH, high nitrogen content and ammonia oxidizing microbes in soil (Senechkin et al., 2014; Bruggen et al., 2015). Physical factors of soil such as low aggregate stability and high clay content primarily contributes to the infection of *P. parasitica* (Phytophthora root rot disease) in tomatoes of organic field while high nitrogen content in soil and tomato tissue leads to severe *P. persici* infection (corky root disease) in tomatoes in organic farms (Bruggen et al., 2003; Letourneau et al., 2006). In organic farms, corky root and other root diseases are controlled and reduced by longer crop rotation and organic content management. Disease suppression in organic farms is done by the use of cover crops, and manure, increasing water penetrability and by lowering the nitrogen content (Letourneau et al., 2006). Significant dominance of *R. Solani* and *Verticillium dahlia* were found in conventional treatments compared to organic ones (Bruggen et al., 2003).

R. suberifaciens (corky root disease) infections are less common in organic lettuce compared to lettuce from conventional farm fields (Bruggen et al., 2015). This infection of *R. suberifaciens* is enhanced by low microbial activity in soil, the use of herbicides and inorganic nitrogen in conventional farming ((Bruggen et al., 2015). A study from Egypt and The Netherlands determined the rate of survival of *R. solanacearum* race 3 biovar 2 which is a causal agent of brown rot in potatoes and reported shorter survival time in organic fields compared to conventional fields (In Egypt but not in The Netherlands (Messiha et al., 2009). Organic tomato producers of Nepal consider root-knot nematodes a negligible problem while conventional tomato producers cannot neglect this disease (Shrestha et al., 2014).

Range of pathogens, that is, fungi and nematodes to oomycetes are known to cause root diseases in perennial crops. *Pythium* spp. is more commonly observed in conventional apple orchards in comparison to organic farms (Mazzola et al., 2002). In another study by (Manici et al., 2003), reduced colonization of *Pythium* spp. and *Rhizoctonia* spp. were observed in organic apple farms compared to conventional apple farms. Root rot disease caused by *Phytophthora* spp. (*P. cinnamomi* in avocados) are commonly found suppressed on older fruit trees in OF systems which may be due to mulching for many years leading to enhanced microbial activity and lysis of *Phytophthora hypae* (Bruggen 1995; Yang et al., 2001). The incidence of root rot disease due to fungus in strawberries is similar in recently converted organic farms, conventional farms, and conventional fields that underwent soil fumigation (Bruggen and Termorshuizen 2003; Njoroge et al., 2009). However, in old and well-established organic fields these root diseases can be reduced by appropriate soil management practices (Bruggen and Termorshuizen 2003). Increased and regular use of soil fumigants resulted in the ineffectiveness of their action against nematodes in Banana and coffee, while organic amendments are suggested to be the best way to control nematodes-associated root diseases (Finch et al., 2015).

5.2.2 AIR BORNE PLANT DISEASES/FOLIAR DISEASES

Comparing soil-borne diseases in organic and conventional farming systems is straightforward but comparing airborne diseases among both the farming systems are quite complex. Scale differences in both systems are hard to consider which may be because organic farms are in small fractions as compared to conventional farms. It might be possible that the pesticides used in neighboring conventional farms may exchange aerial inoculum and vectors to the organic farms which may provide a profit of low inoculum to organic farms. In a spatial survey, it was demonstrated that conventional farms benefit more from organic farms as a greater number of predators and parasites approach greatly toward organic farms (Gomiero et al., 2011).

Despite various difficulties, we compared the foliar diseases in OF system as well as conventional farming system and presented in Table 5.2 in this chapter.

TABLE 5.2 Air-Borne/Foliar Plant Diseases and Their Management in OF System.

Plant pathogen	Plant disease	Management strategy in organic farms
Ustilago maydis	Common smut in corn	Producing winter cover crops, use of composted manure, and practicing wider rotations in organic farm fields
Puccinia striiformis	Stripe rust in wheat and Rye	Practicing longer rotations of organic productions, use of organic fertilizers, and low nitrogen content in the soil
Puccinia recondite	Leaf rust	Practicing longer rotations of organic productions, use of organic fertilizers, and low nitrogen content in the soil
Erysiphe graminis	Powdery mildew in cereals	Practicing longer rotations of organic productions, use of organic fertilizers, and low nitrogen content in the soil
Mycosphaerella graminicola	Leaf blotch in cereals	Practicing Longer rotations of organic productions, use of organic fertilizers, and low nitrogen content in the soil
Fusarium spp.	Head blight in cereals	Practicing longer rotations of organic productions, use of organic fertilizers, and low nitrogen content in the soil
Leptosphaeriano dorum	Leaf spots and glumes blotch in cereals	Practicing longer rotations of organic productions, use of organic fertilizers and low Nitrogen content in soil
Rhynchosporium secalis	Scald in cereals	Practicing Longer rotations of organic productions , use of organic fertilizers and low Nitrogen content in soil
Metopolophium dirhodum	Rose-grain/ rose-grass Aphid, vector of Barley Yellow Dwarf Virus	Practicing less tillage, lower Nitrogen content, and no synthetic pesticides in organic farms
Alternaria brassicae	Leaf blight in Brassicaceae member	Practicing mixed cropping, management of crop residue, and use of botanical pesticides in organic farms

TABLE 5.2 *(Continued)*

Plant pathogen	Plant disease	Management strategy in organic farms
Xanthomonas campestris	Black rot in Brassicaceae member	Practicing mixed cropping, management of crop residue, and use of botanical pesticides in organic farms
Pseudoperonospora cubensis	Downy mildew in *Cucurbit* spp.	Practicing mixed cropping, management of crop residue, and use of botanical pesticides in organic farms
Phytophthora capsici	Blight in *Cucurbit* spp.	Practicing mixed cropping, management of crop residue, and use of botanical pesticides in organic farms
Colletotrichum orbiculare	Anthracnose in *Cucurbit* spp.	Practicing mixed cropping, management of crop residue, and use of botanical pesticides in organic farms
Phytophthora infestans	Late blight of potato	Management of organic farms as per EU guidelines and use of copper fungicides
Septorialyco persici	Septoria leaf spot in tomato	Practicing mixed cropping, management of crop residue, and use of botanical pesticides in organic farms
Pseudomonas tomato	Bacterial speck in tomato	Production of winter cover crops and use of composted manures
Xanthomonas vesicatoria	Bacterial leaf spot in tomato	Production of winter cover crops and use of composted manures
Colletotrichum coccodes	Reduced anthrac-nose in tomato	Use of cannery waste compost
Leveillulataurica	Powdery mildew in tomato	Treatment of solarized soil in plastic green-houses (organic treatment); use of humic acid as fertilizer and use of organic pesticides, spray of sulfur and copper in organic farm fields
Oidium spp.	Powdery mildew in tomato	Treatment of solarized soil in plastic green-houses (organic treatment); use of humic acid as fertilizer and use of organic pesticides, spray of sulfur and copper in organic farm fields
Alternaria solani	Early blight in tomato	Treatment of solarized soil in plastic green-houses (organic treatment); use of humic acid as fertilizer and use of organic pesticides, spray of sulfur and copper in organic farm fields
P. infestans	Late blight in tomato	Treatment of solarized soil in plastic green-houses (organic treatment); use of humic acid as fertilizer and use of organic pesticides, spray of sulfur and copper in organic farm fields
Tomato ringspot virus	In tomato	Treatment of solarized soil in plastic green-houses (organic treatment); use of humic acid as fertilizer and use of organic pesticides, spray of sulfur and copper in organic farm fields

TABLE 5.2 *(Continued)*

Plant pathogen	Plant disease	Management strategy in organic farms
Tomato bushy stunt virus	In tomato	Treatment of solarized soil in plastic greenhouses (organic treatment); use of humic acid as fertilizer and use of organic pesticides, spray of sulfur and copper in organic farm fields
Tomato spotted wilt virus	In tomato	organic treatments of soil, use of extracts from plants, and use of copper fungicides
Solanum lycopersicum	Leaf spots in tomato	organic treatments of soil, use of extracts from plants, and use of copper fungicides
Xanthomonas vesicatoria	Leaf spots in tomato	organic treatments of soil, use of extracts from plants, and use of copper fungicides
A. solani	Early blight and fruit rot in tomato	organic treatments of soil, use of extracts from plants, and use of copper fungicides
Stigmina carpophila	Shot hole in almonds	Practicing mixed cropping, no use of fertilizers or synthetic pesticides
Venturia inaequalis	Apple scab in apple	Organic amendments in organic productions, growing cover crops, and using sulfur sprays
Cercospora coffeicola	Brown eyespot in coffee	organic production under shade, requirement of foliar nutrients, use of composted manure, and spray of inorganic copper and zinc
Botrytis cinerea	Fruit rot in strawberry	Mulch of straw, no use of fumigation or fungicides

5.3 PLANT DISEASE MANAGEMENT IN ORGANIC FARMING SYSTEM

Plant diseases require interaction between host and pathogen including environmental factors that offer favorable habitats for disease establishment. The management strategies for plant diseases aim to disturb this host-pathogen interaction in organic farms and mainly depend on the stages of the crop growth cycle.

5.3.1 DISEASE-RESISTANCE CROP VARIETIES

This method includes genetic modifications in plant seeds to breed disease-resistant plant varieties. A plant disease is established and spread by host-pathogen gene–gene interaction. The disease-resistant plants remain disease-free for a longer time and cause some useful morphological modification. To develop varieties of resistance to multiple diseases, care should be taken to lower the selection pressure on the pathogen. However, this strategy of developing disease-resistant crop varieties is very specific and can tackle only one or two diseases at a time.

5.3.2 EXCLUSION OF PATHOGEN

Use of disease-free seeds, fresh organic matter, sanitized planting material, and anaerobic soil disinfestations (ASD involves the use of a plastic tarp to cover the soil, applying carbon sources to the soil, and irrigation to field capacity) is a common technique to prevent host-pathogen interaction in plants which lowers the disease incidence significantly (van Bruggen et al., 2016; Khulbe, 2000).

5.3.3 ORGANIC AMENDMENTS

OF system requires healthy soil. Healthy soil means soil with high microbial diversity which improves the biological activity of soil by suppressing weeds, pests, and plant pathogens (IFOAM, 1998). This biological activity is improved by using some common techniques which include winter cover crops, and green and animal manures. For instance, neem cakes are used to control nematodes and other pathogens in soil.

5.3.4 CULTURAL CONTROL

Good agricultural practices are cultural controls which include selection of the right plantation date, weed management, reduction in load of inoculums, and favorable habitat for organism growth. One of the most important forms of cultural control is crop rotation, which minimizes the pathogen infectivity and prevents the harm from pests and weeds. Strategies used in cultural control methods to protect plants from pests are preventive rather than curative (Litterick et al., 2002). To maintain the population of beneficial organisms in the farm fields, reduced interaction between host and virulent pathogens is required. This can be achieved by variation in crop spp., maintaining the cropped and un-cropped areas and the temporal and spatial pattern of the crop rotations. A well-designed crop rotation technique is required for successful organic farming (Stockdale et al., 2000).

5.3.5 ORCHARD BIOINTENSIFICATION

This method provides an effective balance of pests and beneficial organisms. Biointensification includes the development of biologically active soils, the development of entomophagous parks to enhance the beneficial insects, weed strips, hedge rows, wind breaks, intercrops and conservation of insect biodiversity (Singh and Srinivas, 2016).

5.3.6 PHYSICAL METHODS

Soil solarization and hot water treatment are commonly used practices to reduce soil-borne inoculums and diseases; in many crops (Cohen et al., 2005), such as mango plants, hot water treatment is commonly used to reduce anthracnose incidence (Srinivas et al. 2012).

5.3.7 BIOCONTROL AGENTS

Microbial biocontrol agents take competitive and antagonistic approaches to prevent plant diseases. These agents work on the principle of antibiosis, by increasing nutrient uptake and by inducing host resistance (Kloepper et al. 1997). These biocontrol agents are natural, safe, eco-friendly, pathogen-specific, harmless, and have a negligible effect on other biotic systems (Srinivas and Ramakrishna, 2005). In organic farms, the use of biofungicides acts as a best way of pest management. These biofungicides are available commercially as powder (seed treatment), granulars (soil application), and suspension (root drenches and foliar spray) and is composed of beneficial living organisms. Species of *Trichoderma*, *Pseudomonas*, and *Bacillus* are some common examples of biological control agents.

5.3.8 MINERAL BASED FUNGICIDES

Copper based fungicides and Bordeaux mixture strongly adheres to the plant parts and resists the heavy rains. These mixtures are usually used on ornamental plants, vegetables, and fruits. These mixtures exhibit both antibacterial and antifungal properties. Prophylactic sprays especially sulfur-based sprays are used to prevent spore germination and used against mildew diseases.

5.3.9 OTHER COMMON TRADITIONAL METHODS

Neem oil spray, cow urine, panchgavya, buttermilk spray, cow dung ferments, baking soda, soft soap solutions, and horticulture-grade oils are some most predominant methods utilized for the control of pests and diseases in organic farms of India. Baking soda controls plant diseases such as mildew and rust in all weathers except during extreme summer seasons. Buttermilk sprays are commonly used for inhibiting blight diseases in plants, mildew and rust disease, plant disease caused due to mosaic viruses and other fungal as well as viral plant diseases. Soft or light soap solutions and neem-based oils are cost-effective for use against viral vectors like aphids. Ferments from cow dung (amrit paani) have high cellulose-degrading bacterial population, a high range of nitrogen fixers, elevated phosphorous solubilizers, higher plant growth promoters, and antifungals (Venkateswarlu et al., 2008) and it helps in promoting crop production and disease management.

5.4 RECOMMENDATIONS

Many field surveys, studies, experiments, and bioassays have compared the incidence of plant pathogens and associated disease severity in both the farming system, that is, organic and conventional. From all the previous studies it can be easily concluded that root diseases in plants are less common in organic fields while foliar diseases are either less or more severe in organic fields as compared to conventional ones. Several root diseases, rust diseases, powdery mildew diseases, or aphids are promoted by high content of nitrogen in plant tissues which is more problematic in conventional farming while viral diseases are more common in OF. Organic farms residing in humid areas do not have effective control measures against multiple-cycle foliar pathogens. Further quantitative research is required to determine the major factors responsible for crop production loss in organic farms.

KEYWORDS

- **farming**
- **plant diseases**
- **control methods**
- **management**

REFERENCES

Abbasi, P. A.; Al-Dahmani, J.; Sahin, F.; Hoitink, H. A. J.; Miller, S. A. Effect of Compost Amendments on Disease Severity and Yield of Tomato in Conventional and Organic Production Systems. *Plant Dis.* **2002,** *86*, 156–161.

Aparna, K.; Pasha, M. A.; Rao, D. L. N.; Krishnaraj, P. U. Organic Amendments as Ecosystem Engineers: Microbial, Biochemical and Genomic Evidence of Soil Health Improvement in A Tropical Arid Zone Field Site. *Ecol. Eng.* **2014,** *71*, 268–277.

Berkelmans, R.; Ferris, H.; Tenuta M.; van Bruggen, A. H. C. Effects of Long-Term Crop Management on Nematode Trophic Levels Other Than Plant Feeders Disappear After One Year of Disruptive Soil Management. *Appl. Soil Ecol.* **2003,** *23*, 223–235.

Bernard, E.; Larkin, R. P.; Tavantzis, S.; Erich, M. S.; Alyokhin, A.; Gross, S. Rapeseed Rotation, Compost, and Biocontrol Amendments Reduce Soilborne Diseases and Increase Tuber Yield in Organic and Conventional Potato Production Systems. *Plant Soil* **2014,** *374*, 611–627

Bettiol, W.; Ghini, R.; Galv˜ao, J. A. H.; Siloto, R. C. Organic and Conventional Tomato Cropping Systems. *Sci. Agricola.* **2004,** *61*, 253–259.

Birkhofer, K.; Bezemer, T. M.; Bloem, J.; Bonkowski, M.; Christenen, S.; Dubois, D. et al. Long-Term Organic Farming Fosters Below and Aboveground Biota: Implications for Soil Quality, Biological Control and Productivity. *Soil Biol. Biochem.* **2008,** *40*, 2297–2308.

Bulluck, L. R.; Ristaino, J. B. Effect of Synthetic and Organic Soil Fertility Amendments on Southern Blight, Soil Microbial Communities, and Yield of Processing Tomatoes. *Phytopathology* **2002,** *92*, 181–189.

Cohen, M. R.; Yamasaki, H.; Mazzola, M. *Brassica napus* Seed Meal Soil Amendment Modifies Microbial Community Structure, Nitric Oxide Production and Incidence of Rhizoctonia Root Rot. *Soil BiolBiochem.* **2005,** *37*, 1215–1227.

Darby, H. M.; Stone, A. G.; Dick, R. P. Compost and Manure Mediated Impacts on Soilborne Pathogens and Soil Quality. *Soil Sci. Soc. Am. J.* **2006,** *70*, 347–358.

Datnoff, L. E.; Elmer, W. H.; Huber, D. M., Eds. *Mineral Nutrition and Plant Disease*; APS Press: St Paul, 2007; pp 278.

Finckh, M. R.; van Bruggen, A. H. C. General Introduction. In: *Plant Diseases and Their Management in Organic Agriculture*; Finckh, M. R., van Bruggen, A. H. C., Tamm, L., Eds.; APS Press: St Paul, 2015; pp 3–11.

Finckh, M. R.; Wolfe, M. S. Biodiversity Enhancement. In: *Plant Diseases and Their Management in Organic Agriculture*; Finckh, M. R.; van Bruggen, A. H. C., Tamm, L., Eds.; APS Press: St Paul, 2015; pp 153–174.

Finckh, M. R.; van Bruggen, A. H. C.; Tamm, L., Eds. *Plant Diseases and Their Management in Organic Agriculture*; APS Press: St. Paul, 2015.

Gomiero, T.; Pimentel, D.; Paoletti, M. G.; Environmental Impact of Different Agricultural Management Practices: Conventional vs. Organic Agriculture. *Crit. Rev. Plant Sci.* **2011,** *30*, 95–124.

Gosme, M.; de Villemandy, M.; Bazot, M.; Jeuffroy M.-H. Local and Neighbourhood Effects of Organic and Conventional Wheat Management on Aphids, Weeds, and Foliar Diseases. *Agric. Ecosyst. Environ.* **2012,** *161*, 121–29.

Grunwald, N. J.; Hu, S.; van Bruggen, A. H. C. Short-Term Cover Crop Decomposition in Organic and Conventional Soils: Characterization of Soil CN, Microbial and Plant Pathogen Dynamics. *Eur. J. Plant Pathol.* **2000,** *106*, 37–50.

He, M.; Ma, W.; Tian, G.; Blok, W.; Khodzaeva, A., et al. Daily Changes of Infections by *Pythium ultimum* After a Nutrient Impulse in Organic Versus Conventional Soils. *Phytopathology* **2010,** *100*, 593–600.

He, M.; Tian, G.; Semenov, A. M.; van Bruggen, A. H. C. Short-Term Fluctuations of Sugar-Beet Damping Off by *Pythium ultimum* in Relation to Changes in Bacterial Communities After Organic Amendments to Two Soils. *Phytopathology* **2012,** *102*, 413–420.

Hiddink, G. A.; Termorshuizen, A. J.; van Bruggen, A. H. C. Mixed Cropping and Suppression of Soilborne Diseases, a Review. In: *Genetic Engineering, Biofertilisation, Soil Quality and Organic Farming*; Lichtfouse, E., Ed.; Springer: Dordrecht, pp 119–146

Hiddink, G. A.; van Bruggen, A. H. C.; Termorshuizen, A. J.; Raaijmakers, J. M.; Semenov, A. V. Effect of Organic Management of Soils on Suppressiveness to *Gaeumannomyces graminis* var. Tritici and Its Antagonist, Pseudomonas Fluorescens. *Eur. J. Plant Pathol.* **2005,** *113*, 417–135.

IFOAM. In "IFOAM Basic Standards for Organic Production and Processing"; IFOAM Publications, Tholey-Tholey, 1998.

Khulbe, D. *Soil Solarization-Effect on Soil Nutrients, Microorganisms and Plant Growth Response*, PhD thesis; GBPUAT, Pantnagar, 2000.

Kloepper, J. W.; Tuzun, S.; Zehnder, G. W.; Wei, G. Multiple Disease Protection by Rhizobacteria That Induce Systemic Resistance—Historic Precedence. *Phytopathology* **1997,** *87* (2), 136–137.

Letourneau, D.; van Bruggen, A. H. C. Crop protection. In: *Organic Agriculture: A Global Perspective*; Kristiansen, P., Taji, A., Reganold, J., Eds.; CSIRO Publ: Clayton, 2006; pp 93–121.

Litterick, A. M.; Watson, C. A.; Atkinson, D. Crop Protection in Organic Agriculture—A Simple Matter? In *Proceedings of the UK Organic Research 2002 Conference*; Powell, J. et al., Eds.; Organic Centre Wales, Institute of Rural Studies, University of Wales Aberystwyth, 2002; , pp 203–206.

Liu, B.; Tu, C.; Hu, S.; Gumpertz, M, Beagle Ristaino, J. Effects of Organic, Sustainable, and Conventional Management Strategies in Grower Fields on Soil Physical, Chemical, and Biological Factors and the Incidence of Southern Blight. *Appl. Soil Ecol.* **2007,** *37*, 202–214.

Manici, L. M.; Ciavatta, C.; Keldere, M.; Erschbaumer, G. Replant Problems in South Tyrol: Role of Fungal Pathogens and Microbial Population in Conventional and Organic Apple Orchards. *Plant Soil* **2003,** *256*, 315–324.

Mansour, A.; Al-Banna, L.; Salem, N.; Alsmairat, N. Disease Management of Organic Tomato Under Greenhouse Conditions in the Jordan Valley. *Crop Prot.* **2014,** *60*, 48–55.

Marzano, S. Y. L.; Villamil, M. B.; Wander, M. W.; Ugarte, C. M.; Wen, L.; Eastburn, D. M. Organic Transition Effects on Soilborne Diseases of Soybean and Populations of Pseudomonadaceae. *Agron. J.* **2015,** *107*, 1087–1097.

Mazzola, M.; Andrews, P. K.; Reganold, J. P.; Levesque, C. A. Frequency, Virulence, and Metalaxyl Sensitivity of Pythium spp. Isolated from Apple Roots Under Conventional and Organic Production Systems. *Plant Dis.* **2002,** *86*, 669–675.

Messiha, N. A. S.; van Bruggen, A. H. C.; Franz, E.; Janse, J. D.; Schoeman-Weerdesteijn, M. E. et al. Effects of Soil Type, Management Type and Soil Amendments on the Survival of the Potato Brown Rot Bacterium *Ralstonia solanacearum*. *Appl. Soil Ecol.* **2009,** *43*, 206–215.

Messiha, N. A. S.; van Bruggen, A. H. C.; van Diepeningen, A. D.; de Vos, O. J.; Termorshuizen, A. J. et al. Potato Brown Rot Incidence and Severity Under Different Management and Amendment Regimes in Different Soil Types. *Eur. J. Plant Pathol.* **2007,** *119*, 367–381.

Njoroge, S. M. C.; Kabir, Z.; Martin, F. N.; Koike, S. T.; Subbarao, K. V. Comparison of Crop Rotation for Verticillium Wilt Management and Effect on Pythium Species in Conventional and Organic Strawberry Production. *Plant Dis.* **2009,** *93*, 519–527.

Pankhurst, C. E.; McDonald, H. J.; Hawke, B. G.; Kirkby, C. A. Effect of Tillage and Stubble Management on Chemical and Microbiological Properties and the Development of Suppression Towards Cereal Root Disease in Soils from Two Sites in NSW, Australia. *Soil Biol. Biochem.* **2002,** *34*, 833–834.

Schillinger, W. F.; Paulitz, T. C. Natural Suppression of *Rhizoctoniab* are Patch in a Long-Term No-Till Cropping Systems Experiment. *Plant Dis.* **2013,** *98*, 389–394.

Senechkin, I. V.; van Overbeek, L.; van Bruggen, A. H. C. Greater Fusarium Wilt Suppression After Complex Than After Simple Organic Amendments as Affected by Soil pH, Total Carbon and Ammonia-Oxidizing Bacteria. *Appl. Soil Ecol.* **2014,** *73*, 148–155.

Shrestha, G.; Prajapati, S.; Mahato, B. N. Plant Diseases and Their Management Practices in Commercial Organic and Conventional Vegetable Farms in Kathmandu Valley. *Nep. J. Agric. Sci.* **2014,** *12*, 129–141.

Singh, H.S.; Srinivas, P. Concept of Bio-Intensification of Orchard for Pest Management (BIOPM) in Mango. *Int. J. Innov. Hortic.* **2015,** *5* (2), 81–90.

Souza, A. G. C.; Maffia, L. A.; Silva, E. F.; Mizubuti, E. S. G.; Teizeira, H. A Time Series Analysis of Brown Eye Spot Progress in Conventional and Organic Coffee Production Systems. *Plant Pathol.* **2015,** *64*, 157–166.

Srinivas, P., Singh, H. S.; Mandal, S. Incidence of Postharvest Diseases in Different Mango Varieties in Coastal Odisha. Oral Presentation in National Symposium Plant Pathology in Genomic Era organized by Indian Phytopathological Society and Department of Plant Pathology, Indira Gandhi Krishi Vishwavidyalaya, Raipur, Chhattisgarh during May 26–28, 2014; 2012.

Srinivas, P.; Ramakrishna, G. Biological Management of Rice Seed Borne Pathogens by Native Biocontrol Agents. *Ann. Plant Protect. Sci.* **2005,** *13* (2), 422–426.

Stockdale, E. et al. Agronomic and Environmental Implications of Organic Farming Systems. *Adv. Agron.* **2000,** *70*, 261–327.

Tuck, S. L.; Winqvist, C.; Mota, F.; Ahnström, J.; Turnbull, L. A.; Bengtsson, J. Land-Use Intensity and the Effects of Organic Farming on Biodiversity: A Hierarchical Meta-Analysis. *J. Appl. Ecol.* **2014,** *51*, 746–755.

van Bruggen, A. H. C.; Finckh, M. R. General Principles of Organic Plant Production. In *Plant Diseases and Their Management in Organic Agriculture*; Finckh, M. R., van Bruggen, A. H. C., Tamm, L., Eds.; APS Press: St Paul, 2015; pp 15–23.

van Bruggen, A. H. C.; Finckh, M. R. Plant Diseases and Management Approaches in Organic Farming Systems. *Annu. Rev. Phytopathol.* 2016; *54*, 25–54.

van Bruggen, A. H. C.; Semenov, A. M. Soil Health and Soilborne Diseases in Organic Agriculture, in Plant Diseases and Their Management in Organic Agriculture; Finckh, M. R., van Bruggen, A. H. C.; Tamm, L., Eds.; APS Press: St Paul, 2015; pp 67–89.

van Bruggen, A. H. C.; Termorshuizen, A. J. Integrated Approaches to Root Disease Management in Organic Farming Systems. *Australas. Plant Pathol.* 2003; *32*, 141–156.

van Bruggen, A. H. C.; Francis, I.; Krag, R. The Vicious Cycle of Lettuce Corky Root Disease: Effects of Farming System, Nitrogen Fertilizer and Herbicide. *Plant Soil* **2015,** *388*, 119–132.

van Bruggen, A. H. C.; Sharma, K.; Kaku, E.; Karfopoulos, S.; Zelenev, V. V.; Blok, W. J. Soil Health Indicators and Fusarium Wilt Suppression in Organically Managed Greenhouse Soils. *Appl. Soil Ecol.* 2015; *86*, 192–201.

vanBruggen, A. H. C. Plant Disease Severity in High-Input Compared to Reduced-Input and Organic Farming Systems. *Plant Dis.* **1995,** *79*, 976–984.

Venkateswarlu, B., Balloli, S. S.; Ramakrishna, Y.S. *Organic Farming in Rainfed Agriculture: Opportunities and Constraints*; Central Research Institute for Dryland Agriculture: Hyderabad, 2008; p 185.

Willer, H.; Lernoud, J., Eds. The World of Organic Agriculture. Statistics and Emerging Trends FiBL-IFOAM Report. Research Institute of Organic Agriculture (FiBL), Frick, Sitzerland/IFOAM, Organics International, Germany, Bonn, 2015.

Yang, C. H.; Crowley, D. E.; Menge, J. A. 16S rDNA Fingerprinting of Rhizosphere Bacterial Communities Associated with Healthy and Phytophthora Infected Avocado Roots. *FEMS Microbiol. Ecol.* **2001,** *35*, 129–136.

Yogev, A.; Laor, Y.; Katan, J.; Hadar, Y.; Cohen, R. et al. Does Organic Farming Increase Soil Suppression Against Fusarium Wilt of Melon? *Org. Agric.* **2011,** *1*, 203–216.

CHAPTER 6

Sclerotium rolfsii: Pathogen Biology, Symptomatology, and Host Range of the Devastating Soil-borne Pathogen

MALAVIKA RAM AMANTHRA KELOTH[1], SEWETA SRIVASTAVA[1], RAVINDRA KUMAR[2], POONAM PANDURANG SHETE[1], RAGHAVENDRA REDDY MANDA[3], AKASH JANGID[1], and MEENAKSHI RANA[1]

[1]*School of Agriculture, Lovely Professional University, Phagwara, Punjab, India*

[2]*ICAR-Indian Agricultural Research Institute, Regional Station, Karnal, Haryana, India*

[3]*Wageningen University and Research, Wageningen, The Netherlands*

ABSTRACT

Sclerotium rolfsii Sacc. [*Athelia rolfsii* (Curzi) Tu Kimbrough] is a soil-borne fungal phytopathogen having a wide host range. Though belonging to Basidiomycotina order, this fungus produces sterile mycelium without asexual spores. Its peculiarity in producing hard resting structures called sclerotia has made its management difficult. Various management practices are carried out which include cultural, physical, chemical, biological control, etc. Few botanicals have also shown promising result. Biological control of the pathogen has caught attention due to the strong inhibition of the pathogen by *Trichoderma* spp. and other bacterial bioagents like *Pseudomonas*, *Bacillus* etc. Endophytic actinomycetes have also been reported for significant control of this pathogen. It can be managed by the integration of all available control measures. Therefore, keeping in view the above facts this

Plant Diseases and Their Management: A Sustainable Approach.
Deepa Srivastava, Rajarshi Kumar Gaur, and Ajay K. Tiwari (Eds.)

chapter was designed and written on the integration of different management practices for the control of this devastating pathogen.

6.1 INTRODUCTION

Sclerotium rolfsii Sacc. is a fungal phytopathogen of global significance due to its wide host range. This destructive and widespread fungus produces sterile mycelia and sclerotia as well as oxalic acid and several cell-wall degrading enzymes (Punja, 1985). Since 1892, when the pathogen was first reported by Rolfs in tomato, several research works are being conducted worldwide (Aycock, 1966). Saccardo in 1913 described the genus *Sclerotium* with its characteristic features (Punja and Damiani, 1996). It was first reported in India from the Mysore and Mandya districts of Karnataka during 2012–2014, causing southern blight and leaf spot in common bean (*Phaseolus vulgaris*) (Mahadev Kumar et al., 2015). Tropic and subtropic conditions are favorable for its growth and development where temperature is high like southern and south eastern parts of US (Punja, 1985). Warm regions in the world having high temperature are optimal for the mycelial growth and sclerotial production of the fungi, wherein it is maximum at 27–30°C and dry weight production and hyphal extension is favored at 8–40°C (Mathur and Sarbhoy, 1977; Sharma and Kushal, 1979; Zoberi, 1980). The shape and size of the sclerotia may vary according to the variation in temperature (Aycock, 1966). Different isolates showed varied linear growth rates of hyphae, ranging from 0.85 to 0.97 mm h^{-1} at 27°C in agar media (Kliejunas and Ko 1975; Punja et al., 1985). The most suitable media for the colony growth of *S. rolfsii* was V8 agar medium (Punja and Damiani, 1996). Moreover, in sterile soil also they showed a similar growth rate at the optimal temperature (Mustafa and Chattopadhyay, 1971). With increasing moisture content mycelial growth retarded progressively (Mustafa and Chaltopadhyay, 1971a), and disease incidence increased in well-drained sandy soils (Weerapat and Schroeder, 1966) and in soil with moisture content below the saturation level (Ramarao and Raja, 1980).

Contrarily, soils containing high silt content which can retain moisture are also found preferred for higher disease incidence (Shew et al., 1984). In cultures subjected to continuous light (especially blue light), *S. rolfsii* exhibited higher growth and sclerotia formation than that in darkness (Humpherson-Jones and Cooke, 1977; Miller and Liberta, 1977; Trevethick and Cooke, 1973). Similarly, more sclerotial production was noticed at the soil surface than deep in the soil (Punja, 1985). *Sclerotium rolfsii* when grew on PDA subjected to 35°C incubation produced 1043 sclerotia in number with an average of 0.6 mm diameter (Punja and Damiani, 1996). Fresh culture about 1–7 days old, which were grown in dark, produced more sclerotia when subjected to light than 8–14 days old cultures (Humpherson-Jones and Cooke, 1977; Miller and Liberta, 1977). *S. rolfsii* has been regarded as a highly aerobic pathogen, but its growth rate on exposure to CO_2 was not much bad as expected. The linear growth rate of the hyphae when exposed to 11.5% CO_2 (Griffin and Nair, 1968) and the mycelial dry weight of the pathogen when exposed to 15% CO_2 (Mitchell and Mitchell, 1973) resulted in only 50% reduction. A constant mycelial growth rate was observed when the fungi was exposed to 3–21% O_2 concentration, whereas the growth decreased when they were exposed to CO_2 concentration above 0.03%. When they were subjected to O_2 concentration below 6% as well as CO_2 concentration above 10%, the rate of sclerotial germination got decreased and no sclerotia formation was reported at O_2 concentration below 15% and at CO_2 concentration above 4% (Griffin and Nair, 1968). Low O_2 concentration showed a reduced linear extension of mycelia in nonsterile soil than in the agar media (Punja and Jenkins, 1984). Agar media maintained at pH ranging from 2 to 5 gave maximum sclerotial germination. Inorganic salts were also tested for their inhibiting ability of sclerotia, out of which K^+, Na^+, and Ca^{2+} showed less inhibition than NH_4^+ and Li^+. At higher pH (8.6 approx.) the H ion concentration of the agar media increased the toxicity of the salts to sclerotia, while at low (6 approx.) pH there was no effect. Moreover, due to the presence of free CO_3^{2-}, HCO_3^-, and NH_3 in the salts of carbonate, bicarbonate, and ammonium respectively at high pH, these salts were found having fungicidal effect against sclerotia (Punja and Grogan, 1982).

Remoistening of dried plant tissues which were undecomposed, resulted in higher rate of growth of the mycelium in soil due to the release of several volatile compounds (Beute and Rodriguez-Kabana, 1979a; Beute and Rodriguez-Kabana, 1979b; Linderman and Gilbert, 1969; Linderman and Gilbert, 1973; Punja and Grogan, 1981a). Volatiles caused increase in the amount of aerial mycelium, whereas they delayed the formation of sclerotial initials (Punja et al., 1984; Wheeler, 1972). No improvement in dry weight of the mycelia was reported due to these compounds (Punja et al., 1984). Also, their presence has resulted in directional growth of the fungal mycelium (Linderman and Gilbert, 1969; Punja and Grogan, 1981b). It has been found that a wide range of carbon sources have aided the growth of *Sclerotium rolfsii* (Mathur and Sarbhoy, 1976; Maxwell and Bateman, 1968; Sharma and Kaushal, 1979; Zoberi, 1980) and many organic and inorganic nitrogen compounds have also supported its growth (Mathur, 1977; Narasimhan, 1969; Sharma and Kaushal, 1979; Zoberi, 1980). Media that provided extensive growth of the pathogen yielded large quantity of sclerotia and also the carbon (Bahadur et al., 1976; Mathur and Sarbhoy, 1976, Sharma and Kaushal, 1979; Zoberi, 1980) and nitrogen (Mathur, 1977) sources along with optimum pH (Sharma and Kaushal, 1979) have manipulated the number and size of sclerotia. The formation of sclerotia was also facilitated by the external application of potassium (Wheeler and Sharan, 1965; Zoberi, 1980). The submerged cultures of the fungus have led to the production of an extracellular polysaccharide, β-1,3 glucan (Okon et al., 1974; Kritzman et al., 1979). Besides, the variation in the presence of soluble proteins and enzymes amid the formation of sclerotia has led to nucleic acid and protein production, a build-up in cyclic adenosine monophosphate (cAMP) levels and a rise in translocation of compounds (Chet et al., 1972; Hadar et al., 1983; Igwegbe et al., 1977; Okon et al., 1973).

Inhibition of sclerotial production has been noted at initial, developmental, and maturation stages (Henis and Chet, 1968; Willetts, 1972) due to less concentration of various unrelated chemicals (Chet et al., 1966; Chet and Henis, 1968; Bozarth and Tweedy, 1971; Melhuish and Bean, 1971; Trevethick and Cooke, 1971; Okon et al., 1972; Wheeler, 1972; Ercegovich et al., 1973; Okon et al., 1973; Christias, 1975; Tourneau, 1976; Igwegbe et al., 1977; Tourneau, 1978; Fellman et al., 1983). There are many other factors that caused modification in the pattern and rate of growth of the sclerotia like, partially inhibiting hyphal extension (Kritzman et al., 1977) and increasing hyphal branching (Henis et al., 1973; Kritzman et al., 1977), adding staling compounds (Humpherson-Jones and Cooke, 1977), threonine (Kritzman et al., 1976; Kritzman et al., 1977; Liu and Wu 1971), lactose (Okon et al., 1972; Okon et al., 1975), iodoacetic acid (Chet et al., 1966; Henis and Chet, 1968), or the sodium salt of ethylenediaminetetracetic acid (Chet and Henis, 1968; Henis and Chet, 1968) to the medium, tearing or cutting the mycelium (Henis, 1965; Hadar et al., 1981a) or imposing restrictive growth conditions (Henis, 1965; Wheeler and Waller, 1965) or nutrient deprivation (Christias and Lockwood, 1973; Hadar et al., 1983). A reverse effect was shown by glucose, cysteine, or cupperic ions when lactose, iodoacetic acid, or sodium salt of ethylenediaminetetracetic acid were replaced respectively (Chet and Henis, 1968; Chet et al., 1966; Okon et al., 1975). As such, there are many other factors responsible for the formation, development, or inhibition of sclerotia of *S. rolfsii* which indicates the necessity to accept only those factors particular to the isolate obtained. Rising moisture content was not in favor with the growth of the fungi; instead, it showed progressive decrease (Mustafa and Chaltopadhyay, 1971a).

Depending on the different concentration gradient, age of culture, light interaction, direction and distance of translocation of glucose (Okon et al., 1973) and ^{32}P (Wilcoxson and Subbarayudu, 1968) through the hyphae of *S. rolfsii*, the number of substances accumulated in sclerotia varied (Wilcoxson and Subbarayudu, 1968). The efficiency and distance of translocation can be improved through the formation of cords by coalescing and clustering of hyphae in soil, which is also found in dual culture with *Trichoderma harzianum*, having unknown significance (Hadar et al., 1981).

6.2 STRUCTURE AND COMPOSITION OF SCLEROTIA

The size of sclerotia of *Sclerotium rolfsii* ranges from 0.5 to 2 mm in diameter (Punja, 1985). The formation of sclerotia was observed laterally from the main hyphal strands (Willetts, 1972) and their growth pattern was examined through fluorescent brighteners (Edelstein et al., 1983; Kritzman et al., 1978). The sclerotial surface was witnessed with drops of exudates comprising carbohydrates, oxalic acids, cations, proteins, enzymes, and amino acids (Colotelo, 1978; Christias, 1980) after the formation of the sclerotia and during their development. This exudate may be formed due to the excess amount of nutrients, water, and other materials that are translocated in order to maintain improved metabolic activity in the developing sclerotia (Punja, 1985). Further, it was also observed that the sclerotia when exposed to ambient air in culture (Punja and Grogan, 1981a) and those formed in soil were devoid of the exudate droplet (Punja, 1985). Two to four cell-layered outer melanized rind (Chet and Henis, 1968a; Chet and Henis, 1969; Chet et al., 1967; Liu et al., 1977), an underlying cortical layer, and an innermost medullary region comprised of loosely interwoven hyphae (Chet et al., 1969; Nair et al., 1969; Chet, 1975; Tu and Kimbrough, 1975; Punja et al., 1984) contribute to the composition of sclerotia. Around the rind there exists a cuticle or skin (Chet et al., 1969; Tu and Kimbrough, 1975; Willetts, 1971) that could be the remnants of membranous-like material enclosing the exudation droplets (Colotelo, 1978; Christias, 1980) or a viscous film formed by the polymerization of compounds in the exudates (Punja and Grogan, 1981a). This outer layer was recognized to be nonessential and nonsignificant in terms of biological aspects (Punja and Grogan,

1981a). The layers comprise of chitin, laminarin, and β-1,3 glucan (Chet and Henis, 1968; Jones et al., 1972) with glucans, sugars (Chet et al., 1967; Mathur and Sarbhoy, 1977), and lipids (Chet et al., 1967; Howell and Fergus, 1964; Tu and Kimbrough, 1975) as their foremost storage reserves along with amino acids (Chet et al., 1967; Christias, 1980; Mathur and Sarbhoy, 1977) and fatty acids. The rind thickness and relative proportions of cortex and medulla vary according to the substrate where sclerotia formation occurs (Punja et al., 1984). The thickness of the cortex ranged from 4–10 layers. The medullary region comprised highest concentration of carbohydrates and metachromatic granules, followed by cortical region and predominantly the cytoplasm on the basis of description of histochemical staining of sclerotia (Punja and Damiani, 1996). The fungi developed clamp connections on the leading hyphae and possessed 1–6 nuclei in the hyphal cells with an average of 2 per cell. Also, PDA with charcoal showed the presence of basidial stage of the fungi (Punja and Damiani, 1996).

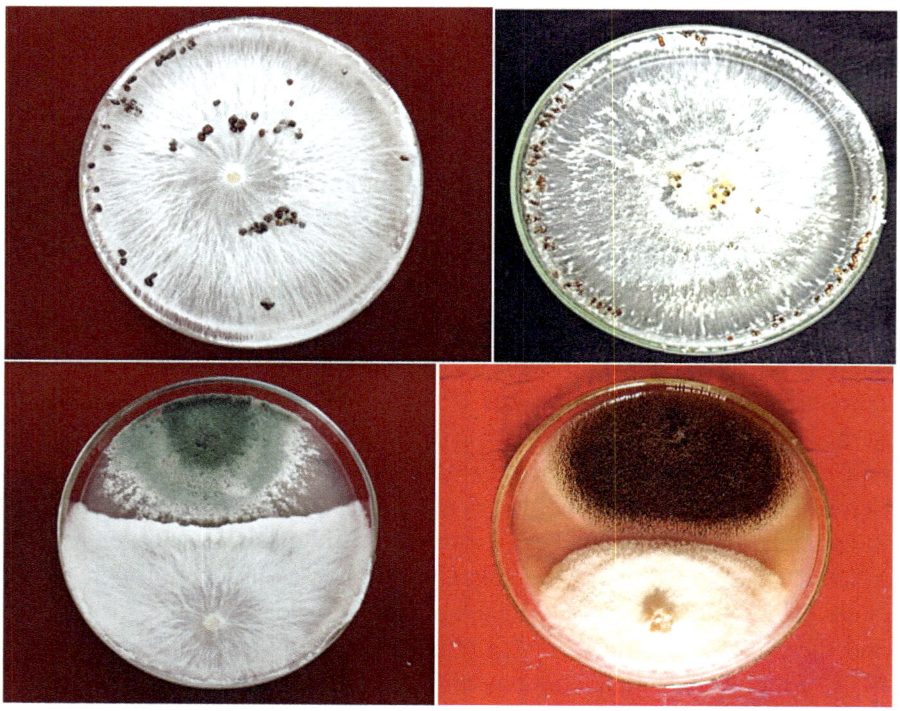

FIGURE 6.1 Pure cultures of *Sclerotium rolfsii* (a and b), and biocontrol of *Sclerotium rolfsii* with *Trichoderma harzianum* (c), and *Aspergillus niger* (d).

Source: Reprinted with permission from Kumar (2010).

6.3 GERMINATION OF SCLEROTIA

Germination in *S. rolfsii* was observed to be of two forms: hyphal and eruptive (Punja and Grogan, 1981a). In hyphal germination, the hyphae originate from the cells of medulla (Chet et al., 1977) and grow as individual strands from the sclerotium surface. While in eruptive germination, a plug(s) or aggregates of mycelium burst out through the sclerotial rind and this phenomenon takes place only once as the internally stored materials are being utilized during the development of the mycelium. Thus, it was only in the case of hyphal germination that the sclerotia was capable of germinating more than once (Smith, 1972; Javed and Coley-Smith, 1973). The widespread growth of the fungal mycelium through hyphal germination takes place only when there is external availability of nutrients (Punja and Grogan, 1981a). Many cases reported the formation of secondary sclerotia, but generally in eruptive germination rind is only that remains after the eruption (Smith, 1972). The eruptive germination of the sclerotia could be triggered off through sclerotial drying (Smith, 1972), their exposure to volatile compounds (primarily alcohols and aldehydes) (Linderman and Gilbert, 1969; Linderman and Gilbert, 1973; Beute and Rodriguez-Kabana, 1979a; Beute and Rodriguez-Kabana, 1979b), or through treatment with sodium hypochlorite (Linderman and Gilbert, 1973; Punja and Grogan, 1981a). A brief dormancy period was observed in mature sclerotia and that tends them to germinate promptly. Familiarity regarding the discrimination in sclerotial germination is thus very significant as it helps in knowing the amount of mycelial growth and the energy available for infection, describing the extent of disease incidence and its further progress in the hosts (Punja and Grogan, 1981a; Punja and Grogan, 1981b).

6.4 FACTORS AFFECTING SCLEROTIAL GERMINATION

The eruptive sclerotial germination is affected by several factors in which few of them stimulate and few inhibit the germination (Lane et al., 2019). Alcohols and acetaldehydes with concentrations ranging from 1.6 to 6.1 μL per Petri dish stimulated the sclerotial germination eruptively on oat kernels. The compounds like isopropyl and butyl alcohol stimulated germination at a higher rate with increased linear growth rate of the mycelium than other volatiles evolved from dried and remoistened alfalfa leaf tissues, with reduced effect on the dry weight. Also, it was noted that eruptive germination was inhibited on the addition of carbon source like 20 mM sucrose or glucose into Noble water agar, whereas it stimulated the hyphal germination. Supplementing glucose has caused reduction in the leakage of amino compounds and carbohydrates and also brought down the dry weight loss from the sclerotia. Moreover, change in the behavior and composition of the sclerotial rind and cortical layer were detected in accordance with the change in the culture media such as PDA, oat, or soil and different treatments they are exposed to, like alcohols and NaClO. Soil-borne sclerotia experienced low and variable germination due to the manifestation of contaminating microorganisms on the surface of sclerotia (Punja et al., 1984).

6.5 BEHAVIORAL VARIATION IN DIFFERENT ISOLATES

Variation in the behavior of germination was also examined on the basis of different isolates that are obtained. Two isolates of *S. rolfsii*, SR_4 and SR_7, isolated from tomato in Ceylon and groundnut in Gold Coast gave distinguishing characteristics on the basis of morphology, physiology, and growth behaviors studied. SR_4 gave numerous amounts of spherical sclerotia while SR_7 produced irregularly shaped sclerotia few in number (Abeygunawardena and Wood, 1957). Luxuriant mycelial growth was observed in carrot extract and Weindling's liquid media while poor growth in maize meal extract and Brown's liquid media. Linear growth of the mycelium was rapid in agar media, at 30°C and optimum temperature for sclerotial production was 10°C for SR_7 isolate (Abeygunawardena and Wood, 1957). Adequate growth of the two isolates in liquid media brought changes in the initial pH values. Initial pH values ranging from 2.1 to 4.2 gave best growth results for both isolates (Abeygunawardena and Wood, 1957). The time taken for the production of sclerotia in PDA was compared among isolates from *Physalis*, *Parthenium*, and groundnut, in which sclerotial production occurred at 120, 84, and 72 h, respectively, after inoculation into media. The average size of sclerotia of *Physalis* isolates (159.48 µm) was observed to be greater than that of *Parthenium* isolates (138.97 µm) and lesser than that of groundnut isolates (286.03 µm). *Parthenium* isolate in PDA media showed much rapid growth rate than the other two isolates. On confirmation regarding cross infectivity, *Physalis* isolates showed varied degrees of noncompatibility with the other two isolates (Nandi et al., 2017).

Thin films of distilled water gave better environment for sclerotial germination. Aeration of soil or water facilitated the germination of all sclerotia, while sclerotia present in or on soil and those submerged in water gave poor germination results (Abeygunawardena and Wood, 1957).

6.6 INTEGRATED MANAGEMENT PRACTICES

Although the disease can be effectively controlled by fungicides, we need to limit their large-scale use because of phytotoxic effect and fungicidal residues causing human health hazards and environmental contamination (Ons et al., 2020). Moreover, by integrated management, very low quantities of different materials are required compared to chemical treatment alone. Keeping in view the current requirement and trend of sustainable development in agriculture, this section of the present chapter was undertaken for the management of diseases caused by *Sclerotium rolfsii* by combining cultural, physical practices along with organic matters, biocontrol agents, and fungicides.

Supplementing the soil with organic amendments or residues symbolizes the eco-friendly management of soil-borne pathogens by safeguarding the soil micro-biota and ecosystem. But inappropriate dosage and untimely application may hinder management practices to avoid favorable conditions for the growth of the pathogen. Many other cultural practices like soil solarization and chemical applications are also used. Biological control of the pathogen has caught attention due to the strong inhibition of the pathogen by *Trichoderma* spp. (Sood et al., 2020) and other bacterial bioagents like *Pseudomonas*, *Bacillus* etc. (Fira et al., 2018; Biessy et al., 2021). Endophytic actinomycetes have also been reported for causing significant control of this pathogen (Pavithra et al., 2020).

6.7 CROP ROTATION

Crop rotation is a conventional and preferred technique of disease control; however, it is ineffective in reducing *S. rolfsii* due to its wide host range and the capacity of sclerotia to survive in the soil. Rotating with nonsusceptible crops like grasses and grains like jowar, bajra, corn, or wheat, on the other hand, may help reduce disease incidence in subsequent years by lowering initial inoculum (Mullen, 2001; Leona et al., 2020).

6.8 SOIL SOLARIZATION

Soil solarization is a nonchemical disease management approach for managing weeds, nematodes, and soil-borne pathogens, and it can be used in combination with chemical and biological disease control (Katan, 1981; Katan, 1998). This has been reported to be an effective method of managing Sclerotium rolfsii-related illnesses (Charirak et al., 2018). Katan et al. (1976) and Katan (1980) were the first to show that solarized soil suppressed the population of soil-borne fungus such as *F. oxysporum, R. solani*, and *Sclerotium rolfsii* by 62–100% at 5–25 cm depths. Soil solarization has proven to be an effective and environmentally friendly alternative to chemicals for the management of a variety of soil-borne pests, including phytopathogens and weeds, in a variety of agricultural circumstances (Yaqub and Shahzad, 2009). During the hot summer months, when soil temperatures rise, soil solarization is particularly effective for the soil pathogen *S. rolfsii*, which kills many key soil-borne fungal and bacterial plant diseases (Dwivedi and Prasad, 2016). Soil solarization is a hydrothermal process that captures solar radiation in the soil using a transparent sheet. Solarization is effective against soil pathogens in India's hot dry regions, especially between the months of April and June (Lodha, 2011). Soil solarization should not be seen as a general strategy, but rather as an extra alternative to consider and incorporate into IPM programmes as needed (Kumar et al., 2017).

6.9 EFFECT OF PLANT EXTRACTS AGAINST SCLEROTIUM ROLFSII

In comparison to synthetic pesticides, plant-based pesticides (botanicals) appear to be one of the best solutions for the management of plant diseases because they are known to cause fewer environmental concerns and pose a lower risk to consumers (Verma and Dubey, 1999). Antibacterial, antifungal, and insecticidal activities have been documented in plant extracts from a variety of higher plants (Fawzi et al., 2009). These botanical pesticides are important because they reduce crop losses; they are eco-friendly, bio-degradable and are cheap and hence affordable. Several researchers have shown that extracts of many plants may control different fungal plants pathogens. Some also have reported various plant extracts to have antimicrobial activity against *S. rolfsii* Sacc. infecting different crops, with positive outcomes (Wavare et al., 2017). Plant extracts were more or less inhibitory to mycelial growth and sclerotia formation of the *Sclerotium rolfsii* is mentioned in Table 6.1.

TABLE 6.1 Effective Botanicals Used for the management of *S. rolfsii* Causing Important Diseases in Crops Plants.

Botanical	Source of extraction	Disease	Reference
Agave americana	Bulb	Collar rot	Sesha Kiran et al. (2006)
Andrographis paniculata	Leaves	Collar rot	Mahato et al. (2018)
Allium cepa	Bulb	Collar rot	Pandey et al. (2005)
Allium sativum	Cloves	Collar rot	Cavalcanti et al. (2018)
Alstonia scholaris	Leaves	Collar rot	Butt et al. (2016)
Annona squamosa	Leaves	Root rot	Quadri and Fatima (2016)
Azadirachta indica	Leaves	Collar rot	Khan et al. (2020)
Brassica juncea	Seeds	Collar rot	Garain et al. (2021)
Calotropis gigantea	Leaves, flower	Collar rot	Gupta et al. (2012)
Cannabis sativa	Leaves	Collar rot	Gupta et al. (2012)
Carum carvi	Seeds	Root rot	El-Mougy and Alhabeb (2009)
Carum copticum	Seeds	Root rot	Farooq et al. (2010)
Cassia fistula	Flower	Collar rot	Ferdosi et al. (2022)
Cassia occidentalis	Fruit	Collar rot	Sesha Kiran et al. (2006)
Catharanthus roseus	Leaves	Collar rot	Mahato et al. (2018)
Chenopodium album	Root	Root rot	Ali et al. (2017)
Chrysanthemum frutescens	Flower	Collar rot and damping-off	Derbalah et al. (2012); Wavare et al. (2017)
Clerodendron inerme	Leaves, flower	Collar rot	Sesha Kiran et al. (2006)
Curcuma longa	Rhizome	Foot and root rot	Amin et al. (2013)
Datura metel	Bark, leaf, root, stem, seed, fruit	Collar rot, stem rot, and soft rot	Jabeen et al. (2014); Hosen et al. (2016)
Datura stramonium	Bark, leaf, root, stem, seed, fruit	Collar rot	Maji and Nath (2016)
Eucalyptus tereticornis	Leaves, bark	Collar rot	Sesha Kiran et al. (2006)
Eupatorium odoratum	Leaves	Sclerotium Wilt	Shwetha and Hegde (2012)
Foeniculum vulgare	Fruit	Root rot	Farooq et al. (2010)
Gaillardia sp.	Flower	Collar rot	Wavare et al. (2017)
Lantana camara	Leaves, flower	Collar rot	Bapat et al. (2016)
Lawsonia inermis	Leaves	Sheath blight	Ansari (1995)
Leucaena leucocephala	Leaves	Collar rot	Sesha Kiran et al. (2006)

TABLE 6.1 *(Continued)*

Botanical	Source of extraction	Disease	Reference
Maesa lanceolata	Stem bark	Collar rot	Okemob (2003)
Melia azedarach	Fruit	Collar rot	Khan and Javaid (2013)
Mentha piperita	Leaves	Root rot	El-Mougy and Alhabeb (2009)
Nerium indicum	Roots, stem, leaves	Collar rot	Siddiqui et al. (2016)
Nicotiana tabacum	Leaves	Foot and root rot	Amin et al. (2013)
Ocimum sanctum	Leaves, inflorescence	Collar rot	Mahato et al. (2018)
Polyalthia longifolia	Leaves	Dry root rot	Bindu Madhavi et al. (2011)
Pongamia pinnata	Leaves, seeds	Collar rot	Akram et al. (2016)
Prosopis juliflora	Leaves	Collar rot	Saleh et al. (2021)
Ricinus communis	Leaves, fruit	Collar rot	Sesha Kiran et al. (2006)
Sida acuta	Root, stem, leaves, buds	Collar rot	Jindal et al. (2012)
Tagetes erecta	Flower	Collar rot	Wavare et al. (2017)
Thespesia populnea	Seed	Damping-off	Derbalah et al. (2012)
Vinca rosea	Leaf, flower, stem and root	Collar rot	Narain and Satapathy (1977)
Zingiber officinale	Rhizome	Collar rot	Gupta et al. (2012)

6.10 BIOLOGICAL CONTROL OF SCLEROTIUM ROLFSII

According to the studies conducted with inhibitory effects of strains and mutants of *Talaromyces flavus* against *Sclerotium rolfsii*, it was found that benomyl-resistant mutant Ben^{R}TF1-R6^{x} was capable of producing a higher number of extracellular enzymes like chitinase and showed an improved antagonistic activity against the pathogen than the other mutant Ben^{R}TF1-R3^{x} and ten other wild-type strains tested. The pathogen was inhibited from causing stem rot disease in bean crop by mycoparasitism on sclerotia and antagonistic activity of cell wall-degrading chitinase enzyme produced by *Talaromyces flavus* (Madi et al., 1997).

High amounts of ferulic acid were detected in the stem portions above the infected collar region of chickpea (*Cicer arietinum*) crop through high-performance liquid chromatography analysis. The phenolic compound was minimal in amount in the roots of healthy seedlings. Ferulic acid exhibited complete inhibition of the fungal mycelial growth at 1000 μg mL^{-1} and even at lower concentrations like 250, 500, and 750 μg mL^{-1}, it gave satisfactory results with compact growth of the colony. Such phenolic compounds are present in increased amounts in the leaves and stems of *Sclerotium rolfsii*-infected seedlings than the healthy plants (Sarma and Singh, 2003).

On the basis of laboratory experiments, it was discovered that the leaf extracts of *Thespesia populnea* var. *acutiloba* (Wild tulip-tree) and *Chrysanthemum frutescens* (Marguerite daisy) were most efficient in suppressing *S. rolfsii* than other leaf extracts from *Euonymus japonicas* (Winged spindle), *Bauhinia purpurea* (Purple camel's foot), *Cassia fistula* (Golden shower tree), *Caesalpinia gilliesii* (Bird of paradise), and *Cassia senna* (Senna plant). Also, their efficiency in preventing the damping-off disease severity in sugarbeet crop was confirmed through seed treatment by coating or soaking the seeds under greenhouse conditions. Interestingly, none of the leaf extracts, even at highest concentration, exhibited phytotoxic effects on sugar beet leaves and were also less toxic to rats. Few bioactive compounds were found responsible for the inhibitory effect in the pathogen through gas chromatography-mass spectrometry analysis. *Chrysanthemum frutescens* extract was slightly more efficient under greenhouse conditions than the extract of *Thespesia populnea* var. *acutiloba*, which was estimated to be due to the higher levels of bioactive compounds like 1,8 cineole (33.33%), linalool (8.18%), and terpinyl acetate (21.6%) (Derbalah et al., 2012).

The cell wall of *S. rolfsii* was used as a substrate for the growth of *Serratia marcescens* along with its components, chitin and laminarin as a source for carbon. Different substrates including cell wall of *S. rolfsii*, its washed mycelium, and dried mycelium were used to incubate a culture filtrate having chitinolytic activity, in which the fungal hyphae of *S. rolfsii* was degraded due to the release of N-acetyl-D-glucosamine at 1–5.2 chitinase units. Abrupt swelling of the hyphal tips of the fungi was reported due to the presence of crude chitinase through light and scanning electron microscopy, along with the lysis of 63% of hyphal tip cells (Ordentlich et al., 1988).

The first report of the isolation of endophytic actinomycetes from various medicinal plants and its efficiency to control *S. rolfsii* in chickpea was presented (Singh and Gaur, 2016). *Streptomyces diastaticus*, *Streptomyces fradiae*, *Streptomyces olivochromogenes*, *Streptomyces collinus*, *Streptomyces ossamyceticus*, and *Streptomyces griseus* were few of the selected strains which showed antagonistic activity against *S. rolfsii*. Efficiency of actinomycetes strains was also evaluated against *S. rolfsii* infecting chilli (Pattanapipitpaisal and Kamlandharn, 2012). The production of chitinase enzyme by the actinomycetes has resulted in chitin degradation of the fungal cell wall. They are phosphate solubilizers and also produce siderophores and phytohormones. Actinomycetes generally produce hydroxamate and catecholate-type siderophores that are capable of competing for iron with phytopathogens (Khamna et al., 2009). Also, the production of phytohormones and the mechanism of phosphate solubilization help to modulate plant defence mechanism (Goudjal et al., 2014; Singh et al., 2016, Singh et al., 2017a). Highest amount of siderophore production and IAA production was observed in actinomycete isolated from *Azadirachta indica* (Verma et al., 2009). A report was made on the role of *Streptomyces* spp. as a biocontrol agent against stem and root rot causing *S. rolfsii* that caused a very destructive disease in chilli pepper in Thailand. Among the 265 isolates, *S. philanthi* RM-1-138, *S. philanthi* RL-1-178, and *S. mycarofaciens* SS-2-243 were showing antifungal and antibacterial property, but *S. philanthi* RM-1-138 inhibited seed germination and seedling growth of chilli pepper crop under laboratory conditions. In greenhouse, *S. philanthi* RL-1-178 had similar effects as that shown by *Trichoderma harzianum* NR-1-52 or that by carboxin treatment. Moreover in field conditions *S. philanthi* RL-1-178 protected the chilli pepper plants against *S. rolfsii* more efficiently than *S. mycarofaciens* SS-2-243 and *Trichoderma harzianum* NR-1-52 (Boukaew et al., 2011). Maximum growth inhibition (57.03%) of *Sclerotium rolfsii* was obtained in dual culture with *Trichoderma harzianum* at 96 hai in the dual culture method, whereas mean colonial diameter of *Sclerotium rolfsii* was found minimum by the effect of cultural filtrates and volatile metabolites of *Trichoderma harzianum*, *Aspergillus niger*, and *Penicillium citrinum* among all the tested fungi (Kumar et al., 2014). *Trichoderma harzianum* had very significant effects over the sclerotia production of this pathogen. After 21 days of incubation, *Trichoderma harzianum* caused the maximum reduction in sclerotia of *S. rolfsii*, *viz*., 86.44%, 95.42, and 87.46%, respectively, under dual culture test, volatile metabolites test, and nonvolatile metabolites tests (Kumar et al., 2017a).

The effect of *Trichoderma harzianum* in inhibiting the growth and development of *S. rolfsii* was assessed both under lab and greenhouse condition, wherein both gave promising results in reducing pathogen's virulence (Elad et al., 1980). High pentachloronitrobenzene concentrations, high pH levels, and low temperatures along with mycelial invasion by *Trichoderma harzianum* accorded adverse conditions to the pathogen development in vitro. Under greenhouse condition, both conidial suspension and wheat bran inoculum preparation of *Trichoderma harzianum* were performed against *S.*

rolfsii in beans, cotton, and tomato, in which the wheat bran preparation exhibited more efficacy (Naeimi et al., 2020).

On the basis of studies conducted and data obtained, it was recorded that mycelial growth and sclerotium germination of *S. rolfsii* were inhibited by the application of high amounts of benzaldehyde at the rate of 0.4 kg^{-1} of soil and dried powder of velvet-bean at the rate of 100 g kg^{-1} (Blum and Rodríguez-Kábana, 2004). Stimulation of mycelial growth and sclerotium germination was noticed at low amount of application of benzaldehyde at the rate of 0.1 mL kg^{-1} of soil and dried powders of kudzu (*Pueraria lobata*) at the rate of 25 g kg^{-1} of soil, and pine-bark (*Pinus elliottii* and *P. taeda*) at the rate of 25 g kg^{-1} of soil. Sclerotial formation was inhibited by Kudzu at 25–100 g kg^{-1} and velvet-bean at 25–100 g kg^{-1}. Sclerotial formation was stimulated by the application of benzaldehyde at 0.1 and 0.2 mL kg^{-1}. The colonization of sclerotia of the fungus by *Trichoderma* sp. was aided by the presence of Kudzu at 50 and 100 g kg^{-1} and pine-bark at 50 g kg^{-1}. Thus, dried powders of kudzu (50 and 100 g kg^{-1}), velvetbean (50 and 100 g kg^{-1}), and pine-bark (50 g kg^{-1}) used in amending the soil helped to reduce disease severity in soybean (*Glycine max*) crop and soil amended with kudzu or velvetbean (30 and 35 g kg^{-1}) and pine-bark (35 g kg^{-1}) reduced disease severity in tomato (*Lycopersicon esculentum*) crop.

Arbuscular mycorrhizal fungi (AMF) aid host plants in a variety of ways, including providing protection from fungal plant diseases (Wu et al., 2013). *Glomus fasciculatum*, an arbuscular mycorrhizal fungus, inhibits the detrimental effects of *Sclerotium* stem rot disease. *S. rolfsii*, which causes groundnut stem rot, is greatly inhibited by arbuscular mycorrhizal fungi (*Glomus fasciculatum*) (Doley et al., 2015).

Efficacy of native strains of *Pseudomonas fluorescens* against *S. rolfsii* causing root rot and stem rot in peanut was evaluated both in vitro and greenhouse condition (Ganesan and Gnanamanickam, 1987). Inhibition of mycelial growth was observed under in vitro condition with 10–20% germination loss of sclerotia after 1 h of immersion in bacterial suspension. Under greenhouse condition, 10^8 cfu mL^{-1} of *Pseudomonas fluorescens* were inoculated into peanut pots after which *S. rolfsii* was inoculated. All peanut plants without treatment with the bacteria wilted in 7 days due to *S. rolfsii* infection, while 99% of plants treated with *P. fluorescens* survived. Another isolate of the bacteria named *Pseudomonas cf. monteilii 9 showed increased antagonistic activity against S. rolfsii* in groundnut causing stem rot (Rakh et al., 2011). Under in vitro condition, the rhizospheric soil bacterium exhibited 94% inhibition on the basis of dry weight in dual culture technique by the production of siderophores, volatile metabolites, diffusible antibiotics, and hydrogen cyanide. There was 45.5–66.67% decrease in disease incidence in *Pseudomonas cf. monteilii 9 treated seeds in pot assay. Also, the bacterium-excreted protease that plays a role in the growth inhibition of the pathogen which was confirmed due to the appearance of clear halo region in Pseudomonas cf. monteilii 9 inoculated skim milk agar plates.* Two different strains of *Pseudomonas* (*Pseudomona fluorescens* CA05 and *Pseudomonas putida* CA28) and *Bacillus subtilis* CA32 were tested separately and in combination against *S. rolfsii* in *Capsicum annum* under greenhouse condition (Abeysinghe, 2009). The bioagents were used as seed treatment and as root treatment before transplanting. Better results were obtained with the combination of *Bacillus* strain with any of the *Pseudomonas* strains, while lower effects were obtained when the two strains of *Pseudomonas* alone were combined. When examined separately *Bacillus* showed highest inhibition rate followed by *Pseudomona fluorescens* CA05 and then *Pseudomona putida* CA28. But when each *Pseudomonas* strain was combined with *Bacillus*, both gave similar results. The sclerotia of the fungi is capable of overwintering in soil and the inoculum emerges to cause infection in the following season, which makes this pathogen difficult to control (Thiessen and Woodward, 2012).

TABLE 6.2 Control of Pathogen, *S. rolfsii* by Using Antagonist Microbial Inoculants in Different Crops.

Name of crop	Name of plant disease	Fungal/bacterial antagonist	References
Brinjal	Collar rot	*Trichoderma* spp. and *Bacillus* spp.	Abinaya et al. (2020)
		Trichoderma viride + castor oil	Murmu et al. (2021)
Chickpea	Collar rot	*Trichoderma viride, T. harzianum, T. koningii, T. reesei, T. pseudo-koningii, T. aureo-viride*, and *T. hamatum*	Ali and Javaid (2015)
		Trichoderma harzianum GKVK isolate, *Pseudomonas fluorescens*	Nagamma and Nagaraja (2015)
Chilli	Foot rot	*Trichoderma harzianum* and *Chaetomium globosum*	Sultana et al. (2012)
	Southern blight	*Bacillus megaterium* and *Pseudomonas fluorescens*	Sharf et al. (2021)
Elephant foot yam	Collar-rot or foot-rot disease	*Trichoderma viride, T. harzianum*, and *T. hamatum*	Srinivasulu et al. (2005)
Finger millet	Foot rot	*Trichoderma harzianum* (GKVK) isolate	Manu et al. (2012)

TABLE 6.2 *(Continued)*

Name of crop	Name of plant disease	Fungal/bacterial antagonist	References
Ground nut	Stem rot	*Trichoderma harzianum, T. viride, Pseudomonas fluorescens, Chaetomium* ND35	Ganesan (2004); Ganesan and Sekar (2004); Manjula et al. (2004)
Groundnut	Stem rot	Combined application of *Rhizobium* and *Trichoderma harzianum* (ITCC – 4572)	Ganesan et al. (2007)
Groundnut	Stem rot	*Trichoderma* spp., *Pseudomonas fluorescens* and *Bacillus subtilis*	Darvin and Kumari (2013)
Groundnut	Stem-rot	Actinomycetes	Jacob et al. (2016)
Jute	Anthracnose and soft-rot	*Aspergillus niger* and *Trichoderma viride*	Hosen et al. (2016)
Lentil	Foot-rot and root-rot	*Trichoderma viride* compost, *T. harzianum* compost, provax 200	Khalequzzaman (2016)
Maize	Stem rot	Viable spore *Chaetomium*	Soytong (1991)
Peanut	Root and stem rot	*Pseudomonas fluorescens*	Ganesan and Gnanamanickam (1987)
	Stem rot	*Trichoderma* spp.	Suheri et al. (2014)
	Stem rot	*Streptomyces* sp. RP1A-12	Jacob et al. (2018)
Soybean	Root rot	*Bradyrhizobium japonicum* and *Trichoderma harzianum*	Chakraborty et al. (2003)
	Foot and collar rot	*Trichoderma harzianum*	Bankoliya et al. (2015)
Stevia rebaudiana	Sclerotium wilt	*Trichoderma harzianum, T. viride*	Shwetha and Hegde (2012)
Sugar beet	Root Rot	*Trichoderma* sp., *P. fluorescens* and *Bacillus subtilis*	Rasu et al. (2013)
Sunflower	Root and collar-rot	*Pseudomonas* spp., *Pseudomonas fluorescens, Alcaligenes odorans, P. putida, Streptomyces* spp., and *Bacillus pantothenticus*	Rangeshwaran and Prasad (2000)
Tomato	Damping-off	*Trichoderma koningii*	Tsahouridou and Thanassoulopoulos (2002)
Turfgrass	Southern blight	*Bacillus cereus* 44bac and *Stenotrophomonas rhizophila* 88bpf	Ünal et al. (2019)
Water Melon	Southern blight rot	Bamboo rhizospheric microbe, that is, *Bacillus subtilis*	Darma et al. (2016)
Wheat	Root rot	*Trichoderma* spp.	Elad et al. (1980)

6.11 CHEMICAL MANAGEMENT

A considerable effect of fungicides on the growth of *S. rolfsii* has been proved by several researchers (Chowdary et al., 1998; Hanumantha and Kannan, 2002; Bhat and Srivastava, 2003; Cilliers et al., 2003; Prabhu and Hiremath, 2003; Thakur et al., 2004). Some fungicides like Bavistin 50 WP, Topgan 50 WP, Dithane M-45, and the Bordeaux mixture have been reported to be useful in controlling the *Sclerotium rolfsii*-caused foot and root rot disease of eggplant (Patil et al., 1986; Tiwari and Ashok, 2004; Dasgupta and Maiti, 2008). Table 2.3 shows the effect of certain significant chemicals on *Sclerotium rolfsii* in the above-mentioned context.

TABLE 6.3 Chemicals Evaluated Against *Sclerotium rolfsii* Causing Diseases in Different Crops.

Chemicals	Crop	Disease	References
Combination of the systemic fungicides etridiazole +quintozene found effective	*Amaryllis*	Bulb rot and leaf sheath blight	Akarapisan et al. (2017)
Pentachloronitobenzene (PCNB) and Banodenil	Apple	Southern blight and root rot	Conway et al. (1996)
Brassicol, Manzate, Parasan, DM-145, Dithane M-45, and Captan	Brinjal	Foot rot	Chaurasia et al. (2014)
Hexaconazole 5% EC, Propiconazole 25% EC, and combo products Tubaconzole 50% + Trifloxystrobin 25% WG, Captan 70% + Hexaconazole 5% WP, Propiconazole 13% + Difenoconazol and Carboxin 37.5% + Thiram 37.5%	Chickpea	Collar rot	Shirsole et al. (2019)
Flusilazole and Tebuconazole showed better performance under both in vitro and in vivo conditions			Sravani and Chandra (2021)
Bavistin 50 WP, Topgan 50 WP, and Ridomil Gold	Eggplant	Foot and root rot	Siddiqui et al. (2016)
Hexaconazole, vitavax, propiconazole, and difenoconazole	Finger millet	Foot/stem rot	Manu et al. (2012); Sahana et al. (2020)
Quintozene used as a dip and added to soil	Grapevine	Root rot	Keyser and Ferreira (1988)
Propiconazole and Tricyclazole	Groundnut	Stem rot	Rangarani et al. (2017)
Benlate, Captan and Ridomil at 500 ppm applied as seed treatment and soil drench respectively	Lentil	Collar rot	Shahid et al. (1990)

TABLE 6.3 *(Continued)*

Chemicals	Crop	Disease	References
Pentachloronitobenzene (PCNB)	Maize	Basal rot	Soytong and Soytong (1997)
Bavistin, Brassicol, Captan, Dithane M-45, DM-145, Fytolan, Manzate, Parasan and Sulfex		Root rot	Prasad and Dwivedi (2020)
Tilt 250 EC (Propiconazole) found most effective	Sesame	Stem rot	Shamsi and Naher (2020)
Thiram and Mancozeb @ 0.1% concentration	Tomato	Collar rot	Dutta and Das (2002)
Potassium salt, salicylic acid, and sorbic acid		Root rot	El-Mohamedy et al. (2014)
Calixin, Captan, Carbendazim, Indofil M-45, and Thiodan	Tea	Sclerotial blight	Bhagat and Chakarborty (2013)
Plantvax [oxycarboxin] and Vitavax [carboxin], as seed treatments at 2 g kg^{-1}	Wheat	Foot rot	Reddy et al. (1976)

Fungicide difenoconazole was applied in combination with *Trichoderma harzianum*, a biocontrol agent to prevent the growth rate of *S. rolfsii* in groundnut crops. Integrating cultural control through lower plant density and use of inversion plough to cultivate infected fields avoided disease incidence and improved the quality of produce respectively. Inversion plough cultivation did not contribute in increasing yield. Difenoconazole had better efficiency than other fungicides like carbendazim, flusilazole, and chlorothalonil which were used to compare and also difenoconazole did not have any impact on the growth rate of *Trichoderma harzianum* (Cilliers et al., 2003). In vitro evaluation of fungicides against *S. rolfsii* infecting chilli causing dry root rot was carried out and examined, in which tebuconazole and combination of carbendazim + mancozeb gave positive results in inhibiting mycelial growth by 94.1%, followed by difenconazole (93.3%) (Madhavi and Bhattiprolu, 2011). In inhibiting sclerotial germination and growth of the pathogen, propiconazole and tebuconazole were most effective followed by the combination of carbendazim and mancozeb. Seedling dip of chilli in fungicide or soil drenching with fungicides can be performed under in vivo condition at 1000, 2000, and 3000 ppm concentrations. Wokocha (1990) concluded that the addition of *T. viride* with PCNB gave excellent disease control and plants treated with the combination showed no symptoms of basal stem rot of tomato caused by *Sclerotium rolfsii* during both dry- and wet-season trials. The use of hymexazol in combination with the tested isolates of *T. harzianum* against *S. rolfsii* is a promising approach in integrated control of sugarbeet root rot (Khattabi et al., 2001). As compared to alone application *Trichoderma harzianum* or *Pseudomonas fluorescens* or Hexaconazole integrated application of *Trichoderma harzianum* or *Pseudomonas fluorescens* and Hexaconazole resulted in minimum mortality and maximum plant growth and yield of chickpea under field condition while managing the collar rot (Singh et al., 2017).

6.12 CONCLUSION

Production of surplus amount of sclerotia poses a greater challenge to the management of the pathogen. The present chapter concluded that soil solarization, biological, plant extract, and chemicals inhibit the growth of *S. rolfsii*. Exposing of soil to natural sunlight and application of bioagents and well-decomposed organic matter during bed preparation can be an eco-friendly option for environment and farmer. Giving importance to proper pH and soil conditions can also be useful in preventing the occurrence and germination of sclerotial bodies. Nonhazardous nature of bioagents and botanicals, their simple treatment, systemic effect they induce, and also their low-cost mark them as an effective, feasible, and eco-friendly alternative to conventional chemical control that is within the means of even small farmers. As host cultivars resistant to *S. rolfsii* are rarely available, and the option for conventional chemical control is also limited and expensive, the present approach, when further developed and standardized, may turn out to be particularly suitable for the control of this highly aggressive pathogen with wide host range.

KEYWORDS

- **Sclerotium rolfsii**
- **soil-borne pathogen**
- **symptomatology**
- **integrated disease management**

REFERENCES

Abeygunawardena, D. V. W.; Wood, R. K. S. Factors Affecting the Germination of Sclerotia and Mycelial Growth of *Sclerotium rolfsii* Sacc. *Trans. Br.. Mycol. Soc.* **1957,** *40* (2), 221–231.

Abeysinghe, S. Efficacy of Combine Use of Biocontrol Agents on Control of *Sclerotium rolfsii* and *Rhizoctonia solani* of *Capsicum annuum*. *Arch. Phytopathol. Pflanzenschutz.* **2009,** *42* (3), 221–227.

Abinaya, A.; Kalpana, K.; Ebenezar, E. G.; Thiruvudainambi, S.; Theradimani, M.; Arunkumar, R Evaluation of Antagonistic Potential of *Bacillus* spp. and *Trichoderma* spp. Against *Sclerotium rolfsii* Sacc. Causing Collar Rot Disease in *Solanum melongena* L. *Int. J. Curr. Microbiol. App. Sci.* **2020,** *9* (12), 694–703.

Akarapisan, A.; Kumvinit, A.; Ruamrungsri, S. Bulb Rot of *Amaryllis* Caused by *Sclerotium rolfsii* and Effect of Fungicides on In Vitro Inhibition of Mycelial Growth. *Asian J. Plant Pathol.* **2017,** *11*, 95–101.

Akram, A.; Shaheen, I.; Akhund, S.; Nayyar, B. G.; Seerat, W. In Vitro Antifungal Activity of *Pongamia pinnata* Against Collar Rot Pathogen (*Sclerotium rolfsii*) of Chickpea. *Pure Appl. Biol.* **2016,** *5* (3), 520–528.

Ali, A.; Javaid, A. Screening of *Trichoderma* Species for Their Biological Control Potential Against *Sclerotium rolfsii*, the Cause of Collar Rot Disease of Chickpea. *Mycopath* **2015,** *13* (2), 93–96.

Ali, A.; Javaid, A.; Shoaib, A. GC-MS Analysis and Antifungal Activity of Methanolic Root Extract of *Chenopodium album* Against *Sclerotium rolfsii*. *Planta Daninha* **2017,** *35*, e017164713.

Amin, R.; Sarker, B. C.; Adhikary, S. K.; Sultana, S.; Zubair, T. Effect of Some Botanical Extracts and Cow's Urine on *Sclerotium rolfsii* Causal Agent of Foot and Root Rot of Betel Vine. *Int. J. Eng. Sci.* **2013,** *2* (9), 77–82.

Ansari, M. M. Control of Sheath Blight of Rice by Plant Extracts. *Indian Phytopathol.* **1995,** *48*, 268–270.

Aycock, R. Stem Rot and Other Diseases Caused by *Sclerotium rolfsii*; North Carolina Agricultural Experiment Station, 1966; pp 136–202.

Bahadur, P.; Sinha, V. C.; Upadhyaya, Y. M. Effect of Certain Sugars and Nitrogen Sources on the Growth of *Sclerotium rolfsii*. *Indian Phytopathology* **1976,** *29*, 426–427.

Bankoliya, M. K.; Kumar, S.; Ashish, S. Effect of intercropping on the fungicides of Foot and Collar Rot in Soybean. *Int. J. Plant Protect.* **2015,** *8* (2), 375–378.

Bapat, U. C.; Prabha, S.; Kumar, J. Antifungal Activity of Ethanolic and Petroleum Ether Extracts of Some Medicinal Plants Against the Plant Pathogenic Fungus *Sclerotium rolfsii* Sacc. *Int. J. Bioassays* **2016,** *5* (7), 4714–4719.

Beute, M. K.; Rodriguez-Kabana, R. Effect of Volatile Compounds from Remoistened Plant Tissues on Growth and Germination of Sclerotia of *Sclerotium rolfsii*. *Phytopathology* **1979a,** *69* (8), 802.

Beute, M. K.; Rodriguez-Kabana, R. Effect of Wetting and the Presence of Peanut Tissues on Germination of Sclerotia of *Sclerotium rolfsii* Produced in Soil. *Phytopathology* **1979b,** *69* (8), 869.

Bhagat, I.; Chakraborty, B.; Efficacy of Fungicides Against Sclerotial Blight of Tea Plant. *Our Nat* **2013,** *11*, 208–210.

Bhat, N. M.; Srivastava, L. S. Evaluation of Some Fungicides and Neem Formulations Against Six Soil-Borne Pathogens and Three *Trichoderma* spp. in vitro. *Plant Dis. Res.* **2003,** *18* (1), 56–59.

Biessy, A.; Novinscak, A.; St-Onge, R.; Léger, G.; Zboralski, A.; Filion, M. Inhibition of Three Potato Pathogens by Phenazine-Producing *Pseudomonas* spp. Is Associated with Multiple Biocontrol-Related Traits. *mSphere* **2021,** *6* (3), e0042721.

Bindu Madhavi, G.; Bhattiprolu, S. L.; Reddy, V. B. Effect of Various Plant Extracts on Dry Root Rot of Chillies Caused by *Sclerotium rolfsii*. *J. Hortl. Sci.* **2011,** *6* (2), 156–158.

Blum, L. E.; Rodríguez-Kábana, R. Effect of Organic Amendments on Sclerotial Germination, Mycelial Growth, and *Sclerotium rolfsii*-Induced Diseases. *Fitopatologia Brasileira* **2004,** *29* (1), 66–74.

Boukaew, S.; Chuenchit, S.; Petcharat, V. Evaluation of *Streptomyces* spp. for Biological Control of *Sclerotium* Root and Stem Rot and *Ralstonia* Wilt of Chilli Pepper. *BioControl* **2011,** *56* (3), 365–374.

Bozarth, G. A.; Tweedy, B. G. Effect of Pesticides on Growth and Sclerotial Production of *Sclerotium rolfsii*. *Phytopathology* **1971,** *61*, 1140–1142.

Butt, B. Z.; Sana, N.; Butt, S. Manangement of *Sclerotium rolfsii* Through Methanolic Leaf Extract of *Alstonia scholaris* (L.) R. Br. and *Azadirachta indica* L. *J Bioresour. Manage* **2016,** *3* (2), 1–8.

Cavalcanti, V. P.; Araújo, N. A. F.; Schwanestrada, K. R. F.; Pasqual, M.; Dória, J. *Athelia* (*Sclerotium*) *rolfsii* in *Allium sativum*: Potential Biocontrol Agents and Their Effects on Plant Metabolites. *An. Acad. Bras. Ciênc.* **2018,** *90* (04), 3949–3962.

Chakraborty, U.; Sarkar, B.; Chakraborty, B. N. Combined Application of *Bradyrhizobium japonicum* and *Trichoderma harzianum* on Root Rot Disease of Soybean. *J. Mycol. Plant Pathol.* **2003,** *33*, 21–25.

Charirak, P.; Saksirirat, W.; Jogloy, S.; Saepaisan, S. Integration of Soil Solarization with Chemical and Biological Control of Stem Rot Disease of Jerusalem Artichoke. *J. Pure Appl. Microbiol.* **2018,** *10* (4), 2531–2539.

Chaurasia, A. K.; Chaurasia, S.; Chaurasia, S, Chaurasia, S. In Vitro Efficacy of Fungicides Against the Growth of Foot-Rot Pathogen (*Sclerotium rolfsii* sacc.) of Brinjal. *Int. J. Curr. Microbiol. Appl. Sci.* **2014,** *3* (12), 477–485.

Chet, I. Ultrastructural Basis of Sclerotial Survival in Soil. *Microb. Ecol.* **1975,** 2 (3), 194–200.

Chet, I.; Henis, Y. The Control Mechanism of Sclerotial Formation in *Sclerotium rolfsii* Sacc. *Microbiology* **1968,** *54* (2), 231–236.

Chet, I.; Henis, Y. X-ray Analysis of Hyphal and Sclerotial Walls of *Sclerotium rolfsii* Sacc. *Can. J. Microbiol.* **1968a,** *14* (7), 815–816.

Chet, I.; Henis, Y. Effect of Catechol and Disodium EDTA on Melanin Content of Hyphal and Sclerotial Walls of *Sclerotium rolfsii* Sacc. and the Role of Melanin in the Susceptibility of These Walls to β- (1→ 3) Glucanase and Chitinase. *Soil Biol. Biochem.* **1969,** *1* (2), 131–138.

Chet, I.; Henis, Y.; Kislev, N. Ultrastructure of Sclerotia and Hyphae of *Sclerotium rolfsii* Sacc. *Microbiology* **1969,** *57* (2), 143–147.

Chet, I.; Henis, Y.; Mitchell, R. The Morphogenetic Effect of Sulphur-Containing Amino Acids, Glutathione and Iodoacetic Acid on *Sclerotium rolfsii* Sacc. *Microbiology* **1966,** *45* (3), 541–546.

Chet, I.; Henis, Y.; Mitchell, R. Chemical Composition of Hyphal and Sclerotial Walls of *Sclerotium rolfsii* Sacc. *Can. J. Microbiol.* **1967,** *13* (2), 137–141.

Chet, I.; Retig, N.; Henis, Y. Changes in Total Soluble Proteins and in Some Enzymes During Morphogenesis of *Sclerotium rolfsii*. *Microbiology* **1972,** *72* (3), 451–456.

Chet, I.; Timar, D.; Henis, Y. Physiological and Ultrastructural Changes Occurring During Germination of Sclerotia of *Sclerotium rolfsii*. *Can. J. Bot.* **1977,** *55* (9), 1137–1142.

Chowdary, K. A.; Reddy, D. R.; Rao, K. C. Efficacy of Systemic (triazoles) and Non-Systemic Fungicides Against Sclerotial Wilt of Bell Pepper Caused by *Sclerotium rolfsii* Sacc. *Indian J. Plant Protect.* **1998,** *26* (2), 125–130.

Christias, C. Specific Inhibition of Sclerotium Formation by 2-Mercaptoethanol and Related Sulfhydryl Compounds in *Sclerotium rolfsii*. *Can. J. Microbiol.* **1975,** *21,* (10), 1541–1547.

Christias, C. Nature of the Sclerotial Exudate of *Sclerotium rolfsii* Sacc. *Soil Biol. Biochem.* 12 (2), 199–201.

Christias, C.; Lockwood, J. L. Conservation of Mycelial Constituents in Four Sclerotium-Forming Fungi in Nutrient-Deprived Conditions. *Phytopathology* **1973,** *63*, 602.

Cilliers, A. J.; Pretorius, Z. A.; Van Wyk, P. S. Integrated Control of *Sclerotium rolfsii* on Groundnut in South Africa. *J. Phytopathol.* **2003,** *151* (5), 249–258.

Colotelo, N. Fungal Exudates. *Can. J. Microbial.* **1978,** *24* (1), 173–81.

Conway, K. E.; Tomasino, S.; Claypool, P. L. Evaluations of Biological and Chemical Controls for Southern Blight of Apple Rootstock in Oklahoma Nurseries. *Proc. Okla Acad. Sci.* **1996,** *75*, 9–15.

Darma, R.; Purnamasari, I. M.; Agustina, D.; Pramudito, T. E.; Sugiharti, M.; Suwanto, A. A Strong Antifungal-Producing Bacteria from Bamboo Powder for Biocontrol of *Sclerotium rolfsii* in Melon (*Cucumis melo* var. *amanta*). *J. Plant Pathol. Microbiol.* **2016,** *7* (2), 334.

Darvin, G.; Prasanna Kumari, V. Effect of Bio-control Agents n Radial Growth of *Sclerotium rolfsii* In Vitro. *Inter. J. Appl. Biol. Pharma. Tech.* **2013,** *4* (4), 61–64.

Dasgupta, B.; Maiti, S. Assessment of *Phytophthora* Root Rot of Betel Vine and Its Management Using Chemicals. *Indian J. Mycological. Pl. Pathol.* **2008,** *29*, 91–95.

Derbalah, A. S.; Dewir, Y. H.; El-Sayed, A. E. B. Antifungal Activity of Some Plant Extracts Against Sugar Beet Damping-Off Caused by *Sclerotium rolfsii*. *Ann. Microbiol.* **2012,** *62* (3), 1021–1029.

Doley, K.; Dudhane, M.; Borde, M. Effects of Arbuscular Mycorrhizal Fungi against *Sclerotium rolfsii* in Groundnut (JL-24). *Ind. J. Fundament. Appl. Life Sci.* **2015,** *5* (1), 293–302.

Dutta, P.; Das, B. C. Management of Collar Rot of Tomato by *Trichoderma* spp. and Chemicals. *Indian Phytopathol.* **2002,** *55* (2), 235–237.

Dwivedi, S. K.; Prasad, G. Integrated Management of *Sclerotium rolfsii*: An Overview. *Eur J Biomed Pharma Sci* **2016,** *3* (11), 137–146.

Edelstein, L.; Hadar, Y.; Chet, I.; Henis, Y.; Segel, L. A. A Model for Fungal Colony Growth Applied to *Sclerotium rolfsii*. *Microbiology* **1983,** *129* (6), 1873–1881.

Elad, Y.; Chet, I.; Katan, J. *Trichoderma harzianum*: A Biocontrol Agent Effective Against *Sclerotium rolfsii* and *Rhizoctonia solani*. *Phytopathology* **1980,** *70* (2), 119–121.

El-Mohamedy, R. S. R.; Jabnoun-Khiareddine, H.; Daatmi-Remadi, M. Control of Root Rot Diseases of Tomato Plant Caused by *Fusarium solani, Rhizoctonia solani* and *Sclerotium rolfsii* Using a Different Chemical, Plant Resistance Inducers. *Tunisian J Plant Protect* **2014,** *9*, 45–55.

El-Mougy, N. S.; Alhabeb, R. S. Inhibitory Effects of Powdered Caraway and Peppermint Extracts on Pea Root Rot under Greenhouse Conditions. *J. Plant Prot. Res.* **2009,** *49* (1), 93–96.

Ercegovich, C. D.; Chrzanowski, R. L.; Cole, H.; Herendeen, N.; Witkonton, S. The Relationship Between the Chemical Structure of S-triazines and Fungi Toxicity to *Sclerotium rolfsii*. *Can. J. Microbiol.* **1973,** *19* (3), 329–334.

Farooq, M. A.; Iqbal, U.; Iqbal Sh, M.; Afzal, R.; Rasool, A. In Vitro Evaluation of Different Plant Extracts on Mycelial Growth of *Sclerotium rolfsii* the Cause of Root Rot of Sugar Beet. *Mycopathology* **2010,** *8* (2), 81–84.

Fawzi, E.; Khalil, A. A.; Afifi, A. Antifungal Effect of Some Plant Extracts on *Alternaria alternata* and *Fusarium oxysporum*. *Afr. J. Biotech.* **2009,** *8* (11).

Fellman, J. K.; Tourneau, D. I.; Stiers, D. L. Effect of Some Carboxamide Fungicides on Growth and Sclerotia Formation of *S. rolfsii*. *Trans. Brit. Mycolog. Soc.* **1983,** *80*, 170–172.

Ferdosi, M. F. H.; Ahmed, H.; Khan, I. H.; Javaid, A. Fungicidal Potential of Flower Extract of *Cassia fistula* Against *Macrophomina phaseolina* and *Sclerotium rolfsii*. *J. Anim. Plant Sci.* **2022,** *32* (4), in press.

Fira, D.; Dimkić, I.; Berić, T.; Lozo, J.; Stanković, S. Biological Control of Plant Pathogens by *Bacillus* Species. *J. Biotechnol.* **2018,** *285*, 44–55.

Ganesan, P.; Gnanamanickam, S. S. Biological Control of *Sclerotium rolfsii* Sacc. in Peanut by Inoculation with *Pseudomonas fluorescens*. *Soil Biol. Biochem.* **1987,** *19* (1), 35–38.

Ganesan, S. *Studies on the Biocontrol of Soilborne Plant Pathogens*, PhD Thesis Submitted to Madurai Kamaraj University, Madurai, India, 2004.

Ganesan, S.; Ganesh Kuppusamy, R.; Sekar, R. Integrated Management of Stem Rot Disease (*Sclerotium rolfsii*) of Groundnut (*Arachis hypogaea* L.) using *Rhizobium* and *Trichoderma harzianum* (ITCC–4572). *Turk. J. Agric. Forestry* **2007,** *31*, 103–108.

Ganesan, S.; Sekar, R. Biocontrol mechanism of Groundnut (*Arachis hypogaea* L.) Diseases-*Trichoderma* system. In: *Biotechnological Applications in Environment and Agriculture*; Pathade. G. R., Goel, P. K., Eds.; ABD Pub: Jaipur, 2004; pp 312–327.

Garain, P. K.; Mondal, B.; Dutta, S. Effect of Bio-Fumigation by Indian Mustard (*Brassica juncea* L.) on *Sclerotium rolfsii* Sacc., Causing Collar Rot in Betelvine (*Piper betle* L.). Indian *Phytopathology* 74 (2), 1–11.

Goudjal, Y.; Toumatia, O.; Yekkour, A.; Sabaou, N.; Mathieu, F.; Zitouni, A. Biocontrol of *Rhizoctonia solani* Damping-off and Promotion of Tomato Plant Growth by Endophytic Actinomycetes Isolated from Native Plants of Algerian Sahara. *Microbiol. Res.* **2014,** *169*, (1), 59–65.

Griffin, D. M.; Nair, N. G. Growth of *Sclerotium rolfsii* at Different Concentration of Oxygen and Carbon Dioxide. *J. Exp. Bot.* **1968,** *19* (4), 812–816.

Gupta, R. P.; Singh, S. K.; Mishra, R. S.; Gupta, S. Efficacy of Botanicals Against *Sclerotium rolfsii* Causing Collar Rot of Chick Pea. *Ann. Plant Prot. Sci.* **2012,** *20* (2), 486–487.

Hadar, Y.; Elad, Y.; Chet, I.; Henis, Y. Induction of Macroscopic Strand Formation in *Sclerotium rolfsii* by *Trichoderma harzianum*. *Israel J. Plant Sci.* **1981,** *30* (3), 156–164.

Hadar, Y.; Henis, Y.; Chet, I. The Potential for the Formation of Sclerotia in Submerged Mycelium of *Sclerotium rolfsii*. *Microbiology* **1981a,** *122* (1), 137–141.

Hadar, Y.; Pines, M.; Chet, I.; Henis, Y. The Regulation of Sclerotium Initiation in *Sclerotium rolfsii* by Glucose and Cyclic AMP. *Can. J. Microbiol.* **1983,** *29* (1), 21–26.

Hanumantha, B. T.; Kannan, N. In Vitro Evaluation of Fungicides Against Soft Rot of Coffee Caused by *Sclerotium rolfsii*. *J. Coffee Res.* **2002,** *30* (2), 137–140.

Henis, Y. Nutritional and Mechanical Factors Involved in Mycelial Growth and Production of Sclerotia by *Sclerotium rolfsii* in Artificial Medium and Amended Soil. *Phytopathology* **1965,** *55*, 87–91.

Henis, Y.; Chet, I. Developmental Biology of Sclerotia of *Sclerotium rolfsii*. *Can. J. Bot.* **1968,** *46* (7), 947–948.

Henis, Y.; Okon, Y.; Chet, I. The Relationship Between Early Hyphal Branching and Formation of Sclerotia in *Sclerotium rolfsii*. *Microbiology* **1973,** *79* (1), 147–150.

Hosen, S.; Shamsi, S.; Bashar, M. A. In Vitro Efficacy of Fungicides and Plant Extracts on the Growth of *Colletotrichum gloeosporioides* (Penz.) Penz. & Sacc. and *Sclerotium rolfsii* Sacc. the Causal Organisms of Anthracnose and Soft Rot of Jute. *Dhaka Univ. J. Biol. Sci.* **2016,** *25* (2), 195–199.

Howell, D. M.; Fergus, C. L. The Component Fatty Acids Found in Sclerotia of *Sclerotium rolfsii*. *Can. J. Microbiol.* **1964,** *10* (4), 616–618.

Humpherson-Jones, F. M.; Cooke, R. C. Morphogenesis in Sclerotium-Forming Fungi: I. Effects of Light on *Sclerotinia sclerotiorum*, *Sclerotium delphinii* and *S. rolfsii*. *New Phytol* **1977,** *78* (1), 171–180.

Igwegbe, E. C. K.; King, V.; Salary, J. 6-Methylpurine-Induced Inhibition of Sclerotia Morphogenesis in *Sclerotium rolfsii* and Its Reversal by Adenosine. *Mycopathology* **1977,** *62* (3), 153–159.

Jabeen, N.; Javaid, A.; Ahmed, E.; Sharif, A. Management of Causal Organism of Collar Rot of Bell Pepper (*Sclerotium rolfsii*) by Organic Solvents Extracts of *Datura metel* Fruit. *Pak. J. Phytopathol* **2014,** *26* (1), 13–18.

Jacob, S.; Rao, R. S.; Kumar, K. V. K.; Varshney, R. K.; Sudini, H. K. Assessing the Prospects of *Streptomyces* sp. RP1A-12 in Managing Groundnut Stem Rot Disease Caused by *Sclerotium rolfsii* Sacc. *J. Gen. Plant Pathol.* **2016,** *82* (2).

Jacob, S.; Sajjalaguddam, R. R.; Sudinia, H. K. *Streptomyces* sp. RP1A-12 Mediated Control of Peanut Stem Rot Caused by *Sclerotium rolfsii*. *J. Integr. Agric.* **2018,** *17* (4), 892–900.

Javed, Z. U. R.; Coley-Smith, J. R. Studies on Germination of Sclerotia of *Sclerotium delphinii*. *Trans. Br. Mycol. Soc.* **1973,** *60* (3), 441-IN6.

Jindal, A.; Kumar, P.; Jain, C. Antifungal Activity of Flavonoids of *Sida acuta* Burmf. Against *Candida albicans*. *Int. J. Drug. Dev. Res.* **2012,** *4* (3), 92–96.

Jones, D.; Farmer, V. C.; Bacon, J. S. D.; Wilson, M. J. Comparison of Ultrastructure and Chemical Components of Cell Walls of Certain Plant Pathogenic Fungi. *Trans. Brit. Mycol. Soc.* **1972,** *59* (1), 11-IN3.

Katan, J. Solar Pasteurization of Soils for Disease Control: Status and Prospects. *Plant Dis.* **1980,** *64*, 450–454.

Katan, J. Solar Heating (Solarization) of Soil for Control of Soilborne Pests. *Ann Rev Phytopathol* **1981,** *19*, 211–236.

Katan, J. Soil Solarization: Integrated Control Aspect. In: *Principles and Practice of Managing Soilborne Plant Pathogens*; Hall, R., Ed.; APS Press: St. Paul, 1998; pp 250–278.

Katan, J.; Greenberger, A.; Alon, H.; Grinstein, A. Solar Heating by Polyethylene Mulching for the Control of Diseases Caused by Soilborne Pathogens. *Phytopathology* **1976,** *76*, 683–688.

Keyser, H. A.; Ferreira, J. H. S. Chemical and Biological Control of *Sclerotium rolfsii* in Grapevine Nurseries. *S. Afr. J. Enol. Vitic.* **1988,** *9* (1), 43–44.

Khalequzzaman, K. M. Control of Foot and Root Rot of Lentil by Using Different Management Tools. *ABC J. Adv. Res.* **2016,** *5* (1), 35–42.

Khamna, S.; Yokota, A.; Lumyong, S. Actinomycetes Isolated from Medicinal Plant Rhizosphere Soils: Diversity and Screening of Antifungal Compounds, Indole-3-acetic Acid and Siderophore Production. *World J. Microbiol. Biotech.* **2009,** *25* (4), 649.

Khan, I. H.; Javaid, A. Antifungal Activity of *Melia azedarach* L. Fruit Extract Against *Sclerotium rolfsii*, the Cause of Collar Rot Disease of Chickpea. *Mycopath* **2013,** *11* (1), 9–13.

Khan, I. H.; Javaid, A.; Al-Taie, A. H.; Ahmed, D. Use of Neem Leaves as Soil Amendment for the Control of Collar Rot Disease of Chickpea. *Egypt J Biol Pest Control* **2020,** *30*, 98.

Khattabi, N.; Ezzahiri, B.; Louali, L.; Oihabi, A. Effect of fungicides and *Trichoderma harzianum* on Sclerotia of *Sclerotium rolfsii. Phytopathol. Mediterranean* **2001,** *40*, 143–148.

Kliejunas, J. T.; Ko, W. H. Continuous Versus Limited Growth of Fungi. *Mycologia* **1975,** *67* (2), 362–366.

Kritzman, G.; Chet, I.; Henis, Y. The Relationship Between Rhythmic Hyphal Growth and Circadian Formation of Sclerotia in *Sclerotium rolfsii* Sacc. *Can. J. Microbiol.* **1977,** *23* (8), 959–963.

Kritzman, G.; Chet, I.; Henis, Y. Isolation of extracellular polysaccharides from *Sclerotium rolfsii. Can. J. Bot.* **1979,** *57* (18), 1855–1859.

Kritzman, G.; Chet, I.; Henis, Y.; Huttermann, A. The Use of the Brightener "Calcofluor White M2R New" in the Study of Fungal Growth. *Israel J. Bot.* **1978,** *27*: 138–146.

Kritzman, G.; Okon, Y.; Chet, I.; Henis, Y. Metabolism of L-threonine and Its Relationship to Sclerotium Formation in *Sclerotium rolfsii. Microbiology* **1976,** *95* (1), 78–86.

Kumar, A.; Azad, C. S.; Kumar, R.; Kumar, R.; Imran, M. Soil Solarization: An Eco-Friendly and Effective Disease Management Approach for Soil Borne Diseases. *J. Pharmacogn. Phytochem.* **2017,** *SP1*, 513–515.

Kumar, R. *Studies on Decomposing Fungi of Sesbania aculeata L. in Soil and Its Effects on Soil Borne Plant Pathogens*; Ph. D. Thesis. Banaras Hindu University, Varanasi, 2010.

Kumar, R.; Sinha, A.; Srivastava, S.; Mahajan, G. Evaluation of Biocontrol Potential of Some Fungal Decomposers of *Sesbania aculeata* L. Green Manure Against Some Soil-Borne Plant Pathogens. *J. Environ. Biol.* **2017a,** *38*, 37–45.

Kumar, R.; Sinha, A.; Srivastava, S.; Singh, S. Evaluation of Substrates for Mass Multiplication of Green Manure Associated Fungi for Biological Control of Soil Borne Phytopathogens. *Indian Phytopathol.* **2014,** *67* (4), 396–401.

Lane, D.; Denton-Giles, M.; Derbyshire, M.; Kamphuis, L. G. Abiotic conditions governing the myceliogenic germination of *Sclerotinia sclerotiorum* allowing the basal infection of Brassica napus. *Austral. Plant Pathol.* **2019,** *48*, 85–91.

Leona, G.; Sudhakar, R.; Uma Devi, G.; Uma Maheswari, T. Management of Stem Rot of Groundnut caused by *Sclerotium rolfsii* Sacc. with Actinomycetes. *Int. J. Curr. Microbiol. App. Sci.* **2020,** *9* (12), 3587–3601.

Linderman, R. G.; Gilbert, R. G. Stimulation of *Sclerotium rolfsii* in Soil by Volatile Components of Alfalfa Hay. *Phytopathology* **1969,** *59*, 1366–1372.

Linderman, R. G.; Gilbert, R. G. Behaviour of Sclerotia of *Sclerotium rolfsii* Produced in Soil or in Culture Regarding Germination Stimulation by Volatiles, Fungistasis, Sodium Hypochlorite Treatment. *Phytopathology* **1974,** *63*, 500–503.

Liu, T. M.; Yu, P. H.; Wu, L. C. The Isolation, Purification, and Identification of Dark Pigment in Sclerotia of *Sclerotium rolfsii* Sacco. *Plant Prot. Bull. (Taiwan)* **1977,** *19*, 223–37.

Liu, T. M. E.; Wu, L. C. Effect of Amino Acids on the Growth and Morphogenesis of *Sclerotium rolfsii* Sacc. *Plant Prot Bull Taiwan* **1971,** *13*, 87–96.

Lodha, S. Soil Solarisation: An Eco-friendly Approach to Manage Soil Borne Plant Pathogens. *CAZRI (Den News)* **2011,** *13* (1), 1–4.

Madhavi, G. B.; Bhattiprolu, S. L. Integrated Disease Management of Dry Root Rot of Chilli Incited by *Sclerotium rolfsii* (Sacc.). *Int. J. Plant Anim. Environ. Sci.* **2011,** *1* (2), 31–37.

Madi, L.; Katan, T.; Katan, J.; Henis, Y. Biological Control of *Sclerotium rolfsii* and *Verticillium dahliae* by *Talaromyces flavus* Is Mediated by Different Mechanisms. *Phytopathology* **1997,** *87* (10), 1054–1060.

Mahadeva kumar, S.; Tejaswini, G. S.; Janardhana, G. R.; Yadav, V. First Report of *Sclerotium rolfsii* Causing Southern Blight and Leaf Spot on Common Bean (*Phaseolus vulgaris*) in India. *Plant Dis.* **2015,** *99* (9), 1280.

Mahato, A.; Biswas, M. K. and Patra, S. Efficacy of Medicinal Plant Extracts Against Collar Rot of Tomato Caused by *Sclerotium rolfsii* (Sacc.). *Int. J. Microbiol. Res.* **2018,** *10* (5), 1224–1227.

Maji, A.; Nath, R. Management Through Eco-Friendly Approach of Collar Rot Disease of Sunflower Caused by *Sclerotium rolfsii* Sacc in Sundarbans of West Bengal. *Int. J. Agric. Sci. Res.* **2016,** *6* (4), 9–12.

Manjula, K.; Kishore, G. K.; Girish, A. G.; Singh, S. D. Combined Application of *Pseudomonas fluorescens* and *Trichoderma viride* Has an Improved Biocontrol Activity Against Stem Rot in Groundnut. *Plant Pathol. J.* 20 (1), 75–80.

Manu, T. G.; Nagaraja, A.; Chetan, Janaward, S.; Hosamani, V. Efficacy of Fungicides and Biocontrol Agents Against *Sclerotium rolfsii* Causing Foot Rot Disease of Finger Millet, Under In Vitro Conditions. *Global J. Biol. Agric. Health Sci.* **2012,** *1* (2), 46–50.

Mathur, S. B. Physiological studies on *Sclerotium rolfsii*-Il. Utilization of Inorganic and Organic Nitrogen Sources. *Indian Phytopathol* **1977,** *30*, 131–132.

Mathur, S. B.; Sarbhoy, A. K. Physiological Studies on *Sclerotium rolfsii* Causing Root Rot of Sugarbeet. *Indian Phytopathol.* **1976,** *29*, 454—455.

Mathur, S. B.; Sarbhoy, A. K. Physiological Studies on *Sclerotium rolfsii* Causing Root Rot of Sugarbeet. *Indian Phytopathol.* **1977,** *30*, 384–387.

Maxwell, D. P.; Bateman, D. F. Influence of Carbon Source and pH on Oxalate Accumulation in Culture Filtrates of *Sclerotium rolfsii*. *Phytopathology* **1968,** *58* (10), 1351.

Melhuish Jr, J. H.; Bean, G. A. Effect of Dimethyl Sulfoxide on the Sclerotia of *Sclerotium rolfsii*. *Can. J. Microbiol.* **1971,** *17* (3), 429–431.

Miller, R. M.; Liberta, A. E. The Effects of Light and Tyrosinase During Sclerotium Development in *Sclerotium rolfsii* Sacc. *Can. J. Microbiol.* **1977,** *23* (3), 278–287.

Mitchell, D. J.; Mitchell, J. E. Oxygen and Carbon Dioxide Concentration Effects on the Growth and Reproduction of *Aphanomyces euteiches* and Certain Other Soil-Borne Plant Pathogens. *Phytopathology* **1973,** *63*, 1053–1059.

Mullen, J. Southern Blight, Southern Stem Blight, White Mold. http://www.apsnet.org/edcenter/intropp/lessons/fungi/Basidiomycetes/Pages/SouthernBlight.aspx

Murmu, J. J.; Tiwari, S.; Jennifer, G. M. J.; Singh, S. Eco-Friendly Management of *Sclerotium rolfsii* Causing Collar Rot of Brinjal (*Solanum melongena* L.). *Int. J. Curr. Microbiol. App. Sci.* **2021,** *10* (03), 373–381.

Mustafa, T. P.; Chaltopadhyay, S. B. Effect of Soil Moisture on the Growth of *Macrophomina phaseoli, Sclerotium rolfsii*, and *Fusarium solani* in Soil. *Indian J. Microbiol* 1971a, *11*, 77–82.

Mustafa, T. P.; Chattopadhyay, S. B. Effect of Soil Temperature on the Growth of *Macrophomina phaseoli* and *Sclerotium rolfsii* in Soil. *Indian J. Microbiol.* **1971,** *11*, 83–86.

Naeimi, S.; Khosravi, V.; Varga, A.; Vágvölgyi, C.; Kredics, L. Screening of Organic Substrates for Solid-State Fermentation, Viability and Bio-efficacy of *Trichoderma harzianum* AS12–2, a Biocontrol Strain Against Rice Sheath Blight Disease. *Agronomy* **2020,** *10*, 1258.

Nagamma, G. and Nagaraja, A. Efficacy of Biocontrol Agents Against *Sclerotium rolfsii* Causing Collar Rot Disease of Chickpea, Under In Vitro Conditions. *Int. J. Plant Protect.* **2015,** *8* (2), 222–227.

Nair, N. G.; White, N. H.; Griffin, D. M.; Blair, S. Fine Structure and Electron Cytochemical Studies of Sclerotium rolfsii Sacc. *Austral. J. Biol. Sci.* **1969,** *22* (3), 639–652.

Nandi, S.; Hembaram, S.; Adhikari, A.; Tiwari, B. K.; Dutta, S. Host Infection Beyond the Traditional Range of *Sclerotium* (*Athelia*) *rolfsii* with *Physalis minima*. *Bioinformation* **2017,** *13* (10), 333.

Narain, A.; Satapathy, J. N. Antifungal Characterstics of *Vinca rosea* Extracts. *Indian Phytopath.* **1997,** *30*, 36–40.

Narasimhan, R. Physiological Studies on the Genus *Sclerotium*. II. Utilization of Inorganic Nitrogen Sources by *Sclerotium rolfsii* (Sacc.) and *Sclerotium oryzae* (Calt.) Under Protracted Incubation. *Proc. Ind. Acad. Sci.* **1969,** *49*, 42–54.

Okemob, P. O.; Bais, H. P.; Vivancoa, J. M. In Vitro Activities of *Maesa lanceolata* Extracts Against Fungal Plant Pathogens. *Fitoterapia* **2003,** *74*, 312–316.

Okon, Y.; Chet, I.; Henis, Y. Lactose-Induced Synchronous Sclerotium Formation in *Sclerotium rolfsii* and Its Inhibition by Ethanol. *Microbiology* **1972,** *71* (3), 465–470.

Okon, Y.; Chet, I.; Henis, Y. Effects of Lactose, Ethanol and Cycloheximide on the Translocation Pattern of Radioactive Compounds and on Sclerotium Formation in *Sclerotium rolfsii*. *Microbiology* **1973,** *74* (2), 251–258.

Okon, Y.; Chet, I.; Henis, Y. The Effect of Glucose and Lactose on β-d-Galactosidase Activity and Formation of Sclerotia in *Sclerotium rolfsii*. *Can. J. Microbiol.* **1975,** *21* (7), 1123–1126.

Okon, Y.; Chet, I.; Kislev, N.; Henis, Y. Effect of Lactose on Soluble-Glucan Production and on the Ultrastructure of *Sclerotium rolfsii* Sacc. Grown in Submerged Culture. *Microbiology* **1974,** *81* (1), 145–149.

Ons, L.; Bylemans, D.; Thevissen, K.; Cammue, B. P. A. Combining Biocontrol Agents with Chemical Fungicides for Integrated Plant Fungal Disease Control. *Microorganisms* **2020,** *8* (12), 1930.

Ordentlich, A.; Elad, Y.; Chet, I. The role of chitinase of *Serratia marcescens* in biocontrol of *Sclerotium rolfsii*. *Phytopathology* **1988,** *78* (1), 84–88.

Pandey, M. K.; Singh, D. P.; Singh, U. P. Phenolic Acid Changes in Mycelia of *Sclerotium rolfsii* After Garlic and Onion Supplementation in a Broth Medium. *Mycobiology* **2005,** *33* (3), 137–141.

Patil, M. R.; Waghe, S. V.; Wangikar, P. D.; Khune, N. N. Chemical Control of Wilt of betel vine. *Pesticides* **1986,** *20*, 28–29.

Pattanapipitpaisal, P.; Kamlandharn, R. Screening of Chitinolytic Actinomycetes for Biological Control of *Sclerotium rolfsii* Stem Rot Disease of Chilli. *Songklanakarin J. Sci. Technol.* **2012,** *34* (4), 387–393.

Pavithra, G.; Bindal, S.; Rana, M.; Srivastava, S. Role of Endophytic Microbes Against Plant Pathogens: A Review. *Asian J. Plant Sci.* **2020,** *19*, 54–62.

Prabhu, H. V.; Hiremath, P. C. Bio-Efficacy of Fungicides Against Collar Rot of Cotton Caused by *Sclerotium rolfsii* Sacc. *Karnataka J. Agric. Sci.* **2003,** *16* (4), 576–579.

Prasad, G.; Dwivedi, S. K. Efficacy of Bioagents, Plant Extracts and Chemical Fungicides Against Root Rot Disease in *Zea mays* Caused by *Sclerotium rolfsii*. *Asia Pac. J. Multidisc. Res.* **2020,** *8* (1), 47–55.

Punja, Z. K. The Biology, Ecology, and Control of *Sclerotium rolfsii*. *Ann. Rev. Phytopathol.* *23* (1), 97–127.

Punja, Z. K.; Damiani, A. Comparative Growth, Morphology, and Physiology of Three *Sclerotium* Species. *Mycologia* **1996,** *88* (5), 694–706.

Punja, Z. K.; Grogan, R. G. Eruptive Germination of sclerotia of *Sclerotium rolfsii*. *Phytopathology* **1981a,** *71* (10), 1092–1099.

Punja, Z. K.; Grogan, R. G. Mycelial Growth and Infection Without a Food Base by Eruptively Germinating Sclerotia of *Sclerotium rolfsii*. *Phytopathology* **1981b,** *71* (10), 1099–1103.

Punja, Z. K.; Grogan, R. G. Effects of Inorganic Salts, Carbonate-Bicarbonate Anions, Ammonia, and the Modifying Influence of pH on Sclerotial Germination of *Sclerotium rolfsii*. *Phytopathology* **1982,** *72* (6), 635–639.

Punja, Z. K.; Huang, J. S.; Jenkins, S. F. Relationship of Mycelial Growth and Production of Oxalic Acid and Cell Wall Degrading Enzymes to Virulence in *Sclerotium rolfsii*. *Can. J. Plant Pathol.* **1985,** *7* (2), 109–117.

Punja, Z. K.; Jenkins, S. F. Influence of Temperature, Moisture, Modified Gaseous Atmosphere, and Depth in Soil on Eruptive Sclerotial Germination of *Sclerotium rolfsii*. *Phytopathology* **1984,** *74* (6), 749–754.

Punja, Z. K.; Jenkins, S. F.; Grogan, R. G. Effect of Volatile Compounds, Nutrients, and Source of Sclerotia on Eruptive Sclerotial Germination of *Sclerotium rolfsii*. *Phytopathology* **1984,** *74* (11), 1290–1295.

Quadri, U.; Fatima, S. Eco-Friendly Management of *Sclerotium rolfsii* Causing Root Rot Disease in Chilli. *Int. J. Adv. Res.* **2016,** *4* (5), 672–678.

Rakh, R. R.; Raut, L. S.; Dalvi, S. M.; Manwar, A. V. Biological Control of *Sclerotium rolfsii*, Causing Stem Rot of Groundnut by *Pseudomonas* cf. *monteilii* 9. *Recent Res. Sci. Technol.* **2011,** *3* (3), 26–34.

Ramarao, P.; Raja, U. Effect of Soil Moisture on Development of Foot and Root Rot of Wheat and on Other Soil Microflora. *Indian J. Mycol. Plant Pathol.* **1980,** *10* (1), 17–22.

Rangarani, A.; Rajan, C. P. D.; Harathi, P. N.; Bhaskar, B.; Sandhya, Y. Evaluation of Fungicides and Heribicides on *Sclerotium rolfsii*, Incitant of Stem Rot Diseases in Groundnut (*Arachis hypogeal* L.). *Int. J. Pure App. Biosci.* **2017,** *5* (3), 92–97.

Rangeshwaran, R.; Prasad, R. D. Biological Control of *Sclerotium*-Rot of Sunflower. *Indian Phytopathol.* **2000,** *53* (4), 444–449.

Rasu, T.; Sevugapperumal, N.; Thiruvengadam, R.; Ramasamy, S. Biological Control of Sugarbeet Root Rot Caused by *Sclerotium rolfsii*. *Int. J. Biol. Ecol. Environ. Sci.* **2013,** *2* (1), 7–10.

Reddy, H. R.; Kulkarni, B. G. P.; Hegde, R. K. Studies on Chemical Control of Foot Rot on Wheat in Karnataka. *Mysore J. Agric. Sci.* **1976,** *10* (3), 428–431.

Sahana, B.; Manjunatha Reddy, T. B.; Mushrif, S. K.; Anjaneya Reddy, B.; Doddabasappa, B. Evaluation of Fungicides Against Stem Rot of Capsicum Caused by *Sclerotium rolfsii* Sacc. *Int. J. Chem. Stud.* **2020,** *8* (4), 306–312.

Saleh, I.; Abu-Dieyeh, M. H. Novel *Prosopis juliflora* Leaf Ethanolic Extract as Natural Antimicrobial Agent Against Food Spoiling Microorganisms. *Sci. Rep.* **2021,** *11*, 7871.

Sarma, B. K.; Singh, U. P. Ferulic Acid May Prevent Infection of *Cicer arietinum* by *Sclerotium rolfsii*. *World J. Microbiol. Biotech.* **2003,** *19*, 123–127.

Sesha Kiran, K.; Lingaraju, S.; Adiver, S. S. Effect of Plant Extracts on *Sclerotium rolfsii*, the Incitant of Stem Rot of Groundnut. *J. Mycol. Plant Pathol.* **2006,** *36* (1), 77–79.

Shahid, M. A.; Ahmad, M.; Khan, M. A. Chemical Control of Collar Rot of Lentil Caused by *Sclerotium rolfsii* Sacc. *Sarhad J. Agric.* **1990,** *6* (5), 503–507.

Shamsi, S.; Naher, N. In Vitro Management of *Sclerotium rolfsii* Sacc.—The Causal Agent of Stem Rot of Sesame (*Sesamum indicum* L.). *Int. J. Bioinfo. Biol. Sci.* **2020,** *8* (2), 1–6.

Sharf, W.; Javaid, A.; Shoaib, A.; Khan, I. H. Induction of Resistance in Chili Against *Sclerotium rolfsii* by Plant-Growth-Promoting Rhizobacteria and *Anagallis arvensis*. *Egypt J. Biol. Pest Control* **2021,** *31*, 16.

Sharma, S. L.; Kaushal, B. R. Cultural and Physiological Studies with Sunflower Isolate of *Sclerotium rolfsii*. *Indian J. Mycol. Plant Pathol.* **1979,** *9*, 105–107.

Shew, B. B.; Beute, M. K.; Campbell, C. L. Spatial Pattern of Southern Stem Rot Caused by *Sclerotium rolfsii*. *Phytopathology* **1984,** *74*, 730–735.

Shirsole, S. S.; Khare, N.; Lakpale, N.; Kotasthane, A. S. Evaluation of Fungicides Against *Sclerotium rolfsii* Sacc. Incitant of Collar Rot of Chickpea. *Pharma Innov. J.* **2019,** *8* (12), 310–316.

Shwetha, G. S.; Hegde, Y. R. Management of *Sclerotium* Wilt of *Stevia rebaudiana* Through Biorationals. *Int. J. Plant Protect*. **2012,** *5* (2), 248–251.

Siddiqui, I.; Bokhari, N. A.; Perveen, K. Antifungal Ability of *Nerium oleander* Against *Fusarium oxysporum*, *Sclerotium rolfsii* and *Macrophomina phaseolina*. *J. Anim. Plant Sci.* **2016,** *26* (1), 269–274.

Singh, S.; Nirmalkar, V. K.; Tiwari, R. K. S.; Jangre, A.; Kumar, P. Integration of *Trichoderma*, *Pseudomonas* and Fungicides for the Control of Collar Rot Disease of Chickpea (*Cicer arietinum* L.). *Int. J. Agric. Environ. Biotech.* **2017,** *10* (1), 125–131.

Singh, S. P.; Gaur, R. Evaluation of Antagonistic and Plant Growth Promoting Activities of Chitinolytic Endophytic Actinomycetes Associated with Medicinal Plants Against *Sclerotium rolfsii* in Chickpea. *J. Appl. Microbiol.* **2016,** *121* (2), 506–518.

Singh, S. P.; Gupta, R.; Gaur, R.; Srivastava, A. K. *Streptomyces* spp. Alleviate *Rhizoctonia solani*-Mediated Oxidative Stress in *Solanum lycopersicon*. *Ann. Appl. Biol.* **2016,** *168* (2), 232–242.

Singh, S. P.; Gupta, R.; Gaur, R.; Srivastava, A. K. Antagonistic Actinomycetes Mediated Resistance in *Solanum lycopersicon* Mill. Against *Rhizoctonia solani* Kühn. *Proc. Natl. Acad. Sci. India Sect. B Biol. Sci.* **2017,** *87* (3), 789–798.

Smith, A. M. Drying and Wetting Sclerotia Promotes Biological Control of *Sclerotium rolfsii* Sacc. *Soil Biol. Biochem.* **1972,** *4* (2), 119–120.

Sood, M.; Kapoor, D.; Kumar, V.; Sheteiwy, M. S.; Ramakrishnan, M.; Landi, M.; Araniti, F.; Sharma, A. *Trichoderma*: The "Secrets" of a Multitalented Biocontrol Agent. *Plants (Basel)* **2020,** *9* (6), 762.

Soytong, K. Species of *Chaetomium* in Thailand Soils. *Thai Phytopathol.* **1991,** *11* (3–4), 86–94.

Soytong, K.; Soytong, K.; Chaetomium as a New Broad Spectrum Mycofungicide. In: *Proceedings of the 1st International Symposium on Biopesticides*; pp 124–132.

Sravani, B.; Chandra, R. In Vitro and In Vivo Evaluation of Chemical Fungicides against *Sclerotium rolfsii* Causing Collar Rot of Chickpea. *Biol. Forum: Int. J.* **2021,** *13* (2), 10–16.

Srinivasulu, B.; Krishna Kumar, K. V.; Aruna, K.; Krishna Prasadji, J.; Rao, D. V. R. In Vitro Antagonism of Three *Trichoderma* spp. Against *Sclerotium rolfsii* Sacc., a Collar-Rot Pathogen in Elephant Foot Yam. *J. Biol. Control.* **2005,** *19* (2), 167–171.

Suheri, H.; Isnaini, M.; Rohyadi, A. Application of Organic Amendment and *Trichoderma* sp. to Control Basal *Sclerotium rolfsii* on Peanut Grown on Partially Degraded Land. *J. Deg. Mining Lands Manag.* **2014,** *1* (3), 131–136.

Sultana, J. N.; Pervez, Z.; Rahman, H.; Islam, M. S. In Vitro Evaluation of Different Strains of *Trichoderma harzianum* and *Chaetomium globosum* as Biological Control Agent's Seedling Mortality of Chilli. *Bangladesh Res. Pub. J.* **2012,** *6* (3), 305–310.

Thakur, K. S.; Keshry, P. K.; Tamrakar, D. K.; Sinha, A. K. Studies on Management of Collar Rot Disease (*Sclerotium rolfsii*) of Chickpea by Use of Fungicides. *Adv. Plant Sci.* **2004,** *17* (2), 553–555.

Thiessen, L. D.; Woodward, J. E. Diseases of Peanut Caused by Soilborne Pathogens in the Southwestern United States. *Int. Scholarly Res. Netw. ISRN Agron.* **2012,** *2012*, ID 517905.

Tiwari, R. K.; Ashok, S. Efficacy of Fungicides on *Rhizoctonia solani* and *Sclerotium rolfsii* and Their Effect on *Trichoderma harzianum* and *Rhizobium leguminosarum*. *J. Mycol. Plant Pathol.* **2004,** *34* (2), 482–485.

Tourneau, D. L. Effects of Phenylthiourea on Growth and Sclerotial Formation of *Sclerotium rolfsii* and *Whetzelinia sclerotiorum*. *Can. J. Microbiol* **1976,** *22* (3), 379–383.

Tourneau, D. L. Inhibition of Sclerotial Formation of *Sclerotium rolfsii* by Cadmium. *Mycologia* **1978,** *70* (4), 849–852.

Trevethick, J.; Cooke, R. C. Effects of Some Metabolic Inhibitors and Sulphur-Containing Amino Acids on Sclerotium Formation in *Sclerotium rolfsii, S. delphinii* and *Sclerotinia sclerotiorum*. *Trans. Br. Mycol. Soc.* **1971,** *57* (2), 340–342.

Trevethick, J.; Cooke, R. C. Non-Nutritional Factors Influencing Sclerotium Formation in Some *Sclerotinia* and *Sclerotium* Species. *Trans. Br. Mycol. Soc.* **1973,** *60* (3), 559–566.

Tsahouridou, P. C.; Thanassoulopoulos, C. C. Proliferation of *Trichoderma koningii* in the Tomato Rhizosphere and the Suppression of Damping off by *Sclerotium rolfsii*. *Soil Biol. Biochem.* **2002,** *34* (6), 767–776.

Tu, C. C.; Kimbrough, J. W. Morphology, Development, and Cytochemistry of the Hyphae and Sclerotia Of Species in the *Rhizoctonia* Complex. *Can. J. Bot.* **1975,** *53* (20), 2282–2296.

Ünal, F.; Aşkın, A.; Koca, E.; Yıldırır, M.; Bingöl, MU Mycelial Compatibility Groups, Pathogenic Diversity and Biological Control of *Sclerotium rolfsii* on Turfgrass. *Egypt J. Biol. Pest Control* **2019,** *29*, 44.

Verma, J.; Dubey, N. K. Prospective of Botanical and Microbial Products as Pesticides of Tomorrow. *Curr. Sci.* **1999,** *76*, 172–179.

Verma, V. C.; Gond, S. K.; Kumar, A.; Mishra, A.; Kharwar, R. N.; Gange, A. C. Endophytic Actinomycetes from *Azadirachta indica* A. Juss.: Isolation, Diversity, And Anti-Microbial Activity. *Microbial. Ecol.* **2009,** *57* (4), 749–756.

Wavare, S. H.; Gade, R. M.; Shitole, A. V. Effect of Plant Extracts, Bio Agents and Fungicides against *Sclerotium rolfsii* Causing Collar Rot in Chickpea. *Indian J. Pharm. Sci.* **2017,** *79* (4), 513–520.

Weerapat, P.; Schroeder, H. W. Effect of soil Temperature on Resistance of Rice to Seedling Blight Caused by *Sclerotium rolfsii*. *Phytopathology* **1966,** *56* (6), 640–644.

Wheeler, B. E. J. Effect of Ethanol on Production of Sclerotia by *Sclerotium rolfsii*. *Trans. Br. Mycol. Soc.***1972,** *59* (3), 453–461.

Wheeler, B. E. J.; Sharan, N. The Production of sclerotia by *Sclerotium rolfsii*: I. Effects of Varying the Supply of Nutrients in an Agar Medium. *Trans. Br. Mycol. Soc.* **1965,** *48* (2), 291-IN14.

Wheeler, B. E. J.; Waller, J. M. The Production of Sclerotia by *Sclerotium rolfsii*: II. The Relationships Between Mycelial Growth and Initiation of Sclerotia. *Trans. Br. Mycol. Soc.* **1965,** *48* (2), 303–314.

Wilcoxson, R. D.; Subbarayudu, S. Translocation to and Accumulation of Phosphorus-32 in Sclerotia of *Sclerotium rolfsii*. *Can. J. Bot.* *46* (1), 85–88.

Willetts, H. J. The Survival of Fungal Sclerotia Under Adverse Environmental Conditions. *Biol. Rev.* **1971,** *46* (3), 387–407.

Willetts, H. J. The Morphogenesis and Possible Evolutionary Origins of Fungal Sclerotia. *Biol. Rev.* **1972,** *47* (4), 515–536.

Wokocha, R. C. Integrated Control of *Sclerotium rolfsii* Infection of Tomato in the Nigerian Savanna: Effect of *Trichoderma viride* and Some Fungicides. *Crop Protect.* **1990,** *9* (3), 231–234.

Wu, F.; Wang, W.; Ma, Y.; Liu, Y.; Ma, X.; An, L.; Feng, H. Prospect of Beneficial Microorganisms Applied in Potato Cultivation for Sustainable Agriculture. *Afr. J. Microbiol. Res.* **2013,** *7* (20), 2150–2158.

Yaqub, F.; Shahzad, S. Effect of Solar Heating by Polyethylene Mulching on Sclerotial Viability and Pathogenicity of *Sclerotium rolfsii* on Mungbean and Sunflower. *Pak. J. Bot.* **2009,** *41* (6), 3199–3205.

Zoberi, M. H. Some Nutritional Factors Regulating Formation of Sclerotia of *Sclerotium rolfsii*. *Can. J. Bot.* **1980,** *58* (23), 2484–2490.

CHAPTER 7

Management of Disease-Transmitting Insect Vectors

K. M. G. CHANCHALA

Sugarcane Research Institute, Sri Lanka

ABSTRACT

Transmission of plant pathogens from a host to another host is a vital step in the disease cycle as the immobility of plants, the strong pectin barrier, and cellulose barrier surrounding cells. Phytoplasma and virus species mostly uses insect vectors for exit, transfer, and access from one host to another, since microbes develop within the plant hosts and pursue for the same resources as herbivorous insects. Aphids, leaf hoppers, white flies, and mealy bugs are the major groups of insect vectors and occasionally mites and nematodes also transmit the pathogens. Plant pathogens, especially viruses and phytoplasma, are capable of affecting physiology and behavior of vectors to enhance their probabilities of transmission directly or through the modification of the host plant. Severity of the plant disease depends basically on two factors viz., presence and viruliferous nature of vector and inoculum density of the pathogen. Therefore, management of the available vector population is at uppermost importance in disease management. Several studies have been conducted on disease vector management during the previous years. Based on vector biology, ecology and tritrophic relationship among plant, vector and pathogen, several management strategies have been developed throughout the world viz., resistant plants for vector, clean seeds/propagation materials, roguing infected materials, habitat management, weed management, mulching, physical control, natural enemies, genetic manipulations, and pesticide application. This chapter discusses the importance of the vector management, its economics, and important vector management strategies.

Plant Diseases and Their Management: A Sustainable Approach.
Deepa Srivastava, Rajarshi Kumar Gaur, and Ajay K. Tiwari (Eds.)

7.1 INTRODUCTION

Transmission of plant pathogens from host to host is a vital step in the disease cycle as the immobility of plants, the strong pectin barrier, and cellulose barrier surrounding cells. Phytoplasma and virus species mostly uses insect vectors for exit, transfer, and access from one host to another, since microbes develop within the plant hosts and pursue for the same necessities as herbivorous insect vector. Aphids, leaf hoppers, white flies, and mealy bugs are the major groups of insect vectors and occasionally mites and nematodes also transmit the pathogens. Plant pathogens, especially viruses and phytoplasma, are capable of affecting physiology and behavioral aspects of vectors to increase their probabilities of transmission, either directly or through the alteration of the host plant.

Among all plant pathogens, viruses signify a prevalent and diverse group, and the transmission of many plant viruses is obligately reliant on insect vectors (Nault, 1997; Hogenhout et al., 2008). Phytoplasmas are noncultivable, degenerate, Gram-positive prokaryotes, causing numerous diseases in plants (Lee et al., 2000; Bertaccini, 2007). They are phloem-limited and thereby, they are spread by specific phloem-feeding Auchenorrhyncha and Sternorrhyncha. It was found that some genera within certain tribes seem to be virus vectors (e.g., Cicadulina species), while others appear only to contribute in the transmission of phytoplasma. The spread and development of plant diseases in general are influenced by inoculum density, the presence of a vector and movements of vectors in relation to healthy and disease-infected plants. Those factors are theoretically key influences of pathogen spread (Sisterson, 2009).

Vector-borne diseases are globally associated with numerous plant species (McCoy et al., 1989). It has been identified as pathogens in several plant categories and in some cases, it has triggered severe epidemics in key crops such as grapevine, sugarcane, coconut, ornamentals, vegetables, and fruit trees (Marcone et al., 1999; Lee et al., 2000). Some of these diseases, specifically those of woody plants, are fatal. The numbers of plants and insects identified to harbor these vector-borne pathogens are constantly increasing (Seemüller et al., 1998, 2002; Lee et al., 2000). Managing vectors in economically important plants varies for annual versus perennial crops, as harms in annual crops are characteristically seasonal, while losses in perennial plant cropping systems accumulate over the years. This chapter discusses the management strategies of disease-transmitting insect vectors comprehensively.

7.2 DISEASE VECTORS

7.2.1 VECTOR?

Vector is an organism that transmits pathogens to plant hosts that the pathogen can then infect. The various types of organisms that can act as vectors viz., Aphids, leaf hoppers, white flies, mealy bugs and with the vector species pathogen or disease transmitting can vary. Several species of the order Homoptera viz., aphids, leafhoppers, plant hoppers, whiteflies, and mealy bugs are important vectors of viruses, while leafhoppers are also the key vectors of phytoplasmas. Depending on the insect spp., the relationship between the pathogen and the insect and the prevailing weather conditions, especially wind, insects can transmit pathogens over short or long expanses. Dispersion of the pathogen depends entirely, or to a large extent, on the performance of apposite vectors.

7.2.2 METHODS OF TRANSMISSION BY VECTORS

Based on the duration of pathogen retained in the vector, Fereres and Raccah (2015) reviewed three transmission methods for vector-borne pathogens.

7.2.2.1 NONPERSISTENT/STYLET-BORNE VECTORS (NP)

Nonpersistent vectors retain pathogen for short period "less than the time the pathogen persists in leaf extracts." They were known as stylet-borne vectors. NP pathogens are acquisitioned and inoculated within brief periods of probing (Fig. 7.1a). They do not necessitate a latent period in the vector and are transferred by several species of aphids, mostly by those not inhabiting the crop.

7.2.2.2 SEMI-PERSISTENT VECTORS (SP)

Some pathogens specify an intermediary retaining in their vector. SP pathogens prerequisite longer periods (hours) for pathogen acquisition and transmission than do NP pathogens. They also have a finer range of vector species. However, they do not necessitate latent period and are absent when the vectors are moulting.

7.2.2.3 PERSISTENT VECTORS (P)

These are persistent for prolonged retention, often for whole life period. In persistent viruses, vectors needed several hours or even days for efficient acquisition and inoculation. They have a narrow range of vectors, which mostly colonies in the crop pass through moults and prerequisite a latent period. Two prime modes of transmission occurred circulative (CV) or internal, where the virus crosses gut barriers and enters the circulatory system of the insect and accumulates inside the salivary glands, and non-circulative (NC) or external, where the virus deposited attached to the cuticle, mouth-parts, or foregut (Fig. 7.1b) of the insect and did not pass the gut barriers.

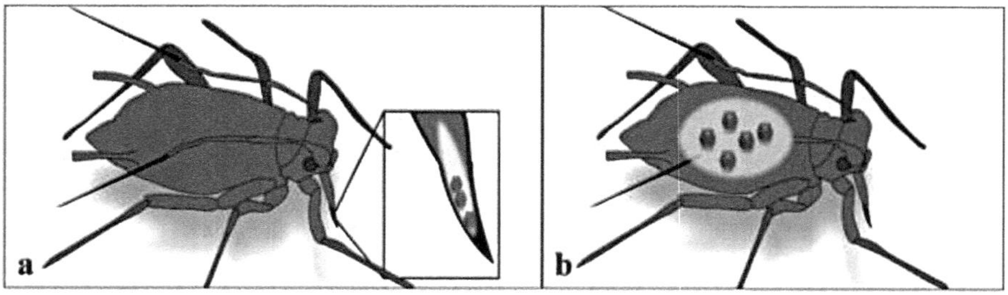

FIGURE 7.1 (a)Nonpersistent /stylet-borne vectors (NP) and (b) persistent vectors (P).

7.2.3 MAJOR INSECT GROUPS VECTOR THE VIRUS AND PHYTOPLASMA

Aphids

Having piercing and sucking mouthparts aphids test the sap in the epidermal cells of several leaves at site selection for feeding. Then feed on the selected plant leaves, cells in the phloem or xylem. During the process of feeding they secrete saliva into the feeding sight releasing virus inoculum into the host plant. Aphids transmit viruses belong to few genera, that is, *Potyvirus*, *Cucumovirus*, and *Luteovirus* in each manner of transmission.

Whiteflies

Whiteflies are small insects with piercing and sucking mouthparts, which feed on crops which are main sources of carbohydrate and protein. Whiteflies mostly transmit diseases with economic importance viz., tomato yellow leaf curl virus, chili curl virus, lettuce infectious yellow virus, cassava African mosaic virus, and bean viruses in tropical areas. Occasionally, they also transfer some viruses in subtropical and temperate areas. Viruses transmitted by whiteflies often cause yellow mosaic, yellow veins, and leaf distortion. They are transferred in a persistent manner but not in a circulative manner.

They transmit viruses from many genera but mostly those in the genus *Begomovirus*.

Hoppers

Leafhoppers, plant hoppers, and treehoppers are also important as virus vectors. Hoppers also have piercing and sucking mouth parts and acquire virus particles from the phloem during feeding process. Viruses transferred by hoppers cause yellowing and leaf-rolling in general. Method of disease transmission by several hopper vectors varies. In their group some are nonpersistent vectors while others act semi-persistent or persistent vectors. Rice tungro virus and maize chlorotic viruses are transmitted by leafhoppers in a semi-persistent manner. Hopper spp also transmit phytoplasmas in a persistent or semi persistent manner.

Thrips

Thrips mostly transmit viruses that belong to genus *Tospovirus*. It is found that tomato spotted wilt virus (TSWV), a tospovirus, is transmitted in a persistent manner.

Mealy Bugs

Mealy bugs feed on the phloem cells of the host plant and transmit viruses in semi-persistent manner. Mealy bugs are not very efficient vectors as they crawl from plant to plant, unless they are dispersed by wind or ants. As a result, mealy bug-borne viruses spreading is slow (Cacao swollen shoot virus, West Africa).

7.2.4 CHARACTERISTICS OF EFFICIENT VECTORS

Hanboonsong et al. (2006) and Weintraub and Beanland (2006) recorded that to become efficient vector, particular vectors should possess several characteristics viz.,

i. Abundant in the relevant crop cultivating areas
ii. Populations dynamics match with the cycle of the plant disease, that is, at peak when the disease is most prevailing
iii. Vector must be hemi-metabolous (their nymphs and adults feed same way and in the same physical locality and often both juvenile and adults can transmit phytoplasma)

iv. Specific and selective feeding on definite plant tissues makes them efficient vectors of pathogens that exist in those tissues. Also, their feeding is nondestructive, promoting efficacious inoculation of the plant vascular system without injuring conductive tissues and causing cautious responses
v. Having a propagative and persistent relationship with pathogen
vi. Having obligate symbiotic prokaryotes that are transmitted to the offspring by transovarial transmission, which is the same mechanisms that permit the transovarial transmission of phytoplasmas

7.3 IMPORTANCE OF MANAGEMENT

Economic threshold level for vectors takes lower number comparatively to the normal pest as even one viruliferous vector would be able to transmit disease efficiently in a favorable environmental condition.

Vector transmitted plant pathogens may cause severe losses to several annual and perennial crop species. On occasion, insects are accountable for substitution from a nonspreading form to the epidemic form of diseases. Outbreaks of disease caused by insect vectors are demonstrated in both perennial and annual crops. In perennials, citrus industry faced a total destruction in the 1930s in Argentina and Brazil due to the aphid *Toxoptera citricida*. Also in annual crops, outbreaks of *Tomato spotted wilt virus* (TSWV) or begomoviruses are caused to the spread of the thrips *Frankliniella occidentalis* and the whitefly complex, *Bemisia tabaci*, correspondingly.

Most vector-borne pathogens require their vectors for subsistence and spread (Mauck et al., 2012) and they have evolved and adapted to their insect vectors to upsurge their chances of spread from host plant to plant.

Usually, plant viruses and phytoplasmas depend on vectors for their survival for two key factors. Plants are rooted and immobile which make it difficult for pathogen to transmit from one plant to another and plant epidermis with impermeable cuticle coats, averting entry of pathogens and vectors are well adjusted to their role as vectors by their capacity to penetrate the epidermis and deposit the pathogens in the cytoplasm without endangering the integrity of the plant cell. Consequently, many plant pathogens depend on insect vectors for transportation among host plants.

There are several cases in which both pathogen and vectors advantage from their communal association by changes induced in their mutual host plant after pathogen infection. Typically, the rate of vector population rises in infected plants as pathogen manipulate vector behavior to augment their own spread (Chanchala et al., 2021; Fereres and Raccah, 2015).

Pathogen effects on fitness of the vector such as growth rate, longevity, and reproduction in several vectors (Kuno and Chang, 2005), as well as on feeding behaviors, facilitate transmission of pathogen (Platt et al., 1997).

Not-virulified aphid, *Rhopalosiphum padi,* prefers to colonize and feed on wheat infested with Barley Yellow Dwarf Virus (BYDV) (Laura et al., 2012). *Rhopalosiphum padi* and *Myzus persicae,* major aphid vectors of Leafroll Virus in Potato (PLRV), incidentally induce changes in the host selection behavior when they feed on PLRV-infected plants (Eigenbrode et al., 2002; Werner et al., 2009).

Shorter nymphal period of *Macrosiphum granarium*: vector of Yellow Dwarf Virus in barley have been observed by Miller and Coon (1964). Similar results in leaf hopper vector *Hishimonous phycitis* on eggplants are infected by little leaf disease reported by Srinivasan and Chelliah (1980). Reduced mortality in nymphs which fed on infected plants was reported by Jayaraj and Seshadri (1967) and Miller and Coon (1964). Preference of vector on diseased plants in *Emilia sanchifolia*, sugarbeet, and *Seasumum* was recorded by Srinivasan and Chelliah (1980).

Thereby, it is vital to manage plant vectors to evade or minimize the epidemics caused by plant pathogens transmitted by vector.

7.4 VECTOR MANAGEMENT STRATEGIES

Prevention is the best strategy for the vector-borne virus and phytoplasmal diseases. There are several control strategies to manage vector-borne virus and phytoplasmal diseases including producing resistant plants, insecticide application, habitat management, natural enemies, physical control, rogueing and genetic manipulation which come under three major categories (Fig. 7.2); avoiding pathogen acquisition, transmission by vector, and keeping vector population in minimized condition which should be integrated in a successful vector management program.

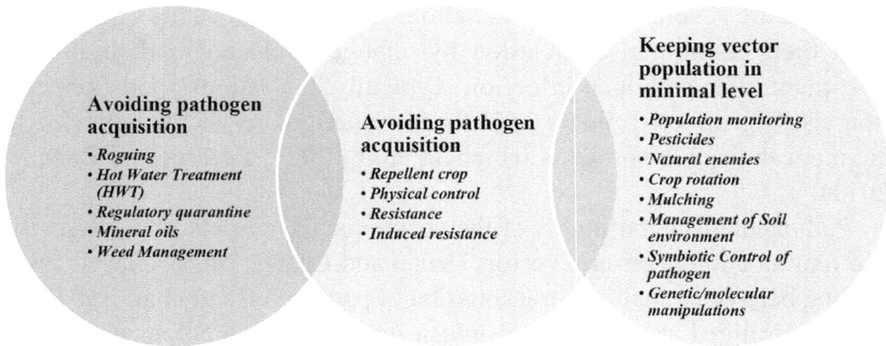

FIGURE 7.2 Strategies for successful disease-transmitting vector management program.

7.4.1 AVOIDING PATHOGEN ACQUISITION BY VECTOR

7.4.1.1 ROGUEING

The elimination of diseased plants is the most effective management strategy in orchards where the trees are noncontiguous. Infected plants can be rogued, removed, or ratooned, only symptomatic shoots are removed.

X-disease of chokecherry (*Prunus virginiana* L.) is a classic example of rogueing to successful management of the disease by the removal of the phytoplasma reservoir host (Parker et al., 1963). Uyemoto et al. (1998) developed a new technique of rogueing combined with insecticide application. They found a significant decline in disease spread around trees consequently removed.

Typically, pathogen affects the fitness of the vector such as existence, growth rate, reproduction as well as on feeding behaviors and facilitating pathogen transmission. Several authors have definite influence on viral or phytoplasma infected host plants persuading the secondary transmission of the pathogen by vector (Miller and Coon, 1964; Jayaraj and Seshadri, 1967; Srinivasan and Chelliah, 1980; Platt et al., 1997; Eigenbrode et al., 2002; Kuno and Chang, 2005; Werner et al., 2009; Laura et al., 2012).

Chanchala et al. (2021) confirmed that WLD-infected sugarcane plants are enhancing its vector performance by attracting, increasing survival and population build-up which induce the transmission of WLD phytoplasma itself. Thereby, they have suggested the regular assessment and rogueing out of diseased clumps which could cause a great impact on secondary transmission of WLD by its vector.

7.4.1.2 HOT WATER TREATMENT (HWT)

Hot water treatment is very popular in the management of seed-borne pathogen and it is also an effective method in the management of the insect vectors who lay eggs in the bark of the host plant or and in trash attached to the vegetative propagation materials as stem cuttings.

It has been identified that the insect vector of grape vine yellows lays eggs under the barks of grape vine. After identification, both phytoplasma and the eggs of the vector were simply eliminated by the treatment with heat (Cauudwell, 1966). The current practice is treating grape vines with hot water (Mannini, 2007).

7.4.1.3 REGULATORY QUARANTINE

The first and principal method for managing vectors is to confirm that they are not transported to new region (Bonfils and Schvester, 1960). In Taiwan, sugarcane is a prohibit material to import as per plant quarantine regulations; a firm inspection is needed for the local WLD to limit further spreading of the disease. This cooperative exercise has contributed for successful control of WLD (Leu, 1983).

7.4.1.4 MINERAL OILS

Mineral oils are hydrophobic materials that impede with pathogen acquisition and retention by vectors. Mineral oil with right viscosity and un-sulfonated remnants was found effective to decline the efficiency of transmission by vectors. This is a popular method in nurseries for protection against NP vectors. The mode of action seems to be by interference with vector requisite by probing vectors. The leaf surface must be fully covered: full coverage requires high pressure regular applications (up to twice a week) of large volume. Successful combination of oil with insecticides that have insect-repelling qualities (e.g., pyrethroids) was described in countries, that is, Israel and England (Raccah, 1986). Blanc et al. (2014) documented that novel molecules viz., *peptides* could be considered to outcompete with capsule proteins of virus and thus affect transmission.

7.4.1.5 WEED MANAGEMENT

Weeds can significantly influence vector infection intensities and distribution, but plant density also affects vector spreading. Thereby, management of alternative host plants can be used to diminish the vector populations and the disease incidences in some cropping systems (Weintraub and Wilson,

2010). In the world many phytoplasma diseases transmitted by vectors (Leaf hoppers) have managed successfully by managing alternative host species. Stark-Urnau and Kast (2008) indicated that management of weed reduces the vector population too by trapping in crops and using pesticides to those crops, removal of habitats, and keeping as repellent crops.

Bois noir disease of grapevine (grapevine yellows) is phytoplasma disease vectored by *Hyalesthys obsoletus,* which is the sole known vector to date (Maixner et al., 1995). *H. obsoletus* prefers not to feed and develop on grapevine and grapevine is considered an "accidental" plant of the *H. obsoletus.* Level of infestation of *H. obsoletus* and disease is reliant on the predominant weedy host plants in the grapevine plantations viz., *C. arvensis, Ranunculus* spp. (Maixner, 2007) and *Urtica dioica* (Bressan et al., 2007). Vector populations in grapevine yards were reduced successfully by using glyphosate to manage weeds (Stark-Urnau and Kast, 2008).

X-disease of stone fruits and Pierce's disease of grapevines are two important vector-borne diseases of trees and vines. It illustrates the impact of weeds in the spread of crop diseases. Pierce's disease is caused by the bacterium Xylella fastidiosa and X-disease Pytoplasma. Pierce's disease and X-disease, weeds and neighboring natural vegetation cover serve as reservoirs of insect vectors or/and vector-borne pathogens. Thereby, habitats of insect vectors can be eradicated through weed control, which may prevent disease spread by eliminating vector population buildup.

The use of a "pull" strategy has been proposed for the management of *Hyalesthys obsoletus* through the use of its favored host plant, *Vitex agnus castus* (Zahavi et al., 2007). By placing *V. agnus castus* 10, 50, and 70 m from vineyards, it was expected that *H. obsoletus* would be haggard to those plants and away from the grapevines.

This two new diseases have become obvious—characterized by severe stunting, profuse tillering, and lethal yellowing—and have been recognized as phytoplasmas (Arocha et al., 2009) and the leafhopper, *Recilia banda* Kramer, has been confirmed as a vector (Obura et al., 2009). They found phytoplasmas in two grass species in the region and recommended disease management by weed removal.

7.4.2 AVOIDING PATHOGEN TRANSMISSION BY VECTOR

7.4.2.1 HABITAT MANAGEMENT

Land use change also plays a significant role in the epidemiology of vector-transmitted diseases. Change of habitat may influence the multiplication habitats and biodiversity of the vectors or/and harboring hosts. As pointed out by Weintraub and Beanland (2006), the surrounding vegetation composition of a field/orchard/vineyard has a significant effect on the presence and spreading of vectors.

a. *Repellent crop*

   Cleome viscosa (Cleomaceae) acts as a repellent to the *D. menoni* (Chanchala et al., 2015). Laboratory studies revealed the potential of using *Cleome viscosa* extract to repel WLD vector from sugarcane plants for 24 hrs period. According to the literature, most of the plant species which belong to order Brasicales are having characteristic odor and eventually repel insects from their presence. *Brassica juncea* (L.) Czern & Coss (Bracicaceae) is also an oil crop with potential to repel many insect pest species. SRI (2020) confirmed the ability of *B. juncea* to repel *D. menoni* in sugarcane plantations by intercropping *B. juncea* with sugarcane and scattering around the sugarcane field at the early growth stage of the sugarcane crop.

b. *Maintaining gap among plots*

   Thein et al. (2012) proposed that the disease risk declines with increasing distance from infected fields. Thein et al. (2012) have experienced the influence of wind direction for the dispersal of WLD vectors *M. hiroglyphicus* and *Y. flavovittatus*. Also, they reported that the infection risk of WLD is increased on the leeward (downwind) side of a disease-infected field than on the windward (upwind) side.

   Larsen and Whalon (1988) also denoted that nearby habitats and floral communities associated with the host crop could influence the dispersal distance. Thereby, weed community and the nearby floral communities to the sugarcane ecosystem also could be an influence in flying distance.

   SRI (2019) suggested that keeping minimum 75 m distance between commercial and nurseries fields would reduce the vector transmission of the WLD to newly planted sugarcane fields.

Thereby flying distance of the vector, wind direction, and nearby habitats and floral communities associated with the host crop may decide the gap among two crop fields to minimize the vector transmission of the disease.

7.4.2.1.1 Physical Control

Dispersal is a key component in the population ecology, dynamics of insects, and secondary transmission by insect vectors. Thereby, the efficient way of managing phytoplasma vectors is by disturbing or avoiding their invasion into the crop. There are several means of physical controlling of the insect vectors viz., barrier crops, insect-exclusion screening (IES), barrier sprays, and maintaining distance among fields.

a. *Barrier crop*

   Several authors propose that high, tall, barrier crops may act basically as mechanical barriers that inhibit aphid colonization of the protected crop.

   When tall barrier crops such as sunflower, maize, or guinea corn (*Sorghum bicolor*) present, aphids incline to alight less commonly on the protected crop and more frequently on the barrier. Since nonpersistent pathogens are reserved for only few hours by the vectors, they drop the inoculum soon after settling or probing on the barrier crop. Also, reports on the effect of sunflower barriers on the spread of *Soybean mosaic virus* (SMV) in soybean fields noted that most aphid vectors fly higher in plots with barriers and, thereby, the barriers altered the pattern of SMV infection in the field.

   However, taller barriers can assuredly redistribute the aphids landing on the protected crop and alter the spatial pattern of pathogen spread in the field.

   Sometimes, the erratic behavior of border crops could be related to variances in the height and development of the borders at times of high risk of pathogen spread (usually related to the ultimate density of alates landing on the crop).

b. *Insect-exclusion screening (IES)*

   Insect-exclusion screening is one of the most trustworthy means of regulating disease-transferring vectors by covering the crop. But there are few constraints in using IES for large-scale agriculture crops such as sugarcane, maize, rice, grapes etc. However, IES has become popular with fruit trees such as banana and papaya.

   There are some success stories on 100% controlled vector-transmitted diseases by covering the trees with IES. Papaya cultivation in Australia is associated with three different phytoplasma diseases: dieback, yellow crinkle, and mosaic (Elder et al., 2002). When considering the effectiveness of IES compared to systemic insecticide treatments and a nontreated control, covered trees had significantly less phytoplasma symptoms.

According to Mannini (2007), IES is the only approach to uphold phytoplasma-free vineyards. It also recommended to constantly cover the clonal or mother plants use to produce new generations of daughter plants with IES to maintain the plants free from pathogens. IES is shown effective over the *Homalodisca vitripennis* (Germar), a sharpshooter vector of *Xylella fastidiosa*, to protect high-value vineyards and nursery stock confirming the leafhopper behavior altered, in that they moved away from the barrier to adjacent plants; few actually flew over the barrier (Blua et al. 2005).

c. *UV-absorbing plastics and nets*

An effective way to interfere with vector landing and steering is the use of UV-absorbing plastics and nets. They absorb UV-light interfere with insect vision and accordingly, were found to cause great reduction of pathogen transmission. Significant reduction of insect alighting was recorded for whiteflies, aphids, and thrips (Antignus, 2012).

d. *Barrier sprays*

Barrier sprays are novel and improved form of an old type of inorganic chemical control: whitewashing or dusting.

Kaolin is one of the typically used barrier sprays which is a nonabrasive, fine-grained aluminosilicate mineral sprayed as a particle film. It acts in a couple of ways by killing an insect by suffocating or coating the plant and barricading feeding and oviposition sites.

Also, kaolin protected grape plants from feeding and oviposition by leafhoppers, by physically coating the plant with a mineral film (Puterka et al., 2003) and Tubajika et al. (2007) reported the fewer leafhoppers in grapevine field treated with kaolin.

Water availability in the sight is the most influencing factor to the efficacy of kaolin as overhead irrigation or heavy rain cause washing away of kaolin. And it has been a very effective mean of pest management in dry areas.

e. *Color traps*

Insects (e.g., aphids) are typically repelled from reflective surfaces and their target can be rehabilitated by the attraction of insects to colors. This effect led to the use of metallic reflective surfaces, straw mulches, or kaolin particle films.

7.4.2.1.2 Resistance

Resistant cultivars are used as one of the major pest management approaches in Integrated Pest Management (IPM) Programs. According to Agrawal (1959) and Sanghera and Kumar (2018) making clones resistant to insects is a continuous and time-taking process. But, in general, it is compatible with other insect control strategies (Khan and Saxena, 1998) and it provides more practical tactic in leafhopper management.

Plant breeding is mostly oriented toward the developing of varieties that are resistant to disease rather than to the vectors. Breeding of plants that are resistance to vectors has been endeavored as a strategy to control diseases and it has been successful in many crops viz., breeding for mite resistance to control virus disease of wheat (Martin, 1983), and breeding for plant hoppers resistance to control virus disease of rice (Heinrichs, 1985). However, there are many incidences of developing disease-resistant varieties as a by-product of breeding varieties for insect pests of a crop, which are later identified as vectors of diseases. Chanchala (2021) has identified some vector-resistant sugarcane varieties (SL 8306, SL 92 5588) for the *Deltocephalus menoni* (Hemiptera; Cicadellidae); vector of sugarcane white leaf disease (WLD).

7.4.2.1.3 Induced Resistance

Osler et al. (2003) reported the induce resistance in certain wine grape varietals for phytoplasma disease. They reported that some wine grape varietals on various rootstocks, vines often go into spontaneous remission from phytoplasma symptoms and symptoms present only for a few years and then abruptly disappear. Also, in Italy higher amount of wine grape plants of the Prosecco cultivar recovered within 1–4 years from a phytoplasma infection after the first appearance of symptoms. As there are no symptoms for 3 successive years, it was considered that plant has recovered from phytoplasmas (Maixner, 2006). Romanazzi and Murolo (2008) stated that stress condition plants faced due to severe pruning, pollarding, uprooting or partial uprooting have promoted the induced resistance in certain grape wines.

Other than to agro technical methods, applications of various chemicals are known to increase recovery from various plant pathogens. Application of the auxins indole-3-acetic acid (IAA) and indole-3-butyric acid (IBA) to phytoplasma-infected periwinkle shoots shows improved recovery (Curkovic Perica, 2008).

7.4.2.2 KEEPING VECTOR POPULATION IN MINIMAL LEVEL

7.4.2.2.1 Population Monitoring

Since the level of vector population is abundant in the cropping systems, it influences the severity of the disease, disease spread, and management. Prasad and Prabhakar (2012) denoted that, pretreatments of seed cane should be decided considering the number of vectors present in the sugarcane cultivation since the presence of vectors in the field causes secondary transmission of the disease. When the disease inoculum presents within the field and nearby fields level of transmission is also high with higher the vector number. Herein, accurate measurement on population level of vector has become vital in designing effective program for vector management.

Most of the insect vectors are however challenging to monitor due to their minute size and flying and jumping capabilities. Therefore, accurate measurement on population level of vector could only be achieved through using appropriate insect population monitoring technique. It also affects making cost-effective and environmentally sound management decisions. Insect population monitoring is considered as the procedure to regularly observe, measure, and record number of insects within a given area and diverse insect sampling techniques and equipment used to discover insects in the air, on plants, and even on and underneath soil according to the insect type.

Frequently using monitoring techniques for insect pest populations in the field are: in-situ counts, knockdown, suction sampling, sweep netting, trapping (sticky traps, light traps, wing traps, bucket traps, pan traps, pheromones traps, and pitfall traps), and visual inspection. From available monitoring techniques, sticky traps, light traps, and sweep nets are the frequently used techniques to monitor leafhoppers populations.

Sticky traps: Sticky traps have been used successfully in monitoring leaf hopper population in cotton, sugarcane, apple, cow pea, chili, tea, and cranberry (Mensah, 1996; Atakan and Canhilal, 2004; Hanboonsong et al., 2006; Wallis and Shaw, 2008; Rodriguez- Saona et al., 2012; Bian et al., 2014; Tang et al., 2015; Toshova et al., 2017; Said et al., 2017).

Sweep nets: Sweep nets is one of the most efficient monitoring techniques generally, especially in early stages. This technique is low cost, speed, and simple, thereby common tool that growers use to assess strategies to control the insect pests in potato, egg plants, pepper, zucchini, okra, Alfa Alfa, maize, sugarcane, and rice (Senevirathne, 2008; Wakeil., et al., 2013).

Light traps: Light traps are commonly used monitoring tool for insects and many leaf and planthoppers which are attracted to light, in large numbers (Labonne, et al., 1998). Sometimes, the light traps also attract non-native insect species which are moving by the particular fields.

7.4.2.2.2 Pesticides

Regardless of the numerous insecticides available, chemical control is not the preferred solution to avert vector activity. Many pathogens are transmitted into crops by visiting insects that transmit pathogen during their first probing. Vectors for NP pathogens oblige only comparatively short inoculation period considerably shorter than the time needed for insecticides to kill. In addition, insecticides can enhance impatience in insects, with the result that they make more inoculation efforts compared to the calm insects. Exclusions are vectors that colonize the crop and transfer phloem-restricted pathogens, for which insecticides may diminish vector-borne pathogen spread.

Many vector-borne viral and phytoplasma diseases do not have any pertinent means to cure them. In here insecticide application to retard the spread of vector is the only way of managing the spread of these diseases at the point of vector has been identified.

Application of insecticide to manage spread of pathogen by vector is well known in managing viral and phytoplasma disease in many crop species including fruit crops and sugarcane.

In controlling sugarcane WLD several insecticides have been used namely, Malathion, carbosulfan, carbomates, Organophosphorus, and Thiomethoxam. But in most cases it had become unsuccessful due to numerous reasons.

Spraying malathion at the young age of sugarcane plant has been unsuccessful to control the vector and the disease incidence (Leu, 1974). In the sugarcane WLD management, over 3.2 hectares of each sugarcane plantation has been comprehensively treated with carbosulfan insecticide by-weekly from April to June before the insect vector population could reach its systematic peak in July and August. But the disease has not been successfully managed by this insecticide application (Wongkaew, 1999). As confirmed by Chen (1980) application of carbomate and Organophosphorus insecticides to the setts before planting or to young plants has been ineffectual in controlling the WLD but reduces the vector population. The Sri Lankan recommendation for foliar application includes the application of Thiomethoxam 25% (Actara); 5 g/16 L (15 mL/tank) to control WLD, also reported to be ineffective for disease management (SRI, 2007) even though it reduces the vector population within the particular fields.

Jaraush and Jaraush (2010) reviewed that no chemical applications had been effective in controlling vectors of pear decline (PD), apple proliferation (AP), and European stone fruit yellows (ESFY).

Pear psyllids, *Cacopsylla pyri*, and *Cacopsylla pyricola* are vectors of PD (*Ca.* Phytoplasma pyri) and initially the management of pear psyllids depends upon a system of IPM where the chemical control was a major component in the system. With the time psyllids have a definite capability to produce resistance against insecticides; several organophosphates and pyrethroids due to their host specificity and the high reproductive ability.

Cacopsylla picta and *Cacopsylla melanoneura*; vectors of AP (*Ca.* Phytoplasma mali) were controlled earlier using several insecticides viz., Ethofenprox, organophosphates, and neonicotinoids. Subsequently, both the re-migrants and the new generations are available in the cultivations evidencing the requirement of new management approaches.

It has been confirmed that no chemical applications had been effective in controlling *Cacopsylla pruni*, vector of ESFY (*Ca.* Phytoplasma prunorum).

Accordingly, conventional insecticides, even when regularly used (e.g., Wally et al., 2004), will not manage the disease appearance, because pathogen transmission occurs more rapidly than insecticides can act and there is often a continuous influx of new vectors from adjacent habitats. Therefore, other management strategies have to be used in combination with insecticide control.

Saracco et al. (2008) have developed an approach for the use of insecticides, namely, to protect plants from virulified migratory vectors use systemics such as neonicotinoids, and to subdue vectors use organo-phosphate insecticides. Pilkington et al. (2004) have confirmed that total vector population control can be accomplished and disease occurrence retarded by just treating crop borders with systemic insecticides. This strategy targets only the vector population, without extensive insecticide application. Accordingly, border treatment with systemic pesticide is efficient on managing the vector population.

In managing the WLD vector population in the sugarcane fields of Sri Lanka, border treatment with systemic insecticide is effective (SRI, 2017). The schedule of application is, treating soil with Fipronil (phenyl pyrozole/ Prince GR) at the rate of 18 kg/ha at the point of plant germination; second application includes treating 2 m wide border of the field from the outer boundary line from the monthly intervals until crop reach to 6-month age at low and moderate disease pressures in the surrounding cropping fields.

However, use of pesticide is hindered for various reasons: insecticide resistance of vector species, low abundance of the vectors, missing homologation of appropriate chemicals, and not least environmental protection considerations.

Therefore, other than the conventional pesticides, alternative strategies should be considered in insect vector management.

7.4.2.2.3 Natural Enemies

Typically, Hemipterans are attacked by an array of predators and parasitoids; spiders, Mirids, Carabids, coccinellids, etc. (Denno and Perfect, 1994; Perfect and Cook, 1994). When it comes to the pathogen-transmitting vectors, use of natural enemies to manage vectors has a mixed assessment.

Economic threshold in vector management is lower as compared to the pest. Sometimes, minimum numbers of vectors are not achievable with natural enemies. Since NEs are more sensitive for pesticides, spraying another pest may even cause to retard the NE population and resulting incensement of disease transmission. Also presence of natural enemies induces the scatter responses of vector causing increased numbers of test bites in feeding which may cause increase in pathogen transmission. Thereby, relining only on natural enemies on vector control is risky for most perennial and ratoon crops.

There is a better opportunity of using NEs for vector management in island ecosystems comparatively to the more complex settings. It was remarked that the introduction of psyllid parasitoids into Runion Island significantly reduced the spread of the bacteria that cause citrus greening disease while same parasitoids were unsuccessful in Viet Nam.

Predators affect the behavior of prey and they possibly diminish pathogen spread by a vector. Arthropod predators have been displayed to reduce the ingestion of plant biomass by pest herbivores, but their surging nonconsumptive influence on vector insects' feeding and subsequent pathogen transmission is also confirmed.

When considering the homopteran vectors, the leafhopper changed its feeding behavior in reaction to predation risk. Phloem ingestion, the feeding stage when pathogen acquisition occurs, was delayed and less common. The phase when pathogen inoculation takes place, via the secretion of pathogen-infested saliva of vector, was shorter when predator was in vicinity.

In addition to predation it has been confirmed that predators are capable of influencing various physiological, morphological, and behavioral characters of their prey through various nonconsumptive effects as vigilance, refuge seeking, fleeing, growth, and mating success. In herbivorous insects the incidental trait-mediated effects of predators have been frequently shown to reduce herbivory and subsequent plant damage; the examples of this comprise the effect of spiders on the praying of grasshoppers, beetles, and caterpillars.

Hence, those predators and parasitoids are vital in agriculture ecosystems as sustainable hemipteran pest control strategy.

a. *Parasitoids*

There are many specialist parasitoids allied with Auchenorrhyncha (Waloff and Jervis, 1987). Hymenopterans (Dryinidae) are the furthermost common nymphal and adult parasites of the leafhoppers and planthoppers. They deposit eggs into the nymph or larvae and the parasitic larvae develop within the hopper. Mymaridae (Hymenoptera), "fairy flies" also infrequently attack the egg stage of leaf and plant-hoppers. Pipunculidae (Diptera) are almost exclusively parasitoids of Auchenorrhyncha. They lay eggs into the adult and nymphal phases and frequently in some Delphacidae species.

b. *Predators*

Predators are capable in transmission of plant pathogens solely through manipulating the feeding behavior of vectors. It was confirmed by studying on predator-mediated pathogen spread mechanisms using the plant pathogen Wheat Dwarf Virus that is transmitted by the leafhopper, *Psammotettix alienus* (Tholt et al. 2017).

Sap feeding invertebrate herbivores are expert in feed deep into the plant's tissues. Basically they belong to the Homoptera and are the most vital vectors of various plant diseases. Most of the known plant viruses (90%) are transferred by them. Retrieving deeper plant tissues consumes longer time and more energy. The removal of the stylets from deeper tissues takes longer period; therefore, feeders will not be able to respond quickly to predator during the penetration. Thereby, they have to assure predator evasion against their least nutritional requirements while feeding.

Generally, this proposes that longer feeding times upsurge exposure to predators, while delaying or shortening feeding periods increases the risks of dehydration and starvation. In the case of all circulative pathogens, Homopteran insects act stylet-borne vectors. There is a strong relationship between the pathogen and its vector. Pathogen transmission occurred only with the vector's specific mode of feeding. In here, different stages of leafhopper penetration are closely associated with the pathogen transmission process. Infections are often restricted to certain tissues and the mode of feeding on these tissues affects transmission of pathogen. In the case of a phloem-restricted pathogen, only the watery saliva secretion of a virulified vector into the phloem can cause pathogen inoculation, and vectors can only attain the pathogen by ingesting phloem sap from a disease-infected host plant. When complex sap feeding behavior is responsible for the acquisition and inoculation of plant pathogens, then the predators have an important role in the process, and it suppresses the spread and epidemic of vector-borne plant diseases.

Leafhoppers are attacked by an array of predators. Spiders are very important predators of both adults and nymphs sometimes may be key egg predators. Coccinelid beetles predate energetically on the eggs in soil and plant debris. Moreover specialist parasitoids attach on Auchenorrhyncha vectors.
Most ground beetle spp. feed on animal food either throughout their individual life or only at some stage of their development. These carnivores include predators attacking larvae, pupae, eggs of insects, and snails. Therefore, they are considered insects useful in agriculture ecosystems.

Coccinelid beetle have been known global as a predator of a number of insects and they are dispersed throughout the world in many countries of Asia. They are economically vital as predaceous both in their larval and adult phases on various important soft bodied crop pests (Hippa et al, 1978; Kring et al, 1985).

c. *Entomopathogenic fungi*

Entomopathogenic fungi, *Metarhizium anisopliae* shows a best control over *M. hiroglyphicus* in Taiwan (Kernasa et al, 2018).

d. *Wolbachia*

Wolbachia is a bacterium ubiquitous in insects which inherited maternally with a potential as insect pest-control agent. Wangkeeree et al. (2020) have detected *Wolbachia* and characterized *Wolbachia* in *Yamatotettix flavovittatus* Matsumura (Cicadellidae: Hemiptera), insect vector of WLD phytoplasma in sugarcane. Moreover, *Wolbachia* was recognized in the eggs and nymphs of *Y. flavovittatus*. These findings are valuable information for the future development of *Wolbachia* as a bio-control agent for disease-transmitting leafhopper vectors.

7.4.2.2.4 Crop Rotation

Crop rotation provides a great opportunity to evade the vector population from the fields as there are no host plants in the particular area for a reasonable time and it has been a successful disease management strategy which can be achieved through managing the vector population present in the field.

It has been a very successful strategy in WLD management in sugarcane which is a devastative phytoplasma disease. In areas under rain-fed cultivation, a switch of sugarcane planting with green manure plants in deferent seasons is highly recommended. The most applicable strategies are likely the sanitation, the disease-free cane setts utilization, and the crop rotation with economic green manure plants. WLD disease could be successfully managed in a defined research area at Udornthani Province of Northeast Thailand during 1989–1991 using pigeon pea (Cajanus canja L.) and sword bean (Canavaliagla diata Jacq.) as green manures for crop cycle and healthy seed material were further used for replacement of old variety (Kusalwong and Ouvanich, 1993).

Similarly, an incidence was occurred in Siyabalanduwa, Sri Lanka during 1995–1997. There was an outbreak of WLD and it spread over 2500 ha of sugarcane fields providing the inoculam to the nearby areas. Eradication of disease plants in the area was done with weedicides and fire. Later on the affected varieties were replaced with hot water treated seedcane of variety SL 83 06, which is resistant to WLD. But disease was reappeared later on due to less attention on continuation of rogueing out programs (Senevirathne, 2008).

In tropical countries there is no much fluctuation of temperature during the year. It is insignificant for changing planting time in relation to temperature. As the population of *M. hiroglyphicus* insect vector and the proliferation of WLD phytoplasma are restricted by low temperature during winter season, changing planting time has been implemented in Taiwan to minimize disease spread (Ling and Yang 1963; Pan and Yang 1970).

This can be improved by exclusion of sources of infection in and around the crop, removal of plant remains from previous season and, if necessary, creation of a time gap between crops and space gap among plots. These operations can retard the numbers of viruliferous insects that reach the crop. Merging these actions with the use of disease-free cane setts strategies has made Taiwan completely manage the disease to a harmless condition (Leu and Kusalwong, 2000). Although the losses have been efficiently reduced through these approaches, the WLD relocated sooner due to ignoring of such proper cultural practices (Wongkaew 1999).

7.4.2.2.5 Mulching

One means of controlling the habitat is the application of several organic and synthetic mulches. Synthetic mulches, such as plastic sheeting, can physically avert the entering of vectors into the soil (*H. obsoletus, D. menoni lay* eggs at or just below soil surface). Vector-borne diseases of several crops have been managed effectively by disturbing the lifecycle of the vectors by manipulating their egg-laying habitations (Howard and Oropeza, 1988; Summers and Stapleton, 2002; Mannini, 2007).

The use of reflective mulches works in a different manner: rather than physical control, the vector is repelled from the plant. Summers and Stapleton (2002) achieved better management of the maize leafhopper vector, *Dalbulus maidis* (DeLong and Wolcott), and increased maize yield with plastic reflective mulch than with insecticide treatments.

Comparable effects were achieved with the management of the aster yellows phytoplasma vector *Macrosteles quadrilineatus* (Forbes) on carrots using aluminum foil mulch. However, Setiawan and Ragsdale (1987) found that the efficiency of the mulch diminished as plant canopy expands. Instead, the type of mulching ingredients used around coconut trees affects the abundance of the planthopper vector of lethal yellows, *Myndus crudus* (Cixiidae). Smaller quantity of nymphs are found around trees mulched with coarse materials such as pine bark nuggets (Howard and Oropeza, 1988).

An integrated approach developed to control the X-disease of stone fruits suggests that the management of ground cover in the orchards is an effective cultural measure for reducing leafhopper vector populations and feeding damage to peach (Douglas and McClure, 1988).

The insect vector of bois noir disease of grape vine *Hyalesthes obsoletus* oviposit at or just below the soil surface, and vector movement into the soil has been physically prevented by using synthetic mulches such as plastic sheeting (Maixner, 2007).

Summer and Stapleton (2002) reported that better control of maize leaf hopper *Dulbulus maidis*, a vector of maize stunt, leads to a higher maize yield. The control has been achieved by using plastic reflective mulch in which oviposition in soil had been prevented. Similarly, the aster yellow phytoplasma vector on carrot, *Macrosteles quadrilineatus,* has been controlled by laying aluminum foil mulch to prevent the oviposition by the vector (Stewan and Ragsdale, 1987).

7.4.3 MANAGEMENT OF SOIL ENVIRONMENT

Several hemipterance insects lay eggs in soil and on trash near to the soil. Management of soil preferred for egg laying also facilitates minimal population of vectors.

As an example, there is a significant difference in ovipositional preference of the WLD vector *D. menoni* between different soil groups, between different soil amendments. And there is high preference of *D. menoni* for oviposition in sandy loam and fine-sand soils. Therefore, further attention should be given to manage WLD in the sugarcane fields with sandy loam and fine-sand soils in the areas with high incidence of this disease (SRI, 2016).

7.4.4 SYMBIOTIC CONTROL OF PATHOGEN

Symbiotic control of pathogen is an effective mean of vector-borne pathogen management. Most symbiotic microbes are present in the mid gut and ovaries of the vector throughout the all phases of the life cycle including eggs, nymphs, and adults. Enhancing mutual benefits symbiotic microbes rescind the vector transmitted pathogen ingest with vector feeding. Wangkeeree et al. (2012) identified bacterium associated with *M. hiroglyphicus* as a potential candidate for symbiotic control of WLD.

7.4.5 GENETIC/MOLECULAR MANIPULATIONS

Weintraub (2007) has reviewed genetic manipulations. They include the modification of plant lectins in host plants, which affects several physiological functions of vectors, including blocking the absorption of free amino acids and sugars. Also plant lectins show efficiency in phytoplasma vectors: the snowdrop lectin (*Galanthus nivalis* agglutinin, GNA) and *Allium sativum* leaf lectin (ASAL). These lectins have been shown to cause mortality in phytoplasma vectors in rice. These approaches are anticipated to be popular in future for most precise management of vector-borne diseases.

KEYWORDS

- **virus**
- **phytoplasma**
- **vectors**
- **transmission**
- **management**
- **resistance**
- **pesticides**
- **habitat management**
- **natural enemies**

REFERENCES

Agarwal, R. A. Morphological Characteristics of Sugarcane and Insect Resistance. *Entomol. exp. et applicata* **1959,** *12*, 767–776.

Antignus, Y. Control Methods of Virus Diseases in the Mediterranean Basin. *Adv. Virus Res.* **2012,** 84, 533–556.

Arocha, Y.; Zerfy, T.; Abebe, G.; Proud, J.; Hanson, J.; Wilson, M.; Jones, P.; Lucas, J. Identification of Potential Vectors and Alternative Plant Hosts for the Phytoplasma Associated with Napier Grass Stunt Disease in Ethiopia. *J. Phytopathol.* **2009,** *157*, 126–132.

Atakan, E.; Canhilal, R. Evaluation of Yellow Sticky Traps at Various Heights for Monitoring Cotton Insect Pests. *J. Agric. Urban Entomol.* **2004,** *21* (1), 15–24.

Bertaccini, A. Phytoplasmas: Diversity, Taxonomy, and Epidemiology. *Front. Biosci.* **2007,** 12, 673–689.

Bian, L.; Sun, X. L.; Luo, Z. X.; Zhang, Z. Q.; Chen, Z. M.; Design and Selection of Trap Color for Capture of the Tea Leafhopper, *Empoasca vitis*, by Orthogonal Optimization. *Entomologia Experimentalis et Applicata* **2014,** *151* (3), 247–258.

Blanc, S.; Drucker, M.; Uzest, M. Localizing Viruses in Their Insect Vectors. *Annu. Rev. Phytopathol.* **2014,** *52*, 402–425.

Blua, M. J.; Campbell, K.; Morgan, D. J. W.; Redak, R. A. Impact of a Screen Barrier on Dispersion Behaviour of *Homalodisca coagulata* (Hemiptera: Cicadellidae). *J. Econ. Entomol.* **2005,** *98*, 1664–1668.

Bonfils, J.; Schvester, D. Les cicadelles (Homoptera Auchenorrhyncha) dans leurs rapport avec la vigne dans le sud-ouest de la France. *Annales des Epiphyties* **1960,** *11*, 325–336.

Bressan, A.; Turata, R.; Naixner, M.; Spiazzi, S.; Boudon-Padieu, E.; Girolami, V. Vector Activity of *Hyalesthes obsoletus* Living on Nettles and Transmitting Astolbur Phytoplasma to Grapevines: A Case Study. *Ann. Appl. Biol.* **2007,** *150*, 331–339.

Caudwell. A. Inhibition of flavescence Doree Virus In Vivo by Means of Heat Treatment. *Annales des Epiphyties* **1966,** *17*, 61–66.

Chanchala, K. M. G. The Plant Vector Relationship Between Sugarcane Plant and Vector of Sugarcane White Leaf Disease (WLD), *Deltocephalus menoni* (Hemiptera: Cicadellidae) to Identify Criteria for Varietal Evaluation; PhD Thesis. University of Peradeniya.

Chanchala, K. M. G.; Wanasinghe V. K. A. S. M.; Hemachandra K. S. Survival of the *Deltocephalus menoni* (Homoptera: Cicadellidae), the Vector of Sugarcane White Leaf Disease in Sri Lanka on Alternative Host Plants. *J. Sugarcane Sri Lanka* **2015,** *2*, 36–43.

Chanchala, K. M. G.; Wanasinghe, V. K. A. S. M.; Kulasekara, B. R.; Hemachandra, K. S.; Nugaliyadde, L.; Witharama, W. R. G. Tri-Trophic Relationship Among Sugarcane White Leaf Disease Phytoplasma (WLDP), *Deltocephalus menoni* (Homoptera: Cicadellidae) and Sugarcane Plant in Secondary Transmission of the Disease. *J. Agric. Sci.—Sri Lanka* **2021,** *16* (3), 443–451. DOI: 10.4038/jas.v16i03.9470.

Chen, C. T. Sugarcane White Leaf Disease in Taiwan; *Taiwan Sugar* March–April 1980; pp 56–61.

Curkovic Perica, M. Auxin-Treatment Induces Recover of Phytoplasma-Infected Periwinkle. *J. Appl. Microbiol.* **2008,** *105*, 1826–1834.

Denno, R. F.; Perfect, T. J. *The Planthoppers: Their Ecology and Management*; Chapman and Hall: London, 1994.

Douglas, S. M.; McClure, M. S. New Intergraded Approach for Controlling X-Disease of Stone Fruits; Connecticut Agricultural Experimental Station; New heaven. Bulletin 854, 1988.

Eigenbrode, S. D.; Ding, H.; Shiel, P.; Berger, P. H. Volatiles from Potato Plants Infected with Potato Leaf Roll Virus Attract and Arrest the Virus Vector, *Myzus persicae* (Homoptera: Aphididae). *Proc. R. Soc. B: Biol. Sci.* **2002,** *269*, 455–460.

Elder, R. J.; Milne, J. R.; Reid, D. J.; Guthrie, J. N.; Persley, D. M. Temporal Incidence of Three Phytoplasma-Associated Diseases of *Carica papaya* and Their Potential Hemipteran Vectors in Central and South-East Queensland. *Austral. Plant Pathol.* **2002,** *31*, 165–176.

Fereres, Al.; Raccah, B. Plant Virus Transmission by Insects. In: *eLS*; John Wiley and Sons, Ltd: Chichester, 2015. DOI: 10.1002/9780470015902.a0000760.pub3.

Foth, H. D. *Fundamentals of Soil Science*, 6th edn.; John Wiley and Sons: New York, 1978; pp 25–62.

Hanboonsong, Y.; Ritthison, W.; Choosai, C.; Sirithorn, P. Transmission of Sugarcane White Leaf Phytoplasma by Yamatotettix flavovittatus, a New Leafhopper Vector. *J. Econ. Entomol.* **2006,** *99* (5), 1531–1537.

Heinrichs, E. A.; Medrano, F. G.; Rapusas, H. R. *Genetic Variation for Insect Resistant in Rice*; Los Banos: Laguna, IRRI, 1985; pp 18–28.

Hippa, H.; Kepeken, S. D.; Laine, T. *On the Feeding Biology of Coccinella hieroglyphica L. (Coleoptera: Coccinellidae). Kevo-subaretitic Res. Station* **1978,** *14*, 18–20.

Hogenhout, S. A.; Ammar, E. D.; Whitfield, A. E.; Redinbaugh, M. G. Insect Vector Interactions with Persistently Transmitted Viruses. *Annu. Rev. Phytopathol.* **2008,** *46*, 327–359.

Howard, F. W.; Oropeza, C. Organic Mulch as a Factor in the Nymphal Habitat of *Myndus crudus* (Hemiptera: Auchenorrhyncha: Cixiidae). *FL Entomol.* **1998,** *8*, 92–97.

Jarausch, B.; Jarausch, W. Psyllid Vectors and Their Control. In: *Phytoplasmas Genomes, Host Plants and Vectors*; Weintraub, P. G., Jones, P., Eds.; CAB International: Whiltshire, 2010; pp 250–271.

Jayaraj, S.; Seshadri, A. R. Preference of the Leafhopper, *Empoasca kerri* Pruthi (Homoptera, Jassidae) to Pigen Pea, *cajanus cajan* (L.) Millsp. Plants Infected with Sterility Mosaic Virus. *Curr. Sci.* **1967,** *36*, 353–355.

Kernasa, N.; Jamrutsri, p.; Kumpiro, S.; Korpaditskul, R.; and Uraichuen, S. Efficiency of Entomopathogenic Fungi to Sugarcane White Leafhopper, *Matsumuratettix hiroglyphicus* (Matsumura) (Hemiptera: Cicadellidae). *Agric. Res. Technol.* **2018,** *14*, 5.

Khan, Z. R.; Saxena, R. C. Host Plant Resistance to Insects. In: *Critical Issues in Insect Pest Management*; Dhaliwal, G. S., Heinrichs, E. A., Eds.; Commonwealth Publishers: New Delhi, 1998; pp 118–154.

Kring, T. J.; Gilstrap, F. E.; Michels, G. J. Role of Indegenous Coccinellid in Regulating Green Bugs (Homoptera: Aphididae) on Texas *Grain sorghum. J. Econ. Entomol.* **1985,** *78* (1), 269–273.

Kuno, G.; Chang, G. J. Biological Transmission of Arboviruses: Re-examination of and New Insights into Components, Mechanisms, and Unique Traits as Well as Their Evolutionary Trends. *Clin. Microbiol. Rev.* **2005,** *18*, 608–637.

Kusalwong, A.; Ouvanich, W. Example Model for Sugarcane White Leaf Disease Control. In: *Proceeding of the 31st Kasetsart University Conference*, 3–6 February, Bangkok, Kasetsart University, Thailand, 1993; pp 348–352.

Labonne, G., Quiot, J. B.; Lauriaut, F. Searching for Leafhopper Vectors of Phytoplasmas: The Choice of Trapping Methods. Acta Hortic. **1998**. https://www.cabi.org/isc/abstract/19991105332.

Larsen KJ, Whalon ME Dispersal of Paraphlepsius Irroratus (Say) (Homoptera: Cicadellidae) in Peach and Cherry Orchards. *Environ. Entomol.* **1988,** *17*, 842–851.

Laura L. Ingwell, Sanford D. Eigenbrode and Nilsa A. Bosque-Pe′rez. Plant Viruses Alter Insect Behaviour to Enhance Their Spread. *Sci. Rep.* **2012,** *2*, 578. DOI: 10.1038/srep00578.

Leu, L. S. An Insectary Method for Testing Sugarcane Varieties for Resistance to White Leaf Disease. In: *Proceedings of the XV International Society of Sugarcane Technologists Congress*, 13–29 June, Durban, the International Society of Sugarcane Technologists, South Africa, 1974; pp 266–274.

Leu, L. S. White Leaf Disease. In: *Disease of Sugarcane*; Agnihothri, V. P., Ed.; Oxford and IBH Publishing Co: Delhi, 1983; pp 250–267.

Leu, L. S.; Kusalwong, A. Control of Sugarcane White Leaf Disease in Taiwan and Thailand. In: *Proceeding of the 6th International Society of Sugarcane Technologists Pathology Workshop*, 16–23 July, Petchaburi, The International Society of Sugarcane Technologists, Thailand, Ph-1, 2000.

Ling, K. C.; Chaun-Yang, C. Studies on the White Leaf Disease of Sugarcane, III. Evidence of Spreading of the Disease Under Natural Condition. *Rep. Republic Taiwan Sugar Exp. Station* 1963. *30*, 91–94.

Maixner, M. Temporal Behavior of Grapevines Infected by Type II of Vergilbungskrankheit (Bois Noir). In: *Extended Abstracts 15th Meeting of the International Council for Study of Virus and Virus-like Diseases of the Grapevine*, Stellenbosch, South Africa, 2006; pp 223–224.

Maixner, M. Biology of *Hyalesthes obsoletus* and Approaches to Control This Soil Borne Vector of Bois Noir Disease. *IOBC/WPRS Bull.* **2007,** *30* (7), 3–9.

Maixner, M.; Ahrens, U.; Seemuller, E. Detection of the German Grapevine Yellows (Vergilbungskrankheit) MLO in Grapevine, Alternative Hosts and a Vector by a Specific PCR Procedure. *Eur. J. Plant Pathol.* **1995,** *101*, 241–250.

Mannini, F. Hot water treatment and field coverage of Mother Plant Vineyards to Prevent Propagation Material from Phytoplasma Infections. *Bull. Insectol.* **2007,** *60*, 311–312.

Marcone, C.; Neimark, H.; Ragozzino, A.; Lauer, U.; Seemuller, E. Chromosome Sizes of Phytoplasmas Composing Major Phylogenetic Groups and Subgroups. *Phytopathology* **1999,** *89*, 805–810.

Martin-Tanguy, J. Conjugated Polyamines and Reproductive Development Biochemical, Molecular and Physiological Approaches. *Physiologia Plantarum* **1997,** *100*, 675–688.

Mauck, K.; Bosque-Pe´rez, N. A.; Eigenbrode, S. D.; De Moraes, C. M.; Mescher, M. C. Transmission Mechanisms Shape Pathogen Effects on Host–Vector Interactions: Evidence from Plant Viruses. *Funct. Ecol.* **2012,** *26*, 1162–1175.

Mensah, R. K. Evaluation of Coloured Sticky Traps for Monitoring Populations of *Austroasca viridigrisea* (Paoli) (Hemiptera: Cicadellidae) on Cotton Farms. *Austral. J. Entomol.* **1996,** *35* (4), 349–353.

Miller, J. W.; B. F. Coon. The Effect of Barley Yellow Dwarf Virus on the Biology of Its Vector the English Grain Aphid, *Macrosiphum granarium*. *J. Econ. Entomol.* **1964,** *57* (6), 970–974.

Nault, L. R. Arthropod Transmission of Plant Viruses: A New Synthesis. *Ann. Entomol. Soc. Am.* **1997,** *90*, 521–541.

Obura, E.; Midega, C. A. O.; Masiga, D. Pickett, J. A.; Hassan, M.; Koji, S.; Khan, Z. R. *Recilia banda* Kramer (Hemiptera: Cicadellidae) a Vector of Napier Stunt Phytoplasma in Kenya. Naturwissenschafen Published Online, 2009. DOI: 10.1007/500114–009–0578-x 4.

Osler, R.; Carraro, L.; Ermacora, P.; Ferrini, F.; Loi, N. Loschi, A.; Martini, M.; Mutton, P. B.; Refatti, E. Roguing: A Controversial Practice to Eradicate Grape Yellows Caused by Phytoplasmas. In: *14th Meeting of the International Council for the Study of Virus and Virus-like Diseases of Grapevine* (ICVG), Locorotondo, Italy, 2003; p 68.

Pan, Y. S.; Yang, S. L. Studies on the Bionomics of *Matsumuratettix hiroglyphicus* (Matsumura), an Insect Vector of Sugarcane White Leaf Disease. I. Seasonal Abundance and Dispersal of the Adult. *Rep. Republic Taiwan Sugar Exp. Station* **1970,** *50*, 65–71.

Parker, K. G.; Palmiter, D. H.; Gilmer, R. M.; Hickey, K. D. X-Disease of Peach and Cherry Trees and Its Control. *Cornell Extension Bull.* **1963,** *1100*, 1–12.

Perfect, T. J.; Cook, A. G. Rice Planthopper Population Dynamics: A Comparison Between Temperate and Tropical Regions. In: *The Planthoppers: Their Ecology and Management*, Denno, R. F., Perfect, T. J., Eds.; Chapman and Hall: London, 1994; pp 282–301.

Pilkington, L. J.; Gurr, G. M.; Fletcher, M. J.; Elliott, E.; Nikandrow, A.; Nicol, H. I. Reducing the Immigration of Suspected Leafhopper Vectors and Severity of Australian Lucerne Yellows Disease. *Austral. J. Exp. Agric.* **2004,** *44*, 983–992.

Platt, K. B.; Linthicum, K. J.; Myint, K. S. A.; Innis, B. L.; Lerdthusnee, K.; Vaughn, D. W. Impact of Dengue Virus Infection on Feeding Behavior of *Aedes aegypti. Am. J. Trop. Med. Hygiene* **1997,** *57*, 119–125.

Prasad, Y.; Prabhakar, M. *Pest Monitoring and Forecasting. Integrated Pest Management: Principles and Practice*; Cabi: Oxfordshire, 2012; pp 41–57.

Puterka, G. J.; Reinke, M.; Luvisi, D.; Ciomperlik, M. A.; Bartels, D.; Wendel, L.; Glenn, D. M. Particle film, Surround WP, Effects on Glassy-Winged Sharpshooter Behavior and Its Utility as a Barrier to Sharpshooter Infestations in Grapes. *Plant Health Progress Online*, 2003. DOI: 10.1094/PHP-2003-0321-RS.

Raccah, B. Nonpersistent Viruses: Epidemiology and Control. *Adv. Virus Res.* **1986,** *31*, 387–429.

Rodriguez-Saona, C. R.; Byers, J. A.; Schiffhauer, D. Effect of Trap Color and Height on Captures of Blunt-Nosed and Sharp-Nosed Leafhoppers (Hemiptera: Cicadellidae) and Non-Target Arthropods in Cranberry Bogs. *Crop Protect.* **2012,** *40*, 132–144.

Romanazzi, G.; Murolo, S. Partial Uprooting and Pulling to Induce Recovery in Bois Noir-Infected Grapevines. *J. Phytopathol.* **2008,** *156*, 747–750.

Rypstra, A. L.; Buddle, C. M. (2013).Spider Silk Reduces Insect Herbivory. *Biol. Lett.* **2013,** *9*, 20120948. DOI: 2012094810.1098/rsbl.2012.0948.

Said, A. E.; Fatahuddin, A.; Nasruddin, A. Effect of Sticky Trap Color and Height on the Capture of Adult Oriental Fruit Fly, *Bactrocera dorsalis* (Hendel) (Diptera:Tephritidae) on Chili Pepper. *Am. J. Agric. Biol. Sci.* **2017,** *12* (1), 13–17.

Sanghera, G. S.; Kumar, R. Prospects of Non-Conventional Approaches for Sugarcane Improvement. In: *Biotechnology to Enhance Sugarcane Productivity and Stress Tolerance*; Sengar, K., Ed.; Boca Raton: CRC Press (Taylor and Francis), 2018.

Saracco, P.; Marzachi, C.; Bosco, D. Activity of Some Insecticides in Preventing Transmission of Chrysanthemum Yellows Phytoplasma ('*Candidatus* Phytoplasma asteris') by the Leafhopper *Macrosteles quadripunctulatus* Kirschbaum. *Crop Protect.* **2008,** *27*, 130–136.

Schmitz, O. J.; Suttle, K. B. Effects of Top Predator Species on Direct and Indirect Interactions in a Food Web. *Ecology* **2001,** *82*, 2072–2081.

Seemüller, E.; Garnier, M.; Schneider, B. Mycoplasmas of Plants and Insects. In: *Molecular Biology and Pathogenicity of Mycoplasmas*; Razin, S., Herrmann, R., Eds.; Kluwer Academic/Plenum Publishers: Dordrecht, 2002; pp 91–116.

Seemüller, E.; Marcone, C.; Lauer, U.; Ragozzino, A.; Giischl, M. Current Status of Molecular Classification of the Phytoplasmas. *J. Plant Pathol.* **1998,** *80*, 3–26.

Senevirathne, J. A. U. T. *An Investigation of the Secondary Transmission of Sugarcane White Leaf Disease in Sri Lanka*; PhD Thesis, unpublished, 2008.

Setiawan, D. F.; and Ragsdale, D. W. Use of Aluminum-Foil and Oat-Straw Mulches for Controlling Aster Leafhopper, *Macrosteles fascifrons* (Homoptera: Cicadellidae), and Aster Yellows in Carrots. *Great Lakes Entomol.* **1987,** *20*, 102–109.

Sisterson M. S. Transmission of Insect-Vectored Pathogens: Effects of Vector Fitness as a Function of Infectivity Status. *Environmental Entomolo*gy **2009,** *38*, 345–355. DOI: 10.1603/022.038.0206.

SRI. *Progress Report: Division of Crop Protection*; Publication of Sugarcane Research Institute: Sri Lanka, 2007.

SRI. *Progress Report: Division of Crop Protection*; Publication of Sugarcane Research Institute: Sri Lanka, 2016.

SRI. *Progress Report: Division of Crop Protection*; Publication of Sugarcane Research Institute: Sri Lanka, 2017.

SRI. *Progress Report: Division of Crop Protection*; Publication of Sugarcane Research Institute: Sri Lanka, 2019.

SRI. *Progress Report: Division of Crop Protection*; Publication of Sugarcane Research Institute, Sri Lanka, 2020.

Srinivasan, K.; Chelliah, S. The Mechanism of Preference of the Leaf Hopper Vector, *Hishimonous phycitis* (Distant) for Egg Plants Infected with Little Leaf Disease. *Proc. Indian Natl. Sci. Acad.* **1980,** *6*, 786–796.

Stark-Urnau, M.; Kast, W. K. Mabnahmen zur Eindammung des Brennesseltyps der Schwarzholzkrankheit bei Weinreben (*Vitis vinifera*). *Gesunde Pfl anzen* **2008,** *60*, 85–89.

Summers, C. G.; Stapleton, J. J. Management of Corn Leafhopper (Homoptera: Cicadellidae) and Corn Stunt Disease in Sweet Corn Using Reflective Mulch. *J. Econ. Entomol.* **2002,** *95*, 325–330.

Tang, L. D.; Zhao, H. Y.; Fu, B. L.; Han, Y.; Liu, K.; Wu, J. H. Colored Sticky Traps to Selectively Survey Thrips in Cowpea Ecosystem. *Neotrop. Entomol.* **2016,** *45* (1), 96–101.

Thaler, J. S.; Griffin, C. A. M. Relative Importance of Consumptive and Non-Consumptive Effects of Predators on Prey and Plant Damage: The Influence of Herbivore Ontogeny. *Entomol. Exp. App.* **2008,** 128, 34–40, https://doi.org/10.1111/j.1570-7458.2008.00737.x.

Thein, M, Jamjanya. T.; Kobori, Y.; Hanboonsong, Y. Dispersal of the Leafhoppers *Matsumuratettix hiroglyphicus* and *Yamatotettix flavovittatus* (Homoptera: Cicadellidae), Vectors of Sugarcane White Leaf Disease. *Appl. Entomol. Zool.* **2012,** *47*, 255–262.

Tholt, G.; Samu, F.; Kiss, B. Feeding Behaviour of a Virus-Vector Leafhopper on Host and Non-Host Plants Characterised by Electrical Penetration Graphs. *Entomol. Exp. Appl.* **2015,** *155*, 123–136, https://doi.org/10.1111/eea.12290.

Toshova, T. B.; Velchev, D. I.; Abaev, V. D.; Kalushkov, P. K.; Orgován, E.; Lohonyai, Z.; Tóth, M.; Koczor, S. Nontarget Coleoptera Species Captured in Coloured Sticky Traps in Maize Crops in Bulgaria. *Acta Zoologica Bulgarica* **2017,** 237–246.

Tubajika, K. M.; Civerolo, E. L.; Puterka, G. J.; Hashim, J. M.; Luvisi, D. A. The Effects of Kaolin, Harpin, and Imidacloprid on Development of Pierce's Disease in Grape. *Crop Protect.* **2007,** *26*, 92–99.

Uyemoto, J. K.; Bethell, R. E.; Kirkpatrick, B. C.; Munkvold, G. P.; Marois, J. J.; Brown, K. W. Eradication as a Control Measure for X-Disease in California Cherry Orchards. *Acta Hortic.* **1998,** *472*, 715–721.

Wakeil.; N. E. E.; Gaafar, N. M.; Abdellah S. H.; Moniem, A. Diversity and Abundance of Leafhoppers (Hemiptera, Cicadellidae) in Different Crops in Egypt. *Adv. Appl. Agric. Sci.* **2013,** *03* (04), 08–15.

Wallis, D. R.; Shaw, P. W. Evaluation of Coloured Sticky Traps for Monitoring Beneficial Insects in Apple Orchards. *New Zealand Plant Protect.* **2008,** *61*, 328–332.

Wally, O.; Daayf, F.; Khadhair, A. H.; Adam, L.; Elliott, B.; Shinners-Carnelley, T.; Iranpour, M.; Northover, P.; Keyworth, S. Incidence and Molecular Detection of Yellows-Type Disease in Carrots, Associated with Leafhoppers in southern Manitoba, Canada. *Can. J. Plant Pathol.* **2004,** 26, 498–505.

Waloff, N.; Jervis, M. A. Communities of Parasitoids Associated with Leafhoppers and Planthoppers in Europe. *Adv. Ecol. Res.* **1987,** *17*, 281–402.

Wangkeeree, J.; Tewaruxsa, P.; Roddee, J.; Y. Hanboonsong. Wolbachia (Rickettsiales: Alphaproteobacteria) Infection in the Leafhopper Vector of Sugarcane White Leaf Disease *J. Insect Sci.* **2020,** *20* (3), 20; 1–8 DOI: 10.1093/jisesa/ieaa053.

Weintraub, P. G. Insect Vectors of Phytoplasmas and Their Control—An Update. *Bull. Insectol.* **2007,** *60* (2), 169–173.

Weintraub, P. G.; Beanland, L. Insect Vectors of Phytoplasmas. *Annu. Rev. Entomol.* **2006,** *51*, 91–111.

Weintraub, P. G.; Wilson, M. R. Control of Phytoplasma Diseases and Vectors. In: *PHYTOPLASMAS Genomes, Host Plants and Vectors*; Weintraub, P. G., Jones, P., Eds.; CAB International: Whiltshire, 2010; pp 233–249.

Werner, B. J.; Mowry, T. M.; Bosque-Pe′rez, N. A.; Ding, H.; Eigenbrode, S. Changes in Green Peach Aphid Responses to Potato Leaf Roll Virus-Induced Volatiles Emitted During Disease Progression. *Environ. Entomol.* **2009,** *38*, 1429–1438.

Wongkaew, P. *Sugarcane White Leaf Disease Management*; Thailand Research Fund, Pimpatana Press: Khon Kaen, 1999; p 228.

Zahavi, T.; Peles, S.; Harari, A. R.; Soroker, V.; Sharon, R. Push and Pull Strategy to Reduce *Hyalesthes obsoletus* Population in Vineyards by *Vitex agnus castus* as Trap Plant. *Bull. Insectol.* **2007,** *60* (2), 297–298.

CHAPTER 8

Management of Fungal Diseases

AHMED NAMSI, SIHEM BEN MAACHIA, and MONIA JEMNI

Regional Research Center in Oasis Agriculture, Degache, Tunisia

ABSTRACT

To combat widespread fungal pathogen infections and postharvest illnesses, it is critical to create long-term approaches and instruments. For infections caused by fungi, the use of synthetic fungicides has been the most common control technique. The fungicides employed were hazardous to biological systems, killing not just the target fungi but also beneficial organisms. To preserve biodiversity, it is becoming increasingly vital to rethink our strategies and find different ways to handle fungal diseases such as: microbiol biocontrol agents, naturally derived botanicals (bioactive phytocompunds, plant extract, volatil aromatic compounds, and essential oils), nanotechnology, genetic engineering (transgenic approaches), physical control of postharvest fruit (hot water treatment, fumigation, ionizing radiation), food additives, and coating (food preservatives, fruit skin coating). Combining the above complementing strategies could result in effective fungal disease control.

8.1 DAMAGE AND LOSS CAUSED BY FUNGAL PATHOGENS IN AGRICULTURE

Plant diseases and pests pose a significant danger to global food security. Pathogens attack crop plants continuously during both the preharvest and postharvest periods, resulting in significant production losses. Losses due to plant diseases may be as high as $30–50 billion in cultivated and stored crops

Plant Diseases and Their Management: A Sustainable Approach.
Deepa Srivastava, Rajarshi Kumar Gaur, and Ajay K. Tiwari (Eds.)

and represent approximately 20% of total global food production annually (Giraud et al., 2010). Plant diseases cost the global economy roughly $220 billion per year, according to the Food and Agriculture Organization (FAO, 2020). Diseases caused by pathogenic microorganisms and animal pests are estimated to account for up to 38.2% of overall yield losses in rice and 36.5% of total yield losses in potatoes. Diseases caused by plant pathogenic fungi have continued to cause considerable losses in harvestable yield and have reduced the visual value and storage life of agricultural crops. Fungi are thought to be responsible for roughly 30% of new diseases (Giraud et al., 2010). These are the primary causes of the most serious plant diseases, which result in major production and crop quality reductions as well as enormous economic losses around the world. Fungi are thought to be the greatest hazard to plant-host species, accounting for around 65% of pathogen-driven host loss (Fisher et al., 2012). Fungal plant pathogens, in particular, are drawn to nutrient-dense plant tissues. As a result, the risk of losing wheat output owing to fungal infections rises from less than 10% with an achievable yield of 2 tons/ha to more than 20% when production intensity rises to 12 tons/ha (Oerke; 2019). The greatest losses were discovered in food-deficit countries with rapidly expanding populations (Agrios, 2005; Savary et al., 2019). Postharvest illnesses generate significant economic losses because microbial diversity in fresh produce rises during transport from the field to the consumer. According to the United Nations' Food and Agriculture Organization (FAO), approximately 20–40% of harvested fruits and vegetables are lost due to decay, though damage can also be caused by physical injury during postharvest handling, especially in developing countries, as a result of poor storage and transportation facilities (Droby, 2006; Braghini et al., 2009a). Fruit postharvest infections, caused by opportunistic bacteria and fungi that cause rotting of the product, have been recognized as the leading source of horticulture commodity losses (Munhuweyi et al., 2016).

In pome and stone fruits, fungal diseases such as *Penicillium*, *Alternaria spp.*, *Aspergillus*, and *Fusarium* create mycotoxins such as patulin, aflatoxin, and citrinin, which can be harmful to one's health. Postharvest infections can have an impact on the quality and shelf life of horticulture fresh food along the supply chain. When a contaminated fruit is used in processing, the likelihood of mycotoxin generation rises. As a result, mycotoxin exposure is regarded as a significant danger factor. The Food and Agriculture Organization (FAO) is the United Nations' food and agriculture agency (FAO, 2020). According to estimates, mycotoxins impact almost a quarter of all global commodities each year (Magan et al., 2003; Atanda et al., 2011). Many fungal diseases have become more common as a result of global warming and concomitant climate changes (Garcia-Solache and Casadevall, 2010; Casadevall, 2017). One of the key difficulties of the twenty-first century is to improve the volume and availability of food for a growing global population (Godfray et al., 2010). To ensure food security and safety, reducing or eliminating crop losses due to disease through well-organized plant disease management and control in agriculture will make a significant contribution to the quest for global food security (Sarrocco and Vannacci, 2018).

8.2 PLANT AND POSTHARVEST FUNGAL PATHOGENS

8.2.1 PLANT PATHOGENIC FUNGI

Fungi account for the majority of plant diseases. Approximately 8000 fungus and oomycetes species have been connected to plant diseases (Horst, 2008; Fisher et al., 2020). Plant fungal pathogens cause devastating diseases in crops and are the leading cause of plant death in agricultural settings (Agrios, 2009; Doehlemann et al., 2017). The majority of fungus exist as threadlike hyphae composed of cells with chitin-rich cell walls. Wind, water splash, dirt, insects, other invertebrates, animals, workers, machinery, tools, seedlings, and other plant material all help to distribute fungal spores. Pathogenic fungi can infect plants at every stage of development, from seedling through seed maturation, in natural environments, either alone or in combination with other phytopathogens (Iqbal et al., 2018; Narayanasamy, 2011). Plant fungal pathogens can assault and infect their hosts in a variety of ways. Fungi get into plants through natural holes like stomata, as well as wounds caused by pruning, harvesting, hail, insects, various diseases, and mechanical damage. These organisms produce enzymes (chemical activity) and employ physical pressure to build entry holes for hyphae to colonize and rob nutrients from plants. Plants are harmed by pathogenic fungus because they kill cells and/or cause stress. Anthracnose, Blight, Botrytis rots; Canker, Damping off, Dieback, Downy mildews; Fusarium rots; Gall, Leaf spot, White blister, Powdery mildews; Rusts; Root rot (Rhizoctonia rots, Pythium rots; Sclerotium rots.), Coils Scab, and Wilt are the most prevalent diseases produced by plant pathogenic fungus. Others, such as Clubroot (Plasmodiophorabrassicae) in brassicas, Leaf blight (*Alternaria dauci*) in carrots, and Red root complex in beans, are peculiar to a particular crop group. The majority of fungal pathogenic species are Ascomycetes (genera: *Alternaria*, *Fusarium*,

Verticillium) and Basidiomycetes (genera: *Sclerotium, Rhizoctonia*) (Iqbal et al., 2018; Hussain and Usman, 2019; Jain et al., 2019).

8.2.2 FUNGAL PATHOGENS OF POSTHARVEST DISEASE

Fungi that cause postharvest diseases can be divided into two groups based on the source and features of the infection: (i) those that poison the field's fruit, while the fruit is still linked to the mother plant and remains dormant until it is harvested and (ii) those that infect the fruit by microscopic wounds in the rind or injuries sustained during harvest, transportation, postharvest handling, and marketing (Palou et al., 2015). Blue or green molds on numerous significant fruit commodities are caused by *Penicillium spp.* Blue mold, produced by the *Penicillium expansum L.* species, can cause severe postharvest losses in apple, pear, stone fruits, kiwifruit, numerous berries, pomegranate, persimmon, and other subtropical and tropical fruits, for example (Errampalli, 2014). Penicillium expansum infections in apples and pears are also becoming a health problem due to their production of patulin (Neri et al., 2010). Green and blue molds, produced by *Penicillium digitatum* and *Penicillium italicum*, respectively, are the two postharvest diseases that cause the most significant losses of citrus fruits over the world (Palou, 2014). *Geotrichum spp.*, which cause sour rot on a variety of horticulture products, and *Rhizopus spp.*, which are severe postharvest infections that induce soft rot or *Rhizopus* rot on a variety of fruit hosts, are two other major wound pathogens. *Geotrichum candidum L.*, which attacks stone fruits such as peach, nectarine, plum, and sweet cherry, as well as tomato and other fruit-like vegetables, and Geotrichumcitri-aurantii, which causes sour rot in citrus fruits, are the two most important species. The most *Botryos phaeriaceae* common cause of fruit soft rot is *Rhizopus stolonifer* (Barkai-Golan, 2001). *Botrytis cinerea*, which forms gray mold, is the principal cause of postharvest degradation in grapes, strawberries, blueberries, and other tiny berries, kiwifruit, pomegranate, and figs, among other fruits and vegetables, but it also infects many other fruits and vegetables (Romanazzi and Feliziani, 2014; Moorman, 2014: Lennox et al., 2003; Hua el al., 2018; Spotts et al., 2008). Many subtropical and tropical fruits, such as citrus, avocado, persimmon, banana, papaya, mango, and pineapple, are infected with *Colletotrichum spp.*, which causes anthracnose. *Colletotrichum gloeosporioides* (mango, papaya, and *Botryos phaeriaceae* avocado), *Colletotrichum musae* (banana), *Colletotrichum acutatum*, and *Colletotrichum horii* are the most important species (Siddiqui and Ali, 2014). Black spot-caused by Alternaria spp., as well as Lasiodiplodiatheobromae that cause stem-end rot and Botryos phaeriaceae pathogens that cause stem-end (Zhang, 2014). *Alternaria alternata* is a postharvest disease that affects pome fruits, persimmons, loquats, and pomegranates, among other fruits (Troncoso-Rojas and Tiznado-Hernánde, 2014). Monilinia species, particularly *M. fructicola, M. laxa,* and *M. fructigena*, cause field and postharvest brown rot in stone fruits and are widely regarded as one of the most significant yield-limiting factors for these crops. In peaches and other stone fruits, *Monilinia* spp. infect the fruit immediately before harvesting; in other words, the symptom may not have appeared yet during the preharvest stage but may become more noticeable during postharvest storage. *Monilinia* spp. can also cause pome fruit postharvest disease (Martini and Mari, 2014).

8.3 ALTERNATIVES APPROACHES OF CONTROLLING FUNGAL DISEASES

Various ways have been used to prevent, alleviate, or control fungal illnesses over the years (Tyagi et al., 2017). Several ways have been studied for the management of fungal diseases and maintaining fresh product quality, including standard synthetic fungicides, natural chemicals, chemical elicitors of induced natural host defenses, biological control, and physical means such as ionizing radiation (Hallman, 2011).

8.3.1 CHEMICAL CONTROL

Due to fungal infection, significant losses in crop production and postharvest storage defy modern agricultural output (Agrios, 2005). The use of fungicides to prevent fungal infections is frequently regarded as essential for ensuring global food security, and has thus become a vital aspect of agriculture (Strange and Scott, 2005; Agrios, 2005; Fisher et al., 2012). To reduce food losses caused by fungal diseases, farmers today rely heavily on synthetic fungicides. Fungicides are primarily used to manage fruits, vegetables, and postharvest fungal deterioration, accounting for more than 35% of the global pesticide industry. Geographically, Europe is the leading market for fungicides, with key applications in grains and cereals (e.g., wheat), fruits (with viticulture being particularly important), and vegetables. Fungicide sales (based on mass) account for more than 40% of total pesticide sales in the European Union (EU), with synthetic, organic fungicides accounting for roughly 60% of all fungicides. Fungicides can account for more than 90% of all pesticide treatments in wine-growing regions. In the United States (US), this accounts for 80% of overall fungicide use. In North America, fungicides account for less than 10% of total pesticide application (including inorganic fungicides). Major fungicide categories include dithiocarbamates, chloronitriles, demethylation inhibitors (DMIs), and strobilurins, which represent for roughly 65, 12, 7, and 6% of total synthetic fungicide mass used in the United States, respectively (Carvallo, 2017; Russell, 2005).

8.3.1.1 USE OF FUNGICIDES IN CROPS

In comparison to herbicides and insecticides, fungicides are the smallest group of pesticides, but they show a good growth rate and future potential. The classical protective organic fungicides are the most commonly used, although use of the highly active, systemic compounds with curative action is increasing (Baibakova et al. 2019). Fungicides are classified into five chemical groups. Triazoles are the most common type. This class of fungicides has been used to combat pathogens that cause a variety of diseases in fruit and vegetable crops (Table 8.1).

i. Nonsystemic (Contact) Fungicides: These fungicides prevent mycelia from growing and developing within plant tissues by killing or suppressing fungi and fungal spores (Oliver and Hewitt, 2014), However, after the fungus has invaded or colonized host tissue, they have little or no effect. Furthermore, while nonsystemic fungicides tend to stay on the surface of plants, they can be phytotoxic and cause damage if absorbed (Lesemann et al., 2006). Dithiocarbamates acid, sulfur-based agents, copper-based agents, and other derivatives have contact action. As a result, these fungicides can only be used as preventatives. To reduce the risk of infection, it is also vital to apply them to certain plants before the known infection period begins. Zineb, polycarbonate, copper oxychloride, sulfur, mancozeb, bordeaux liquid, and other contact agents are unable to heal infected plants. Inorganic copper compounds such as Bordeaux mixture, copper carbonate, and inorganic sulfur in the form of elemental sulfur and lime sulfur are examples of primary contact fungicides.

Organic contact fungicides (e.g., thiram, ferbam, and ziram) are more effective and less toxic than inorganic chemicals and hence play an essential role in the comprehensive control of plant diseases.

ii. Fungicides that are absorbed by the plant and transferred to the infection site are known as systemic fungicides. As a result, after the mycelia have penetrated the parenchyma of the plant tissue, these substances can kill the fungus, halting the progress of infection (Oliver and Hewitt, 2014). Some systemic fungicides only go a short distance within the plant from the point of penetration. Local-systemic fungicides are what we are talking about. This group includes dicarboximide fungicides, for example (González and Caetano, 2017). Iprodione, procymidone, vinclozolin, chlozolinate, metomeclan, and dicarboximide fungicides are particularly promising for the management of plant diseases caused by *Botrytis*, *Sclerotinia*, *Monilinia*, *Alternaria*, *Sclerotium*, and *Phoma* species (Dias, 2012). The suppression of triglyceride production in fungus appears to be the mechanism of action for these drugs. Some locally systemic fungicides pass the leaf plate from one leaf surface to the next, but they do not penetrate the plant. These fungicides are known as translaminar fungicides, such as trifloxystrobin (Mueller, 2006). Xylemmobile or acropetal systemic fungicides move inside the water-conducting tissue (xylem), which raises them in the transpiration flow; nonetheless, mobility within the plant is limited. DMI fungicides, for example, are somewhat mobile within plants. Others are quite mobile and can travel around the xylem with ease. Thiophanate-methyl and mefanox are two examples of systemic fungicides that are mobile in the xylem (Baibakova et al. 2019). A phloem-mobile system is the third form of systemic fungicide. The compound circulates in phloem out of the sheet, where it is deposited upwardly to the other leaves and downward to the roots (Lesemann et al., 2006).

8.3.1.2 SYNTHETIC FUNGICIDES FOR CONTROLLING POSTHARVEST DISEASES OF FRESH FRUITS

Fruit postharvest infections and the economic losses they cause have been controlled for many years over the world thanks to postharvest applications of synthetic chemical fungicides. The mainstay of postharvest disease

TABLE 8.1 Main Classes of Fungicides (Baibakova et al., 2019).

Chemical class	Mechanism of action	Fungi
Triazoles	Inhibit sterol biosynthesis	Botrytis, Ustilago, Cercospora, TilletiaZymoseptoria, Fusarium, Cochliobolus, Erysiphe, Altemaria, Puccinia, Septoria, Pythium, Rhynchosporium, Cladosporium, Epicoccum, Phoma
Phenylpyrroles	Inhibit micellic development, decrease glucose phosphorylation during cell respiration, and alter cell membrane function.	Tilletia, Fusarium, Ascochyta, Altemaria, Fusarium, Aspergillus, Rhizoctonia Helminthosporium,
Strobilurins	Block electron transport in the cytochrome b and c1 chains to stop mitochondrial respiration.	Puccinia, Septoria, Pyrenophora, Alternaria, Cladosporium, Epicoccum, Botrytis, Rhynchosporium, Drechslera, Fusarium, Rhizoctonia, Ustilago, Erysiphe
Benzimidazoles	Disrupt the fungal cell's life activity by inhibiting ergosterol production.	Fusarium, Botrytis, Sclerotmia, Septoria, Uncinula, Erysiphe
Morpholines (cinnamic acid derivatives)	Prevent the growth of mycelium and the decrease of the double compound C-C, as well as the synthesis of ergosterol.	Erysiphe, Uncinula, Septoria, Puccinia

management has been synthetic fungicide applications (Romanazzi et al., 2016). In general, postharvest management is performed by applying a preharvest fungicide treatment to the fruit and then applying a postharvest dip or drench treatment to the fruit after harvesting, depending on the disease. To minimize infection, synthetic fungicides such as benomyl, carbendazim, propineb, and difenoconazole are routinely used (Ali et al., 2013). In conventional horticulture, active compounds such as imazalil (IMZ), pyrimethanil (PYR), fludioxonil (FLU), thiabendazole (TBZ), and others are widely employed as cost-effective ways of postharvest decay control. In banana plants, two chemical fungicides, IMZ and TBZ, showed strong antifungal efficacy against *Colletotrichum musae* (Khan et al., 2001). These fungicides, according to Bhutia (2015), are the principal means of controlling postharvest infections of banana fruits: IMZ and TBZ. During the rainy seasons, the fungicide guazatine is often used as a prestorage drench treatment to combat sour rot (Geotrichum candidum) in citrus fruit. A systemic fungicide known as imazalil is used to reduce green mold rot in citrus fruit during the postharvest chain. In South Africa, New Zealand, and Australia, postharvest infections in avocados are controlled economically by the use of prochloraz fungicide treatments (Smith et al., 2011).

To manage a wide range of fungal diseases, numerous efficient "reduced risk" fungicides belonging to distinct chemical classes have recently been created. A number of studies have been conducted to investigate the potential of these fungicides to control the primary postharvest pathogens of horticultural crops, and some of them have been registered in the United States for postharvest treatments of various horticultural crops (Adaskaveg and Förster, 2010). The widespread and ongoing use of these pesticides in the treatment of postharvest fungal infections is causing serious issues (Murray et al., 2017).

Fungicides (such as imazalil, thiabendazole, and o-phenylphenol) are commonly used, which might result in fungal mutations and resistance to specific fungicides (Deising et al., 2008; Arrebola et al. 2010). Due to the continued use of fungicides, resistance to thiabendazole has been found in P. digitatum and P. italicum. Pseudocercosporella herpotrichoides isolates, on the other hand, have been found to be resistant to prochloraz. *C. gloeosporioides* isolates from mango fruit showed a wide range of sensitivity to prochloraz (Arauz, 2000), while B. cinerea has recently shown resistance to anilinopyrimidines (pyrimethanil). In South Africa, *P. digitatum* showed resistance to imazalil. Moreover, the usage of fungicides is increasingly causing additional serious issues: (i) Fungicide residues on fresh fruit can also be hazardous to the environment and human health, (ii) restricted access to new high-value organic markets or existing export markets that are increasingly requiring pesticide-free products to meet customer needs, (iii) consumer preferences and legal changes obviously promote a decrease in the usage of traditional fungicides (Aktar et al., 2009).

8.3.1.3 CHEMICAL TREATMENT LIMITS AND CHALLENGES

Furthermore, due to changes in climatic conditions, the development of fungicide resistance, and invasive fungus species, fungicide use is expected to grow regionally (Hakala et al., 2011, Boxall et al., 2009; Seagraves and Lundgren, 2012). However, because of their widespread use, fungicide-resistant organisms have emerged. Fungal diseases can develop resistance within a few years of exposure, depending on the pathogen genome and the method of fungicidal activity. This necessitates a continuous search for new fungicides and treatment approaches (Avenot and Michailides, 2007). Continuous use of chemical fungicides might be ineffective and not environment friendly.. The detrimental effects of these fungicides and chemical residues highlight the need to move away from high-dose chemical applications and toward safer, more environmentally friendly alternatives, as well as to apply new methodologies and technology for the control of postharvest fruit disease. However, the extensive use of these concerns has been made about the environment's and human health's long-term impacts. Seed-applied fungicides, for example, are effective against soil-borne diseases, but they have the ability to linger in low quantities in the plant or rhizosphere for several months (Nettles et al., 2016). Currently, fresh produce-importing countries enforce rigorous laws regarding minimal pesticide residual levels in the edible section of the fresh product. The environmental impact of fungicide waste disposal is significant. Regulators have passed legislation that prohibits or restricts their usage by setting lower maximum residue levels (MRLs) (Kim et al., 2017).

The use of various chemical fungicides is not an environmentally friendly approach to fungal diseases because it pollutes the environment and poses a health risk to humans. Crop protection in general, and plant disease control in particular, are critical to satisfying future food demand in terms of both quality and quantity (Strange and Scott 2005). As an alternative to chemical pesticides, new environmentally friendly and sustainable solutions are needed to manage diseases on the world's most valuable crops. Plant disease management employs a variety of biocontrol technologies, strategies, and approaches. To establish long-term disease control measures, the pathogens' genetic diversity will need to be thoroughly investigated. The detrimental effects of these fungicides and chemical residues highlight the need to move away from high-dose chemical applications and toward safer, more environmentally friendly alternatives, as well as to apply new methodologies and technology for the control of postharvest fruit disease (Yazdani et al., 2011; Sales et al., 2016).

8.3.2 MICROBIAL BIOCONTROL AGENTS (BCAS)

The use of nonpathogenic microbial antagonists or compounds produced from their metabolism, referred to as biological control, is a valid and promising alternative from an ecological standpoint for reducing the activities and controlling populations of target diseases (Singh, 2016). As a result, a biocontrol strategy based on competition for space and nutrients, as well as the induction of plant defenses against virulent infections by attenuated or avirulent pathogens, could be regarded a viable option (Ghorbanpour et al., 2018).

Trichoderma has been identified as the genus with the greatest promise, with 25 biocontrol compounds employed against a variety of plant fungal diseases. *Trichoderma* produces a variety of enzymes that attack fungi's cell walls. *Trichoderma* spp., on the other hand, do not directly attack the plant disease Rhizoctonia solani when fresh bark is utilized in composts. However, as the concentration of readily available cellulose drops in disintegrating bark, the chitinase genes of *Trichoderma* spp. are activated, resulting in the production of chitinase to parasitize *R. solani.* In addition to *Trichoderma*, nine genera, *Alternaria*, *Aspergillus*, *Candida*, *Fusarium*, *Penicillium*, *Pichia*, *Pythium*, *Talaromyces*, and *Verticillium*, are regarded as important, each with five or more known antagonistic species (Thambugala et al., 2020). Plants can be protected from fungal infections using antagonistic microorganisms, which are environmentally friendly (Bloemberg and Lugtenberg 2001). Bacteria inhabiting the root interior have been found to be ideal candidates for use as biological control agents in studies (BCAs) (Sturz et al. 2000; Tjamos et al. 2004). Pseudomonads that produce the antibiotic 2,4-diacetylphloroglucinol (DAPG) may also cause host defenses to get activated (Iavicoli et al. 2003). Furthermore, DAPG-producing bacteria can colonize roots aggressively, a characteristic that may contribute to their ability to limit pathogen activity in the wheat rhizosphere by competing for organic resources (Raaijmakers and Weller 2001). Endophytic bacteria are also known to aid plant growth by producing plant hormones, increasing nutrient availability, and nitrogen fixation (Whipps 2001). Pseudomonas putida WCS358r strains genetically altered to produce phenazine and DAPG have recently been shown to have better abilities to suppress plant diseases in field-grown wheat. Serratia marcescens seems to influence *Sclerotium rolfsii* via chitinase expression, for example. Furthermore, a b-1,3-glucanase contributes greatly to Lysobacterenzymogenes strain C3's biocontrol capabilities. As a result, contemporary agriculture has effectively used commercial products containing microbial BCAs (e.g., Trichoderma-based products and biopesticides based on Bacillus thuringiensis).

Several commercial fungal and bacterial antagonistic products, such as GiloGard, are available (Gliocladium virens seedling diseases of ornamentals and bedding plants), BINAB T (Trichoderma *harzianum*/*Trichoderma polysporum* to reduce wood decay), Gallex or Galltrol (*Trichoderma harzianum* several soil-borne illnesses), F-Stop (Trichoderma harzianum several soil-borne diseases), Gallex or Galltrol (Trichoderma harzianum several soil-borne diseases) (Agrobacterium radiobacter K-84 crown rot), Dagger G (*Pseudomonas fluorescens*, *Rhizoctonia*, and *Pythium* damping-off of cotton), and Kodiac (Bacillus subtillis seed illnesses) are two plant fungal disease treatments that are widely utilized around the world (Agrios, 2005; Pal and McSpadden Gardener, 2006). Similarly, certain microbes can engage in antagonistic behavior by directly competing for nutrients (e.g., iron), forming biofilms on the surface of the host/fruit, or producing volatiles (bioactive metabolites) and toxic proteins, as well as enzymes (glucanases, chitinases, proteases, and hydrolases) that inhibit fungal pathogens (Spadaro and Droby, 2016; Qin et al., 2017; Zhou et al., 2018; Shuping and Eloff , 2017). Because of their great inhibitory activity and propensity to colonize surfaces for lengthy periods of time, yeasts have been widely researched as antagonists (Zhimo et al., 2016). Their antagonistic activity has been attributed to a number of mechanisms. Competition for nutrition and space, pathogen parasitism, antifungal chemical release, induction of host resistance, biofilm development, and, most recently, the role of reactive oxygen species (ROS) in defense response are all examples. Zhimo et al. (2016), for example, tested the biological efficiency of 29 antagonistic yeast isolates against a variety of postharvest fruit diseases, including *Colletotrichum musae, Fusarium oxysporum, Colletotrichum capsici, Colletotrichum gloeosporioides,* and *Aspergillus niger*. The pathogens' mycelial growth was greatly decreased by the yeast isolates, with a mycelial growth inhibition of above 20%. Only four yeast isolates were found to have a broad range of antagonistic activity against the pathogens examined, with an average MGI of 17.3 to 22%. This demonstrates the complication of antagonistic mechanisms. In general, antagonistic bacteria's principal modes of action are similar to those of fungal antagonists (Zhimo et al., 2016; Carmona-Hernandez et al., 2019). Biological control agents (BCA) have been used in a limited number of places around the world, despite the ecological benefits they provide. The spectrum of biological controls, according to Carmona-Hernandez et al. (2019), has several restrictions, such as temperature, humidity, and a limited range of activity. The author also stated that in the field, this control technique is subject to high variability, which is a significant constraint to its practical deployment. Furthermore, commercialization of BCAs will necessitate scale-up from lab to large-scale manufacturing, creation of standard product formulation, adequate biosafety of selected microorganisms, and registration (Nunes, 2012). Regulatory impediments or procedures prior to registration, as well as the expensive expense of scaling up production, are other recurring roadblocks (Mari et al., 2014; Anuagasi et al., 2017).

8.3.3 NATURALLY DERIVED BOTANICALS

8.3.3.1 BIOACTIVE PHYTOCOMPOUNDS

Plant bioactive chemicals, as well as other natural compounds including metabolites, volatiles, and active enzymes from specific microbes, are rich sources of antifungal compounds, making them ideal fungicides for the management of postharvest fungal infections (Bhutia, 2015; Klein et al., 2013; Lopes et al., 2015; Yazdani et al., 2011; Jiménez-Reyes et al., 2019). Many bioactive chemicals have been discovered to date, and some of them have aided in the development of innovative plant-based biopesticides for agricultural use. A safer alternative for sustainable and organic production is the proper selection of biomolecules for the manufacture of biopesticides with diverse modes of action against target pathogens. A thionin from wheat endosperm was the first bioactive chemical identified from plants. *Drechslera teres*, a barley pathogen, and *Thielaviopsis paradoxa* are both inhibited by thionins found in the endosperm and leaf of barley (Montesinos, 2007). Allicin (diallyl thiosulfinate) is an antibacterial and antifungal organosulfur chemical derived from garlic (Curtis et al., 2004). *Phytophthora cryptogea, P. cinnamomi, Fusarium oxysporum f. sp. cyclaminis*, as well as soybean crops (*Phomopsis phaseoli, Fusarium* spp., *Sclarotinia sclerotiorum, Phomaexigua*), have all demonstrated the effectiveness of grapefruit extract in the field against gray mold, fusariosis, and alternariosis (Jamiołkowska et al., 2011). This organic compound has been demonstrated to be effective against fungus and other diseases in studies (Elhadrami et al., 2010).

8.3.3.2 PLANT EXTRACTS

Plant extracts are complex combinations of molecules with bioactive activities from several chemical classes. They are traditionally made from the full plant as well as components of the plant (rhizome, root, stem, leaf, flower, and tissues/cell cultures) (Hao et al., 2015). A solvent is used to extract these complicated combinations of bioactive chemicals. Water, acetone, acetic acid, ethanol, ethyl acetate, methanol, and hexane are the most frequent solvents. It is worth noting that the solvent used and the experimental conditions used can have a big impact on the extraction yield and constituent composition (Cosa et al., 2006; Sasidharan et al., 2011). Secondary metabolites such as alkaloids, phenols, flavonoids, terpenoids, phenolic acids, quinones, flavones, flavonoids, flavonols, tannins, and coumarins are largely responsible for plants' antifungal activity (Bhutia et al., 2015; Gurjar et al., 2012). Leaf extracts from seven different plants, namely *Acacia nilotica* (ethyl alcohol), *Citrus aurantifolia* (ethyl acetate), *Murraya koenigii* (ethyl acetate), *Nerium indicum* (ethyl acetate), *Ocimum gratissimum* (benzene, ethyl acetate), *O. sanctum* (petroleum ether), *Prunus persica* (ethyl acetate), and the bark extract of *A. farnesiana* and *A. nilotica* (ethyl acetate extract) exhibited 100% efficacy against the test fungus. Latex is produced by over 10% of flowering plants in over 40 plant families, including Euphorbiaceae, Moraceae, Cannabinaceae, Apocynaceae, Astereaceae, Papavereceae, Sapotaceae, and Asclepidiaceae (Neri et al., 2009; Agrawal and Konno, 2009). Latex is a clear or variably colored, generally milky emulsion or suspension of many solid particles that emerges when plant tissue is damaged (Dickison, 2000). Inside the laticifers of roots, stems, leaves, and fruits, latex pours (Pickyard, 2008). Plant laticifer cells' subcellular organelles produce a wide range of defensive proteins and secondary compounds that act as a barrier to pathogens (Konno, 2011; Ramos et al., 2007; El-Moussaoui et al., 2001). Plant latex contains a variety of physiologically active chemicals, and its antibacterial properties have been thoroughly documented (Siritapetawee et al., 2012; Kanokwiroon et al., 2008; Wititsuwannakul et al., 2002). Plant latex bioactive substances could be used as an antifungal agent against postharvest infections. Phytochemical analyses revealed the presence of alkaloids, flavonoids, glycosides, phenols, saponins, steroids, tannins, and terpenoids in latex extracts. Various levels of sensitivity were identified across plant species, with *A. terreus* and *P. digitatum* being the most sensitive. The latex extracts examined were moderately susceptible to *F. solani* and *A. fumigatus*. Thevetia peruviana latex (75.2%) and Artocarpus heterophyllus latex (64.8%) were shown to have potential antifungal activity against the isolates, followed by Manilkara zapota latex (51.1%) (Sibi et al., 2013).

8.3.3.3 VOLATILE AROMATIC COMPONENTS AND ESSENTIAL OILS

A variety of components including inducible volatile aromatic and flavor chemicals can be found in fruits and vegetables (Tripathi, 2007). Fruits create these aromatics and flavor components during ripening and provide resilience to the fruits throughout the postharvest stage. Flavor molecules are secondary metabolites with distinctive features such as limited water solubility and volatility. Acetaldehyde, six carbon (C6) aldehydes, benzaldehyde, hexenel, and hexanal are some of the most important volatile aromatic components. The effect of trans-2-hexenel on the control of blue mold disease (*P. expansum*) in the decrease of patulin content and fruit quality improvement of "Conference" pears was assessed, and better decay reduction was obtained (Neri et al., 2006). Jasmonates are naturally occurring plant growth regulators that are found across the plant kingdom and are thought to control a variety of elements of plant development and responses to environmental challenges. Essential oils (EOs) are aromatic oily solutions generated from diverse parts of plants (roots, flowers, buds, leaves, twigs, bark, herbs, wood bark and fiber, fruit, and seeds) (Burt, 2004). Essential oils (EOs) as antifungal agents were thoroughly discussed by Jiménez-Reyes et al. (2019) Kalemba and Kunicka (2003), Tzortzakis (2007a, 2007b). These EOs are volatile, natural, and complex chemicals with a strong odor that are generated as secondary metabolites by aromatic plants (Abdolahi et al., 2010). EOs have been studied for their antifungal properties in a variety of studies. Abdolahi et al. (2010), for example, investigated the antifungal effects of EOs from ajowan (*Carumc opticum L.),* fennel (*Foeniculum vulgare Mill.*), and caraway (*Carum carvi L.*) against *Penicillium digitatum* and *Alternaria alternata*, finding that caraway had poor antifungal activity against both fungal pathogens. This could be due to a variety of variables, including cultivar, storage temperature, and the oils' dependence on time (Lopez-garcia et al., 2003). Several investigations have found carvacrol's efficacy against *Xanthomonas perforans*, *A. tomatophila*, and *Podosphaera xanthii* (Zhang et al., 2019a), as well as eugenol and thymol's antifungal activity against *Botrytis cinerea* (Zhang et al., 2019b), eugenol, and thymol's antifungal (Wang et al., 2010; Olea et al., 2019; Zhang et al., 2019a). Furthermore, oregano EO was discovered to be high in carvacrol and thymol, while cinnamon EO was shown to be high in eugenol and cinnamaldehyde (Avila sosa et al., 2012). Phenylpropanoids produced from these chemicals have antifungal activity that is intimately linked to their chemical structure (Olea et al., 2019). These chemicals have been investigated and shown to have potent antifungal action against a variety of fungi (Elshafie and Camele, 2016). Boyraz and Ozcan (2006) studied the effects of EOs, hydrosols, ground materials, and extracts from summer savory on *A. mali* and *B. cinerea* (*Satureja hortensis*). All doses of the studied extracts completely inhibited the mycelial growth of both harmful fungi, according to the authors. The study proved that *S. hortensis* has the potential to be a powerful antifungal agent. Tripathi et al. (2008) tested the antifungal efficacy of EOs derived from 10 plants against *B. cinerea*: *Chenopodium ambrosioides, Eucalyptus citriodora, Eupatorium cannabinum, Lawsonia inermis, Ocimum canum, O. gratissimum, O. sanctum, Prunus persica, Zingiber cassumunar,* and *Zingiber officinal B. cinerea* growth was completely inhibited, according to the authors. The synergy of these aromatic components enhances their effectiveness and prevents pathogenic fungus from developing resistance (Elshafie and Camele, 2016). Mbili (2015) studied the effectiveness of EOs as a postharvest treatment against *B. cinerea, P. expansum,* and *N. alba* on apples using fumigation, dipping, and thermal fogging (Karpisk, 2020). Seasonal variations, on the other hand, have been found to have a significant impact on the chemical profile and antifungal properties of EOs (de Macêdo et al., 2020; Gourine et al., 2010; Tanoh et al., 2020). De Macêdo et al. (2018), for example, looked at the effect of seasonal variation on the chemical composition of *Psidium salutare* EO, its antifungal activity, and its effect on *C. albicans* morphogenesis. For instance, De Macêdo et al. (2018) investigated the impact of seasonal variability on the chemical composition of the *Psidium salutare* EO, its antifungal activity, and its effect on the *C. albicans* morphogenesis. Chemical differences in *P. myrtoides* EO composition between the dry and rainy seasons considerably influenced the intrinsic minimum fungicide concentration against *Candida albicans* and *Candida krusei*, according to De Macêdo et al. (2020). Fungicidal effects of Eos derived from various plants have been reported to suppress fungal growth and development in vitro and in vivo in various fresh products (Meepagala et al., 2002; Imelouane et al., 2009). Plants produce terpenes, sesquiterpenes, aldehydes, ketones, and phenolic components, which are found in EOs. *B. cinerea, Rhizopus stolonifer*, and *Alternaria alternata* have all been reported to be completely controlled with thyme oil (Plotto et al., 2003) and *C. gloeosporioides* (Sellamuthu et al., 2013a) during in vitro experiments After 30 days at 20°C, thyme oil reduced deterioration by 76% in naturally infected fruit (Liu et al., 2009). Cinnamon oil was also found to successfully inhibit *C. gloeosporioides* in bananas in in vitro studies (Maqbool et al., 2010). The effectiveness of several postharvest decay control strategies, such as spraying or dipping, was investigated in cherries, citrus fruit, apples, peaches, and cabbages. Fumigation of sweet cherries with thymol vapor was found to be beneficial in controlling gray mold rot (*B. cinerea*), brown rot (*M. fructicola*) (Chu et al., 2001), and anthracnose (*C. gloeosporiodes*) in avocados (Sellamuthu et al., 2013a). One of the most appealing aspects of EOs is their bioactivity during the vapor phase, which makes them appealing as probable fumigants for the preservation of stored items while having no effect on the sensory attributes of the food (Anthony et al., 2003; Arrebola et al., 2010; Laird and Phillips, 2011; Antunes and Cavacob, 2010).

8.3.4 NANOTECHNOLOGY: NANOMATERIALS AND NANOPARTICULES

Nanotechnology is a branch of science that focuses on the use of individual atoms, molecules, or submicron particles with sizes ranging from 0.1 to 100 nanometers in chemical, physical, and biological systems. Fungal pathogens are fought with nanosized organic and inorganic particles with a variety of biological functions (Patel et al., 2014). A number of inorganic and organic nanomaterials have been produced and shown to have strong antifungal activity against phytopathogenic microorganisms in vitro, with some of them still causing toxicity in greenhouse and field settings. TiO2, CuO, and other oxides have been used in the past (Hao et al., 2017, 2019; Liu et al., 2017). The toxicity of Zn, ZnO (Xue et al., 2014), carbon nanomaterials, Al, and Si nanoparticles (NPs) against phytopathogenic bacteria and fungus has been documented, reducing disease incidence. Graphene oxide silver nanoparticle (GO-AgNP) nanocomposites can inhibit the growth of hyphae, demonstrating a considerable effect in the control of Fusarium graminearum leaf spot disease (Chen el al., 2020). Agricultural methods have used a variety of nano-pesticides, nano-fertilizers, and nano-sensors, as well as nanomaterials (Roberto et al., 2019). Despite the fact that nanoparticles improve plant health, they also match the growing demand for high output in agriculture (Atiq, 2020). Because phyto-extracts control fungal infections, boost plant development, and successfully eliminate agricultural diseases, they could be used to make environmentally friendly NPs (Abd-Elsalam, 2013; Khan and Rizvi 2014; Nasrollahzadeh and Sajadi 2015). Another methodology, such as the disc diffusion method, can be utilized to test the varied antifungal activity of these NPs. According to various research, the green synthesis of metal NPs results in a wide range of effective and natural reducing agents (Kharissova et al. 2013; Min et al. 2009). Silver nanoparticles have been shown to efficiently inhibit phytopathogens such as *Rhizoctonia solani*, *Sclerotinia sclerotiorum*, and *Sclerotinia minor*. Furthermore, silver nanoparticles can cause severe damage to hyphae by shattering the hyphal wall membrane and causing internal hyphal injury. The rice pathogens *Fusarium verticillioides*, *Bipolaris oryzae*, and *Fusarium fujikuroi* were all resistant to the (SE-MgO) nanocomposite (Sidhu et al. 2020). Carbon nanoparticles have been found to have potent antifungal properties against microorganisms. *Fusarium graminearum* and *Fusarium poae* are two types of Fusarium (Wang et al. 2016). At various concentrations, single-walled carbon nanotubes (SWCNTs), multi-walled carbon nanotubes (MWCNTs), graphene oxide (GO), and reduced graphene oxide (rGO) reduced fungal pathogen biomass, mycelia growth, and prevented spore germination.

8.3.5 GENETIC ENGINEERING

New fungal strains with improved biocontrol efficacy could be created by genetically modifying the metabolic pathways that trigger the manufacture of secreted proteins and secondary chemicals. Previous research has shown that silencing negative regulatory elements, signal-transduction components, or genes from contiguous metabolic networks can produce the same effect, shifting metabolite flow and biosynthesis or suppressing the feedback inhibition that regulates its production. Trichoderma virens tvk1 mutants, for example, showed higher biocontrol ability against *R. solani*, as well as increased expression of mycoparasitism-related genes and lytic enzyme overproduction. Virens has a better ability than the wild type to cause ISR (induced systemic resistance) on corn against *Cochliobolus heterostrophus*. Another case of improved biocontrol abilities was described by Reithner et al. (2005). Δtga1 mutants produced an excessive amount of antifungal secondary metabolites in *Trichoderma atroviride*. Similarly, *Trichoderma atroviride* Δtmk1 mutants overproduced 6-pentyl-pyrone and peptaibols, leading in higher antifungal activity and protection of bean plants against *Rhizoctonia solani* (Reithner et al., 2007). To untangle the involvement of these genes in the establishment and progression of infection processes, genetic transformation techniques have been frequently employed to silence genes putatively implicated in pathogenicity (Johnson et al., 2018).

To develop disease resistance genetically, genetic engineering has the benefit of integrating resistant genes from any species. Currently, a lot of effort is being put into identifying and isolating genes that, when transferred, could make target plants immune to fungus. Some of these initiatives are centered on resistance genes that have been identified through traditional breeding techniques. In addition, there is a large search for genes that encode enzymes involved in the creation of fungus-toxic substances, as well as genes that encode proteins that have a direct inhibitory effect on fungi growth. Transgenic plants have been created using methods such as neutralization of fungal toxins and exploitation of antifungal genes from non-plant sources. In many cases, using these measures has resulted in a significant reduction of fungal diseases (Islam, 2006; Borel, 2017). The involvement of a large number of genes involved in immune responses following fungus infestation, as well as the numerous pathways involved, has been revealed (Melchers and Stuiver, 2000; Rommens and Kishore, 2000). The toxicity of the gene product toward fungal growth or development, as well as the efficacy of individual genes in the illness response pathway, was evaluated in these experiments. Potential compounds for the production of transgenic tolerant plants were identified based on such assessments. As a result, plant pathologists and biotechnologists are working hard to create disease-resistant plants (Wani et al., 2010). The involvement of glutathion-stransferase in the detoxification of toxins, the jasmonic acid signaling pathway, and several effector genes underlying a more general response where the toxins could be recognized as an elicitor for the trunk pathogens was discovered by scientists working on Eutypa dieback disease of grapevine in Switzerland. To make gene testing easier, grapevines were investigated for the penetration of double standard RNA into leaves. In *Puccinia striiformis*, RNAs were found to be effective in suppressing host plant identification. Plants and fungi have been studied for genes that encode for post-transcriptional gene silencing. A transgenic plant's genome contains foreign DNA that was introduced artificially through genetic engineering (Dalmay et al., 2000).

CRISPR-associated protein 9 (clustered regularly interspaced short palindromic repeats—CRISPR Associated Protein 9) is a new gene editing breakthrough. It can be used to leverage plant defense mechanisms against pathogen attack, with the fungal immune system recognizing and degrading invading pathogenic DNA (Knott and Doudna, 2018; Bono et al., 2015). In most cases, the technique's implementation in filamentous fungi was merely a proof of concept for its viability. The system has the advantage of being able to modify the target gene by changing the sgRNA spacer sequence, as well as silencing many genes at once by transforming the cell with different sgRNAs and Cas9 at the same time (Hsu et al., 2015).

The use of CRISPR-Cas not only saves time when performing genomic functional analyses, but it also has the potential to generate new fungal genotypes that could be exploited as potential plant disease competitors or in priming plant defensive responses. Fusarium Head Blight (FHB), one of the most damaging diseases of grain cereal crops worldwide caused by various Fusarium spp., with F. Graminearum and F. culmorumas being the most prevalent and aggressive culprits, could be one viable scenario for the application of CRISPR-Cas9 silenced mutants. While sterility of infected florets causes yield loss, grain quality deterioration is primarily due to the buildup of trichothecenes coded by the fungal trigenes cluster, which are very poisonous to humans and animals (Muñoz et al., 2019). Another benefit of CRISPR-Cas9 is the creation of well-known antagonists with improved biocontrol abilities thanks to genome editing. Trichoderma species, for example, have been regarded as exceptional biocontrol agents capable of reducing disease severity not only by restricting, if not killing, phytopathogen proliferation, but also by triggering plant defensive responses (Muñoz et al., 2019). The production of transgenics has mostly been used to target the host susceptibility genes in order to attain fungal resistance. Despite its accomplishments, the CRISPR/Cas9 system suffers from off-targeting. The ongoing discovery/evolution of novel varieties is addressing this and other difficulties linked with this system. Finally, the fate of employing CRISPR/Cas9 as a versatile tool for building disease resistance in crop plants will be determined by regulatory considerations. A new technology that allows the generation of precise knock-out mutants in a saprotrophic/pathogenic fungus without the use of foreign DNA opens up new avenues for managing plant infections (Das et al., 2019; Muñoz et al., 2019).

The notion is based on the injection of short RNA (dsRNA/siRNA) molecules to pathogenic organisms, which causes the silencing of critical genes, limiting or preventing their growth. The success of RNAi technology in crop protection is determined on the delivery of dsRNAs to a target organism. Insects, nematodes, viruses, and parasitic plants have all been successfully controlled using a new generation of species-specific control approaches that take advantage of a cellular defense mechanism termed RNA interference (RNAi) (Whyard et al., 2009; Alakonya et al., 2012; Papolu et al., 2013; Schmitt-Engel et al., 2015). As a result, RNAi is being studied as an environmentally benign alternative to pest and pathogen control in agriculture. Because of its sequence-dependent method of action, RNAi technology has a wide range of potential applications in plant defense, including plant diseases (Jahan et al., 2015). The notion of cross-kingdom RNAi, in which fungal pathogens use short noncoding RNAs (sRNAs) to quiet host defense-related genes, holds a lot of promise (Weiberg et al., 2013). MicroRNAs (miRNAs) producing sRNAs can be used by some plants to mute fungal virulence genes (Cai et al., 2018). To improve disease susceptibility, oomycete pathogens use effectors that block cross-kingdom RNAi (Hou et al., 2019). Host-induced gene silencing (HIGS) is a biotechnological approach that uses miRNAs that have been shown or expected to protect against fungal infections (Younis etal., 2014; Huang et al., 2019; Iqbal et al., 2016).

8.3.6 PHYSICAL METHODS CONTROL OF POSTHARVEST FRUIT

Physical treatments like cold or heat have been offered as viable treatments, but they have some drawbacks, including expense, effectiveness at nonphytotoxic temperatures, lack of preventive activity, and low durability (Lurie &Pedreschi, 2014).

8.3.6.1 HOT WATER TREATMENT (HWT)

Fruit that has been subjected to hot water treatments (HWT) has been shown to be resistant to postharvest deterioration (Fallik, 2004; Fallik and Lurie, 2007). Hot water treatment (HWT) has been shown to be helpful in preventing postharvest degradation in pears, lemons, papayas, peaches, nectarine, plums, and mangoes (Nafussi et al., 2001; Zhang et al., 2008; Karabulut et al., 2010; Le et al., 2010; Li et al., 2013). After 5 days of storage at 5°C, a short duration hot water treatment for 12 seconds at 63°C showed good results for strawberry decay management with no fruit loss (Wszelaki and Mitcham, 2003). In addition to directly killing fungal spores, the efficacy of HWT is predicated on stimulating transcription and translation of heat shock proteins, of which a subset contains pathogenesis-related proteins (Fallik et al., 2001). Another mechanism implicated in the control impact of HWT in harvested fruit has been reported to be the activation of the host defensive response (Schirra et al., 2000; Fallik, 2004; Palou et al., 2008). Induced pathogenesis-related proteins and antioxidant enzymes, as well as other antifungal substances, are among the defense responses triggered by HWT in harvested fruit (Nafussi et al., 2001; Kim et al., 2007).

8.3.6.2 FUMIGATION

Traditionally, fumigation has been employed to eliminate insect pests, but more recently, fumigation using ethanedinitrile (EDN) or cyanogen has been used to control both insect pests and fungal infection (Ren et al., 2006). It was created as a replacement for methyl bromide in soil fumigation (Mattner et al., 2006). Sulfur dioxide generator sheets are employed in export packages, and when hydrated by water vapor, they constantly generate a low concentration of gas within the packages throughout storage (Droby and Lichter, 2004). EDN is also known as cyanogen (or oxalonitrile) and is commonly used in nurseries as an eco-friendly alternative to methyl bromide to disinfect the soil. EDN is a colorless gas with an almond-like odor that is environmentally friendly. At room temperature, it is a gas that is highly soluble in water. The principal pathway for EDN breakdown is through oxalic acid derivatives (Mattner et al., 2006). Biological fumigation, or biofumigation, utilizing the volatile-producing fungus *Muscodor albus* is a unique method for controlling fungal diseases (Strobel et al., 2001; Strobel, 2006; Mercier et al., 2007). The endophytic fungus *Muscodor albus* was isolated from a cinnamon tree (Strobel et al., 2001). *M. albus* was discovered to create a total of 28 volatile chemicals that inhibited and killed a variety of fungi, oomycetes, and bacteria. The fungus is a sterile mycelium related to Xylaria, an endophytic fungus. Plant pathogens *Rhizoctonia solani*, *Sclerotinia sclerotiorum*, and *Verticillium dahliae* are among the fungi destroyed by *M. albus* volatiles. Infested barley seed was fumigated with volatiles from *M. albus*, which successfully suppressed covered smut (Ustilagohordei) in the emerging barley plants (Strobel et al., 2001). Without requiring direct contact with the treated product, biological fumigation with *M. albus* volatiles could be utilized to inhibit or kill plant diseases in enclosed environments. Furthermore, *Phomopsis spp.*, *Nodulisporium spp.*, and *Hypoxylon spp.* have been reported to generate volatile compounds that control fruit decay and were reported to produce volatile compounds that control fruit decay (Lee et al., 2009).

8.3.6.3 IONIZING RADIATION

The use of ionizing radiation to treat postharvest diseases and extend the storage life of fresh food has recently gained popularity as an environmentally friendly solution. Postharvest ionizing radiation preserves fresh food by inactivating pathogens and preventing potatoes and onions from sprouting (Piri et al., 2011). Ionizing radiation is also employed as a promising phytosanitary therapy to disinfect agricultural commodities from quarantine pests all over the world (Hallman, 2011). Ionizing radiation's ability to penetrate fresh food makes it a more effective way to combat postharvest illnesses. As a result, the use of ionizing radiation to control decay should be examined while maintaining the balance between pathogen sensitivity and host resistance in mind (Li et al., 2010; Liu et al., 2011). The best radiation dosages and types for fresh food, such as gamma rays, X-rays, and electron beams, are mainly determined empirically. For fruits and vegetables, the United States, the European Union, and Asian countries currently allow irradiation dosages of 1 kGy or less (Ghosh, 2014). Many research have looked into the physiological reactions of postharvest pathogens to ionizing radiation, either alone or in combination with other treatments, in order to determine the best doses or circumstances for inactivation (Li et al., 2010; Hallman, 2011; Liu et al., 2011). Ionizing radiation can harm people in two ways: directly and indirectly. Ionizing radiation's major direct effect on a pathogen is damage to its genetic material, either DNA or RNA. The interaction between the radiation and other atoms or molecules within the pathogens is the indirect effect of ionizing radiation on pathogens. Ionizing radiation can potentially harm enzymes or proteins that are required for pathogen development. One technology that can aid in the inactivation of fungal toxins is ionizing radiation. For example, after gamma irradiation doses of 1 and 10 kGy, the toxicity of aflatoxin B1 (AFB1) found in peanut meal was reduced by 75% and 100%, respectively. Furthermore, a 1.5 and 2 kGy electron beam was adequate to remove *B. cinerea* from contaminated vegetable seeds without impairing seed germination. Despite this, due to the expensive costs of irradiation equipment and customer concerns about its safety, postharvest ionizing radiation is not as widely used as other conventional technologies (Ghanem et al., 2008).

8.3.7 FOOD ADDITIVES AND COATING

8.3.7.1 FOOD ADDITIVES

For the control of postharvest diseases of fruits and vegetables, food additives or generally recognized as safe (GRAS) substances are permitted (Palou et al., 2002). Ammonium carbonate, ammonium bicarbonate, sodium bicarbonate, sodium benzoate, potassium sorbate, and sodium carbonate are some of the chemicals that have been successfully used to treat postharvest infections (Smilanick et al., 2008; Karaca et al., 2014; Youssef et al., 2014; Montesinos-Herrero et al., 2016). D'Aquino et al. (2013) reported that 2 g/L potassium sorbate (PS) reduced the incidence of *Penicillium digitatum*, the cause of green mold in citrus fruit, and 12.5 g/L or 15.0 g/L potassium sorbate (PS) reduced the incidence and severity of sour rot in lemon fruits by 76.0% and 78.0%, respectively (Abd-El-Kareem and Saied, 2015; Palou et al. 2002). Green and blue molds on clementine mandarins were reduced by 3% sodium carbonate (SC), and comparable results were seen for other citrus fruits (Youssef et al., 2014). Furthermore, combining food additives with hot water has been shown to improve the additives' efficacy in reducing fruit rot. Palou et al. (2002), for example, discovered that a heated 3% SC solution at 50°C completely controlled green and blue molds better than either a SC solution or HWT alone. Similarly, according to Fadda et al. (2015), 1% PS dramatically decreased apple blue mold, especially when treated at 53°C. Ethanol and sodium bicarbonate are antibacterial agents that are commonly utilized in the food sector. Treatments with postharvest ethanol and sodium bicarbonate have been shown to be successful in treating postharvest diseases in a variety of fruits.

8.3.7.2 FRUIT SKIN COATINGS

Skin coverings can improve fruit storage quality by reducing water loss and delaying ripening and rotting caused by infections. Oils, waxes, and colloidal carboxymethyl cellulose solutions are commonly used for coating. Fruit coating is a typical process in citrus packing houses to replenish natural waxes that are washed away during fruit washing and handling on the packing line. Because the early formulations were based on paraffin wax or a combination of several other waxes such as beeswax or carnauba, citrus commercial coatings are referred to as waxes. Shellac, wood rosin, candelilla wax, carnauba wax, beeswax, polyethylene, or petroleum waxes are typical anionic microemulsions containing resins and/or waxes. Their primary goal is to reduce fruit weight loss, shrinkage, and improve beauty, but they can also help to prevent CI and other citrus rind diseases (Porat et al., 2004; Bajwa and Anjum, 2007). Fruit coating, on the other hand, can have a negative impact on fruit flavor if the coatings impede gaseous exchange excessively during storage due to an overproduction of volatiles associated with anaerobic conditions (Hagenmaier et al., 2002). As a result, synthetic components such as polyethylene wax, ammonia, and morpholine are not used in the formulations. Additional antifungal ingredients (low-toxicity or food-grade preservatives, biocontrol agents, etc.) are incorporated into these biodegradable formulations to replace the use of conventional chemical fungicides for postharvest disease control, giving rise to the concept of antifungal edible coatings. Hydrocolloids (proteins or polysaccharides), lipids (waxes, acylglycerols, or fatty acids), and resins are the basic or matrix components of edible coatings. Polysaccharides or proteins are combined with lipids in composite coatings or mixes. Plasticizers for increased flexibility and extensibility, as well as emulsifiers or surfactants for improved emulsion stability, may be used as matrix components. These matrixes can be utilized directly in foods or as transporters for additional food additives such as antioxidants, nutraceuticals, flavoring agents, and other additives that are added to change the coating's functioning. The extra ingredients in antifungal edible coatings are food-grade antibacterial chemicals of various natures that are useful in preventing or suppressing fungal growth (Palou et al., 2015).

8.4 CONCLUSION AND FUTURE PROSPECTS

The move from cultural techniques to chemical and biological control strategies has resulted from the identification of pathogens, genetic, and pathological diversity in fungal disease. For fungal decay control, a number of nonchemical options have been offered. None of the nonchemical control approaches have been found to reliably provide a consistent level of disease control that supports acceptability as an alternative to synthetic fungicides when employed as stand-alone treatments. Fruits and vegetables are critical components of human diet and wellness. With rising demand for these commodities, intensive farming and environmentally friendly postharvest methods have become critical (Usall et al., 2015). As a result, proper crop/produce management is essential during cultivation, harvest, storage, transportation, and marketing in order to avoid fruit or vegetable spoiling and reduce health concerns (Prusky et al., 2013). According to the Food and Agriculture Organization of the United Nations, integrated pest management (IPM) is described as the best combination of plant disease control measures, taking into account crop productivity, profit, and safety profile (FAO). IPM is a widely established approach for reducing pesticide use in pest management around the world (Peshin et al., 2009). Preventing infection and considering all available plant disease control techniques while considering their economic benefits and toxicity are given special attention. Plant diseases, on the other hand, are caused and controlled by a complex set of events at multiple levels, including the invader (the plant pathogen), the environment, and the plant itself. These types of combinations put IPM ideas into practice by reducing fungicide doses (under MRLs) or treatment frequency while improving disease control (Marian and Shimizu, 2019). Effective disease control will require integrated strategies that take into consideration all aspects affecting disease epidemiology and incidence, including preharvest factors (Wisniewsk et al., 2016). In order to limit decay losses, IPM must be based on pathogen biology and epidemiology and must take into account all preharvest, harvest, and postharvest aspects that can influence disease appearance and incidence. Naturally, all planned efforts, whether in the field or after harvest, should be cost-effective and have no negative impact on fruit quality. These alternative treatments might be physical, chemical, or biological in character, depending on their nature (Palou et al., 2016). Traditional procedures are tedious and time-consuming. As a result of the demand for fresh, quick, and effective control mechanisms, new genetic approaches have been developed. Resistance genotypes can now be designed thanks to recent breakthroughs in the creation of genetic resources. As a result, our next challenge will be to integrate novel and more effective nonchemical control mechanisms into system development. Furthermore, these methods should be used in conjunction with host resistance components that employ genetic tools and various resistance breeding procedures. By eliminating or replacing fungicides, these strategies will allow for the continued development of ecologically friendly disease management. In the near future, accurate diagnostic technologies for detecting plant disease in situ must be created, tested, and perfected. DNA sequencing could be used in new fungal disease detection methods suited for field use. However, sampling and the deployment of this technology in the form of transportable biosensors could be difficult logistically (He el al., 2016; Singh et al., 2018; Pandey et al., 2018; Sarsaiya et al., 2019). To identify purely fungal genetic biomarkers, RNA sequencing and double-stranded RNA may be used. The accuracy of fungal crop disease detection will be improved by combining biosensing mechanisms with DNA sequencing technologies (McLoughlin et al., 2018; Martinelli et al., 2015).

KEYWORDS

- **fungal diseases**
- **postharvest diseases**
- **management**
- **chemical control**
- **biological control**
- **nanotechnology**
- **transgenics**
- **antagonists**
- **IPM**

REFERENCES

Abd-El-Kareem, F.; Saied, N. M. Evaluation of Hot Water and Potassium Sorbate for Controlling Sour Rot Disease of Lemon Fruits. *J. Chem. Pharmaceut. Res.* **2015,** *7,* 505–210.

Abd-Elsalam, K. A. Nanoplatforms for Plant Pathogenic Fungi Management. *Fungal Genom. Biol.* **2013,** *2,* 107.

Abdolahi, A.; Hassani, A.; Ghosta, Y.; Javadi, T.; Meshkatalsadat, M. H. Essential Oils as Control Agents of Postharvest Alternaria and Penicillium Rots on Tomato Fruits. *J. Food Safety* **2010,** *30,* 341–252.

Adaskaveg, J. E.; Förster, H. New Developments in Postharvest Fungicide Registrations for Edible Horticultural Crops and Use Strategies in the United States. In *Postharvest Pathology*; Prusky, D.; Gullino, M. L. Ed.; Springer: New York, 2010, pp 107–216.

Agrawal, A. A.; Konno, K. Latex: A Model for Understanding Mechanisms, Ecology and Evolution of Plant Defense against Herbivory. *Annu. Rev. Ecol. Evol. Syst.* **2009,** *40,* 311–231.

Agrios, G. N. *Plant Pathology*, Ed. Academic Press: London, 2005.

Aktar, M. W.; Sengupta, D.; Chowdhury, A. Impact of Pesticides Use in Agriculture: Their Benefits and Hazards, *Toxicol.* **2009,** *2,* 1–22.

Alakonya, A.; Kumar, R.; Koenig, D.; Kimura, S.; Townsley, B.; Runo, S.; Garces, H. M.; Kang, J.; Yanez, A.; David-Schwartz, R.; et al. Interspecific RNA Interference of Shoot Meristemless-Like Disrupts *Cuscuta pentagona* Plant Parasitism. *Plant Cell.* **2012,** *24,* 3153–3166.

Ali, A.; Zahid, N.; Manickam, S.; Siddiqui, Y.; Alderson, P. G.; Maqbool, M. Effectiveness of Submicron Chitosan Dispersions in Controlling Anthracnose and Maintaining Quality of Dragon Fruit. *Postharvest Biol. Technol.* **2013,** *86,* 147–253.

Anthony, S.; Abeywickrama, K.; Wijeratnam, S. W. The Effect of Spraying Essential Oils of *Cymbopogon nardus, Cymbopogon flexuosus and Ocimumba silicum*on Postharvest Diseases and Storage Life of Embul Banana. *J. Hortic. Sci. Biotech.* **2003,** *78,* 780–285.

Antunes, M. D. C.; Cavacob, A. M. The Use of Essential Oils for Postharvest Decay Control. *Flavour Frag. J.* **2010,** *25,* 351–266.

Anuagasi, C. L.; Okigbo, R. N.; Anukwuorji, C. A.; Okereke, C. N. The Impact of Biofungicides on Agricultural Yields and Food Security in Africa. *Int. J. Agric. Technol.* **2017,** *13,* 953–978.

Arauz, L. F. Mango Anthracnose: Economic Impact and Current Options for Integrated Management. *Plant Dis.* **2000,** *84,* 600–211.

Arrebola, E.; Sivakumar, D.; Bacigalupo, R.; Korsten, L. Combined Application of Antagonist *Bacillus Amyloliquefacien S*and Essential Oils for the Control of Peach Postharvest Diseases. *Crop Prot.* **2010,** *29,* 369–277.

Arrebola, E.; Sivakumar, D. Korsten: Effect of Volatile Compounds Produced by Bacillus Strains on Post-harvest Decay in Citrus. *Biol Cont.* **2010,** *53,* 122–128.

Atanda, S. A.; Pessu, P. O.; Agoda, S.; Isong, I. U.; Adekalu, O. A.; Echendu, M. A.; Falade, T. C. Fungi and Mycotoxins in Stored Foods. *Afr. J. Microbiol. Res.* **2011,** *5,* 4373–4382.

Atiq, M.; Naeem, I.; Sahi, S. T.; Rajput, N. A.; Haider, E.; Usman, M.; Shahbaz, H.; Fatima, K.; Arif, E.; Qayyum, A. Nanoparticles: A Safe Way towards Fungal Diseases. *Arch. Phytopathol. Plant Protec.*, 2020.

Avenot, H. F.; Michailides, T. J. Resistance to Boscalid Fungicide in *Alternaria alternata* Isolates From Pistachio in California. *Plant Dis.* **2007,** *91* (10), 1345–2350.

Avila-Sosa, R.; Palou, E.; Teresa, M.; Munguía, J.; Nevárez-Moorillon, G. V.; Rhode, A. Cruz, N.; López-Malo, A. Antifungal Activity by Vapor Contact of Essential Oils Added to Amaranth, Chitosan, or Starch Edible Films. *Int. J. Food Microbiol.* **2012,** *153*, 66–22.

Baibakovaa, E. V.; Nefedjeva, E. E.; Suska-Malawska, M.; Wilk, M. Sevriukova, G. A.; Zheltobriukhov, V. F. Modern Fungicides: Mechanisms of Action, Fungal Resistance and Phytotoxic Effects. *Annu. Res. Rev. Biol.* **2019,** *32,* 1–26.

Bajwa, B. E.; Anjum, F. M. Improving Storage Performance of Citrus Reticulata Blanco Mandarins by Controlling Some Physiological Disorders. *Int. J. Food Sci. Technol.* **2007,** *42,* 459–201.

Barkai-Golan, R. Postharvest Diseases of Fruit and Vegetables. In *Development and Control*; Elsevier Science: Amsterdam, the Netherlands, 2001.

Bhutia, D. D.; Zhimo, Y.; Kole, R.; Saha, J. Antifungal Activity of Plant Extracts against *Colletotrichum musae*, the Post-harvest Anthracnose Pathogen of Banana cv. Martaman. *Nutrifood Sci.* **2015,** *46* (1), 2–25.

Bloemberg, G. V.; Lugtenberg, B. J. J. Molecular Basis of Plant Growth Promotion and Biocontrol by Rhizobacteria. *Curr. Opin. Plant Biol.* **2001,** *4,* 343–350.

Bono, J. M.; Olesniky, E. C.; Matzkin, L. M. Connecting Genotypes, Phenotypes and Fitness: Harnessing the Power of CRISPR/Cas9 Genome Editing. *Mol. Ecol.* **2015,** *24,* 3810–3822.

Borel, B. CRISPR, Microbes and More are Joining the War Against Crop Killers. *Nature* **2017,** *543,* 302–204.

Boxall, A. B.; Hardy, A.; Beulke, S.; Boucard, T.; Burgin, L.; Falloon, P. D.; Haygarth, P. M.; Hutchinson, T.; Kovats, R. S.; Leonardi, G.; Levy, L. S.; Nichols, G.; Parsons, S. A.; Potts, L.; Stone, D.; Topp, E.; Turley, D. B.; Walsh, K.; Wellington, E. M.; Williams, R. J. Impacts of Climate Change on Indirect Human Exposure to Pathogens and Chemicals from Agriculture. *Environ. Health Perspect.* **2009,** *117* (4), 508–514.

Boyraz, N.; Ozcan, M. Inhibition of Phytopathogenic Fungi by Essential Oil, Hydrosol, Ground Material and Extract of Summer Savory (*Saturejahortensis* L.) Growing Wild in Turkey. *Int. J. Food Microbiol.* **2006,** 107(3), 238–242.

Braghini, R; Pozzi, C. R.; Aquino, S.; Rocha, L. O.; Correa, B. Effects of Gamma Radiation on the Fungus *Alternaria alternata* in Artificially Inoculated Cereal Samples. *Applied. Radiat. Isotopes* **2009,** *67,* 1622–1628.

Brodersen, P.; Voinnet, O. The Diversity of RNA Silencing Pathways in Plants. *Trends Genet.* **2006,** *22,* 268–280.

Burt, S. *Essential Oils: Their Antibacterial Properties and Potential Applications in Foods—A Review. Int. J. Food Microbiol.* **2004,** *94,* 223–253.

Cai, Q.; Qiao, L.; Wang, M.; He, B.; Lin, F. M.; Palmquist, J.; Huang, S. D.; Jin, H. Plants Send Small RNAs in Extracellular Vesicles to Fungal Pathogen to Silence Virulence Genes. *Science* **2018,** *360,* 1126–2129.

Carmona-Hernandez, S.; Reyes-Pérez, J. J.; Chiquito-Contreras, R. G.; Rincon-Enriquez, G.; Cerdan-Cabrera, C. R.; Hernandez-Montiel, L. G. Biocontrol of Postharvest Fruit Fungal Diseases by Bacterial Antagonists: A Review. *Agronomy* **2019,** *9,* 121.

Carvalho, F. P. Pesticides, Environment, and Food Safety. *Food Energy Secur.* **2017,** *6,* 48–60.

Casadevall, A. Don't Forget the Fungi When Considering Global Catastrophic Biorisks. *Health Secur.* **2017,** *15,* 341–342.

Catalanotto, C.; Azzalin, G.; Macino, G.; Cogoni, C. Gene Silencing in Worms and Fungi. *Nature* **2000,** *404,* 245.

Chen, J.; Wu, L.; Lu, M.; Lu, S.; Li, Z.; Ding, W. Comparative Study on the Fungicidal Activity of Metallic MgO Nanoparticles and Macroscale MgO Against Soilborne Fungal Phytopathogens. *Front. Microbiol.* **2020,** *11,* 365.

Chu, C. L.; Liu, W. T.; Zhou, T. Fumigation of Sweet Cherries with Thymol and Acetic Acid to Reduce Postharvest Brown Rot and Blue Mold Rot. *Fruits* **2001,** *56,* 123–230.

Cosa, P.; Vlietinck, A. J.; Berghe, D. V.; Maes, L. Anti-Infective Potential of Natural Products: How to Develop A Stronger In Vitro 'Proof-Of-Concept'. *J. Ethnopharmacol.* **2006,** *106,* 290–302.

Curtis, H.; Noll, U.; Störmann, J.; Slusarenko, A. J. Broad Spectrum Activity of the Volatile Phytoanticipin Allicin in Extracts of Garlic (*Allium sativum* L.) Against Plant Pathogenic Bacteria, Fungi and *Oomycetes. Physiol. Mol. Plant Pathol.* **2004,** *65,* 79–89.

Dalmay, T., Hamilton, A., Mueller, E.; Baulcombe, D. C. Potato Virus × Amplicons in Arabidopsis Mediate Genetic and Epigenetic Gene Silencing. *Plant Cell* **2000,** *12,* 369–379.

Dang, Y.; Yang, Q.; Xue, Z.; Liu, Y. RNA Interference in Fungi: Pathways, Functions, and Applications. *Eukaryot. Cell.* **2011,** *10,* 1148–2155.

D'Aquino, S.; Fadda, A.; Barberis, A.; Palma, A.; Angioni, A.; Schirra, M. Combined Effects of Potassium Sorbate, Hot Water and Thiabendazole against Green Mould of Citrus Fruit and Residue Levels. *Food Chem.* **2013,** *141,* 858–264.

Darras, A. I. Postharvest Disease Management. In *Handbook of Florists Crops Diseases*; Mcgovern, R., Elmer W., Eds.; Springer: Cham, 2016.

Das, A.; Sharma, N.; Prasad, M. CRISPR/Cas9: a Novel Weapon in the Arsenal to Combat Plant Diseases. *Front. Plant Sci.* **2019,** *9,* 2008.

De Macêdo, D. G.; Souza, M. M. A.; Morais-Braga, M. F. B.; Coutinho, H. D. M.; Dos Santos, A. T. L.; Da Cruz, R. P.; Da Costa, J. G. M.; Rodrigues, F. F. G.; Quintans-Junior, L. J.; Almeida, J. R. S.; De Menezes, I. R. A. Effect of Seasonality on Chemical Profile and Antifungal Activity of Essential Oil Isolated From Leaves *Psidium salutare* (Kunth) O. Berg. *PeerJ.* **2018,** *6,* 5476.

De Macêdo, D. G.; Souza, M. M. A.; Morais-Braga, M. F. B.; Coutinho, H. D. M.; Dos Santos, A. T. L.; Machado, A. J. T.; Rodrigues, F. F. G.; Da Costa, J. G. M.; De Menezes, I. R. A. Seasonality Influence on the Chemical Composition and Antifungal Activity of *Psidium myrtoides* O. Berg. *South Afr. J. Bot.* **2020,** *128,* 9–27.

Deising, H. B.; Reimann, S.; Pascholati, S. F. Mechanisms and Significance of Fungicide Resistance. *Braz. J. Microbiol.* **2008,** *39,* 286–295.

Dias, M. C. Phytotoxicity: An Overview of the Physiological Responses of Plants Exposed to Fungicides. *J. Bot.* **2012,** 1–4.

Dickison, W. C. *Integrative Plant Anatomy*; ED; Academic Press: Massachusetts, **2000,** 85.

Doehlemann, G.; Ökmen, B.; Zhu, W.; Sharon, A. Plant Pathogenic Fungi. *Microbiol. Spectr.* **2017,** *5* (1), FUNK-0023-2016.

Droby, S. Improving Quality and Safety of Fresh and Vegetables after Harvest by the Use of Biocontrol Agents and Natural Materials. *Acta Horti.* **2006,** *709,* 45–21.

Droby, S.; Lichter, A. Post-harvest Botrytis Infection: Etiology, Development and Management. In *Botrytis: Biology, Pathology and Control*; Elad, Y.; Williamson, B.; Tudzyinski, P.; Delen, N., Eds.; Kluwer Academic Publishers: Dordrecht, the Netherlands., 2004; pp 349–267.

Dubey, V. K.; Agannadham, J.; Procerain, M. V. A Stable Cysteine Protease from the Latex of *Calotropis Procera. Phytochemistry* **2003,** *62,* 1057–2071.

El Hadrami, A.; Adam, L. R.; El Hadrami, I.; Daayf, F. Chitosan in Plant Protection. *Mar. Drugs.* **2010,** *8,* 968–287.

El-Moussaoui, A.; Nijs, M.; Paul, C.; Wintjens, R.; Vincentelli, J.; Azarkanm.; Looze, Y. Revisiting the Enzymes Stored in the Laticifers of *Carica papaya* in the Context of their Possible Participation in the Plant Defence Mechanism. *Cell Mol. Life Sci.* **2001,** *58,* 556–270.

Elshafie, H. S.; Camele, I. Investigating the Effects of Plant Essential Oils on Post-harvest Fruit Decay. *Fungal Pathogen.* **2016,** *5,* doi: 10.5772/62568.Intechopen.

Errampalli, D. *Penicillium expansum* (Blue Mold). In *Postharvest Decay. Control Strategies*; Bautista-Baños, S., Ed.; Academic Press, Elsevier Inc: London, UK, 2014, pp 189–231.

Fadda, A.; Barberis, A.; D'Aquino, S.; Palma, A.; Angioni, A.; Lai, F.; Schirra, M. Residue Levels and Performance of Potassium Sorbate and Thiabendazole and their Co-application against Blue Mold of Apples when Applied as Water Dip Treatments at 20 or 53°**C**. *Postharvest Biol. Technol.* **2015,** *106,* 33–23.

Fallik, E. Prestorage Hot Water Treatments (Immersion, Rinsing and Brushing). *Postharvest Biol. Technol.* **2004,** *32* (2), 125–234.

Fallik, E.; Lurie, S. Thermal Control of Fungi in the Reduction of Postharvest Decay. In *Heat Treatment for Postharvest Pest Control: Theory and Practice*; Tang, J.; Mitcham, E.; Wang, S.; Lurie, S.; Ed.; CAB International: Wallingford, U.K., 2007; pp 162181.

FAO. New Standards to Curb the Global Spread of Plant Pests and Diseases [Online]. http://www.fao.org/news/story/en/item/1187738/icode/ (Accessed September 1, 2020).

Fisher, M. C.; Gurr, S. J.; Cuomo, C. A.; Blehert, D. S.; Jin, H.; Stukenbrock, E. H., et al. Threats Posed by the Fungal Kingdom to Humans, Wildlife, and Agriculture. mBio, **2020,** *11* (3), 420–449.

Fisher, M. C.; Henk, D. A.; Briggs, C. J.; Brownstein, J. S.; Madoff, L. C.; Mccraw, S. L.; Gurr, S. J. Emerging Fungal Threats to Animal, Plant and Ecosystem Health. *Nature* **2012,** *484* (7393), 186–194.

Garcia-Solache, M. A.; Casadevall, A. Global Warming Will Bring New Fungal Diseases for Mammals. *mBio* **2010,** *1,* 00061–20.

Ghanem, I., Orfi, M.; Shamma, M. Effect of Gamma Radiation on the Inactivation of Aflatoxin B_1 in Food and Feed Crops. *Braz. J. Microbiol.* **2008,** *39,* 787–291.

Ghorbanpour, M.; Omidvari, M.; Abbaszadeh-Dahaji, P.; Omidvar, R.; Kariman, K. Mechanisms Underlying the Protective Effects of Beneficial Fungi Against Plant Diseases. *Biol. Cont.* **2018,** *117,* 147–257.

Ghosh, D. Food Safety Regulations in Australia and New Zealand Food Standards. *J. Sci. Food Agric.* **2014,** *94,* 1970–2973.

Giraud, T.; Gladieux, P.; Gavrilets, S. Linking the Emergence of Fungal Plant Diseases with Ecological Speciation. *Trends Ecolevol.* **2010,** *25,* 387–295.

Godfray, H. C. J.; Beddington, J. R.; Crute, I. R.; Haddad, L.; Lawrence, D.; Muir, J. F.; Pretty, J.; Robinson, S.; Thomas, S. M.; Toulmin, C.; Food Security: The Challenge of Feeding 9 Billion People. *Science* **2010,** *327,* 812–218.

González M., Caetano P. SME. Testing Systemic Fungicides for Control of Phytophthora Oak Root Disease. *Forest Pathol.* **2017,** *47* (4), 1–3.

Gourine, N.; Yousfi, M.; Bombarda, I.; Nadjemi, B.; Gaydou, E. Seasonal Variation of Chemical Composition and Antioxidant Activity of Essential Oil From *Pistaciaatl antica* desf. Leaves. *J. Am. Oil Chem. Soc.* **2010**, *87* (2), 157–266.

Gurjar, M. S.; Ali, S.; Akhtar, M.; Singh, K. S. Efficacy of Plant Extracts in Plant Disease Management. *Agric. Sci.* **2010,** *3* (3), 425–233.

Gurr, S. Fungi Pose Major Threat to Global Food Security (Emerging Technologies for Global Food Security Conference). University of Exeter. Harvest Fruit Decay. In *Fungal Pathogen;* 2016, Chapter 5. http://dx.doi.org/10.5772/62568

Hagenmaier, R.; Goodner, K.; Roussef, R.; Dou, H. Storage of 'Marsh' Grapefruit and 'Valencia' Oranges with Different Coatings. *Proc. Fla. State. Hort. Soc.* **2002,** *115,* 303–308.

Hakala, K.; Hannukkala, A. O.; Huusela-Veistola, E.; Jalli, M.; Peltonen-Sainio, P. Pests and Diseases in a Changing Climate: A Major Challenge for Finnish Crop Production. *Agric. Food Sci.* **2011,** *20,* 3–24.

Hallman, G. J. Phytosanitary Applications of Irradiation. *Compr. Rev. Food Sci. Food Safety* **2011,** *10,* 143–251.

Hao, D. C.; Gu, X. J.; Xiao, P. G. *Medicinal Plants: Chemistry, Biology and Omics*; Woodhead Publishing Series in Biomedicine: Number *73,* Elsevier Ltd: Oxford, UK, 2015, p 217.

He, D. C.; Zhan, J. S.; Xie, L. H. Problems, Challenges and Future of Plant Disease Management: From an Environment Point of View. *J. Integr. Agric.* **2016,** *15* (4), 705–215.

Hodges, R. J.; Buzby, J. C.; Bennett, B. Foresight Project on Global Food and Farming Futures. Postharvest Losses and Waste in Developed and Less Developed Countries: Opportunities to Improve Resource Use. *J. Agric. Sci.* **2011,** *149,* 37–45.

Horst, R. *Classification of Plant Pathogens in Westcott's Plant Disease Handbook*; Horst, R., Ed.; Springer: Dordrecht, 2008; pp 43–79.

Hou, Y.; Zhai, Y.; Feng, L.; Karimi, H. Z.; Rutter, B. D.; Zeng, L.; Choi, D. S.; Zhang, B.; Gu, W.; Chen, X. et al. A Phytophthora Effector Suppresses Trans-kingdom RNAi to Promote Disease Susceptibility. *Cell Host Microbe* **2019,** *25,* 153–265.

Hsu, P. D.; Lander, E. S.; Zhang, F. Development and Applications of CRISPR-Cas9 for Genome Engineering. *Cell* **2015,** *157,* 1262–2278.

Hua, L.; Yong, C.; Zhanquan, Z.; Boqiang, L.; Guozheng, Q.; Shiping, T. Pathogenic Mechanisms and Control Strategies of *Botrytis cinerea* Causing Post-harvest Decay in Fruits and Vegetables. *Food Qual Safe.* **2018,** *3,* 111–219.

Huang, C. Y.; Wang, H.; Hu, P.; Hamby, R.; Jin, H. Small RNAs—Big Players in Plant-Microbe Interactions. *Cell Host Microbe.* **2019,** *26,* 173–282.

Hussain, F.; Usman, F. Fungal Biotic Stresses in Plants and Its Control Strategy. In *Abiotic and Biotic Stress in Plants*; De Oliveira, A. B., Ed.; London: Intechopen, 2019.

Iavicoli, A.; Boutet, E.; Buchala, A.; Métraux, J. P. Induced Systemic Resistance in Arabidopsis Thaliana in Response to Root Inoculation with Pseudomonas Fluorescens CHA0. *Mol. Plant Microbe Interact.* **2003,** *16,* 851–258.

Imelouane, B.; Elbachiri, A.; Ankit, M.; Benzeid, H.; Khedid, K. Physico-Chemical Compositions and Antimicrobial Activity of Essential Oil of Eastern Moroccan *Lavandula Dentate*. *Int. J. Agr. Biol.* **2009,** *11,* 113–218.

Iqbal, M. S.; Hafeez, M. N.; Wattoo, J. I.; Ali, A.; Sharif, M. N.; Rashid, B.; Tabassum, B.; Nasir, I. A. Prediction of Host-derived Mirnas with the Potential to Target PVY in Potato Plants. *Front. Genet.* **2016,** *7,* 159.

Iqbal, Z.; Khan, M. A.; Sharif, M.; Shah, J. H.; Rehman, M. H.; Javed, K. An Automated Detection and Classification of Citrus Plant Diseases Using Image Processing Techniques: A Review. *Comput. Electron. Agric.* **2018,** *153,* 12–32.

Islam, A. Fungus Resistant Transgenic Plants: Strategies, Progress and Lessons Learnt. *Plant Tissue Cult. Biotech.* **2006,** *16* (2), 117–238.

Jahan, S. N.; Åsman, A. K. M.; Corcoran, P.; Fogelqvist, J.; Vetukuri, R. R.; Dixelius, C. Plant-Mediated Gene Silencing Restricts Growth of the Potato Late Blight Pathogen *Phytophthora infestans*. *J. Exp. Bot.* **2015,** *66,* 2785–2794.

Jain, A.; Sarsaiya, S.; Wu, Q.; Lu, Y.; Shi, J. A Review of Plant Leaf Fungal Diseases and Its Environment Speciation. *Bioengineered.* **2019,** *10* (1), 409–224.

Jamiołkowska, A. Laboratory Effect of Azoxystrobin (Amistar 250 SC) and Grapefruit Extract (Biosept 33SL) on Growth of Fungi Colonizing Zucchini Plants. *Acta Sci. Polhortoru.* **2011,** *10,* 245–257.

Jiménez-Reyes, M. F.; Carrasco, H.; Olea, A. F.; Silva-Moreno, E. Natural Compounds: Asustainable Alternative to the Phytopathogens Control. *J. Chilean Chem. Soc.* **2019,** *64* (2), 4459–2465.

Johnson, E. T.; Proctor, R. H.; Dunlap, C. A.; Busman, M. Reducing Production of Fumonisin Mycotoxins in Fusarium Verticillioides by RNA Interference. *Myc. Res.* **2018,** *34,* 29–37.

Kalemba, D.; Kunicka, A. Antibacterial and Antifungal Properties of Essential Oils. *Curr. Med. Chem.* **2003,** *10,* 813–229.

Kanokwiroon, K.; Teanpaisan, R.; Witistuwannakul, D.; Hooper A. B.; Witistuwannakul, R. Antimicrobial Activity of Protein Purified from the Latex of *Hevea brasiliensis* on Oral Microorganisms. *Mycoses* **2008,** *51,* 301–207.

Karabulut, O. A.; Smilanick, J. L.; Crisosto, C. H.; Palou, L. Control of Brown Rot of Stone Fruits by Brief Heated Water Immersion Treatments. *Crop Protect.* **2010,** *29,* 903–206.

Karaca, H.; Pérez-Gago, M. B.; Taberner, V.; Palou, L. Evaluating Food Additives as Antifungal Agents Against *Monilinia fructicolain vitro* and in Hydroxypropyl Methylcellulose–Lipid Composite Edible Coatings for Plums. *Int. J. Food Microbiol.* **2014,** *179,* 72–29.

Karpińsk, T. M. Essential Oils of Lamiaceae Family Plants as Antifungals. *Biomolecules* **2020,** *10,* 103.

Khan, M. R.; Rizvi, T. F. Nanotechnology: Scope and Application in Plant Disease Management. *Plant Pathol. J.* **2014,** *13* (3), 214–231.

Khan, S. H.; Aked, J.; Magan, N. Control of the Anthracnose Pathogen of Banana (*Colletotrichum musae*) Using Antioxidants Alone and in Combination with Thiabendazole or Imazalil. *Plant Pathol.* **2001,** *50,* 601–208.

Kharissova, O. V.; Dias, H. R.; Kharisov, B. I.; Perez, B. O.; Perez, V. M. J. The Greener Synthesis of Nanoparticles. *Trends Biotechnol.* **2013,** *31* (4), 240–248.

Kim, K.; Kabir, E.; Ara, S. Exposure to Pesticides and the Associated Human Health Effects. *Sci. Total Environ.* **2017,** *575,* 525–535.

Klein, M. N.; Silva, A. C.; Lopes, M. R.; Kupper, K. C. Application of Microorganisms, Alone or in Combination, to Control Postbloom Fruit Drop in Citrus. *Tropi. Pl. Pathol.* **2013,** *38,* 505–512.

Knott, G. J.; Doudna, J. A. CRISPR-Cas Guides the Future of Genetic Engineering. *Science* **2018,** *361,* 866–869.

Konno, K. Plant Latex and Other Exudates as Plant Defense Systems: Roles of Various Defense Chemicals and Proteins Contained Therein. *Phytochem.* **2011,** *72,* 1510–2530.

Laird, K.; Phillips, C. Vapour Phase: A Potential Future Use for Essential Oils as Antimicrobials? *Lett. Appl. Microbiol.* **2011,** *54,* 169–274.

Lazarovits, G.; Turnbull, A; Johnston-Monje, D. Plant Health Management: Biological Control of Plant Pathogens. In *Encycloped Agriculture and Food Systems*; Van Alfen, N. K., Ed.; New York, NY: Academic Press, 2014; pp 388–399.

Le, T.; Shiesh, C.; Lin, H. Effect of Vapor Heat and Hot Water Treatments on Disease Incidence and Quality of Taiwan Native Strain Mango Fruits. *Int. J. Agric. Biol.* **2010,** *12,* 673–278.

Lee, Y. S.; Kim, J.; Shin, S. C.; Lee, S. G.; Park, I. K. Antifungal Activity of Myrtaceae Essential Oils and their Components Against Three Phytopathogenic Fungi. *Flavourfragr. J.* **2008,** *23,* 23–28.

Lennox, C. L.; Spotts, R. A.; Cervantes, L. A. Populations of *Botrytis Cinerea* and *Penicillium* spp. on Pear Fruit, and in Orchards and Packinghouses, and Their Relationship to Post-harvest Decay. *Plant Dis.* **2003,** *87,* 639–244.

Lesemann, S. S.; Schimpke, S.; Dunemann, F.; Deising, H. B. Mitochondrial Heteroplasmy for the Cytochrome B Gene Controls the Level of Strobilurin Resistance in the Apple Powdery Mildew Fungus Podosphaeraleucotricha (Ell. & Ev.) E.S. Salmon. *J. Plant Dis/ Prot.* **2006,** *113* (6), 259–266.

Li, J.; Li, H.; Ji, S.; Chen, T.; Tian, S.; Qin, G. Enhancement of Biocontrol Efficacy of *Cryptococcus laurentii* by Cinnamic Acid Against *Penicillium italicum* in Citrus Fruit. *Postharv. Biol. Technol.* **2019,** *149,* 42–29.

Li, J.; Zhang, Q.; Cui, Y.; et al. Use of UV-C Treatment to Inhibit the Microbial Growth and Maintain the Quality of Yali Pear. *J. Food Sci.* **2010,** *75,* 503–507.

Li, W.; Zhu, Z.; Chern, M.; Yin, J.; Yang, C.; Ran, L.; Cheng, M.; He, M.; Wang, K.; Wang, J. et al. A Natural Allele of a Transcription Factor in Rice Confers Broad-Spectrum Blast Resistance. *Cell* **2017,** *170,* 114–226.

Li, X.; Zhu, X.; Zhao, N.; Fu, D.; Li, J.; Chen, W.; Chen, W. Effects of Hot Water Treatment on Anthracnose Disease in Papaya Fruit and Its Possible Mechanism. *Postharvest. Biol. Technol.* **2013,** *86,* 437–246.

Liu, C., Cai, L.; Han, X.; Ying, T. Temporary Effect of Postharvest UV-C Irradiation on Gene Expression Profile in Tomato Fruit. *Gene.* **2011,** *15,* 56–24.

Liu, X.; Wang, L. P.; Li, Y. C.; Li, H. Y.; Yu, T.; Zheng, X. D. Antifungal Activity of Thyme Oil Against *Geotrichumcitri-aurantii In Vitro and In Vivo. J. Appl. Microbiol.* **2009,** *107,* 1450–2456.

Lopes, M. R.; Klein, M. N.; Ferraz, L. P.; Silva, A. C.; Kupper, K. C. *Saccharomyces cerevisiae*: A Novel and Efficient Biological Control Agent for *Colletotrichumacutatum* during Pre-harvest. *Microb. Res.* **2015,** *175,* 93–29.

López-Garcia, B.; Veyrat, A.; Pérez-Payá, E.; González-Candelas, L.; Marcos, J. F. Comparison of the Activity of Antifungal Hexapeptides and the Fungicides Thiabendazole and Imazalil Against Postharvest Fungal Pathogens. *Int. J. Food Microbiol.* **2003,** *89,* 163–270.

Lurie, S.; Pedreschi, R. Fundamental Aspects of Postharvest Heat Treatments. *Hortic. Res.* **2014,** *1,* 14030.

Magan, N.; Hope, R.; Cairns, V.; Aldred, D. Post-harvest Fungal Ecology: Impact of Fungal Growth and Mycotoxin Accumulation in Stored Grain. *Eur J Plant Pathol.* **2003,** *109,* 723–730.

Maqbool, M.; Ali, A.; Alderson, P. G. Effect of Cinnamon Oil on Incidence of Anthracnose Disease and Postharvest Quality of Bananas during Storage. *Int. J. Agr. Biol.* **2010,** *12,* 516–220.

Mari, M.; Di Francesco, A.; Bertolini, P. Control of Fruit Postharvest Diseases: Old Issues and Innovative Approaches. *Stewart Postharve Rev.* **2014,** *1,* 1–2.

Marian, M.; Shimizu, M. Improving Performance of Microbial Biocontrol Agents Against Plant Diseases. *J. Gen. Plant Pathol.* **2019,** *85,* 329–336.

Martinelli, F.; Scalenghe, R.; Davino, S.; et al. Advanced Methods of Plant Disease Detection. A Review. *Agron. Sustain. Dev.* **2015,** *35* (1), 25.

Martini, C.; Mari, M. Moniliniafructicola, Monilinialaxa (Monilinia Rot, Brown Rot). In *Postharvest Decay. Control Strategies*; Bautista-Baños, S., Ed.; Academic Press. Elsevier Inc.: London, UK, 2014; pp 233–265.

Mattner, S. W.; Gounder, R. K.; Mann, R.; Porter, I. J.; Matthiessen, J. N.; Ren, Y. L.; Sarwar, M. Ethanedinitrile (C2N2)-A Novel Soil Fumigant for Strawberry Production. *Acta Hortic.* **2006,** *708,* 197–204.

Mbili, N.; Vries, F.; Opara, U. L.; Lennox, C. Chemical Composition and Antifungal Activity of Citrus and Lemongrass Essential Oils in Combination with Cold Storage Regimes Against *Botrytis cinerea. Int. Organ. Biol. Integr. Control-West Palaearctic Regional Sec. Bull.* **2015,** *110,* 165–268.

Mcloughlin, A. G.; Wytinck, N.; Walker, P. L. Identification and Application of Exogenous ds RNA Confers Plant Protection Against *Sclerotinia Sclerotiorum* and *Botrytis Cinereal. Sci Rep.* **2018,** *8,* 7320.

Meepagala, K. M.; Sturtz, G.; Wedge, D. E. Antifungal Constituents of the Essential Oil Fraction of *Artemisia Drancunculus* L. var. Dracunculus. *J. Agric. Food Chem.* **2002,** *50,* 6989–2992

Melchers, L. S.; Stuiver, M. H. Novel Genes for Disease Resistance Breeding. *Curr. Opin. Plant Biol.* **2000,** *3,* 147–252.

Min, J. S.; Kim, K. S.; Kim, S. W.; Jung, J. H.; Lamsal, K.; Kim, S. B.; Jung, M.; Lee, Y. S. Effects of Colloidal Silver Nanoparticles on Sclerotium-Forming Phytopathogenic Fungi. *Plant Pathol. J.* **2009,** *25,* 376–380.

Montesinos, E. Antimicrobial Peptides and Plant Disease Control. *FEMS Microbiol. Lett.* **2007,** *270,* 1–21.

Montesinos, E.; Bardaji, E. Synthetic Antimicrobial Peptides as Agricultural Pesticides for Plant-Disease Control. Chem. Biodivers. **2008,** *5, 1225–2237.*

Montesinos-Herrero, C.; Moscoso-Ramírez, P. A.; Palou, L. Evaluation of Sodium Benzoate and Other Food Additives for the Control of Citrus Postharvest Green and Blue Molds. *Postharvest. Biol. Technol.* **2016,** *115,* 72–20.

Mueller, D. S. Fungicides: Terminology. *Integr. Crop Manag.* **2006,** 120–123.

Munhuweyi, K.; Lennox, C. L.; Meitz-Hopkins, J. C.; Caleb, O. J.; Opara, U. L. Major Diseases of Pomegranate (*Punicagranatum* L.), their Causes and Management—A Review. *Scientia Horticult.* **2016,** *211,* 126–239.

Muñoz, I. V.; Sarrocco, S.; Malfatti, L.; Baroncelli, R.; Vannacci, G. CRISPR-Cas for Fungal Genome Editing: A New Tool for the Management of Plant Diseases. *Front. Plant Sc.* **2019,** *10,* 135.

Murray, K.; Wu, F.; Shi, J.; Xue, S. J.; Warriner, K. Challenges in the Microbiological Food Safety. *Food Qual Safety.* **2017,** *1* (4), 289–201.

Nafussi, B.; Ben-Yehoshua, S.; Rodov, V.; Peretz, J.; Ozer, B. K.; Dhallewin, G. Mode of Action of Hot-Water Dip in Reducing Decay of Lemon Fruit. *J. Agric. Food Chem.* **2001,** *49,* 107–213.

Nagrale, D. T.; Sharma, L.; Kumar, S.; Gawande, S. P. Recent Diagnostics and Detection Tools: Implications for Plant Pathogenic Alternaria and Their Disease Management. In *Current Trends in Plant Disease Diagnostics and Management Practices*; Springer: Switzerland, 2016; pp 111–164.

Narayanasamy, P. Diagnosis of Fungal Diseases of Plants. In *Microbial Plant Pathogens-Detection and Disease Diagnosis*; Springer: Dordrecht, 2011; pp 273–284.

Nasrollahzadeh, M.; Sajadi, S. M. Green Synthesis of Copper Nanoparticles Using Ginkgo Biloba L. Leaf Extract and Their Catalytic Activity for the Huisgen Cycloaddition of Azides and Alkynes at Room Temperature. *J. Colloid. Interface Sci.* **2015,** *457,* 141–147.

Neri, F.; Donati, I.; Veronesi, F.; Mazzoni, D.; Mari, M. Evaluation of *Penicillium expansum* Isolates for Aggressiveness, Growth and Patulin Accumulation in Usual and Less Common Fruit Hosts. *Int. J. Food Microbiol.* **2010,** *143* (3), 109–217.

Neri, F.; Mari, M.; Brigati, S.; Bertolini, P. Control of *Neofabraea alba* by Plant Volatile Compounds and Hot Water. *Postharvest Biol. Technol.* **2009,** *51,* 425–230.

Nettles, R.; Watkins, J.; Ricks, K.; Boyer, M.; Licht, M.; Atwood, L. W.; Peoples, M.; Smith, R. G.; Mortensen, D. A.; Koide, R. T. Influence of Pesticide Seed Treatments on Rhizosphere Fungal and Bacterial Communities and Leaf Fungal Endophyte Communities in Maize and Soybean. *Appl. Soil Ecol.* **2016,** *102,* 61–69.

Nunes, C. A. Biological Control of Postharvest Diseases of Fruit. *Eur. J. Plant Pathol.* **2012,** *133* (1), 181–296.

Oerke, E. C. Crop Losses to Pests. *J. Agric. Sci.* **2006,** *144,* 31–43.

Oerke, E. C. Crop Losses to Pests. *J. Agric. Sci.* **2019,** *144,* 31–43.

Olea, A. F.; Bravo, A.; Martínez, R.; Thomas, M.; Sedan, C.; Espinoza, L.; Zambrano, E.; Carvajal, D.; Silva-Moreno, E.; Carrasco, H. Antifungal Activity of Eugenol Derivatives Against *Botrytis cinerea*. *Molecules* **2019,** *24,* 1239.

Oliver, R. P.; Hewitt, HG. *Fungicides in Crop Protection*; 2nd ed.; CAB International, 2014.

Pal, K. K.; B. Mcspadden Gardener. Biological Control of Plant Pathogens. *Plant Health Instructor.* 2006.

Palou, L.; Usall, J.; Muñoz, J. A.; Smilanick, J. L.; Viñas, I. Hot Water, Sodium Carbonate, and Sodium Bicarbonate for the Control of Postharvest Green and Blue Molds of Clementine Mandarins. *Postharvest Biol. Tec.* **2002,** *24,* 93–26.

Palou, L.; Usall, J.; Smilanick, J. L.; Aguilar, M. J.; Viñas, I. Evaluation of Food Additives and Low-toxicity Compounds as Alternative Chemicals for the Control of *Penicillium digitatum* and *Penicillium italicum* on Citrus Fruit. *Pest Manag. Sci.* **2002,** *58,* 459–266.

Palou, L. *Penicillium digitatum, Penicillium italicum* (Green Mold, Blue Mold). In *Postharvest Decay. Control Strategies*; Bautista-Baños, S., Ed.; Academic Press: Elsevier Inc: London, UK, 2014; pp 45–102.

Palou, L.; Ali, A.; Fallik, E.; Romanazzi, G. GRAS, Plant- and Animal-derived Compounds as Alternatives to Conventional Fungicides for the Control of Post-harvest Diseases of Fresh Horticultural Produce. *Postharvest Biol. Technol.* **2016,** *122,* 41–52.

Palou, L.; Smilanick , J. L.; Crisosto, C. H.; Mansour, M.; Plaza, P. Ozone Gas Penetration and Control of the Sporulation of *Penicillium digitatum* and *Penicillium italicum* within Commercial Packages of Oranges During Cold Storage. *Crop Prot.* **2003,** *22,* 1131–2134.

Palou, L.; Valencia-Chamorro, S. A.; Pérez-Gago, M. B. Antifungal Edible Coatings for Fresh Citrus Fruit: A Review. *Coatings* **2015,** *5,* 962–986.

Pandey, A. K.; Burlakoti, R. R.; Kenyon, L. Perspectives and Challenges for Sustainable Management of Fungal Diseases of Mungbean [*Vigna radiata* (L.) R. Wilczek Var. Radiata]: A Review. *Front. Environ. Sci.* **2018,** *6,* 53.

Papolu, P. K.; Gantasala, N. P.; Kamaraju, D.; Banakar, P.; Sreevathsa, R.; Rao, U. Utility of Host Delivered RNAi of Two FMRF Amide Like Peptides, flp-14 and flp-18, for the Management of Root Knot Nematode, *Meloidogyne incognita. PLoS One.* **2013,** *8,* 1–26.

Patel, N.; Desai, P.; Patel, N.; Jha, A.; Gautam, H. K. Agro Nanotechnology for Plant Fungal Disease Management: A Review. *Int. J. Curr. Microbiol. App. Sci.* **2014,** *3* (10), 71–24.

Peshin, R.; Bandral, R. S.; Zhang, W.; Wilson, L.; Dhawan, A. K. Integrated Pest Management: A Global Overview of History, Programs and Adoption. In *Integrated Pest Management: Innovation-Development Process*; Peshin, R., Dhawan, A. K., Eds.; Springer Dordrecht, the Netherland, 2009; pp 1–50. ISBN 978–2-4020–2992–2.

Pickyard, W. F. Laticifers and Secondary Ducts: Two other Tube Systems in Plants. *New Phytol.* **2008,** *177,* 877–288.

Piri, I.; Babayan, M.; Tavassoli, A.; Javaheri, M. The Use of Gamma Irradiation in Agriculture. *African J. Micro. Res.* **2011,** *5,* 5806–2811.

Plotto, A.; Roberts, R. G.; Roberts, D. D. Evaluation of Plant Essential Oils as Natural Postharvest Disease Control of Tomato (*Lycopersicum esculentum*). *Acta Hortic.* **2003,** *628,* 737–245.

Porat, R.; Weiss, B.; Cohen, L.; Daus, A.; Aharoni, N. Reduction of Postharvest Rind Disorders in Citrus Fruit by Modified Atmosphere Packaging. *Postharvest Biol. Technol.* **2004,** *33,* 35–43.

Prusky, D.; Alkan, N.; Mengiste, T.; Fluhr, R. Quiescent and Necrotrophic Lifestyle Choice during Postharvest Disease Development. *Ann. Rev. Phytopathol.* **2013,** *51,* 155–276.

Qin, X.; Xiao, H.; Xue, C.; Yu, Z.; Yang, R.; Cai, Z.; Si, L. Biocontrol of Gray Mold in Grapes with the Yeast *Hanseniaspora Uvarum* Alone and in Combination with Salicylic Acid or Sodium bi Carbonate. *Posthar. Biol. Technol.* **2015,** *100,* 160–267.

Raaijmakers, J. M.; De Bruijn, I.; De Kock, M. J. D. Cyclic Lipopeptide Production by Plant-Associated Pseudomonass sp: Diversity, Activity, Biosynthesis, and Regulation. Mol. Plant Microbe Interact. **2006,** *19, 699–210.*

Ramos, M. V.; Freitas, C. D. T.; Staniscuaski, F.; Macedo, L. L. P.; Sales, M. P.; Sousa, D. P.; Carlini, C. R. Performance of Distinct Crop Pests Reared on Diets Enriched with Latex Proteins from *Calotropis procera*: Role of Laticifer Proteins in Plant Defense. *Plant Sci.* **2007,** *173,* 349–257.

Reithner, B.; Schuhmacher, R.; Stoppacher, N.; Pucher, M.; Brunner, K.; Zeilinger, S. Signaling via the *Trichoderma atroviride* Mitogen-activated Protein Kinase Tmk 1 Differentially Affects Mycoparasitism and Plant Protection. 2007.

Reithner, B.; Brunner, K.; Schuhmacher, R.; Peissl, I.; Seidl, V.; Krska, R.; Zeilinger, S. The G Protein Alpha Subunit Tga1 of *Trichoderma atroviride* is Involved in Chitinase Formation and Differential Production of Antifungal Metabolites. *Fungal Genet. Biol.* **2005,** *42,* 749–260.

Ren, Y. L.; Wang, Y. J.; Barak, A. V.; Wang, X.; Liu, Y.; Dowsett, H. A. Toxicity of Ethanedinitrile to *Anoplophora glabripennis* (Coleoptera: Cerambycidae). *J. Econ. Entomol.* **2006,** *99,* 308–212.

Roberto, R. S.; Youssef, K.; Hashim, A. F.; Ippolito, A. Nanomaterials as Alternative Control Means Against Postharvest Diseases in Fruit Crops. *Nanomaterials* **2019,** *9* (12), 1752.

Romanazzi, G.; Feliziani, E. *Botrytis cinerea* (Gray Mold). In *Postharvest Decay. Control Strategies*; Bautista-Baños, S., Ed.; Academic Press, Elsevier Inc.: London, UK, 2014; pp 131–246.

Romanazzi, G.; Smilanick, J. L.; Feliziani, E.; Droby, S. Integrated Management of Postharvest Gray Mold on Fruit Crops. *Postharv. Biol. Technol.* **2016,** *113,* 69–26.

Rommens, C. M.; Kishore, G. M. Exploiting the Full Potential of Disease-Resistance Genes for Agricultural Use. *Curr. Opin. Biotechnol.* **2000,** *11,* 120–225.

Russell, P. E. A Century of Fungicide Evolution. *J. Agric. Sci.* **2005,** *143,* 11–25.

Sales, M. D. C.; Costa, H. B.; Fernandes, P. M. B.; Ventura, J. A.; Meira, D. D. Antifungal Activity of Plant Extracts with Potential to Control Plant Pathogens in Pineapple. *Asian Pacific J. Trop. Biomed.* **2016,** *6,* 26–21

Sankarana, S.; Mishraa, S.; Ehsania, R.; Davis, C. A Review of Advanced Techniques for Detecting Plant Diseases. *Comput. Electron Agr.* **2010,** *72,* 1–23.

Sarrocco, S.; Vannacci, G. Preharvest Application of Beneficial Fungi as a Strategy to Prevent Postharvest Mycotoxin Contamination: A Review. *Crop Prot.* **2018,** *110,* 160–270.

Sarsaiya, S.; Jia, Q.; Fan, X.; et al. First Report of Leaf Black Circular Spots on *Dendrobium nobile* Caused by *Trichoderma longibrachiatum* in Guizhou Province. *China. Plant Dis.* **2019,** 1–2.

Sasidharan, S.; Chen, Y.; Saravanan, D.; Sundram, K. M.; Yoga Latha, L. Extraction, Isolation and Characterization of Bioactive Compounds from Plants' extracts. *Afr. J. Tradit. Complement Altern. Med.* **2011,** *8,* 1–20.

Savary, S.; Ficke, A.; Aubertot, J. N.; Hollier, C. Crop Losses due to Diseases and their Implications for Global Food Production Losses and Food Security. *Food Secur.* **2012,** *4,* 519–237.

Savary, S.; Willocquet, L.; Pethybridge, S. J.; Esker, P.; Mcroberts, N.; Nelson, A. The Global Burden of Pathogens and Pests on Major Food Crops. *Nat. Ecol. Evol.* **2019,** *3,* 430–239.

Scherm, B.; Orrù, M.; Balmas, V.; Spanu, F.; Azara, E.; Delogu, G.; Hammond, T. M.; Keller, N. P.; Migheli, Q. Altered Trichothecene Biosynthesis in TRI6-Silenced Transformants of Fusarium Culmorum Influences the Severity of Crown and Foot Rot on Durum Wheat Seedlings. *Mol. Plant Pathol.* **2011,** *12,* 759–771.

Schirra, M.; 'D'Hallewin, G.; Ben-Yehoshua, S.; Fallik, E. Hostpathogen Interactions Modulated by Heat Treatment. *Postharvest Biol. Technol.* **2000,** *21* (1), 71–25.

Schmitt-Engel, C.; Schultheis, D.; Schwirz, J.; Ströhlein, N.; Troelenberg, N.; Majumdar, U.; Dao, V. A.; Grossmann, D.; Richter, T.; Tech, M. The Ibeetle Large-Scale RNAi Screen Reveals Gene Functions for Insect Development and Physiology. *Nat Commun.* **2015,** *6,* 7822.

Seagraves, M. P.; Lundgren, J. G. Effects of Neonicitinoid Seed Treatments on Soybean Aphid and Its Natural Enemies. *J. Pest Sci.* **2012,** *85* (1), 125–232.

Sellamuthu, P. S.; Sivakumar, D.; Soundy, P.; Korsten, L. Enhancing the Defence Related and Antioxidant Enzymes Activities in Avocado Cultivars with Essential Oil Vapours. *Postharvest. Biol. Tec.* **2013,** *81,* 66–22.

Shuping, D.; Eloff, J. N. The Use of Plants to Protect Plants and Food against Fungal Pathogens: A Review. *Afr. J. Tradit. Complement Altern. Med.* **2017,** *14* (4), 120–127.

Sibi, G.; Wadhavan, R.; Singh, S.; Shukla, A.; Dhananjaya, K.; Ravikumar, K. R.; Mallesha, H. Plant Latex: A Promising Antifungal Agent for Postharvest Disease Control. *Pak. J. Biol. Sci.* **2013,** *16* (23), 1737–2743.

Siddiqui, Y.; Ali, A. *Colletotrichum gloeosporioides* (Anthracnose). In *Postharvest Decay. Control Strategies*; Bautista-Baños, S., Ed.; Academic Press, Elsevier Inc.: London, UK, 2014; pp 337–371.

Sidhu, A.; Bala, A.; Singh, H.; Ahuja, R.; Kumar, A. Development of MgO-Sepoilite Nanocomposites Against Phytopathogenic Fungi of Rice (*Oryzae sativa*): A Green Approach. *ACS Omega*, **2020,** *5,* 13557–13565.

Singh, A. K.; Ganapathysubramanian, B.; Sarkar, S.; et al. Deep Learning for Plant Stress Phenotyping: Trends and Future Perspectives. *Trends Plant Sci.* **2018,** *23* (10), 883–298.

Singh, D.; Sharma, R. R. *Post-harvest Disinfection of Fruits and Vegetables*; Chapter 1. Academic Press, 2018, pp 1–22.

Singh, S. Role of Nonpathogenic Fungi in Inducing Systemic Resistance in Crop Plants Against Phytopathogens. *Micr. Inoc. Sust. Agric. Prod.* **2016,** 69–23.

Siritapetawee, J.; Thammasirirak, S.; Samosornsuk, W. Antimicrobial Activity of A 48-kDa Protease (AMP48) from *Artocarpusheterophyllus* Latex. *Eur. Rev. Med. Pharmacol. Sci.* **2012,** *16,* 132–237.

Smilanick, J. L.; Mansour, M. F. Control of Citrus Postharvest Green Mold and Sour Rot by Potassium Sorbate Combined with Heat and Fungicides. *Postharvest Biol. Technol.* **2008,** *47,* 226–238.

Spadaro, D.; Droby, S. Development of Biocontrol Products for Postharvest Diseases of Fruit: The Importance of Elucidating the Mechanisms of Action of Yeast Antagonists. *Trends Food Sci. Tech.* **2016,** *47,* 39–29.

Spotts, R. A.; Wallis, K. M.; Serdani, M.; O'Gorman, D. T.; Sholberg, P. L. Methodology for Determining Relationships between Inoculum Concentration of *Botrytis cinerea* and *Penicillium expansum* and Stem End Decay of Pear Fruit. *Plant Dis.* **2008,** *92,* 451–255.

Strange, R. N.; Scott, P. R. Plant Disease: A Threat to Global Food Security. *Annu. Rev. Phytopathol.* **2005,** *43,* 83–216.

Strobel, G. A. *Muscodoralbus* and Its Biological Promise. *J. Ind. Microbiol. Biotechnol.* **2006,** *33,* 514–222.

Strobel, G. A.; Dirkse, E.; Sears, J.; Markworth, C. Volatile Antimicrobials from *Muscodoralbus*, a Novel Endophytic Fungus. *Microbiology* **2001,** *147,* 2943–2950.

Sturz, A. V.; Christie, B. R.; Nowak, J. Bacterial Endophytes: Potential Role in Developing Sustainable Systems of Crop Production. *Crit. Rev. Plant Sci.* **2000,** *19,* 1–20.

Tanoh, E. A.; Boué, G. B.; Nea, F.; Genva, M.; Wognin, E. L.; Ledoux, A.; Martin, H.; Tonzibo, Z. F.; Frederich, M.; Fauconnier, M.-L. Seasonal Effect on the Chemical Composition, Insecticidal Properties and other Biological Activities of *Zanthoxylum leprieurii* guill. & Perr. Essential Oils. *Foods* **2020,** *9,* 550.

Thambugala, K. M.; Daranagama, D. A.; Phillips, A. J. L.; Kannangara, S. D.; Promputtha, I. Fungi vs. Fungi in Biocontrol: An Overview of Fungal Antagonists Applied Against Fungal Plant Pathogens. *Front. Cell. Infect. Microbiol.* **2020,** *10,* 604923.

Tjamos, E. C.; Tsitsigiannis, D. I.; Tjamos, S. E.; Antoniou, P.; Katinakis, P. Selection and Screening of Endorhizosphere Bacteria from Solarised Soils as Biocontrol Agents Against *Verticillium dahliae* of Solanaceous Hosts. *Eur. J. Plant Pathol.* **2004,** *110,* 35–24.

Tripathi, D.; Raikhy, G.; Kumar, D. Chemical Elicitors of Systemic Acquired Resistance-Salicylic Acid and Its Functional Analogs. *Curr. Plant Biol.* **2019,** *17,* 48–29.

Troncoso-Rojas, R.; Tiznado-Hernández, M. E. *Alternaria alternata* (Black Rot, Black Spot). In *Postharvest Decay. Control Strategies*; Bautista-Baños, S., Ed.; Academic Press, Elsevier Inc.: London, UK, 2014, pp 147–287.

Tyagi, S.; Sahay, S.; Imran, M.; Rashmi, K.; Mahesh, S. S. Pre-harvest Factors Influencing the Post-harvest Quality of Fruits: A Review. *Cur J App Sci Technol.* **2017,** *23* (4), 1–22.

Tzortzakis, N. G. Maintaining Postharvest Quality of Fresh Produce with Volatile Compounds. *Innov. Food Sci. Emerg. Technol.* **2007a,** *8,* 111–216.

Tzortzakis, N. G. (B) Methyl Jasmonate-Induced Suppression of Anthracnose Rot in Tomato Fruit, Crop Prot., **2007b,** *26,* 1507–2513.

Usall, J.; Casals, C.; Sisquella, M.; Palou, L.; De Cal, A. Alternative Technologies to Control Postharvest Disease of Stone Fruits. *Stewart Postharv. Rev.* **2015,** *4,* 2.

Wang, X.; Liu, X.; Chen, J.; Han, H.; Yuan, Z. Evaluation and Mechanism of Antifungal Effects of Carbon Nanomaterials in Controlling Plant Fungal Pathogen. *Carbon* **2014,** *68,* 798–806.

Wang, C.; Zhang, J.; Chen, H.; Fan, Y.; Shi, Z. Antifungal Activity of Eugenol Against *Botrytis cinerea. Trop. Plant Pathol.* **2010,** *35* (3), 137–243.

Wang, M.; Weiberg, A.; Lin, F. M.; Thomma, B. P. H. J.; Huang, H. D.; Jin, H. L. Bidirectional Cross-Kingdom RNAi and Fungal Uptake of External RNAs Confer Plant Protection. *Nat. Plants.* **2016,** *2,* 16151.

Wani, S. H.; Haider, N.; Kumar, H.; Singh, N. B. Plant Plastid Engineering. *Curr. Genomics.* **2010,** *11,* 500–212.

Weiberg, A.; Wang, M.; Lin, F. M.; Zhao, H.; Zhang, Z.; Kaloshian, I.; Huang, H. D.; Jin, H. Fungal Small RNAs Suppress Plant Immunity by Hijacking Host RNA Interference Pathways. *Science* **2013,** *342,* 118–223.

Whipps, J. Microbial Interactions and Biocontrol in the Rhizosphere. *J. Exp. Bot.* **2001,** *52,* 487–211.

Whyard, S.; Singh, A. D.; Wong, S. Ingested Double-Stranded RNAs can Act as Species-Specific Insecticides. *Insect. Biochem. Mol. Biol.* **2009,** *39,* 824–232.

Wisniewski, M.; Droby, S.; Norelli, J.; Liu, J.; Schena, L. Alternative Management Technologies for Postharvest Disease Control: The Journey from Simplicity to Complexity. *Postharv. Biol. Technol.* **2016,** *122,* 3–20.

Wititsuwannakul, D.; Chareonthiphakorn, N.; Pace M.; Wititsuwannakul, R. Polyphenol Oxidases From Latex of *Heveabrasiliensis*: Purification and Characterization. *Phytochemistry*, **2002,** *61,* 115–221.

Wszelaki, A. L.; Mitcham, E. J. Effect of Combinations of Hot Water Dips, Biological Control and Controlled Atmospheres for Control of Gray Mold on Harvested Strawberries. *Postharvest. Biol. Technol.* **2003,** *27,* 255–264.

Xu, S. J.; Choi, W. Antifungal Activity of *Paenibacillus kribbensis*strain T-9 Isolated from Soils against Several Plant Pathogenic Fungi. *Plant Pathol. J.* **2014,** *30,* 102–208.

Yazdani, D.; Tan, H.; Abidin, M. A.; Jaganath, I. B. A Review on Bioactive Compounds Isolated from Plants against Plant Pathogenic Fungi. *J. Medi. Plants Res.* **2011,** *5* (30), 6584–2589.

Younis, A.; Siddique, M. I.; Kim, C. K.; Lim, K. B. RNA Interference (RNAi) Induced Gene Silencing: A Promising Approach of Hi-Tech Plant Breeding. *Int. J. Biol. Sci.* **2014,** *10,* 1150–2158.

Youssef, M.; Sanzani, S. M.; Ligorio, A.; Ippolito, A.; Terry, L. A. Sodium Carbonate and Bicarbonate Treatments Induce Resistance to Postharvest Green Mould on Citrus Fruit. *Postharvest. Biol. Technol.* **2014,** *87,* 61–29.

Zhang, H.; Wang, S.; Huang, X.; Dong, Y.; Zheng, X. Integrated Control of Postharvest Blue Mold Decay of Pears with Hot Water Treatment and *Rhodoto rulaglutinis*. *Postharvest Biol. Technol.* **2008,** *49,* 308–213.

Zhang, J. *Lasiodiplodia theobromae* in Citrus Fruit (Diplodia Stem-End Rot). In *Postharvest Decay. Control Strategies*; Bautista-Baños, S., Ed.; Academic Press, Elsevier Inc.: London, UK, 2014; pp 309–235.

Zhang, S.; Zheng, Q.; Xu, B.; Lui, J. Identification of the Fungal Pathogens of Post-harvest Disease Onpeach Fruits and the Control Mechanisms of *Bacillus subtilis* JK-14. *Toxins* **2019,** *5* (11), 322.

Zhimo, V. Y.; Bhutia, D. D.; Saha, J. Biological Control of Postharvest Fruit Diseases Using Antagonistic Yeasts in India. *J. Plant Pathol.* **2016,** *98* (2), 275–283.

Zhou, Y.; Li, W.; Zeng, J.; Shao, Y. Mechanisms of Action of the Yeast *Debaryomyces nepalensis* for Control of the Pathogen *Colletotrichum gloeosporioides* in Mango Fruit. *Biological Control.* **2018,** *123,* 111–219.

CHAPTER 9

Management of Phytopathogenic Fungal Infections: Traditional and Recent Implications

HARIS AHMED KHAN, TEHSIN FATMA, MUHAMMAD ALI, HABIBA SAEED, ZEESHAN ZAFAR, SIDRA FATIMA, AQEEL AHMED, AMBER MALIK and MUHAMMAD FARAZ BHATTI

Atta-ur-Rahman School of Applied Biosciences (ASAB), National University of Sciences and Technology (NUST), Islamabad, Pakistan

GRAPHICAL ABSTRACT

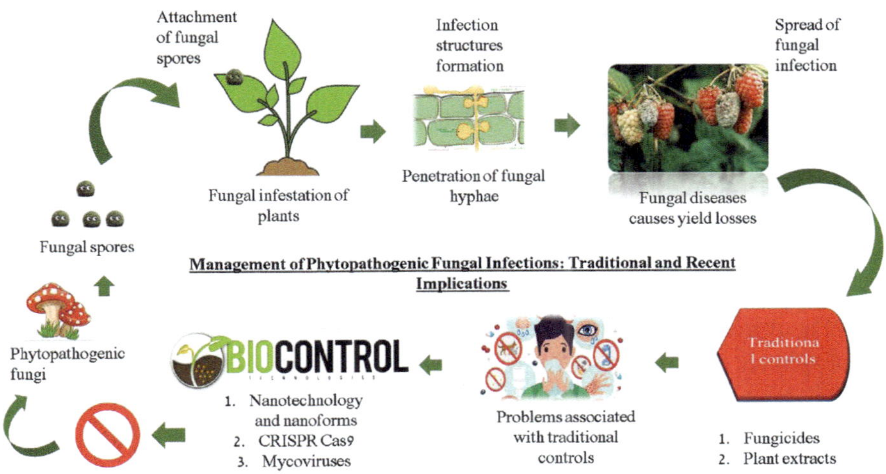

Plant Diseases and Their Management: A Sustainable Approach.
Deepa Srivastava, Rajarshi Kumar Gaur, and Ajay K. Tiwari (Eds.)

ABSTRACT

Fungal infestation causes huge economic losses and poses serious threats to agricultural crops including fruits and vegetables all over the world. The comprehensive knowledge about nature of phytopathonic fungi is important to know for developing novel disease control strategies against these pathogens. Many specialized structures and diverse virulence factors are involved in successful colonization into host tissues by interfering plant defense mechanism. Different virulence factors perform different functions that mediate infection strategy. This chapter highlights brief information regarding phytopathogenic fungi, invasion mechanism into their hosts cells, nutrient acquisition strategies, and associated diseases. Traditionally, different chemicals/fungicides have been used for managing these infections. However, overuse of these fungicides has caused an increase in fungal resistance. Human beings, other living organisms, and beneficial microbial community are greatly affected by excessive fungicides. This extensive overview highlights how traditional, recent, and emerging strategies including plant-based extracts nanomaterials and CRISP-Cas technology can offer greener and effective solutions. This chapter also emphasizes the significance of mycoviruses as biological control agent. Mycoviruses/biocontrol agents (BCAs) can be used as an effective strategy to counter fungal infections by offering environment friendly alternatives against phytopathogenic fungi.

9.1 INTRODUCTION

Members of kingdom Fungi are eukaryotes, which secrete special enzymes to digest food externally and absorb directly via cell wall. They have heterotrophic mode of nutrition. Some of the fungal species absorb nutrients from living source (animal or plants) which are named as biotroph while other fungi acquire nutrients from dead animals or plants and are called saprotrophs (saprophytes or saprobes). But few fungal species infect the living organisms and kill host cells for their nutrients, so these are named necrotrophs (Petrasch et al., 2019; Prusky et al., 2013; van Kan, 2006). They invade both living and dead tissues of plant and stay dormant until conducive circumstances for proliferation arrive (opportunistic) or may develop inside tissues of host plant and cause infection and infest the whole crop (parasites) (Johnston-Monje et al., 2014). According to USEPA survey held in 2005 there are more than 20,000 fungal species that are categorized as parasites and cause several diseases in crop plants. Horticulture rotting during postharvesting is mainly caused by fungi (Gatto et al., 2011) and these fungi are notorious to destroy

10–30% crop yield (Agrios, 2005). These losses were further increased by 50% in tropical regions as well as in developing countries (Tripathi et al., 2004). Fungal spoilages are majorly caused by *Botrytis, Penicillium Fusarium, Geotrichum*, and *Aspergilli* species as they attack on majority of horticulture crops (Agrios, 2005; Dean et al., 2012). *M. oryzae, B. cinerea, P. spp. F. graminearum, F. oxysporum, B. graminis, M. graminicola, Colletotrichum sp., U. maydis*, and *M. lini* have been declared as top 10 fungal pathogens by international community of fungal pathologists (Dean et al., 2012). They are responsible for causing rice blast, gray mold, rust, smut, blight, scab, powdery mildews, blotchs, and many more severe infections respectively in almost all kinds of plants and crops (Dean et al., 2012).

9.1.1 MYCOTOXINS AND POSTHARVEST LOSSES

Mycotoxin is a term used for fungal compound with lower molecular weight which are considered the secondary metabolites. These toxins play a developmental role for fungal growth {Hussein (Hussein and Brasel, 2001), 2001 #40}. These include aflatoxins, patulin, ochratoxin, and citrinin. Majority of them are produced in Aspergillus, Fusarium, Penicillium, and Alternaria species (Sanzani et al., 2012). Mycotoxins are responsible for 25% crop contamination worldwide and one billion tons of food is lost each year approximately, leading to affecting food security and economy adversely (Matny et al., 2015). *P. expansum* is known for producing various fungal toxins most significantly, patulin (Tannous et al., 2018). The molecular biology advanced sequencing technology has promoted the timely identification of fungal gene clusters which encode for the secondary metabolite synthesis (Ballester et al., 2015; Li et al., 2019). Biosynthesis of patulin starts firstly from the synthesis of 6-methylsalicylate synthase and it was also observed that breakdown of this enzyme led to the significant decrease in the biosynthesis of patulin and related infection caused by *P. expansum* on apple (Sanzani et al., 2012) However, it was observed that wild strain without patulin caused similar symptoms of patulin hogenicitrininy as caused by patulin-producing strain (Ballester et al., 2015). Patulin is considered a stimulant which strongly aids *P. expansum* for colonizing different fruit varieties during ripening stage (Snini et al., 2016). Another significant mycotoxin, citrinin, is also produced by *P. expansum* that does not directly cause infection, but it may access *P. expansum* for infecting fruits such as apple (Ballester et al., 2015; Touhami et al., 2018). *P. griseofulvum* contains a virulence gene *gsfR1* responsible for patulin synthesis as well as promotes conidial production on the apple (Valente et al., 2020). Different fungi possess proteins such as VeA and LaeA for the secondary metabolite and to control developmental processes through the formation of nuclear complex. In *A. carbonarius* the above-mentioned proteins are known for the biosynthesis of ochratoxin A and its derivatives in *A. alternata* (Crespo-Sempere et al., 2013; Estiarte et al., 2016).

9.2 FUNGAL INFECTION ESTABLISHMENT IN PLANTS

9.2.1 SPORE ATTACHMENT

Fungal infection initiates from spore attachment on the surface of host body (Neri et al., 2010; van Kan, 2006). The spores, therefore, need to ensure strong adhesion to the surface of the plant, provided spores get dislodged, the germinated spore tubes face greater selective pressure. This ability to selectively adhere to the plant surface varies from species to species. Some species like *Pyricularia oryzae* are having a lower adhesion unless they are hydrated for over 30 minutes (Hamer et al., 1988). Through several electron microscopic analyses, it has been determined that spore tip mucilage (STM) is confined inside their apical spore and is only released during the posthydration event (Braun and Howard, 1994). Other species, like *Collectorichum lindemuthianum*, have several adaptations for increasing adhesion on the surface of the plant. Their ungerminated spores contain adhesive hair-like projections on their surface which increase the strength of the binding to the plant's surface, thereby ensuring that the spores stay on the plant surface (O'Connell et al., 1996).

9.2.2 GERM TUBE FORMATION AND ELONGATION

The initiation of germ-tube formation marks beginning of the germination. The growth of germ-tube grows unidirectional on apical tip by the process of expansion of cell wall, and cell membrane. This expansion is driven through the turgor pressure which is produced due to the accumulation of metabolites in the cytoplasm, which leads to a reduction in the water potential of the cytoplasm, thereby leading to overall osmotic influx of water in the fungal cells (Davis et al., 2000). Some fungi do not require metabolite accumulation for the initial growth of the germ tube, and instead rely on water already stored inside the spore, or by the absorption of moisture in the air (Carver and Bushnell, 1983). While the germ tube is elongating, its surface gets covered by compounds being produced by the fungus and collectively known as the extracellular matrix (ECM), and is speculated to be involved in the adhesion to host surface (Xiao et al., 1994). ECM of fungal germ tubes has been found to have a structurally fibrous nature. The ECM also contains hydrophobins, whose function seems to vary from species to species. They are the surface proteins that are—as their name suggests—hydrophobic in nature and are common to nearly all filamentous fungi. These proteins are classified into two as class I, and class II (Wösten, 2001). Hydrophobins are known to be expressed differentially throughout the different phases of the growth of the fungus. Two hydrophobins of *P. oryzae* are completely characterized: *Mpg1* (class 1) (Talbot et al., 1996), *Mhp1* (class 2) (Kim et al., 2005a). It has been demonstrated that mutants that lack *mpg1* have lower appressorium formation, whereas knocking it down leads to reduced adhesion (Inoue et al., 2015; Kim et al., 2005a). *Mhp1* mutants that have disrupted expression showed reduced pathogenicity, hydrophobicity, and viability of the spores (Kim et al., 2005a). Other fungi, such as *Botrytis cinerea*, have *BcHpb1* which is involved in adhesion but it seems to serve no role in pathogenicity. However, some hydrophobins also seem to downplay the fungal infection, for instance, the HCf-6 in *Cladosporium fulvum* has been found to inhibit the adhesion to the plant surface—something that is not very desirable for fungi (Lacroix et al., 2008). Although the fungal ECM has tested positive for several of the adhesion factors found in the animal ECM, no genomic orthologs have been found for the two.

9.2.3 APPRESSORIUM DEVELOPMENT

In the germ tube, a nucleus from three-nuclei conidium migrates into the germ tube before undergoing mitosis. The daughter nucleus then controls the development of the appressorium whereas the parent nucleus goes back to conidium (Ryder and Talbot, 2015; Veneault-Fourrey et al., 2006). The tri-nuclei conidium is then degraded with the spore. The control of the development of the infectious structures of the fungi is mediated by the cell cycle. When the developing cells enter the S-phase of cell cycle, the appressorium formation starts, and it matures while the cell is in the G2-M part of the cell cycle (Saunders et al., 2010). Initiation of the appressorium is triggered through the DNA damage response cascade, whereas further maturation throughout the S-phase proceeds through a novel pathway which is activated through the generation of turgor by the appressorium and its melanization (Osés-Ruiz et al., 2017). The appressorium nucleus further moves to the penetration peg where it divides continually as the invasive hyphae keep on growing (Deising et al., 2000; Wilson and Talbot, 2009). A melanized dome-like appressorium is formed, which becomes the source of the invasive hyphae. This development of the appressorium is triggered by a lot of factors such as chemical compounds, hydrophobicity, and hardness of the leaf (Gilbert et al., 1996; Liu et al., 2011).

Once appressorium have been matured, it penetrates inside the host's tissues using hydrolytic enzymes, i.e., lipases, cutinases, and polygalacturonases directly (van Kan, 2006; Vilanova et al., 2018). Host cell surface lysis and cell wall hydrolysis initiates different mechanisms related to host or fungus defense or attack as war is initiated between both interacting species (Veloso and van Kan, 2018). The force required to puncture through the host's cell walls is achieved by the turgor pressure generation. Fungus undergoes drastic changes in its metabolism during the infection, and the accumulation of some of the metabolites in the appressorium reduces the water potential in the area (Foster et al., 2017). As a result, water moves into the appressorium down the water potential gradient, via osmosis, which leads to the generation of turgor (Chang et al., 2014). This newly developed turgor, along with some cytoskeletal rearrangement, allows the fungus to make its way into the host (Lo Presti et al., 2015). The spread of the fungus inside the host depends on the lifestyle of the invading fungus. Fungi with a biotrophic lifestyle form structures called haustoria that penetrate the host cells and obtain a part of the host's plasma membrane (Bozkurt and Kamoun, 2020). This host-derived membrane forms an extrahaustorial compartment that separates the haustorium from host's cytoplasm (Panstruga, 2003). The hyphae hemibiotrophic fungi also extract a similar membrane from the host, called the extra invasive hyphae membrane (Kankanala et al., 2007). Lastly, the third type of fungi, the necrotrophic fungi, have their hyphae growing in the intramural space of the host that harbors pectin and cellulose (Bellincampi et al., 2014). The infection process is summarized in Figure 9.1.

9.3 FUNGAL EFFECTORS: KEY PLAYER IN HOST IMMUNE SYSTEM INVASION

The invasion of phytopathogenic fungi also uses compounds called "effectors" that can provide the fungus an edge in evading the host's immune response. These compounds can cause necrosis of the host tissue, or temporarily disable the host's immune system (Hernández-Chávez et al.,

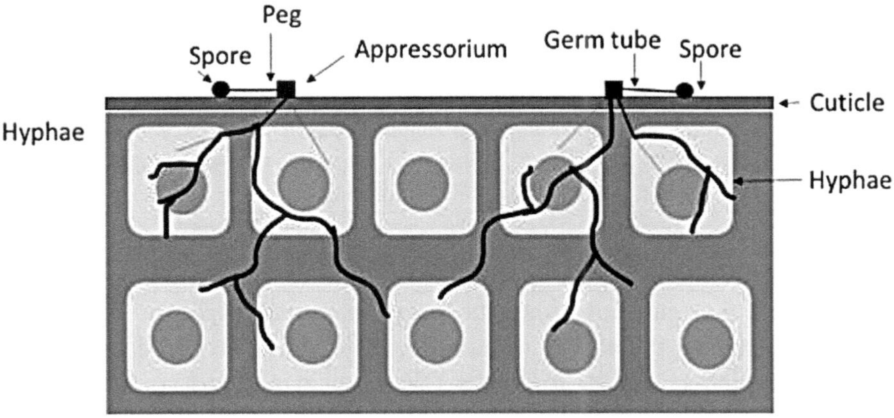

FIGURE 9.1 It shows the fungal infection process adapted from Arya et al. (2021). Spores attach at the surface of their host parts followed by the formation of peg and appressorium development. Hydrolytic enzymes degrade cell wall and membrane; as a result the hyphae penetrates through the pores (wounds) into the living cells.

2017; König et al., 2021; Selin et al., 2016). Sometimes, these toxins can be host-specific, therefore providing the fungi a set range of hosts that they can infect (Nishimura and Kohmoto, 1983). During infection, some fungi like the *P. oryzae* tend to form biotrophic interfacial complexes, structures that are able to produce and release particular effectors into the cytoplasm such as the BAS, Avr-pita, PWL2, and AvrPiz-t (Khang et al., 2010) through the help of t-SNARE and several exocyst components. AvrPiz-t binds to a ubiquitin ligase and a potassium channel in the membranes in rice, to downregulate the host's defense mechanisms against the fungus (Khang et al., 2010). Some effectors, such as the Slp1, instead of being released into the cytoplasm, are being released along the elongating hyphae in the apoplastic region, through the endoplasmic reticulum-Golgi-based secretion network (Giraldo et al., 2013). Slp1 is known to prevent the host from recognizing the chitin in the cell wall of the invading fungi, thereby allowing for a successful infection (Mentlak et al., 2012). Effectors can also be specific, for instance the NIS1 (necrosis-inducing secreted protein 1), is found in *P. oryzae* and nearly all *Colletotrichum* species. It binds the PRR (patter recognition receptor) kinases BIK1 and BAK1, thereby preventing the plant from being able to defend itself using PAMP-PRR-triggered immune response (Irieda et al., 2019). Few effectors are mentioned in Table 9.1.

TABLE 9.1 Showing Some Fungal Effectors and Their Role in Establishing Infection.

Effectors	Function of the effector
AvrPiz-t	Downregulates the host's defense response by binding to ubiquitin ligase
Slp1	Prevents hosts from recognizing the chitin in cell walls of invading fungi
NIS1	Inhibits PAMP-PRR response by binding to a PRR kinase, thereby preventing further downstream signaling

Not surprisingly, sRNA (small RNAs) also serve as a powerful class of effectors that can downplay the hosts' immune response. *B. cinerea* has demonstrated the ability to downregulate genes involved in defense in tomato plants, and *Arabidopsis* (Weiberg et al., 2013). sRNAs are secreted into the apoplastic region by exocytic extracellular vesicles and can then enter cells all around them through direct membrane fusion (Micali et al., 2011).

9.4 MODIFICATION OF HOSTS' INTERNAL ENVIRONMENT

The effectors of phytopathogenic fungi help in infecting the host. The plants itself also possess several mechanisms to counteract against the infection including activation of intracellular defenses, production of ROS (reactive oxygen species), and several different kinds of antimicrobial compounds production (Zachary Nimchuk et al., 2003). However, the fungi can work their way around all these defenses by significantly altering the internal physiological conditions of the host, to their own advantage. For instance, *P. oryzae* uses its enzyme "Abm" to hydroxylate the JA (jasmonic acid) into its hydroxylated form 12-hydroxyjasmonic acid, which suppresses the JA-based immune response of the host (Patkar et al., 2015). It also produces the "Cks1" which can synthesize cytokinin and inhibit the host's defense response (Chanclud et al., 2016; Patkar et al., 2015).

9.5 REACTIVE OXYGEN SPECIES TOLERANCE: A UNIQUE MECHANISM FOR FUNGAL SURVIVAL

Fungi can strive in the face of high concentrations of reactive oxidants. Some of the enzymes required for surviving the oxidative stress are also found to be involved in determining the pathogenicity of the phytopathogenic fungus. *P. oryzae* with mutations in NMO2 (nitronate monooxygenase) are unable to form colonies inside the host plant, even though they can invade the host plants (Marroquin-Guzman et al., 2017). Similarly, mutants with a disrupted glutathione peroxidase gene are not as pathogenic as their wild-type counterparts because the host's defense-related ROS start accumulating at the site where the fungus forms its penetration peg (Huang et al., 2011). While catalases are also important enzymes involved in the detoxification of ROS by the fungi, their disruption mutations do not seem to have a very significant effect on the pathogenicity (Skamnioti et al., 2007).

9.6 MODIFICATIONS TO PAMPS FOR INFECTION ESTABLISHMENT

The fungi are also able to change their pathogen-associated molecular patterns (PAMPs) in a way that the host immune system is unable to recognize them. For instance, the cell wall of the invasive hyphae is changed from β-1,3-glucans to α-1,3-glucans (Fujikawa et al., 2009) and chitin is converted into chitosan by the action of deacetylases (El Gueddari et al., 2002). This allows the fungi to escape the host's pattern recognition receptors (PRRs), thus increasing their chances of successfully colonizing the plant. This leads to the zig–zag model of the plant's immune response. The infection process initiates as the pathogen enters the host and tries to colonize it, but the plant recognizes the PAMPs and launches an immune response using PAMPS-triggered immunity (PTI) (Mogensen, 2009). To counter the increased immune stress, the pathogen releases effectors that allow the pathogen to evade the PTI (Jang et al., 2015). However, the plant manages to recognize the effectors as well and tries to mitigate the damage they are causing through effector-triggered immunity (ETI). This leads to a zig–zag pattern that continues until either the plant is colonized by the pathogen, or the invading pathogen is destroyed by the plant (Jones and Dangl, 2006).

9.7 SYMPTOMS INDUCED BY FUNGAL INFECTIONS

Fungi are known to cause systemic or localized symptoms. Mostly fungal infections lead toward necrotic lesions on the tissues of host and often cause inhibiting, distortions, and abnormalities in tissues and organs of plants (Shoresh et al., 2010). Here are the common symptoms present in fungal infections whether alone or combined with other fungal pathogens. Some of the representative disease/symptoms are mentioned in Table 9.2 and Figure 9.2.

TABLE 9.2 Showing Some Representative Fungal Disease/Symptoms and Their Description.

Disease/symptoms	Description
Leaf spot	Leaf spots are in the form of localized lesions indicating necrotic and collapsed form of tissue. Also, they vary in size and are generally round and concentric, but they can take some other shape as ovoid or elongated at both leaves and stems of the host surface. Characteristic leaf spots will have "bulls-eye-like" appearance comprising of roughly concentric rings which may display zones of various colors such as yellow, purple, or red and will mostly have tan centers. Fungal leaf spot mostly has a dry texture but not dry as dry like paper (Ward et al., 1999).
Anthracnose	It is ulcer-like lesion which can either be necrotic and sunken. Such a type of lesion can appear on flowers, fruit, and stems of host plant—e.g., Apple Anthracnose of the stems and or leaves (*Cryptosporiopsis sp.*). *Colletotrichum* species are mostly known for causing this disease. The favorable elements for anthracnose disease include moist, warm, and humid environment (Than et al., 2008). It also spreads through splash of rain, infested seeds, and humid winds. Infection is progressed by the appearance of lesions in the form of small and dark spots on petiole of stolons. As time passes by, these lesions become brownish in appearance and their diameter also increases ultimately forming rings in concentric pattern. This disease causes the fruit rot and even drop of fruits {Zhang et al., 2017 #71}.
Canker	It is a localized necrotic lesion on the woody tissue that is mostly sunken, e.g., Apple European Canker (*Nectria galligena*) (Das, 2003).
Damping off	It is a rapid collapse and death of the young seedlings. Either seed rots earlier to emergence or the seedling rots at the surface of soil line and falls over and dies off. Some of the most familiar fungi are *Fusarium*, *Rhizoctonia*, and *Pythium* species. This disease can infect the upper and lower ground plant parts; therefore, it is very challenging. The characteristic features of this disease include decomposition of cotyledons before and after the formation of small seedlings, rots of root and lower stems, vegetables rotting during harvesting and cultivation stages, transportation, and storage stages. They can survive in diverse environmental situations. This disease causes the losses in cucurbits, cabbage, capsicum, chili, cauliflower, tomato, ginger, spinach, and brinjal (Sampaio et al., 2020).
Scab	These are the localized lesion formed on the different parts of the host plant including leaves, fruit, or tubers. These infections cause roughened and crust-like area on host surface, e.g., Apple Scab (*Venturia inaequalis*) or Pear Scab (*Venturia pirina*) (Shiller et al., 2015).
Soft/dry root rots	These are disintegration and rots of fleshy roots, leaves, tubers, and fruit, e.g., Rhododendron *Phytophthora* root rot (*P. cinnamomi*), *Armaillaria spp.* Also the root rot of trees and hundreds of other plants are included (Macedo et al., 2017).

TABLE 9.2 *(Continued)*

Disease/symptoms	Description
Blight	Bight is a rapid generalized browning and the death of leaves, branches, stems, and the floral organs. Blight is referred to the both biotic and abiotic disorders, i.e., tomato and Potato Late Blight (*Phytophthora infastans*). *P. infestans* was notorious for great potato famine in Ireland in 1845 (Haverkort et al., 2009). Early blight in tomato is caused by *Alternaria solani* (Chaerani and Voorrips, 2006).
Dieback	Dieback is a slow progressing death of the shoots and twigs that usually start from the tip of infected plant part, e.g., Shoot dieback of apple caused by brown rot of cherry (*Monilinia sp.*) and the Poplar Shoot Dieback (*Venturia populina*). Mostly bacteria are associated with diebacks (*Pseudomonas syringae*) (Malhi et al., 2009).
Mildews	These are manifested by the presence of fruiting bodies, mycelium, and the necrotic tissues such as powdery mildews and downey mildews. One of the most common plant pathogens of plants are powdery mildews (Erysiphales) that belong to ascomycetes. Infects fruits, stems, leaves, and flowers of hundreds of plants, shrubs, herbs. Powdery mildews have some characteristic diagnostic symptoms such as formation of white powdery mass on the infected plant leaves, sometimes browning of leaves along with curling and distortion. The favorable temperature range for this disease is moderate to high; 68–85°F and elevated humidity levels (Glawe, 2008).
Rusts	Infected plants show many tiny lesions on the surface of leaves or stems that are usually rusty color, but it can be black or white. Rusts are commonly caused by Puccina species (Leaf rust is caused by *Puccinia recondite* and *Puccinia striiformis* is responsible for stripe rust). Spring and winter wheat are majorly attacked by rusts, but it also has potential to infect triticale. Acceleration of the disease can result in yellowing and ultimately death of leaves. Significant reduction in yield (more than 50%) can occur due to this disease (Różewicz et al., 2021).
Smuts	This disease is marked by physical presence of mycelium or black spores on seeds, in the form of galls or seeds "replaced" by spores. The word smut is adopted from the Greek language that means dark blackish coloration within reproductive body, the name is given to this disease because of turning of reproductive organs into dark brown to black patches that seem like ash (Kahmann et al., 2000).
Clubroot disease	In this disease club-like galls are formed on the root of the plants due to which it is named clubroot disease. *Plasmodiophora brassicae* (a soil-borne fungus) is known to cause this disease in cruciferous family crops. Most favorable environmental conditions for the spread of this disease are cold, humid, and acidic soil. Desirable temperature range is 50–95°F in acidic soil whereas required temperature range is lower in case of alkaline soils (Akahoshi et al., 2019).

TABLE 9.2 *(Continued)*

Disease/symptoms	Description
Botrytis rot	Known as gray mold rot, caused by *Botrytis cinerea*. Strawberry, apple, tomatoes, and grapes are the hosts. This pathogen is capable of causing both pre- and postharvest losses, as it has the ability to invade in all parts of plants such as fruits, leaves, flowers, seeds, and stem (Hua et al., 2018).
Sclerotinia rot	*Sclerotinia* rot is mainly caused by *Sclerotinia sclerotiorum*. This fungus is widespread, soil borne, and phytopathogenic in nature. A number of factors such as growth stage, number of plants affected and early or late flowering stage has impacts on the yield losses caused by this pathogenic fungus. This disease is the reason of millions dollars of yield losses by decreasing the seed quality and reducing the grain production annually (Sharma et al., 2016).
Fusarium wilt	Diagnostic features of Fusarium wilt comprise browning of vascular tissues, defoliation, stunted growth, epinasty, progressive wilting, and finally the death of infected plant. In roots and stems of infected plants brownish discoloration can be observed when split in cross or vertical sections. Fusarium species cause major loss of various fruits, vegetables, and cereal crops globally such as cotton, chickpea, melon, tomato, sweat pea, banana, wheat, common bean, alfalfa, and many other production (Sampaio et al., 2020).

9.8 CONTROL OF PLANT FUNGAL PATHOGEN

9.8.1 TRADITIONAL CONTROLS

Chemical control (including fungicides/pesticides) and plant extracts are commonly used strategies by farmers for controlling fungal diseases. The description of these methods is given below.

9.8.1.1 PLANT BREEDING FOR IMPROVED RESISTANCE

Since 1990, plant breeding has played an important role in improving the traits of plants and ensured food security (Shiferaw et al., 2013; Tester and Langridge, 2010). This technique can be used to develop plants with desired traits. The process involves combining parental plants to obtain the next generation with the best characteristics. Breeders improve plants by selecting those with the greatest potential based on performance data, pedigree, and more sophisticated genetic information (Ahmar et al., 2020). Plants are improved for food, feed, fiber, fuel, shelter, landscaping, eco-systems services, and a variety of other human activities. Previously, the aim was to improve nutritional values of

FIGURE 9.2 Pictorial representation of disease/symptoms caused by phytopathogenic fungi (images taken from APS press).

crop plants. But recent developments in this field have explored a wide range of possibilities for improving different traits, i.e., wide crosses, introgression of traits from wild relatives by hybrid breeding, mutagenesis, double haploid technology, and some tissue culture-based approaches such as embryo and ovule rescue (to achieve maximum plant regeneration) and protoplast fusion. Resistant varieties offer the cheapest means of disease control and reduce environmental pollution. Environmental factors do not affect the resistance efficiency of these plants (Ahmar et al., 2020).

9.8.1.2 *CHEMICAL CONTROL OF FUNGI (FUNGICIDES)*

Fungicide, also termed as antimycotic, is any toxic constituent that is used to kill or constrain the fungal growth. Parasitic fungi that are major cause for economic loss of crop and risk health of local plants and animals are controlled by fungicides. Agricultural and horticultural fungicides are used as spray or dust to control pathogenic fungus. Before germination of plant seedling, seed fungicides are being used as shield. Systemic fungicides, also termed as chemotherapeutants, are functional to plants, where they become disseminated through the tissue and act to eliminate prevailing disease spread or provide protection against probable disease. Pharmaceutical fungicides are generally used as antifungal creams and oral medications in human and veterinary medicine (Leroux, 2007).

In the 1980s, Botrytis on different plants such as vines, vegetables, and fruits were controlled by different pathogenic fungus such as vinvloxolin, procymidone (Hirooka and Ishii, 2013), benzimidazloles, and iprodione. In later years, Bortyis fungus developed resistance against different fungicides such as benzimidazoles and dicarnoximides, so control of fungus was mediated by different chemicals such as cyprodinil, pryimethanil, and mepanipyrim (Feng et al., 2012; Maeno et al., 1990).Oomycetes are being controlled by different plant pathogenic fungicides such as phenylamines including metalaxyl-M (Hirooka and Ishii, 2013), Quinone inside inhibitor (Qil) fungicides including cyazofamid and amisulbrom (Honda et al., 2007), and carboxylic acid amides (CAA) (Gisi et al., 2019; Miyake et al., 2005) including mandipropamid, dimethomorph, and benthiavalicarb-isopropyl. Some fungicides are also reported whose mode of action is not reported yet such as cymoxanil and fosetyl-aluminum. Broad spectrum fungus fluazinam is being used globally to control many fungal diseases such as gray mold, potato late blight, and downy mildews (Suzuki et al., 1995). Riceblast has been majorly controlled by different pathogenic fungal measures such as host defense inducers and melanin biosynthesis inhibitors (MBIs) in Japan and Korea. Host defense inducers do not directly affect the fungal growth in vitro; however, they provide protection against rice blast pathogen, have protective activity against the rice blast pathogen, *Magnaporthe oryzae*. They develop resistance against plant pathogen and hence provide protection against pathogen (Hirooka, 2004; Iwata, 2001).

9.8.1.3 CONTROL OF FUNGI USING PLANT EXTRACTS

Bio-fungicidal products such as plant extracts, resins, essential oils, and gums are being used as bio-control agent against pathogenic fungus that is a major emerging issue either in vitro or in vivo (Fawzi et al., 2009; Romanazzi et al., 2012). According to the literature idea, it is found that around 25 million higher plant species on globe are being offered as major source of bioactive chemical compounds with diverse applications such as their potential use in agriculture and pharmaceutical industry (Cowan, 1999). Natural products obtained through plant extract show great efficiency against plant pathogenic fungus and mycotoxins including clove, palmarosa, cinnamon oregano, lemongrass oils (Marín et al., 2004), tea tree oil (Burgiel and Smaglowski, 2008), common thyme, aniseed oils (Ćosić et al., 2010), sweet basil, neem, eucalyptus, datura, garlic, and oleander extracts (NAShwA and Abo-ElyouSr, 2013). Carvacrol and Thymol are found to be most effective active ingredients against pathogenic fungal species (Gavaric et al., 2015; Numpaque et al., 2011; Shin et al., 2014; Villanueva Bermejo et al., 2015). An effective way of controlling pathogenic fungus through these chemicals is not known completely, but general mechanism involves the ability of these compounds to dissolve or interrupt cell membrane and fungal cell wall (Isman and Machial, 2006). Most effectively used plant products against plant pathogenic fungus are plant extracts and essential oils. Products obtained through plant source are similar in nature but differ in many characteristics. Essential oils are oily liquids in nature obtained from plant fermentation and steam distillation (Burt, 2004), whereas plant extracts are product of dried plant parts as a result of filtration and evaporation using various source as solvents (Wang et al., 2004). The major compounds that are obtained as plant extracts include phenols, flavonoids, quinones, terpenes, tannins, alkaloids, lectins, polypeptides, saponins, and sterols (Halama and Van Haluwin, 2004). Plant extract products have fungicidal and fungistatic activity on plant pathogens or they may become a reason to create unfavorable for growth and multiplication of pathogenic fungus (Scheuerell et al., 2002).

9.9 PROBLEMS ASSOCIATED WITH TRADITIONAL CONTROL STRATEGIES

Conventional plant breeding also has its limitations. First, breeding can only be done between two plants that can sexually mate with each other. This limits the new traits that can be added to those that already exist in a particular species. Second, when plants are crossed, many traits are transferred along with the trait/s of interest—including those traits that have undesirable effects on yield potential. There is a possibility of resistance breakdown in plants produced by traditional breeding techniques; also this is a time-consuming procedure and high-yielding varieties are very difficult to achieve using these procedures. Chemical pesticides agricultural practice also has drawbacks (Aktar et al., 2009). Major use of chemical pesticides may lead to public health issues and may lead to harmful effects to agriculture sector. Harmful effect of chemical pesticides may include many deadly effects including carcinogenicity, allergic reactions, and sometimes death (Aktar et al., 2009). Chemical pesticides may also lead to development of resistant races of pathogens, nonavailability of pathogenic pesticides. Other problems include development of resistant races of pathogens, a gradual elimination and phasing out of some available pesticides, and the reluctance of some chemical companies to develop and test new chemicals due to the problems with registration process and cost. The need for the development of nonchemical alternative methods to control plant diseases is therefore clear. Some of these products are cancerous to humans and a withdrawal duration is mandatory between the last dosage and the picking. Several fungicides also negatively affect environmental soil organisms, many insects, and pant pollinators. Environmentalists and consumers are giving more attention to fungicides produced naturally and support production of organic food against plant pathogenic fungus (Davis et al., 2000).

9.10 BIOCONTROL OF PLANT PATHOGENIC FUNGUS

Control of plant pathogenic fungus is majorly a cost-effective component in crop production at large scale. Through some traditional practices, plant pathogenic fungus can also be controlled by some genetically modified resistant varieties to avoid the spread environmental pathogenic fungi. However, increase in the use of fungicides potentially affects the health of human and leads to environmental hazards that caught attention of researchers to move toward biological control. Biological control is a broad term being used for control of plant pathogenic fungus through bio-control agent (Sharma, 2021). Biocontrol can be fungus, bacterium, or virus, or a mixture of these to the plant or the soil (Deepa and Sreenivasa, 2019). BCA leads to preventing early infection in plant or establishment of pathogen in plant host. Major lead to use biocontrol is that they are target specific to pathogen and therefore create no harm to non-target species other than pathogen (Costa et al., 2013; O'Brien, 2017).

9.11 RECENT APPROACHES AND RESEARCH TRENDS FOR DISEASE MANAGEMENT

9.11.1 NANOTECHNOLOGY: NEW SOLUTION TO CONTROL OF PHYTOPATHOGENIC FUNGI

Use of nanotechnology to control plant pathogenic fungus is the most suitable option. Nanoparticles act as diagnostic tool for early detection of disease in plant and can also be used into part of the plant that was damaged by pathogenic fungus (Sharon et al., 2010). Nanoparticles are stable in nature, biodegradable and can be used in the production of nano-capsules that can be used for delivery of pesticides, fungicides, herbicides, and some other agrochemicals (Prasad, 2014). Nanoforms of carbon, silver, silica, and aluminosilicates are widely being used to control pathogenic fungus (Singh et al., 2015). The promising forms are discussed below.

9.11.1.1 CARBON NANOPARTICLES

Carbon nanoparticles not only have been used to control plant pathogenic fungus but also have great impact in agricultural sector for increasing growth rate of plants. In Brazil, carbon nanoparticles are being used not only for control of plant pathogenic fungus but also have a great application in agriculture not only for diseases control but also for increasing rate of growth in plants (Patel et al., 2014). Additionally, carbon nanofibers are being used to create nanoparticle for not only improvement of natural fibers including sisal, coconut oil but also control of their release (Patel et al., 2014). Carbon nanomaterials have different shapes such as hollow spheres, cylindrical, ellipsoids, and multi-walled carbon nanotubes (MWCNT)] (Singh et al., 2015). The effect of carbon nanotubes (CNT) has also been seen in enhancing growth in tomato seeds (Khodakovsky et al., 2003) and it was found that enhanced growth of tomato seed is due to high uptake of water due to the penetration of CNT that act as vehicles to deliver desired molecules into seeds during germination of seeds that also protect them during seed germination (Patel et al., 2014).

9.11.1.2 SILVER NANOPARTICLES (SNPS)

Silver is small, ionic nanoparticle has antimicrobial activity. Silver nanosized particles show a great effect in unicellular organism due to enzyme degeneration {Kim (Kim et al., 2005b), 2006 #5}. Control of plant pathogenic fungus by silver nanoparticles has been done against fungal species infecting cereal crops such as *Bipolaris sorokiniana* (Sacc.) (Spot blotch of wheat), *Magnaporthe grisea*, sclerotium-forming species *Rhizoctonia solani*, *Sclerotinia sclerotiorum*, and *Scerotinia minor.* Hyphae exposed to silver nanoparticles leads to severe damage that results in the separation of fungal lining and collapse of fungal hyphae (Min et al., 2009).

9.11.1.3 SILICA NANOPARTICLES

Silica particles are used for specific target delivery into plant cell. These particles deliver DNA and chemicals in the specified tissues of plant cell (Wang et al., 2002). Size of these particles is small and hence increases surface area of the particles (Banik and Sharma, 2011). Mesoporous silica particles had been used for the delivery of DNA and chemicals inside the plant cells and leaves of maize and tobacco. It was analyzed that the DNA-coated Type–II MSN system can be an effective system for the delivery of chemical and protoplast, hence leading to the delivery of DNA as transcription machinery and generating gene expression, resulting in the expression of transgene (Torney et al., 2007). Powdery mildews of cucurbits has been controlled by these nanosized silica–silver particles and resulted in 100% control within three weeks (Park et al., 2006).

9.11.1.4 ALUMINO-SILICATES NANOPARTICLES

In recent times, pesticides are being produced by alumino-silicates nanoparticles. These pesticides are aluminum-silicate nanotubes that can be exposed to plant-affected surface through spray and hence can be easily dropped in insect hairs. These pathogens may then lead to consumption of silicate-filled nanotubes. Pesticides obtained through aluminum-silicates nanoparticles are extra energetic and ecologically harmless (Madhuri et al., 2010; Misra et al., 2013; Patel et al., 2014).

9.12 CONTROLS OF FUNGAL INFECTIONS USING GENOME EDITING TECHNOLOGIES

Countries with highly developed economic rate are used to apply control measure to prevent continuous spread of pathogens and certified seed material after all diagnostic measures. Diagnostic laboratories require highly specific and skilled staff to perform genome editing analyses. For long-term pathogen-resistant crop, new genome editing technologies are developed to produce pathogen resistance in plants. Among the emerging genome editing technique includes CRISPR cas system. CRIPSR/cas9 is one of the most efficient and popular genome editing technologies currently available that allows changing the genes of interest (Ahmad et al., 2018). CRISPR uses two main components that allow it to target the desired regions of genome: a guide RNA (sgRNA), and an endonuclease called the Cas9 (Ahmar et al., 2020). During the process of CRISPR-based genome editing the sgRNA binds with Cas9 and complementary sequence allows the sgRNA to direct Cas9 toward the target sequence. Upon reaching the target sequence, the Cas9 places a double-stranded break (DSB) in the DNA sequence (Janik et al., 2020). This DSB triggers the cell's repair mechanisms and the DSB gets repaired through non-homologous end-joining (NHEJ) and homology direct repair (HDR), which leads to error-free

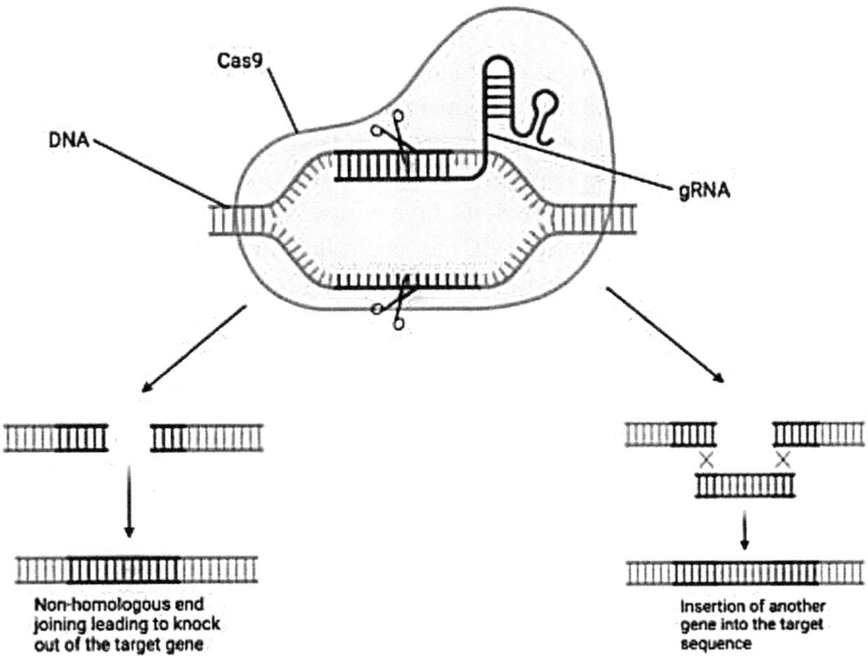

FIGURE 9.3 Mechanism of CRISPR Cas9 system showing different repair systems.

Source: Adapted from Ahmar et al. (2020).

repair, and usually does not induce frameshifts due to indels (insertions or deletions) (Hille and Charpentier, 2016) illustrated in Figure 9.3.

Ectopic expression of the genes encoding inhibitors of fungal enzymes in desired plants can be a potent way to induce resistance. Strengthening the plants' cell walls through increased callose deposition can also help in resisting fungal infections; this can be achieved through the introduction of genes into the plant that code for callose-producing enzymes. A great example of this is when SIMPR4, an ortholog of *Powdery MildewResistance 4* (PMR4), a gene required for callose deposition, was targeted using CRISPR; it was discovered that the loss of function mutants of PMR4 showed greater levels of salicylic acid and hyperactive response, upon being exposed to pathogens (Imam et al., 2016). This led to a significant reduction in the susceptibility of the plants against Powdery Mildew Disease. This can also be processed through various species by manipulation of several different genes (Imam et al., 2016; Tyagi et al., 2021). The process of disease-resistant varieties development is shown in Figure 9.3. Some possible ways to use CRISPR cas9 for biocontrol are shown in Figure 9.4.

9.13 MYCOVIRUSES AS BIOCONTROL AGENTS

Mycoviruses are the best and environment-friendly biological control agents (BCAs) which constitute the most promising area of research. The first mycovirus was identified from a diseased mushroom (Agaricus bisporus –a basidiomycete) in Pennsylvania during the year 1948. The disease was named La France disease (Sinden and Hauser, 1950). Later some mycoviruses have been used for biocontrol first demonstrated by Nuss (1992) for controlling the late blight of chestnut caused by *Chryphonectria parasitica* in Europe. Cryphonectria hypovirus 1 (CHV1) was responsible for this hypovirulence effect (Nuss, 1992).

For considering mycovirus as a biological control agent the knowledge of both host and virus behavior and understanding population biology of mycoviruses is crucial (Pearson et al., 2009). Hypovirulence ability of mycoviruses can be exploited as natural fungicides by diminishing the growth and

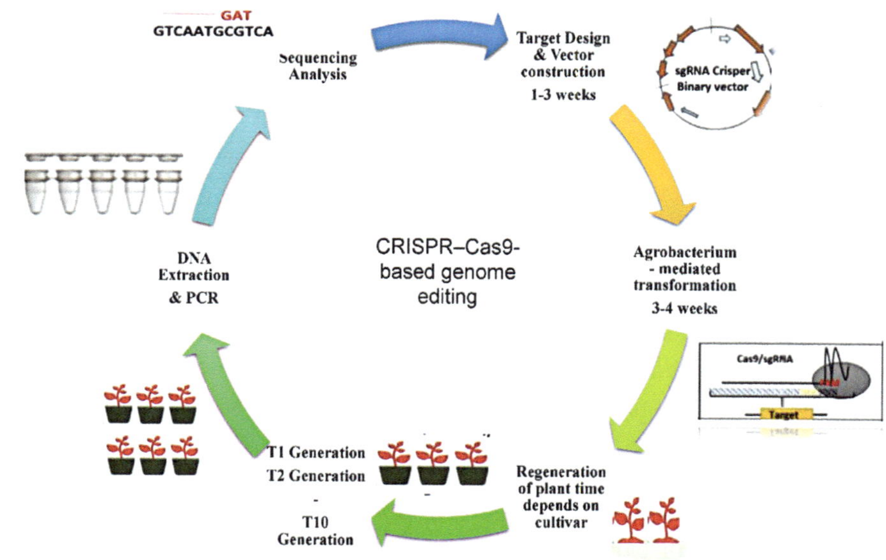

FIGURE 9.4 Schematic representation of CRISPR-Cas9 system integration in plants for developing disease-resistant varieties. It works in three steps: (1) the expression of the nuclear-localized Cas9 protein, (2) the generation of gRNA containing first 20-nt complementary to the target gene, and (3) the NGG PAM site recognition located nearly at the 3' end of the target site. This process is followed by three additional steps: (1) design target and construction of a gene-specific sgRNA (vector), (2) CRISPR–Cas9 sgRNA can be transfected into the plant protoplast through Agrobacterium-mediated transformation, and (3) regenerated plants are screened for mutation via PCR-assay and sequencing.

Source: Adapted from Ahmar et al. (2020).

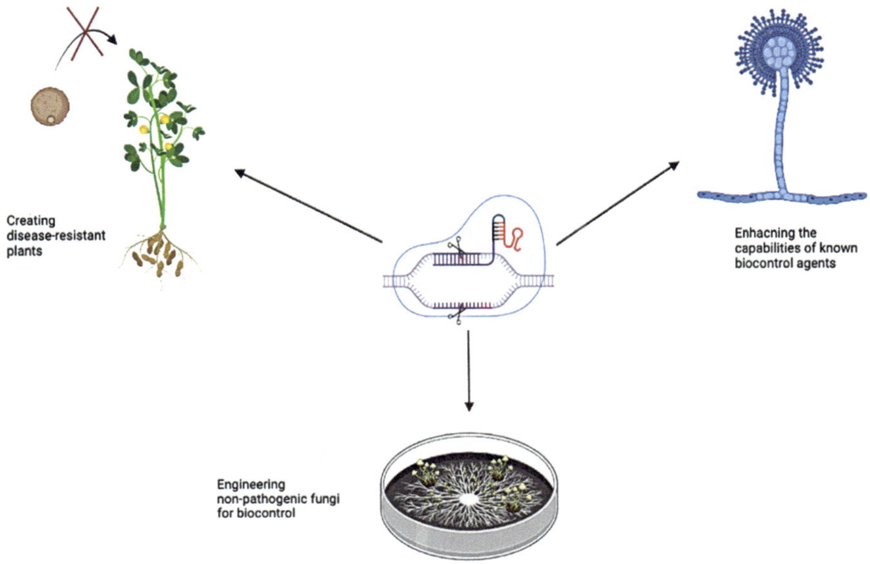

FIGURE 9.5 Showing the application of CRISPR Cas9 in different dimensions.

pathogenicity of phytopathogenic fungi; secondly, infection of hypovirulent strains induces the production of pathogen-associated molecular patterns (PAMPs), which rather than causing disease, activate the host defense mechanism against virulent strains. Many hypovirulent mycoviruses have been reported from different phytopathogenic fungi (including Cryphonectria Fusarium, Botrytis, Sclerotium, Alternaria, Trichoderma, and many others) (Zhang and Nuss, 2016).

Mycoviruses utilize natural transmission means (hyphal anastomosis) for cell-to-cell movement (no extracellular transmission means; ssHADV-1 is an exception). The vegetative incompatibility (VIC) is a significant hurdle in using mycoviruses as BCAs, demonstrated in North America. Same CHV1 from Europe was used but failed to control the chestnut blight disease due to diversity of fungal strains and increased VCGs {Nuss (Nuss, 1992), 1992 #36}. Sclerotinia sclerotiorum debilitation-associated RNA virus (ssDRV) besides showing hypovirulent properties was also unable to cope with these barriers when used as biocontrol. ssHADV-1 (a ssDNA virus) has the ability of extracellular transmission with small host range. The mechanism behind this hypovirulence is still unknown (Nuss, 2011).

Some killer phenotypes have been reported previously in species of yeasts carrying mycoviruses, i.e., *Saccharomyces*, *Hanseniaspora*, *Zygosaccharomyces*, and in *Ustilago maydis* {Schmitt (Bruenn, 2005; Schmitt and Breinig, 2006). This secretes a proteinaceous toxin killer toxin, having killer activity against many different fungal species. The current focus of research is on using killer activity as fungicides against other sensitive (non-killer strains) which helps in eliminating nonbeneficial yeasts. "Super killer" strains have also been constructed by the utilization of these mycoviruses. The process of using mycoviruses as biocontrol is shown in Figure 9.6.

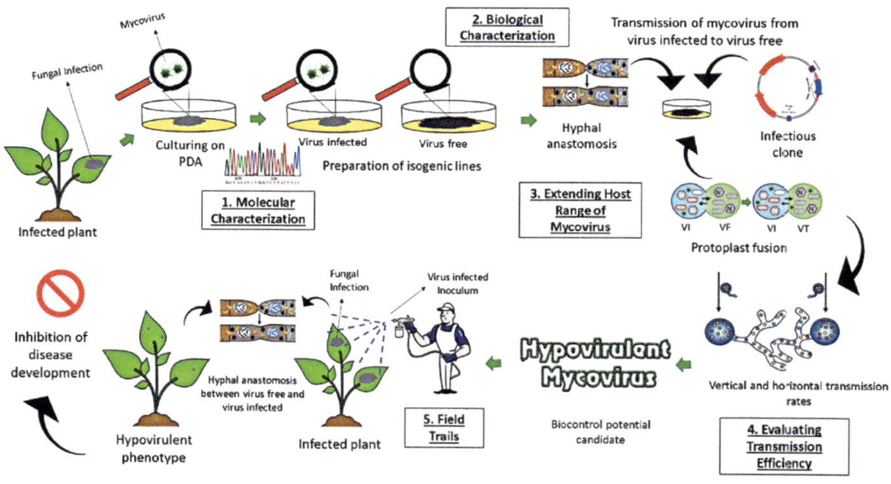

FIGURE 9.6 Schematic representation of major steps in the implementation of fungal disease control utilizing mycoviruses. The first step is the identification of hypovirulent mycovirus (i.e., that reduce the pathogenicity of fungal strains). Culturing fungal strain and studying molecular features is done under lab conditions. Preparation of mycoviral isogenic line (virus free and virus infected) using different available strategies for studying hypovirulence properties of subjected mycovirus. Host range of mycovirus can be extended utilizing different techniques including hyphal anastomosis, constructing infectious clones and protoplast fusion. Mycoviruses usually use intercellular transmission means for their propagation to progeny, which is through spores (vertical transmission) or hyphal anastomosis (Horizontal transmission), which determine their ability to spread under in vivo conditions. This determines their potential for using them as biocontrol. Field trails should be conducted before declaring subjected mycovirus as bio-control agent, but their successful deployment in field conditions poses a great challenge. Mycovirus-infected spore suspension or purified particles (in case can be transferred extracellularly) can directly be applied to infected plants. Uninfected fungal strains causing the disease will get infected and their disease-causing ability will be compromised which eventually stops disease spread.

9.14 FUTURE PROSPECTS

The biopesticide market will keep on growing in the future due to increase in pest resistance problem and high demand of safe and quality food products. The need of the hour is to use green and cheap bio-control methods other than synthetic pesticides. More efforts are required toward integrating the biocontrol's in field, which will ultimately lead toward improved yield. The pace of mycovirus discovery is on the rise and likely to expand more in the next few years, as elegant new tools are becoming increasingly available in this field. This will ultimately give a better understanding of mechanisms of hypovirulence potential of mycoviruses (Fillinger and Elad, 2016). Since mycoviruses reside in fungal cytoplasm, it make them suitable candidate to be used as a vector to transfer non-viral genes into targeted (pathogenic) fungi, i.e., killer yeasts system to kill the pathogenic fungi. VIC-based PCD can also be switched off using CRISPR cas9 by considering the same idea, which is a significant hurdle in using mycoviruses as biocontrol (Kopertekh et al., 2004). Exploring multiple infections, i.e., using combination of mycoviruses which can produce a synergistic hypovirulent effect and favors the horizontal and vertical transmission is also an important aspect to be explored. The population investigations of mycoviruses are in their early stages, which are expected to grow fast with the expanding accessibility of virus-specific identification techniques. Such improvements will support biocontrol opportunities and will provide new avenue into the mycovirology (Pearson et al., 2009). But the real challenge, which one can observe in using mycoviruses as a biocontrol, is to develop field implementation methods and other significant aspects including inoculum sources, infection routes, colonization modes, development of diseases, and characteristics of infected crops. A sufficient understanding of the type, behavior, and ecological conditions of development of the infection is very important in disease control. Plant growth-promoting rhizobacteria (PGPR) such as *Bacillus spp.* and *Pseudomonas spp.* can also be employed as biocontrol agents. The growth, development, and interactions of plants with environment and other microbes can be monitored by using nano biosensors which would be possible in near future. These sensors can give an alert before the onset of a disease so that necessary action can be taken at early stages. Multidisciplinary approaches for designing and synthesizing nanoparticles can be employed in the field of pathology which can open further horizons in phyto nanotechnology. The use of agrochemicals can be reduced and increase in yield can be expected in near future due to the proposed advancements (Das et al., 2021).

KEYWORDS

- **phytopathogenic fungi**
- **fungal infections**
- **antifungal**
- **chemical control biocontrol**
- **plant extracts**
- **mycovirus**
- **nanotechnology**

REFERENCES

Agrios, G. *Plant Pathology*, 5th Edn Amsterdam: Elsevier Academic Press. 2005. [Google Scholar].

Ahmad, H. I.; Ahmad, M. J.; Asif, A. R.; Adnan, M.; Iqbal, M. K.; Mehmood, K.; Muhammad, S. A.; Bhuiyan, A. A.; Elokil, A.; Du, X. J. C. i. i. m. b. A Review of CRISPR-based Genome Editing: Survival. *Evol. Chall.* **2018,** *28,* 47–68.

Ahmar, S.; Gill, R. A.; Jung, K. H.; Faheem, A.; Qasim, M. U.; Mubeen, M.; Zhou, W. Conventional and Molecular Techniques from Simple Breeding to Speed Breeding in Crop Plants: Recent Advances and Future Outlook. *Int. J. Mol. Sci.* **2020,** *21.*

Akahoshi, K.; Motomura-Wages, S.; Shintaku, M.; Hamasaki, R. T. J. P. D. Clubroot Disease of Crucifer Crops. **2019,** *103,* 1–5.

Aktar, M. W.; Sengupta, D.; Chowdhury, A. Impact of Pesticides use in Agriculture: Their Benefits and Hazards. *Interdiscip. Toxicol.* **2009,** *2,* 1–12.

Arya, G. C.; Sarkar, S.; Manasherova, E.; Aharoni, A.; Cohen, H. The Plant Cuticle: An Ancient Guardian Barrier Set Against Long-Standing Rivals. *Front. Plant Sci.* **2021,** *12.*

Ballester, A.-R.; Marcet-Houben, M.; Levin, E.; Sela, N.; Selma-Lázaro, C.; Carmona, L.; Wisniewski, M.; Droby, S.; González-Candelas, L.; Gabaldón, T. J. M. P.-M. I. Genome, Transcriptome, and Functional Analyses of Penicillium Expansum Provide New Insights Into Secondary Metabolism and Pathogenicity. **2015,** *28,* 232–248.

Banik, S.; Sharma, P. J. I. P. Plant Pathology in The era of Nanotechnology. **2011,** *64,* 120–127.

Bellincampi, D.; Cervone, F.; Lionetti, V. Plant cell wall dynamics and wall-related susceptibility in plant-pathogen interactions. *Front. Plant Sci.* **2014,** *5,* 228.

Bozkurt, T. O.; Kamoun, S. The Plant-pathogen Haustorial Interface at a Glance. *J. Cell Sci.* **2020,** *133,* jcs237958.

Braun, E. J.; Howard, R. J. Adhesion of Fungal Spores and Germlings to Host Plant Surfaces. *Protoplasma* **1994,** *181,* 202–212.

Bruenn, J. *The Ustilago Maydis Killer Toxins, Microbial Protein Toxins*; Springer, 2005; pp 157–174.

Burgiel, Z.; Smaglowski, M. J. Z. P. P. N. R. Fungistatyczne własności olejku z drzewa herbacianego. 2008.

Burt, S. Essential Oils: Their Antibacterial Properties and Potential Applications in Foods—A Review. **2004,** *94,* 223–253.

Carver, T. L. W.; Bushnell, W. R. The Probable Role of Primary Germ Tubes in Water Uptake Before Infection by Erysiphe Graminis. *Physiol. Plant Pathol.* **1983,** *23,* 229–240.

Chaerani, R.; Voorrips, R. E. Tomato Early Blight (Alternaria Solani): The Pathogen, Genetics, and Breeding for Resistance. **2006,** *72,* 335–347.

Chanclud, E.; Kisiala, A.; Emery, N. R. J.; Chalvon, V.; Ducasse, A.; Romiti-Michel, C.; Gravot, A.; Kroj, T.; Morel, J.-B. Cytokinin Production by the Rice Blast Fungus Is a Pivotal Requirement for Full Virulence. *PLoS Pathog.* **2016,** 12, e1005457–e1005457.

Chang, H. X.; Miller, L. A.; Hartman, G. L. Melanin-independent Accumulation of Turgor Pressure in Appressoria of Phakopsora Pachyrhizi. *Phytopathology* **2014,** *104,* 977–984.

Ćosić, J.; Vrandečić, K.; Postić, J.; Jurković, D.; Ravlić, M. J. P. In Vitro Antifungal Activity of Essential Oils on Growth of Phytopathogenic Fungi. **2010,** *16,* 25–28.

Costa, F. G.; Zucchi, T. D.; Melo, I. S. Biological Control of Phytopathogenic Fungi by Endophytic Actinomycetes Isolated From Maize (Zea Mays L.). *Technology* **2013,** *56,* 948–955.

Cowan, M. M. Plant Products as Antimicrobial Agents. **1999,** *12,* 564–582.

Crespo-Sempere, A.; Marin, S.; Sanchis, V.; Ramos, A. Transcriptional Factors Regulate Ochratoxin A Biosynthesis in Aspergillus Carbonarius. **2013,** *166,* 479–486.

Das, A. K. Citrus Canker–A Review. **2003,** *5,* 52–60.

Das, K.; Jhan, P. K.; Das, S. C.; Aminuzzaman, F.; Ayim, B. Y. Nanotechnology: Past, Present and Future Prospects in Crop Protection. 2021.

Davis, D. J.; Burlak, C.; Money, N. P. Osmotic Pressure of Fungal Compatible Osmolytes. *Mycol. Res.* **2000,** *104,* 800–804.

Dean, R.; Van Kan, J. A.; Pretorius, Z. A.; Hammond-Kosack, K. E.; Di Pietro, A.; Spanu, P. D.; Rudd, J. J.; Dickman, M.; Kahmann, R.; Ellis, J.; Foster, G. D. The Top 10 Fungal Pathogens in Molecular Plant Pathology. *Mol. Plant Pathol.* **2012,** *13,* 414–430.

Deepa, N.; Sreenivasa, M. Y. Chapter 13 - Biocontrol Strategies for Effective Management of Phytopathogenic Fungi Associated With Cereals. In *New and Future Developments in Microbial Biotechnology and Bioengineering*; Singh, J. S., Singh, D. P., Eds.; Elsevier, 2019; pp 177–189.

Deising, H. B.; Werner, S.; Wernitz, M. J. M. The Role of Fungal Appressoriain Plant Infection. *Infection* **2000,** 2, 1631–1641.

El Gueddari, N. E.; Rauchhaus, U.; Moerschbacher, B. M.; Deising, H. B. Developmentally Regulated Conversion of Surface-exposed Chitin to Chitosan in Cell Walls of Plant Pathogenic Fungi. **2002,** *156,* 103–112.

Estiarte, N.; Lawrence, C. B.; Sanchis, V.; Ramos, A. J.; Crespo-Sempere, A. LaeA and VeA are Involved in Growth Morphology, Asexual Development, and Mycotoxin Production in Alternaria Alternata. *Int. J. Food Microbiol.* **2016,** *238,* 153–164.

Fawzi, E.; Khalil, A. A.; Afifi, A. F. Antifungal Effect of Some Plant Extracts on Alternaria Alternata and Fusarium Oxysporum. **2009,** 8.

Feng, P.; Martino-Catt, S.; Padgette, S.; Krämer, W.; Schirmer, U.; Jeschke, P.; Witschel, M. J. M. C. P. C. Inhibitor of 5-enolpyruvyl Shikimate 3-phosphate Synthase. **2012,** *1,* 406–422.

Fillinger, S.; Elad, Y. *Botrytis-the Fungus, the Pathogen and its Management in Agricultural Systems*. Springer; 2016.

Foster, A. J.; Ryder, L. S.; Kershaw, M. J.; Talbot, N. J. The Role of Glycerol in the Pathogenic Lifestyle of the Rice Blast Fungus Magnaporthe Oryzae. *Environ. Microbiol.* **2017,** *19,* 1008–1016.

Fujikawa, T.; Kuga, Y.; Yano, S.; Yoshimi, A.; Tachiki, T.; Abe, K.; Nishimura, M. Dynamics of Cell Wall Components of Magnaporthe Grisea During Infectious Structure Development. *Mol. Microbiol.* **2009,** *73,* 553–570.

Gatto, M. A.; Ippolito, A.; Linsalata, V.; Cascarano, N. A.; Nigro, F.; Vanadia, S.; Di Venere, D. Activity of Extracts From Wild Edible Herbs Against Postharvest Fungal Diseases of Fruit and Vegetables. *Postharvest Biol. Technol.* **2011,** *61,* 72–82.

Gavaric, N.; Mozina, S. S.; Kladar, N.; Bozin, B. Chemical Profile, Antioxidant and Antibacterial Activity of Thyme and Oregano Essential Oils, Thymol and Carvacrol and Their Possible Synergism. **2015,** *18,* 1013–1021.

Gilbert, R. D.; Johnson, A. M.; Dean, R. A. Chemical Signals Responsible for Appressorium Formation in the Rice Blast FungusMagnaporthe Grisea. *Physiol. Mol. Plant Pathol.* **1996,** *48,* 335–346.

Giraldo, M. C.; Dagdas, Y. F.; Gupta, Y. K.; Mentlak, T. A.; Yi, M.; Martinez-Rocha, A. L.; Saitoh, H.; Terauchi, R.; Talbot, N. J.; Valent, B. Two Distinct Secretion Systems Facilitate Tissue Invasion by the Rice Blast Fungus Magnaporthe Oryzae. *Nat. Commun.* **2013,** *4,* 1996–1996.

Gisi, U.; Lamberth, C.; Mehl, A.; Seitz, T.; Blum, M. J. M. C. P. C. Carboxylic Acid Amide (CAA) Fungicides. **2019,** *2,* 845–869.

Glawe, D. A. The Powdery Mildews: A Review of the World's Most Familiar (Yet Poorly Known) Plant Pathogens. **2008,** *46,* 27–51.

Halama, P.; Van Haluwin, C. J. B. Antifungal Activity of Lichen Extracts and Lichenic Acids. **2004,** *49,* 95–107.

Hamer, J. E.; Howard, R. J.; Chumley, F. G.; Valent, B. A Mechanism for Surface Attachment in Spores of a Plant Pathogenic Fungus. *Science (N.Y.)* **1988,** *239,* 288–290.

Haverkort, A. J.; Struik, P.; Visser, R.; Jacobsen, E. J. P. R. Applied Biotechnology to Combat Late Blight in Potato Caused by Phytophthora Infestans. **2009,** *52,* 249–264.

Hernández-Chávez, M. J.; Pérez-García, L. A.; Niño-Vega, G. A.; Mora-Montes, H. M. Fungal Strategies to Evade the Host Immune Recognition. *J. Fungi. (Basel)* **2017,** *3,* 51.

Hille, F.; Charpentier, E. CRISPR-Cas: Biology, Mechanisms and Relevance. *Philos. Trans. R. Soc. B: Biol. Sci.* **2016,** *371,* 20150496.

Hirooka, T.; Ishii, H. Chemical Control of Plant Diseases. **2013,** *79,* 390–401.

Hirooka, T. J. A. J. Tiadinil. A New Systemic Fungicide Against Rice Blast and Other Diseases. **2004,** *85,* 12–15.

Honda, T.; Hasunuma, N.; Nishioka, M. Amisulbrom (NC-224): Performance of New Fungicide for Potato Late Blight Control. **2007,** 59.

Hua, L.; Yong, C.; Zhanquan, Z.; Boqiang, L.; Guozheng, Q.; Shiping, T. J. F. Q. Safety Pathogenic Mechanisms and Control Strategies of Botrytis Cinerea Causing Post-harvest Decay in Fruits and Vegetables. **2018,** *2,* 111–119.

Huang, K.; Czymmek, K. J.; Caplan, J. L.; Sweigard, J. A.; Donofrio, N. M. HYR1-mediated Detoxification of Reactive Oxygen Species is Required For Full Virulence in the Rice Blast Fungus. *PLoS Pathog.* **2011,** *7,* e1001335.

Hussein, H. S.; Brasel, J. M. Toxicity, Metabolism, and Impact of Mycotoxins on Humans and Animals. *Toxicology* **2001,** *167,* 101–134.

Imam, J.; Singh, P. K.; Shukla, P. Plant Microbe Interactions in Post Genomic Era: Perspectives and Applications. **2016,** 7.

Inoue, K.; Kitaoka, H.; Park, P.; Ikeda, K. Novel Aspects of Hydrophobins in Wheat Isolate of Magnaporthe Oryzae: Mpg1, but not Mhp1, is Essential for Adhesion and Pathogenicity. **2015,** *82,* 18–28.

Irieda, H.; Inoue, Y.; Mori, M.; Yamada, K.; Oshikawa, Y.; Saitoh, H.; Uemura, A.; Terauchi, R.; Kitakura, S.; Kosaka, A.; Singkaravanit-Ogawa, S.; Takano, Y. Conserved Fungal Effector Suppresses PAMP-triggered Immunity by Targeting Plant Immune Kinases. *Proc. Natl. Acad. Sci.* **2019,** *116,* 496.

Isman, M. B.; Machial, C. M. Pesticides Based on Plant Essential Oils: From Traditional Practice to Commercialization. **2006,** *3,* 29–44.

Iwata, M. Probenazole-a Plant Defence Activator. **2001,** *12,* 28–31.

Jang, J.-H.; Shin, H. W.; Lee, J. M.; Lee, H.-W.; Kim, E.-C.; Park, S.H. An Overview of Pathogen Recognition Receptors for Innate Immunity in Dental Pulp. *Mediat. Inflamm.* **2015,** *2015,* 794143.

Janik, E.; Niemcewicz, M.; Ceremuga, M.; Krzowski, L.; Saluk-Bijak, J.; Bijak, M. Various Aspects of a Gene Editing System—CRISPR–Cas9. **2020,** *21,* 9604.

Johnston-Monje, D.; Lazarovits, G.; Turnbull, A. Plant Health Management: Biological Control of Plant Pathogens, 2014; pp 388–399.

Jones, J. D.; Dangl, J.L. The Plant Immune System. *Nature* **2006,** *444,* 323–329.

Kahmann, R.; Steinberg, G.; Basse, C.; Feldbrügge, M.; Kämper, J.. *Ustilago Maydis, the Causative Agent of Corn Smut Disease, Fungal Pathology*. Springer; 2000, pp 347–371.

Kankanala, P.; Czymmek, K.; Valent, B. Roles for Rice Membrane Dynamics and Plasmodesmata During Biotrophic Invasion by the Blast Fungus. *Plant Cell* **2007,** *19,* 706–724.

Khang, C. H.; Berruyer, R.; Giraldo, M. C.; Kankanala, P.; Park, S.-Y.; Czymmek, K.; Kang, S.; Valent, B. Translocation of Magnaporthe Oryzae Effectors Into Rice Cells and Their Subsequent Cell-to-Cell Movement. *Plant Cell* **2010,** *22,* 1388–1403.

Khodakovsky, A.; Litke, N.; Schröder, P. Globally Smooth Parameterizations with Low Distortion. **2003,** *22,* 350–357.

Kim, S.; Ahn, I.-P.; Rho, H.-S.; Lee, Y.-H. MHP1, a Magnaporthe Grisea Hydrophobin Gene, is Required for Fungal Development and Plant Colonization. **2005a,** *57,* 1224–1237.

Kim, S.; Ahn, I. P.; Rho, H. S.; Lee, Y. H. MHP1, a Magnaporthe Grisea Hydrophobin Gene, is Required for Fungal Development and Plant Colonization. *Mol. Microbiol.* **2005b,** *57,* 1224–1237.

König, A.; Müller, R.; Mogavero, S.; Hube, B. Fungal Factors Involved in Host Immune Evasion, Modulation and Exploitation During Infection. *Cell. Microbiol.* **2021,** *23,* e13272.

Kopertekh, L.; Jüttner, G.; Schiemann, J. PVX-Cre-mediated Marker Gene Elimination From Transgenic Plants. **2004,** *55,* 491–500.

Lacroix, H.; Whiteford, J. R.; Spanu, P. D. Localization of Cladosporium Fulvum Hydrophobins Reveals a Role for HCf-6 in Adhesion. *FEMS Microbiol. Lett.* **2008,** *286,* 136–144.

Leroux, P. *Chemical Control of Botrytis and its Resistance to Chemical Fungicides, Botrytis: Biology, Pathology and Control*. Springer, 2007; pp 195–222.

Li, B.; Chen, Y.; Zong, Y.; Shang, Y.; Zhang, Z.; Xu, X.; Wang, X.; Long, M.; Tian, S. Dissection of Patulin Biosynthesis, Spatial Control and Regulation Mechanism in Penicillium Expansum. **2019,** *21,* 1124–1139.

Liu, W.; Zhou, X.; Li, G.; Li, L.; Kong, L.; Wang, C.; Zhang, H.; Xu, J.-R. Multiple Plant Surface Signals are Sensed by Different Mechanisms in the Rice Blast Fungus for Appressorium Formation. *PLOS Pathog.* **2011,** *7,* e1001261.

Lo Presti, L.; Lanver, D.; Schweizer, G.; Tanaka, S.; Liang, L.; Tollot, M.; Zuccaro, A.; Reissmann, S.; Kahmann, R. Fungal Effectors and Plant Susceptibility. *Annu. Rev. Plant Biol.* **2015,** *66,* 513–545.

Macedo, R.; Sales, L. P.; Yoshida, F.; Silva-Abud, L. L.; Lobo, M. Potential Worldwide Distribution of Fusarium Dry Root Rot in Common Beans Based on the Optimal Environment for Disease Occurrence. **2017,** *12,* e0187770.

Madhuri, S.; Choudhary, A.; Rohit, K. Nanotechnology in Agricultural Diseases and Food Safety. **2010,** *2,* 78–82.

Maeno, S.; Miura, I.; Masuda, K.; Nagata, T. Mepanipyrim (KIF-3535), a New Pyrimidine Fungicide, Brighton Crop Protection Conference, Pests and Diseases-1990. *Br. Crop. Prot. Council* **1990,** *2,* 415–422.

Malhi, Y.; Aragão, L. E.; Galbraith, D.; Huntingford, C.; Fisher, R.; Zelazowski, P.; Sitch, S.; McSweeney, C.; Meir, P. Exploring the Likelihood and Mechanism of a Climate-Change-Induced Dieback of the Amazon Rainforest. **2009,** *106,* 20610–20615.

Marín, S., Velluti, A.; Ramos, A. J.; Sanchis, V. Effect of Essential Oils on Zearalenone and Deoxynivalenol Production by Fusarium Graminearum in non-sterilized Maize Grain. **2004,** *21,* 313–318.

Marroquin-Guzman, M.; Hartline, D.; Wright, J. D.; Elowsky, C.; Bourret, T. J.; Wilson, R. A. The Magnaporthe Oryzae Nitrooxidative Stress Response Suppresses Rice Innate Immunity During Blast Disease. *Nat. Microbiol.* **2017,** *2,* 17054.

Matny, O. N.; Al-Warshan, S. H.; Ali, A. M. Antifungal Evaluation of Iraqi Propolis Against Penicillium Expansum and Mycotoxin Production in Apple. **2015,** *4,* 399–405.

Mentlak, T. A.; Kombrink, A.; Shinya, T.; Ryder, L. S.; Otomo, I.; Saitoh, H.; Terauchi, R.; Nishizawa, Y.; Shibuya, N.; Thomma, B. P. H. J., Talbot, N. J. Effector-mediated Suppression of Chitin-triggered Immunity by Magnaporthe Oryzae is Necessary for Rice Blast Disease. *Plant Cell* **2012,** *24,* 322–335.

Micali, C. O.; Neumann, U.; Grunewald, D.; Panstruga, R.; O'Connell, R. Biogenesis of a Specialized Plant–Fungal Interface During Host Cell Internalization of Golovinomyces Orontii Haustoria. **2011,** *13,* 210–226.

Min, J. S.; Kim, K. S.; Kim, S. W.; Jung, J. H.; Lamsal, K.; Bin Kim, S.; Jung, M.; Lee, Y. S. Effects of Colloidal Silver Nanoparticles on Sclerotium-Forming Phytopathogenic Fungi. 2009.

Misra, A. N.; Misra, M.; Singh, R. Sciences, A. Nanotechnology in Agriculture and Food Industry. *Technology* **2013,** *16,* 1.

Miyake, Y.; Sakai, J.; Shibata, M.; Yonekura, N.; Miura, I.; Kumakura, K.; Nagayama, K. Fungicidial Activity of Benthiavalicarb-Isopropyl Against Phytophthora Infestans and its Controlling Activity Against Late Blight Diseases. **2005,** *30,* 390–396.

Mogensen, T. H. Pathogen Recognition and Inflammatory Signaling in Innate Immune Defenses. *Clin. Microbiol. Rev.* **2009,** *22,* 240–273.

NAShwA, S. M.; Abo-ElyouSr, K. A. Evaluation of Various Plant Extracts Against the Early Blight Disease of Tomato Plants Under Greenhouse and Field Conditions. **2013,** *48,* 74–79.

Neri, F.; Donati, I.; Veronesi, F.; Mazzoni, D.; Mari, M. Evaluation of Penicillium Expansum Isolates for Aggressiveness, Growth and Patulin Accumulation in Usual and Less Common Fruit Hosts. *Int. J. Food Microbiol.* **2010,** *143,* 109–117.

Nishimura, S.; Kohmoto, K. Host-Specific Toxins and Chemical Structures From Alternaria Species. *Annu. Rev. Phytopathol.* **1983,** *21,* 87–116.

Numpaque, M. A.; Oviedo, L. A.; Gil, J. H.; García, C. M.; Durango, D. L. Thymol and Carvacrol: Biotransformation and Antifungal Activity Against the Plant Pathogenic Fungi Colletotrichum Acutatum and Botryodiplodia Theobromae. **2011,** *36,* 3–13.

Nuss, D. L. Mycoviruses, RNA Silencing, and Viral RNA Recombination. **2011,** *80,* 25–48.

Nuss, D. L. J. M. R. Biological Control of Chestnut Blight: An Example of Virus-Mediated Attenuation of Fungal Pathogenesis. **1992,** *56,* 561–576.

O'Connell, R. J.; Pain, N. A.; Hutchison, K. A.; Jones, G. L.; Green, J. R. Ultrastructure and Composition of the Cell Surfaces of Infection Structures Formed by the Fungal Plant Pathogen Colletotrichum lindemuthianum. **1996,** *181,* 204–212.

O'Brien, P. A. Biological Control of Plant Diseases. *Aust. Plant Pathol.* **2017,** *46,* 293–304.

Osés-Ruiz, M.; Sakulkoo, W.; Littlejohn, G. R.; Martin-Urdiroz, M.; Talbot, N. J. Two Independent S-phase Checkpoints Regulate Appressorium-Mediated Plant Infection by the Rice Blast Fungus Magnaporthe Oryzae. *Proc Natl Acad Sci U S A* **2017,** *114,* E237–E244.

Panstruga, R. Establishing Compatibility Between Plants and Obligate Biotrophic Pathogens. *Curr. Opin. Plant Biol.* **2003,** *6,* 320–326.

Park, H.-J.; Kim, S.-H.; Kim, H.-J.; Choi, S.-H. A New Composition of Nanosized Silica-Silver for Control of Various Plant Diseases. **2006,** *22,* 295–302.

Patel, N.; Desai, P.; Patel, N.; Jha, A.; Gautam, H. K. Agronanotechnology for Plant Fungal Disease Management: A Review. **2014,** *3,* 71–84.

Patkar, R. N.; Benke, P. I.; Qu, Z.; Chen, Y. Y.; Yang, F.; Swarup, S.; Naqvi, N. I. A Fungal Monooxygenase-Derived Jasmonate Attenuates Host Innate Immunity. *Nat. Chem. Biol.* **2015,** *11,* 733–740.

Pearson, M. N.; Beever, R. E.; Boine, B.; Arthur, K. Mycoviruses of Filamentous Fungi and Their Relevance to Plant Pathology. **2009,** *10,* 115–128.

Petrasch, S.; Silva, C. J.; Mesquida-Pesci, S. D.; Gallegos, K.; van den Abeele, C.; Papin, V.; Fernandez-Acero, F. J.; Knapp, S. J.; Blanco-Ulate, B. Infection Strategies Deployed by Botrytis Cinerea, Fusarium Acuminatum, and Rhizopus Stolonifer as a Function of Tomato Fruit Ripening Stage. 2019, 10.

Prasad, R. Synthesis of Silver Nanoparticles in Photosynthetic Plants. *J. Nanopart.* **2014,** *2014,* 963961.

Prusky, D.; Alkan, N.; Mengiste, T.; Fluhr, R. Quiescent and Necrotrophic Lifestyle Choice During Postharvest Disease Development. **2013,** *51,* 155–176.

Romanazzi, G.; Lichter, A.; Gabler, F. M.; Smilanick, J. L. Recent Advances on the Use of Natural and Safe Alternatives to Conventional Methods to Control Postharvest Gray Mold of Table Grapes. *Technology* **2012,** *63,* 141–147.

Różewicz, M.; Wyzińska, M.; Grabiński, J. The Most Important Fungal Diseases of Cereals—Problems and Possible Solutions. *Agronomy* **2021,** *11.*

Ryder, L. S.; Talbot, N. J. Regulation of Appressorium Development in Pathogenic Fungi. *Curr. Opin. Plant Biol.* **2015,** *26,* 8–13.

Sampaio, A. M.; Araújo, S. d. S.; Rubiales, D.; Vaz Patto, M. C. Fusarium wilt Management in Legume Crops. **2020,** *10,* 1073.

Sanzani, S.; Reverberi, M.; Punelli, M.; Ippolito, A.; Fanelli, C. Study on the Role of Patulin on Pathogenicity and Virulence of Penicillium Expansum. **2012,** *153,* 323–331.

Saunders, D. G.; Dagdas, Y. F.; Talbot, N. J. Spatial Uncoupling of Mitosis and Cytokinesis During Appressorium-mediated Plant Infection by the Rice Blast Fungus Magnaporthe Oryzae. *Plant Cell* **2010,** *22,* 2417–2428.

Scheuerell, S.; Mahaffee, W. J. C. S. Compost Tea: Principles and Prospects for Plant Disease Control. *Utilization* **2002,** *10,* 313–338.

Schmitt, M. J.; Breinig, F. J. N. R. M. Yeast Viral Killer Toxins: Lethality and Self-Protection. **2006,** *4,* 212–221.

Selin, C.; de Kievit, T. R.; Belmonte, M. F.; Fernando, W. G. D. Elucidating the Role of Effectors in Plant-Fungal Interactions: Progress and Challenges. **2016,** *7.*
Sharma, I. Chapter 7 - Phytopathogenic Fungi and Their Biocontrol Applications. In *Fungi Bio-Prospects in Sustainable Agriculture, Environment and Nano-Technology*; Sharma, V. K., Shah, M. P., Parmar, S., Kumar, A., Eds.; Academic Press, 2021; pp 155–188.
Sharma, P.; Verma, P.; Meena, P.; Kumar, V.; Singh, D. Research Progress Analysis of Sclerotinia Rot (Sclerotinia Sclerotiorum) of Oilseed Brassicas Through Bibliography. **2016,** *1,* 45–125.
Sharon, M.; Choudhary, A. K.; Kumar, R. Nanotechnology in Agricultural Diseases and Food Safety. **2010,** *2.*
Shiferaw, B.; Smale, M.; Braun, H.-J.; Duveiller, E.; Reynolds, M.; Muricho, G. Crops That Feed the World 10. Past Successes and Future Challenges to the Role Played by Wheat in Global Food Security. *Food Secur.* **2013,** *5,* 291–317.
Shiller, J.; Van de Wouw, A. P.; Taranto, A. P.; Bowen, J. K.; Dubois, D.; Robinson, A.; Deng, C. H.; Plummer, K. M. A Large Family of AvrLm6-like Genes in the Apple and Pear Scab Pathogens, Venturia Inaequalis and Venturia Pirina. **2015,** *6,* 980.
Shin, M. H.; Kim, J.-H.; Choi, H.-W.; Keum, Y. S.; Chun, S. C. Effect of Thymol and Linalool Fumigation on Postharvest Diseases of Table Grapes. **2014,** *42,* 262–268.
Shoresh, M.; Harman, G. E.; Mastouri, F. Induced Systemic Resistance and Plant Responses to Fungal Biocontrol Agents. **2010,** *48,* 21–43.
Sinden, J.; Hauser, E. J. M. S. Report on Two New Mushroom Diseases. **1950,** *1,* 96–100.
Singh, S.; Singh, B. K.; Yadav, S.; Gupta, A. KApplications of Nanotechnology in Agricultural and Their Role in Disease Management. *Nanotechnology* **2015,** *5,* 1–5.
Skamnioti, P.; Henderson, C.; Zhang, Z.; Robinson, Z.; Gurr, S. J. A Novel Role for Catalase B in the Maintenance of Fungal Cell-wall Integrity During Host Invasion in the Rice Blast Fungus Magnaporthe Grisea. *Mol. Plant-Microbe Interact.: MPMI* **2007,** *20,* 568–580.
Snini, S. P.; Tannous, J.; Heuillard, P.; Bailly, S.; Lippi, Y.; Zehraoui, E.; Barreau, C.; Oswald, I. P.; Puel, O. Patulin is a Cultivar-dependent Aggressiveness Factor Favouring the Colonization of Apples by P enicillium Expansum. **2016,** *17,* 920–930.
Suzuki, K.; Sugimoto, K.; Hayashi, H.; Komyoji, T. J. Biological Mode of Action of Fluazinam, a New Fungicide, for Chinese Cabbage Clubroot. **1995,** *61,* 395–398.
Talbot, N. J.; Kershaw, M. J.; Wakley, G. E.; De Vries, O.; Wessels, J.; Hamer, J. E. MPG1 Encodes a Fungal Hydrophobin Involved in Surface Interactions During Infection-Related Development of Magnaporthe Grisea. *Plant Cell* **1996,** *8,* 985–999.
Tannous, J.; Keller, N. P.; Atoui, A.; El Khoury, A.; Lteif, R.; Oswald, I. P.; Puel, O. Secondary Metabolism in Penicillium Expansum: Emphasis on Recent Advances in Patulin Research. *Nutrition* **2018,** *58,* 2082–2098.
Tester, M.; Langridge, P. Breeding Technologies to Increase Crop Production in a Changing World. *Science (New York, N.Y.)* **2010,** *327,* 818–822.
Than, P. P.; Prihastuti, H.; Phoulivong, S.; Taylor, P. W. J., Hyde, K. D. Chilli Anthracnose Disease Caused by Colletotrichum Species. *J. Zhejiang Univ. Sci. B* **2008,** *9,* 764–778.
Torney, F.; Trewyn, B. G.; Lin, V. S.-Y., Wang, K. Mesoporous Silica Nanoparticles Deliver DNA and Chemicals Into Plants. **2007,** *2,* 295–300.
Touhami, N.; Soukup, S. T.; Schmidt-Heydt, M.; Kulling, S. E.; Geisen, R. Citrinin as an Accessory Establishment Factor of P. Expansum for the Colonization of Apples. **2018,** *266,* 224–233.

Tripathi, P.; Dubey, N. J. P. b. Exploitation of Natural Products as an Alternative Strategy to Control Postharvest Fungal Rotting of Fruit and Vegetables. *Technology* **2004,** *32,* 235–245.

Tyagi, S.; Kumar, R.; Kumar, V.; Won, S. Y.; Shukla, P. J. G. C. Engineering Disease Resistant Plants Through CRISPR-Cas9 Technology. *Food* **2021,** *12,* 125–144.

Valente, S.; Cometto, A.; Piombo, E.; Meloni, G. R.; Ballester, A.-R.; González-Candelas, L.; Spadaro, D. Elaborated Regulation of Griseofulvin Biosynthesis in Penicillium Griseofulvum and its Role on Conidiation and Virulence. **2020,** *328,* 108687.

van Kan, J. A. Licensed to Kill: The Lifestyle of a Necrotrophic Plant Pathogen. *Trends Plant Sci.* **2006,** *11,* 247–253.

Veloso, J.; van Kan, J. A. L. Many Shades of Grey in Botrytis-Host Plant Interactions. *Trends Plant Sci.* **2018,** *23,* 613–622.

Veneault-Fourrey, C.; Barooah, M.; Egan, M.; Wakley, G.; Talbot, N. J. J. S. Autophagic Fungal Cell Death is Necessary For Infection by the Rice Blast Fungus. **2006,** *312,* 580–583.

Vilanova, L.; López-Pérez, M.; Ballester, A. R.; Teixidó, N., Usall, J.; Lara, I.; Viñas, I.; Torres, R.; González-Candelas, L. Differential Contribution of the Two Major Polygalacturonases From Penicillium Digitatum to Virulence Towards Citrus Fruit. *Int. J. Food Microbiol.* **2018,** *282,* 16–23.

Villanueva Bermejo, D.; Angelov, I.; Vicente, G.; Stateva, R. P.; Rodriguez García-Risco, M.; Reglero, G.; Ibañez, E.; Fornari, T. Extraction of thymol from Different Varieties of Thyme Plants Using Green Solvents. *Agriculture* **2015,** *95,* 2901–2907.

Wang, W.; Ben-Daniel, B.; Cohen, Y. J. P. Control of Plant Diseases by Extracts of Inula Viscosa. **2004,** *94,* 1042–1047.

Wang, Y. A.; Li, J. J.; Chen, H.; Peng, X. Stabilization of Inorganic Nanocrystals by Organic Dendrons. **2002,** *124,* 2293–2298.

Ward, J. M. J.; Stromberg, E. L.; Nowell, D. C.; Nutter, F. W. Gray Leaf Spot: A Disease of Global Importance in Maize Production. **1999,** *83,* 884–895.

Weiberg, A.; Wang, M.; Lin, F.-M.; Zhao, H.; Zhang, Z.; Kaloshian, I.; Huang, H.-D.; Jin, H. Fungal Small RNAs Suppress Plant Immunity by Hijacking Host RNA Interference Pathways. *Science (New York, N.Y.)* **2013,** *342,* 118–123.

Wilson, R. A.; Talbot, N. J. Under Pressure: Investigating the Biology of Plant Infection by Magnaporthe Oryzae. *Nat. Rev. Microbiol.* **2009,** *7,* 185–195.

Wösten, H. A. Hydrophobins: Multipurpose Proteins. *Annu. Rev. Microbiol.* **2001,** *55,* 625–646.

Xiao, J.; Ohshima, A.; Kamakura, T.; Ishiyama, T.; Yamaguchi, I. Extracellular Glycoprotein(s) Associated with Cellular Differentiation in Magnaporthe Grisea. **1994,** *7,* 639–644.

Zachary Nimchuk; Thomas Eulgem; Ben F. Holt III, Dangl, J. L. Recognition and Response in the Plant Immune System. **2003,** *37,* 579–609.

Zhang, D.-X.; Nuss, D. L. Engineering Super Mycovirus Donor Strains of Chestnut Blight Fungus by Systematic Disruption of Multilocus vic Genes. **2016,** *113,* 2062–2067.

Zhang, D. D.; Guo, X. J.; Wang, Y. J.; Gao, T. G.; Zhu, B. C. Novel Screening Strategy Reveals a Potent Bacillus Antagonist Capable of Mitigating Wheat Take-all Disease Caused by Gaeumannomyces Graminis var. Tritici. *Lett. Appl. Microbiol.* **2017,** *65,* 512–519.

CHAPTER 10

Management of Bacterial Diseases in Plants

LOVLISH GUPTA[1], MONIKA CHAUHAN[1], AJAY KUMAR[2], and DIWAKAR CHAUHAN[3]

[1]*Division of Forensic Science, School of Basic and Applied Sciences, Galgotias University, Greater Noida, Uttar Pradesh, India*

[2]*Division of Life Science, School of Basic and Applied Sciences, Galgotias University, Greater Noida, Uttar Pradesh, India*

[3]*Division of Chemistry, School of Basic and Applied Sciences, Galgotias University, Greater Noida, Uttar Pradesh, India*

ABSTRACT

Bacteria are a group of prokaryotic organisms that belongs to the kingdom of prokaryotes, also known as Procaryota. This kingdom is commonly called as Monera. Bacteria are unicellular life forms and primarily, microorganisms with least morphological differentiation. Bacterial life forms are most abundant microbes and are ubiquitous in nature, that is, they are found at all places. They can tolerate extreme and strident environmental conditions, such as hot springs, frozen waters, and deserts. Bacteria are coordinated to impartially affect the plant and animal life by causing multitudinous (manifold) diseases to them. For causing various diseases, bacteria show genetic and phenotypic reorganization and variations. There are various molecular and chemical multistationarity that generates stability of the bacterial population. They are expressed in their natural environments and exhibit their core structure that transmits these forces to a cell. The outer membranes of the two types of bacteria, that is, Gram-positive and Gram-negative bacteria potentially

Plant Diseases and Their Management: A Sustainable Approach.
Deepa Srivastava, Rajarshi Kumar Gaur, and Ajay K. Tiwari (Eds.)

transmit mechanical stress to their core structures, such as flagella, pili, or capsule that mediate their physical interactions between the cell, and the surface also bears external forces. There are numerous bacterial diseases in plants, such as seed-borne diseases, as bacterial diseases are extensive but are very difficult to be identified, cured, and controlled. Bacteria are always said to be playing two main roles as our friend and foe—they are responsible in causing various diseases, such as bacterial spots, bacterial wilts, bacterial cankers, leaf blotches, and spots. A recent research has shown that bacteria are playing their vital roles in causing innumerous diseases in potatoes by altering the amylopsin and amylopectin present in them. Plant diseases caused by bacterial pathogens place major constraints on crops and cause significant yearly losses worldwide. Various bacterial diseases are delineated by their ability to favor disease incidences, pathogen dissemination, growth rate of bacteria, which are pathogens that cause disease epidemics.

10.1 INTRODUCTION

Bacterial diseases in plants are becoming the most important concern for all botanists, agriculturists, and various biologists. However, scientists have been involved in coming up with numerous solutions for the management of several diseases caused by bacteria in plants. This is achieved with the better understanding and scientific studies on various forms of bacteria, the nature of bacteria, their functioning, their shapes, sizes, morphological characters, and especially their physiology (Alvarez, 2004). Bacteria were first identified, microbiologists have attempted to standardize their nomenclature and classification. Bacteria are one of the smallest prokaryotic organisms that fall under the kingdom Monera. Monera is a kingdom of prokaryotes; hence, also called prokaryotes. This kingdom mainly includes primitive life forms that have been developed from an early stock known as Progenote, and these organisms are primitive; they have a precise morphological balance on nature since various bacteria are the earliest forms that can survive even in extreme ranges of the environment, such as in icebergs where the temperature is below 0°C; in hot water springs, where the temperature is around or more than 100°C. Ancient bacteria can also survive in extreme saline water as that of oceans; they may survive in driest deserts, in dust particles, inside or outside of the plants, animals and our bodies; the air we breathe contains thousands of bacteria (Hogg, 2005; Koch, 2006). Bacteria were being classified into two subkingdoms called Archaebacteria and Eubacteria, which are responsible for giving rise to the six-kingdom classification (introduced by Carl Woese) after the five-kingdom classification of R. H. Whittaker. Bacteria was first discovered by Anton Von Leeuwenhoek in 1676 (Koch, 2006). Kingdom Monera was first introduced by Copeland in the four-kingdom classifications. Later, the kingdom has played the most vital role in the five-kingdom classifications by R.H. Whittaker (Margulis, 1971).

Bacteria are responsible for causing various diseases in plants as well as in many other organisms. It all depends upon the characteristics of bacteria toward the host based on their motility and non-motile nature. Motile bacteria move toward favorable locations by their chemotactic motility, whereas non-motile bacteria stick to a location randomly more or less to reach their favorable locations and nutrition. This behavior of bacteria holds for animal pathogens, aquatic bacteria, and also seen in bacteria surviving in the environment with limited nutrition. Hence, preventing themselves from being extricated. The proneness of bacteria to cause infection includes the ability to withstand the greater extremities of the environment, such as drought, radiation, and physical removal of plants from the surface of the soil, rain wire, or wind potential. The role of attachment for plant-specific bacteria has mostly been studied about the virulence of plant pathogens for certain bacteria. As a consequence, the association of the bacteria is the most imperative and foremost step to cause infection (Romantschuk, 1992). There are numerous other ways in which bacteria are even beneficial for plants and various organisms. Various bacteria show a symbiotic relationship with the plant and interact with their life cycles. Various endophytic bacteria are associated with plants and facilitate growth and biomass gain. These microbes or endophytic bacteria show mutual interaction in which they interact with various tissues and induce various metabolic influences which then result in pausing physiological and biochemical changes in plants, which help in increasing the nutrition absorption capacity of plants, regulate plant growth, and biomass gain. However, bacteria in turn obtain nutrition from the plant. This type of interaction between endophytic bacteria and plant is most common in almost every plant, most commonly seen in a wide range of bacteria as they live their complete or partial life cycles in plant bodies only. This type of interaction is known as plant–endophyte interaction (Santos et al., 2018). Furthermore, characterization of bacteria is also one of the most important steps in diagnosing the disease caused by bacteria to plants. There are various analytical and instrumental techniques, such as Raman spectroscopy, vibrational spectroscopy, denaturing gradient gel electrophoresis (DGGE) to investigate and identify bacteria (Jarvis and Royston, 2008; Brunvold et al., 2007).

10.2 PHYSIOLOGY OF BACTERIA

10.2.1 MORPHOLOGY OF BACTERIA

Bacteria are a group of prokaryotic organisms or Monera that subdues the cell wall concocted of peptidoglycan; being prokaryotic organisms, the DNA in bacteria is naked and it is in supercoiled stage inside the protoplasm, and also it is attached to the internal extension of the plasma membrane known as mesosome.

The morphology or patterning of bacteria include the succeeding peculiarities:

1. The bacterial cell contains a cell wall made up of peptidoglycan. In eubacteria, the load-bearing structure is peptidoglycan. This is an interlaced monomolecular meshwork that defines the cellular shape. The cell boundaries in eubacteria and Archaebacteria are defined in a generalized way by internal scaffolding or by external scaffolding (Young, 2003).
2. Bacteria are fundamentally unicellular prokaryotic organisms.
3. The outermost covering is a mucilage called glycocalyx, also outlines bacterial cells. It consists of non-cellulosic polysaccharides with or without proteins. If glycocalyx is present in the form of loose sheath, then it is referred to as slime layer, and if it is thick and rigid outside, then the covering is called the capsule.
4. The plasma membrane of the bacterial cell has certain infoldings called mesosomes, which are designed to store reserved food and also to perform the function of respiration. Mesosomes are also known as the villiform specialization of the cell membrane. This also helps in the rapid elongation of the membrane during cell division.
5. The outermost layer glycocalyx, the cell wall, and the innermost plasma membrane together form the cell envelope of the bacterium.
6. The space between the plasma membrane and the cell wall is periplasmic.
7. Have circular DNA folded to form nucleoid or genophore, which is attached to mesosome via an attachment called Y-Fork. This genetic material is also known as prochromosome.
8. Sap vacuoles are absent. Instead, gas vacuoles occur in several cases.
9. Membranous cell organelles, including endoplasmic reticulum, Golgi apparatus, etc. are absent.
10. 70S natural ribosomes are present. These are membrane less; submicroscopic ribonucleoprotein entities having a size of 20 nm× 15 nm. S denotes sedimentation coefficient. The ribosome has two units,

the larger unit is the 50 S sub unit and the smaller unit is the 30 S subunit.

11. In the prokaryotes, there are various types of motile structures associated with the cellular bodies, namely, cilia, fimbriae, pili, and flagellum. Flagella present in most of the bacteria are single-stranded, and are mainly composed of structural protein called flagellin.
12. Bacterial flagella are single-stranded, equivalent to single microtubular fiber. It measures about 20 nm in diameter and 1–7 micrometer in length (Young, 2007).

10.2.2 SHAPES AND SIZES

Bacterial cells are the most abundant organisms and are ubiquitous, found in many places in different shapes and sizes.

- The average size of a bacterium is 2.0–2.6 micrometer in length and 1.0–1.8 micrometer in width.
- The smallest bacterium is *Dialister pneumosintes,* with a size of approximately 0.15 micrometer, the largest bacterial cell belongs to spirochaetes and blue-green algae (BGA), the cell of which may have a size of 500 micrometers.
- Certain bacterial cells are similar in size to eukaryotic cells, for example, *Thiomargartia ramibiensis.*
- Bacteria are also found in various shapes.

10.2.3 CLASSIFICATION OF BACTERIA

10.2.3.1 CLASSIFICATION OF BACTERIA BASED ON THE SHAPE

1. **COCCUS:** The spherical or ovoid outlined bacteria are referred to as coccus.

   a. Coccus may also be grouped, and based on their grouping, various names are given to them: (i) *Monococcus:* when bacteria are not grouped, they are present singly, (ii) *Diplococcus:* when bacteria are present in pair, (iii). *Tetracoccus:* when bacteria are present in the tetrads, (iv). *Streptococcus:* when the long chain of coccus is arranged, (v). *Staphylococcus:* when bacteria are present in clusters like grapes, (vi). *Sarcina:* when coccus is arranged in such a way as it is giving a three-dimensional geometrical form.

2. **BACILLUS:** the rod-like, straight, or cylindrical bacteria with rounded ends are referred to as *Bacillus*. Based on the grouping of *Bacillus*, bacteria are further classified as: (i). *Diplobacillus:* when *Bacillus* bacteria are present together, (ii). *Palisade bacillus:* bacteria are arranged in a stack-like manner, (iii). *Streptobacillus:* when Bacillus bacteria are arranged in a chain.
3. **SPIRILLUM:** The bacterium is called a corkscrew.
4. **VIBRIO:** The body of the bacterium is like a curved rod. Also referred to as comma (,)-shaped bacterium.
5. **STALKED:** A bacterium with the stalk.
6. **BUDDED:** Bacteria give rises to buds hypothetically as their body is swollen in places.

10.2.3.2 CLASSIFICATION OF BACTERIA BASED ON FLAGELLUM

Based on the presence and absence of flagella, bacteria are considered flagellated bacteria (with flagella) and Eflagellated bacteria (without flagella). In the presence of flagella, bacteria are of the following described types:

1. **Atricous:** Bacteria with no flagella
2. **Monotricous:** Bacteria with one flagellum present at one of the ends of the cell.
3. **Amphitricous:** Bacteria with two flagella present at the two ends of the cell.
4. **Cephalotricous:** When several flagella are present at only one end of the cell.
5. **Lophotricous:** When a group or several flagella are present at both the ends of the cell.
6. **Peritricous:** When several flagella are present allover in the periphery of bacterium (Koch, 2006).

There are various molecular mechanisms data accountable for building cells of various diversification. that is, with various shapes and sizes.

The shapes help bacteria to survive through various activities, such as eating, moving, dividing, to survive, or any other activity like attachment of bacteria to a specific body. It can benefit bacteria from adopting an appropriate shape. Bacteria demonstrate ample cellular forms and are capable of changing their shape during the life cycle. Cell shape in most bacteria is examined by the presence of the peptidoglycan cell wall. There are copious distinct cell body shapes of bacteria from space to rods of various curvatures and helizites and more exotic shapes. Cell curvature in bacteria also adds to surface colonization in aquatic environments with moderate flow. Bacteria alter, adapt, and revise their morphology in response to environmental cues or during the cause of pathogenesis. Cell shape in bacteria affects nutrient accession and these nutritional situations create a condition in which one shape of bacteria dominates the other (Whittaker, 1969; Young, 2007).

10.2.3.3 ARCHAEBACTERIA AND EUBACTERIA

1. There are two major groups of monerans referred to as Archaebacteria and Eubacteria.
2. Archaebacteria are ancient bacteria and eubacteria are referred to as modern bacteria or true bacteria.
3. Eubacteria is further divided into two subtypes: bacteria and cyanobacteria.
4. There are some other groups of monerans, called mycoplasma, rickettsiae, and actinomycetes.
5. Archaebacteria can survive in various types of environmental extremities. We can survive in the harshest conditions. Virus Eubacteria can survive in limited environmental extremities. They can only survive in the narrow range of temperature, pressure, etc.
6. Woese and Coworkers described Archaebacteria as a new life form (Yang et al., 2016).
7. Archaebacteria are told to be the closest ancestors of eukaryotic cells. Initially, the Archaebacteria were discovered based upon their 16S rRNA Sequences and phylogeny. Genetically, Archaebacteria are most closely related to eukaryotes rather than eubacteria. Archaebacteria were considered to be different bacteria than any other type of bacteria with a universal ancestry (Woese et al., 1990).
8. Gram stain was discovered by Christian gram in 1884. Gram-positiveness and negativeness.
9. Evolutionary relationships are being set based on molecular structures and sequences. Does the study of natural taxonomy suggests that while moving toward a natural system of the organization are shifting from organismal level to cellular level true molecular level?

This molecular level helps in studying the conventional character of the specific type of organisms and is also responsible for broadly categorizing the typical Kingdoms (Gupta, 1998).

10.2.4 GRAM-POSITIVE AND NEGATIVE BACTERIA

Bacteria are stained with a weak alkaline solution of crystal violet when all of them appeared to be pink ok a blue. They are then treated with 0.5% iodine solution followed by washing with water and then absolute alcohol or acetone. Bacteria that retain blue or purple color are known as Gram-positive bacteria; whereas the bacteria which does not retain the stain and become colorless are called Gram-negative bacteria which can be further stained by safranin.

Characteristics of bacteria	**Gram-positive bacteria**	**Gram-negative bacteria**
1. Staining	Remain in blue to purple color with Gram stain even after washing with absolute alcohol	Do not retain any color when washed with absolute alcohol
2. Wall (cell envelope)	Single-layered	Two-layered
3. Thickness	20–80 nm	8–12 nm
4. Lipid content	Very low	Medium (20–30%)
5. Mucopeptide content	70–80%	10–20%
6. Mesosomes	More prominent	Less prominent
7. Behavior toward antibiotics	More susceptible to antibiotics	More resistant to antibiotics
8. Pathogenicity	Very few of the pathogenic bacteria belong to Gram-positive	Most of the pathogenic bacteria present in the Gram-negative group
9. Cell walls	• Teichoic acid is present • Consist of thick and multilayered peptidoglycan • Contains virtually nonlipopolysaccharide content. • 20–80 nm • Periplasmic space is absent	• Teichoic acids are absent • Consist of thin and unlayered peptidoglycan • Contains high lipopolysaccharide content • 5–10 nm • Periplasmic space is present
10. Examples	*Clostridium*, streptococci, actinomycetes, etc.	*Chlamydia*, *Enterobacter*, *Borrelia*, etc.

Gram-positive bacteria are most susceptible to antibiotics as they lack external membrane, Gram-positive bacteria has a cell wall made up of thick peptidoglycan and teichoic acid which is the lead for the bacteria to get stained with crystal violet dye; however, Gram-negative bacteria have cell wall with thin peptidoglycan and without teichoic acid, that is the reason behind the staining of the bacterial cell wall to be in pink color by safranin (Fox et al., 1980).

10.3 BACTERIA: FRIENDS OR FOE?

Bacteria cannot be simplified in easy terms. Bacteria are of innumerable ranges present all around us, even a handful of soil contains thousands of bacteria. Bacteria have both good effects and ill effects on other living individuals. As bacteria are ubiquitous, it is found in all places where organic matter is present—in water, air, soil, in and on the bodies of various organisms. We can tolerate extreme environments like hot water springs, icebergs, deserts, deep into the seas, alkaline, acidic conditions, etc. Around 130 years ago, a bacterium known as *Erwinia amylovora* was the first bacterium found to be the tenet against plants or the first bacterium to cause plant disease, that is, fire blight disease in pears. Almost every considerable crop is affected by at least one pathogenic bacterium, and certain diseases sometimes even do not have any practical cure. Hence, in this review, we will also be discussing its management via various processes as accounted by molecular, genomic, and post-genomic approaches (Tavares et al., 2020; Berg et al., 2014).

Bacteria also play various beneficial roles in plant disease management as well as in stimulating plant growth and development, as bacteria are an integral part of the management system in crops. Some of the beneficial bacteria and their essential roles are as follows:

1. There is a crowd of certain bacteria called plant growth promoting bacteria (PGPB), these bacteria are highly useful in increasing crop production and yields; some of these bacteria are eco-potent, that is, they can provide benefits to the environment as well.
2. Some groups of PGPB are called biocontrol PGPB. They helps in decreasing the phytopathogenic nature of several other bacteria and microbes to prevent the crop from several diseases.
3. Bacteria like *Azospirillum* increase in bioremediation of wastewater by increasing algal proliferation (Berg et al., 2014; Odelade and Babalola, 2019).
4. Legume symbiont *Rhizobium* helps in preventing soil erosion.
5. Many other bacteria belonging to different genera, such as *Flavobacterium, Frankia, Clostridium, Hydrogenophora, Streptomyces*

help in nitrogen fixation and also help in dissolving various forms of chemicals, such as phosphorus, a process known as phosphate solubilization, as we know that phosphate is present in large amounts inside the soil, but is not available to the plants for their growth and development.

6. Bacteria may act as helper bacteria to promote the growth of economically important aquatic microbes (Berg et al., 2014).
7. Microbial metabolites increase the sustainability of crop production by decreasing the use of chemicals as they are versatile, degradable, and eco-friendly (Bashan et al., 2008).
8. Various bacteria help in preventing soil erosion, help in the restoration of arid lands.
9. In PGPB–plant interaction, various bacteria are commercialized, such as *Agrobacterium radiobacter, Pseudomonas solanacearum, Delftia acidovorans, Bacillus subtilis, Streptococcus griseoviridis* (Odelade and Babalola, 2019; Hayat et al., 2010).
10. Examples of some PGPB: *Erwinia herbicola* multiply in Apple blossoms and provide effective control for fire blight, *Agrobacterium radiobacter* strain 84 provides effective biological control of crown gall of several woody plants and is used throughout the world (Glick, 2012).

10.3.1 ECONOMIC IMPORTANCE

Various bacteria are economically important, some of them are as follows:

1. **Lactic Acid:** Lactic acid is commercially obtained from ammoniated sugar solution through fermentation caused by *Lactobacillus delbrueckii.* Acid is used in food preservation, tanning, and preparation of several drugs.
2. **Butyl Alcohol and Acetone:** *Clostridium acetobutylicum* can produce butyl alcohol, methyl alcohol, and acetone from molasses.
3. **Retting of Fibers:** Stem and leaf fibers are separated from soft tissues by bacterial action of two types—dew retting and anaerobic retting.
4. **Dairy Industry:** Lactic acid bacteria convert milk sugar lactose into lactic acid.

5. **Sulfur Bacteria:** *Beggiatoa* and other sulfur bacteria pick up H_2S released during the putrefaction of proteins.
6. **Antibiotics:** Several antibiotics are obtained from mycelial bacterium *streptomyces.* several antifungal medicines are also produced by the same genus, for example, hamycin, trichomycin, primaricin.
7. **Curing:** Leaves of tea and tobacco are cured of their bitterness with the help of certain bacteria, for example, *Bacillus megatherium.* Beans of coffee and cocoa are similarly cured too.

10.3.2 HARMFUL ACTIVITIES OF BACTERIA

1. Harmful bacteria produce toxins and cause the rotting of fruits, vegetables, and other edible items.
2. Harmful bacteria cause food poisoning that is a complaint by many other diseases in humans.
3. Harmful bacteria also deteriorate domestic articles, such as clothing, wooden articles, and leather.
4. Bacteria like *Thiobacillus* denitrifies the nitrogen by converting nitrates into gaseous nitrogen.
5. For several decades, various harmful bacteria have been responsible for several diseases in humans, animals, and many plants (Burr et al., 1984; Sharma, 2015).

For example, *Phytophthora infestans* causes leaves and stems of tomato and potato plants to be affected with leaf spots, which darkens with time; white fungal growth is seen on the underside of the infected leaves. This disease is specifically called late blight (Rowan et al., 2009).

Puccinici sorghi causes common rust in the leaves and stems of corn (Haverkort et al., 2009).

10.4 BACTERIAL DISEASE IN PLANTS

Stakmann and Harrar (1957) defined disease as a physiological disorder or any structural malfunctioning that is deleterious to the plants or its part or product, that reduces the plant's functioning and its economic value, for example, wilt, potato blight, crown gall, slack rot disease, and Pierce's disease. The plant pathogen interferes with one or more crucial parts of the plant and its functioning. Plant disease is the interaction between the plant host, plant pathogen, and the environment (Sucher et al., 2017).

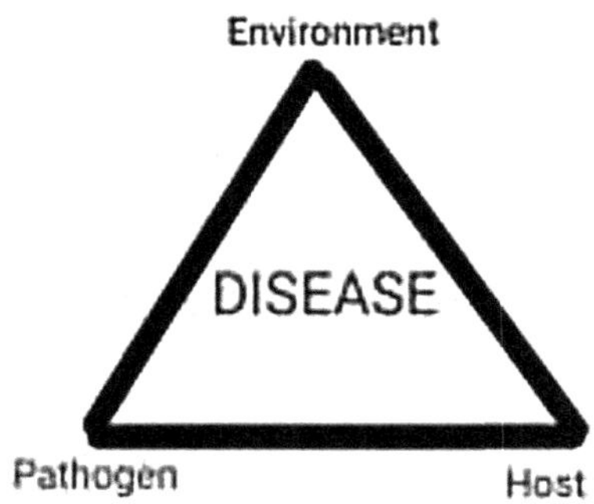

Phytopathology is the term constituted of two terms phyto (= plant) and pathology (=medicinal study), meaning "study of suffering plant" or in other words, phytopathology is the study of biological Sciences that including agricultural and botanical sciences, that includes causation of disease, effects of disease and diagnosis and control of plant disease. Study of the origin, causes, what the reason behind a specific disease, that is, living and non-living, as well as other causes of disorder in plants are referred to as plant etiology. However, the study of the mechanism of a disease being caused includes the integral process of infection and colonization of the host by the pathogen, that is, host-phytopathogen interaction mechanism (HPIM). Some of the normal physiological disease triangle functions of plants include (Wallace, 1978):

1. Utilization and storage of photosynthates.
2. Reproductive growth and activity.
3. Normal cell division, differentiation, and development.
4. Normal metabolic activities such as management of metabolism of metabolic life and synthates.
5. Absorption of water and other minerals from the soil.
6. Adequate amount of translocation, transpiration, guttation, etc. (Shoda, 2000).

The study of interactions between the causative agents and the diseased plant about the environment and ecological conditions, at the population level, is known as Epidemiology (Hirano and Upper, 1983).

In plants, epidemiology is basically of two types:

1. Slow epidemiology, is seen mainly in monoclinic diseases that occur in perennial plants. for example, *Citrus tristeza*, Dutch elm disease.
2. Fast epidemiology, is seen mainly in polyclinic diseases that occur in annual plants/crops, for example, rice blast, potato blight (Davis and Sinclair, 1998).

Epidemics are the dynamics of change in plant disease in time and space. The study of epidemics and the factor influencing them is called epidemiology, for example, the Irish potato famine of 1845–1846 due to infection of potato crops by *Phytophthora* (late blight epidemic of potato), Bengal Famine of 1943 Epidemic of rice. *Cochliobolus* was a major cause of this famine. The subdiscipline of plant pathology is concerned with the factors that cause plant epidemics. Epidemiology has also been useful for predicting the effects of climate change on plant disease; it includes both theoretical and experimental analysis along with mathematical proofs.

10.4.1 MAJOR BACTERIAL DISEASES

Some of the major bacterial diseases include (Tavares et al., 2020):

1. Bacterial wilt
2. Crown gall
3. Citrus canker
4. Bacterial soft rot
5. Pierce's disease
6. Bacterial leaf blight
7. Citrus greening
8. Bacterial fruit blotch
9. *Salmonella* infection in plants
10. Bacterial disease in Leguminosae

10.4.1.1 BACTERIAL WILT

The disease with complexities occurring in plant species, such as Cucurbitaceae and Solanaceae are caused by various types of bacteria. This disease is a vascular tissue disease in plants. When a plant gets infected by the pathogen, the pathogen makes its multiple copies within the xylem, which eventually leads to the mechanical malfunctioning of the water transportation system in plants. The incubation period of these bacteria depends upon the type of bacteria; but in general, it is from 5 days to a week. This is a fatal disease for the plant and causes the death of the plant. Bacterial disease is mostly common in the Solanaceae family, and the most destructive agent that causes the bacterial wilt is *Ralstonia solanacearum*. This bacterium has infected over 200 plant species with the same disease to date. This is a soil-borne pathogen, grows rapidly and more efficiently in hot climates and humid environments. Increasing growth of this bacterium leads to the reduced yields. This bacterium is responsible for causing wilt in many plants such as tomato and potato (Dhingra and Sinclair, 2019). Bacterial wilts were initially reported in Ethiopia on *Ensete*. Because of the infection Banana production has decreased through early ripening and rotting of fruits; this is caused by a bacterium known as *Xanthomonas campestris* PV. *musacearum*. It was observed in a research that the reason for the spread of bacterial wilt is the transmission through insects, as insects were found to be present in large numbers in the infected plants, whereas less number of insects were present in the non-diseased plant (Kim et al., 2016).

Mechanism: The bacterium first enters the root in the elongation zone or through wounds created at the little site. The bacteria then enter the vasculature, and from there it moves from roots to shoot and causes several pathogenicity factors. Water flow is restricted from root to the shoot and results in the shoot wilting that eventually causes the death of the plant (Dhingra and Sinclair, 2019).

Epidemiology of Disease: The main source of the spread of infection is infected tubers, which spread the disease from one plant to another. Wilting increases with certain environmental changes, for example, an increase in temperature decreases the quantity of water inside a plant, hence, causing its wilting. High moisture helps the bacterial pathogen to grow, hence, wilting increases (Tinzaara et al., 2006).

Symptoms

- The disease affects both the shoot and the root system of the plant.
- Infected plants show sudden wilting and complete collapse of plants, causing their death.
- The vascular system becomes brown and this is called vascular browning, it very common in plants suffering from bacterial wilting.
- Bacterial mass ooze out from the cut ends of the stem (Dhingra and Sinclair, 2019; Kim et al., 2016).

Management: Bacterial wilt can be prevented in the following ways

- By using disease-free tubers for planting.
- By avoiding injuries to the tubers during intercultural operations and practices.
- Bacterial wilt can be treated by biocontrol agent too, which belongs to the group of bacteria such as *Bacillus subtilis,* and *Pseudomonas fluorescent* (Tinzaara et al., 2006).

10.4.1.2 CROWN GALL

This is a plant disease caused by a very common soil bacterium, *Agrobacterium tumefaciens.* The disease is represented by the abnormal growth of galls on roots and on the branches that leads to the stunted growth of the plant mainly shrubs. It is a Gram-negative bacterium, bacillus in nature. It gets involved in the competition with the roots of many plants by chemotactic activities. It contains Ti (Tumor-inducing) plasmid. Furthermore, the disease affects many commercially important crops, such as grapes, vines, nut trees, roses, rice crops, and various other shrubs. Mostly, dicotyledon species like herbaceous plants are affected by this disease. Environmentally acquired infectious agents create the challenge of transitioning between extremely different disease and non-disease environments. The transitions between environmental and host body most prominently depend on tight regulatory control of key attributes, including virulence functions, motility, and surface colonization phenotypes, and are shaped by interactions with host defenses and features of the host environment, including the host's microbiome (Hayward, 1991).

Ti plasmid: When the bacterium comes into contact with the plant and interacts with it, through its wounds, the Ti plasmid is expressed. It manifests when the bacteria identify the dimethoxy phenolic compounds found in the plant. Two strains of *Agrobacterium tumefaciens* are known which includes A6 strain and C58 strain. These two strains are responsible for the pathogenicity caused in plants (Barton et al., 2018).

Mechanism: Ti contains T-DNA (Transfer-DNA) that gets transferred from bacterium to the plant genome. Ti plasmid is a megaplasmid, double-stranded circular DNA that contains roughly about 225 kbp. When the bacterium enters the plant through the root, it affects the plant through the Ti plasmid where it releases its T-DNA into the body of the host resulting in its exposure (Barton et al., 2018).

Management: Crops can be protected from disease spread using *Agrobacterium*-resistant crops and also using biochemical agents. Following methods can be employed:

- Replantation of the entire crop field with cereal crops for a long period.
- Use of antibiotic sprays like agrocin, antibiotic fertilizers should be followed on the entire field, as this inhibits the nucleic acid and protein synthesis by abstracting the intake of amino acids by the pathogenic bacteria.
- Injuries or pruning wounds should be avoided, and especially their contact with the soil should be avoided while caring for susceptible plants.
- Tree wrap should be followed to protect against string trimmer damages (Braun, 1962).

10.4.1.3 CITRUS CANKER

Citrus canker is caused by the bacterium *Xanthomonas axonopodis.* The disease is also caused by many other groups of bacteria and affects the citrus species. As a result, it leaves lesions on the leaves, stems, fruits of citrus trees, etc. Hence, this disease is also known as the bacterial spot of citrus (Gelvin, 1990).

Causative Agent: *Xanthomonas citri, Xanthomonas axonopodis* PV. *citri, Xanthomonas Campestris* pv. *citri* and *Xanthomonas citri* subsp. *Citri.*

***Hosts*:** Lime, orange, lemon, kalpi papeda (*Citrus webberi*), Citron (*Citrus medica*), etc. (Gelvin, 1990).

Transmission and Infection Mechanism: The bacterium is transmitted by contaminated equipment, through workers who come in contact with infected trees, or contaminated plants. The bacterium can also be spread through rainwater.

Pathogen generally enters during natural closing and opening of stomata, but sometimes, the pathogen also enters through to the wounds, through the water pressure caused by heavy rains. The pathogen induces hyperplasia of the epidermal and stem cortical tissue resulting in cankers. That circular spots on the abaxial side of the leaf indicate that the pathogen has entered through the stomata, and not from the wound (Kumar et al., 2019; Gottwald et al., 2002).

Symptoms

- The symptoms are shown in both the above and ground parts of the plant.
- The lesions on stems, leaves, and fruits usually appear as raised brownish spots with water-soaked margins and with yellow halos (Zekri et al.).

Management: The best way to prevent infection is to burn the infected and adjacent trees by using prophylactic sprays made up of copper compounds generally.

Quarantine measures are equally impressive to prevent the entry of infected planting materials into canker-free areas (Das, 2003; et al.; Zekri).

10.4.1.4 BACTERIAL SOFT ROT

Bacterial soft rot disease generally results in damaging succulent parts of the plants, such as fruits, tubers, bulbs, and stems of plants. This is the most widespread disease among plants that has caused the greatest and excessive loss of crops. This disease causes a greater total loss of production than other any bacterial disease. The most common occurrence of this disease is on fleshy vegetables such as onions carrots tomatoes, and potatoes,. Soft rots may often develop on fleshy stems. However, almost every fresh vegetable is affected by bacterial soft rot (Danos et al., 1984).

Symptoms

- The appearance of a watery lesion, which is initially smaller but later it is larger in diameter at a faster pace.
- The discoloration of the affected area starts, and the area becomes soft, pulpy, and squishy, due to which a depression starts appearing on the surface.

- The disintegration of tissues into a mass of disorganized plant cells and bacteria starts, due to which the color of the affected tissue within that area turns creamy.
- The outer surface may remain intact while the internal contents have changed into a turbid liquid, which results in the development of crafts due to which a certain mass of this turbid liquid squeezes out of the surface, and when it comes in contact with the air, it turns grayish to dark brown.
- Within a week, the entire fruit containing fleshy part of the plant becomes watery, the fruit or the tuber then becomes odorless until they collapse when secondary bacteria start to grow to decompose the tissue, and it produces a foul smell.
- Lower parts of the stem may also get infected if root crops are affected in the field (Chung et al., 2013; Simeon and Abubakar, 2014).

Causative Agents: *Erwinia carotovora* PV. *carotovora*, *Pseudomonas fluorescens* cause the most common and the most destructive soft rots.

Erwinia crysanthemi affects many hosts and causes many of the soft rot of tropical plants. This bacterium was active over the range of temperatures from 5 to 35°C (Simeon and Abubakar, 2014).

Development of Disease:

Bacteria may first appear in the fields of crops grown from previously infected seeds. The fleshy parts of the plant, such as tubers and bulbs become infected when bacteria come in contact with wounds or lenticels. Later, the growth of the bacterium inside the plant parts is facilitated by the action of insects. Hence, even the plant is resistant to the soft rot bacteria and can prevent its entrance by creating multiple layers on wounds, still can get affected by bacteria because maggots destroy the wound cork/barrier. Bacteria entering inside the broken cells feed and start multiplying and then they produce the pectolytic enzyme that breaks down the pectic substances of the middle lamella and bring about maceration to the tissues. Because of the high osmotic pressure of the macerated tissue, water conduction starts by diffusion water diffuses into the intercellular spaces resulting in plasmolysis, collapsing, and death of the tissue. Similarly, bacteria continue this process with the other cells, and the intercellular spaces resulting in massive destruction of tissues, causing the invaded tissue to be more soft and to get transformed into a slimy mass or bacterial "swimming pool" (Himel et al., 2016; Agyemang et al., 2020).

Management

- Exclusive increase of sanitary and cultural practices.
- Plant wounds with their storage organs should be avoided.
- Debris should be removed from houses, and cleaning of the walls and the area should be followed by disinfecting it using formaldehyde.
- Chemical sprays should be avoided because they may soften the fleshy part of the plant (Himel et al., 2017).
- Crop rotation of susceptible crops to cereals or other non-susceptible crops can help to reduce the pathogen effect (Himel et al., 2017).

10.4.1.5 PIERCE'S DISEASE

Pierce's Disease is mostly prevalent across the United States. This disease affects grapevines (*Vitis vinifera).*

Causative Organism: *Xylella fastidiosa (Xf) , Xylella fastidiosa* subsp. *fastidiosa (Xff.)*

The pathogen is a Gram-negative bacterium. It is atrichous (without flagella) (Su et al., 2013), nutritionally fastidious, and occurs in the vascular system (xylem) of the crops (Hopkins and Purcell, 2002). The bacterium interferes with the normal functioning of the xylem. Xylem is a conducting tissue that helps in the conduction of water and nutrients from roots to other parts of the plants. The bacteria is known for limiting the translocation by producing a chemical substance called tyloses, which blocks the vessels of the xylem, ultimately making it incapable of transporting an adequate amount of water and nutrition. The most common host is the grapevine. *Xf* is endemic and is a threat to the entire population of grapevine (Hopkins and Purcell, 2002).

Symptoms

- Berry softening.
- Sugar accumulation (in red-fruited cultivars).
- Marginal leaf necrosis is a very common symptom, necrotic tissue with yellow and burgundy margins of the signaling leaves edges (Su et al., 2013).
- "Matchstick" is the term used to describe the symptom when the leaf blade abscises and petiole remains attached to the shoot.

- Islands of green, hardening of soft tissues unevenly, resulting in green, vegetative "islands" within brown woody tissue (Himel et al., 2017).

Testing of Sample

Samples can be tested by the following tests:

1. Serological tests like ELISA (enzyme-linked immunosorbent assay).
2. Pathogenicity test and re-isolation of the causal bacterium.
3. Testing under transmission electron microscope.

Mechanism

It is observed that PD is spread by the transmission of pathogen by certain groups of insects known as sharpshooters (Hopkins and Purcell, 2002) and spittles bugs (Kyrkou1 et al., 2018) which served as a vector for the pathogen. This insect introduces the *Xf* into the grapevine. This insect was very frequently seen in the southeastern United States.

Management

- Low temperature helps in reducing insect vector, as this virus is only partially functional in a cold environment.
- Insecticidal treatments can also be used to prevent the vector population.
- Alterations of cropping techniques.
- *Organic approaches:* MIMETIC TECHNOLOGY involves the development of peptides, which closely resemble the natural substrates of proteins necessary for the survival of this pest organism. Hence, the host organism is altered to express mimetic peptides, which proves to be fatal for the pest. As, in the body of the vector protein binds with the midgut microvilli membrane or salivary enzymes of the vectors to bind protein transport in the insect, leading to its death.
- Paratransgenesis is the process of introducing antipathogenic molecules in an insect's gut to block the transmission of bacterial control.
- Fungus infection (thief against thief theory): Entomopathogenic fungus is used to kill the insect; some other fungi used to kill insects include *Pseudogibellula formicarium* and *metarhizium anisopliae.*

- Organic combat against PD: These are natural existing parasites that work against Organic combat against PD: These are natural existing parasites that work against Pierce's disease as they are the natural enemies of the insect population (Kyrkou et al., 2018).

10.4.1.6 BACTERIAL LEAF BLIGHT

Bacterial blight is a vascular disease resulting in tannish gray to white lesions along the veins. The most severe form of this disease is Kresek (wilting) which occurs when the roots and leaves are damaged and infected during transplanting at the seedling stage or tillering stage (Mew et al., 1993). The disease is most prevalent in both irrigated and rained arenas, most prominently in wet seasons.

Causative Organism:
Xanthomonas oryza PV. *oryzae*, the causative bacterium was found to cause disease in rice initially. This is a worldwide disease being caused by this bacterium; the bacterium was most endemic in Asia. In India, it was also endemic in the states of Bihar and Tamil Nadu. This is a perennial bacterium, it can survive in infected rice stubble or plant debris (Nagendran et al., 2013).

Alternate Host: *Leersia oryzoides, Zizania latifolia*

Mechanism
Bacterial blight pathogen enters life primarily through hydathodes. Later, bacteria invade through stomata and enters the xylem vessel only at the late stages of infection when multiplication is limited.

Symptoms

- Symptoms of water-soaked yellowish lesions with waxy margins on the leaf blade.
- Pale yellow color of the leaves is the most commonly seen symptom.
- Appearance of tannish gray to white lesions along the veins.
- Reduced photosynthetic area.
- Empty grains (Mew et al., 1993).

Management

- Pull out the diseased plant, avoid burying diseased plants into paddy soil. Since bacteria can easily survive and can be spread in irrigation water.
- Drain and saturate the soil to reduce humidity.
- Apply a minimal amount of nitrogen fertilizer or withhold further application.

10.4.1.7 CITRUS GREENING

This is the most common disease of citrus fruits also known as Yellow shoot disease, because of its symptom. This disease is a widespread disease persisting since the 19th century. This is the most significant challenge to citrus worldwide. The disease affects the vascular system of the plant (Ghosh et al., 2018).

Causative Agent

Candidatus liberibacter asiaticus is the phloem-limited bacterium, transmitted by psyllid vectors.

Asian Citrus Psyllid (*Diaphorina citri)* is the smallest insect to feed upon leaves. This was first found in Florida in June 1998. The insect vector fields upon healthy trees and injects the bacterium into their phloem. About 3–4 mm in length in adults, 0.3 mm in length in eggs. Their life span is around 30–60 days or even more (Ghosh et al., 2018; da Graca, 1991).

Effects

- Worldwide crop production loss.
- Reduced citrus growth.
- Matured trees when infected, declined in their productivity and become non-productive.
- Young infected trees do not bear fruits (Batool et al., 2007).

Symptoms

- Yellowing of leaves along with the veins.
- The newly formed leaves become small and become upright due to chemotropism.

- Occurrence of nutrition deficiency symptoms.
- Asymmetric blotches, the mottled yellow pattern on leaves.
- Lopsided fruits.
- Color inversion and aborted seeds.
- Premature fruit drop.
- Retention of green color at maturity (Ghosh et al., 2018).

Management

- Control of vectors.
- Healthy propagating materials should be used.
- Quarantined plants (da Graca, 1991).

10.4.1.8 SALMONELLA INFECTION IN PLANTS

- Excretion from a host and subsequent colonization of plants may be part of the life cycle of the bacterium.
- A wide range of biotic and abiotic environmental factors can influence the complex interactions between *Salmonella* and plants.
- Fresh produce contaminated with *Salmonella* can trigger a salmonellosis outbreak.
- *Salmonella* is the main cause of disease, for example, foodborne and gastrointestinal disease.
- *Salmonella* was assumed to survive on plants after a more or less accidental infection.
- Plants may be valuable alternate hosts for enteric pathogens by providing a refuge after excretion from the animal intestinal tract. The ability to colonize edible plants may be an effective survival strategy for *Salmonella* as it provides a direct route from its excretion in the environment back to its numerous herbivorous and omnivorous hosts.
- Plants recognize *Salmonella* by detecting conserved microbial patterns that later activate basal defenses (Batool et al., 2007; Brandl et al., 2013).

10.5 DIAGNOSIS/DETECTION OF PLANT BACTERIAL DISEASE

Diagnosis or detection of the disease in plants is one of the most imperative methods of management of plant bacterial disease. Hence, various methods are employed regarding the detection of diseases.

10.5.1 NEED FOR DIAGNOSIS

- Administration of the cure or to cultivate treatments for particular diseases.
- Initiation and applications for control measures to be in use.
- To apply and understand the preventive measures.
- To understand the historical background of the particular disease.
- To develop an opinion regarding a particular disease to ensure the prognosis.
- To ensure the epidemiology and etiology of diseases (Sobiczewski, 2008).

10.5.2 PRINCIPLES OF DIAGNOSIS

1. Identification of the common disease.
2. Observation of signs and symptoms of a particular disease.
3. Ensuring the presence of the pathogen and the specific disease with the help of various diagnostic techniques.
4. Relating the Koch's postulates with these diagnostic principles.

Koch's Principles

- The suspected pathogen must consistently be associated with diseased plants.
- Isolation of microbes and they should be grown in pure culture.
- Culture and microbes should cause disease when it is introduced into a healthy organism.
- The microorganism must be reisolated from the inoculated plant and should be grown in pure culture. The microbe should have the same character as that of the microbe grown in the original culture (Sobiczewski, 2008).

Symptoms and Signs

Symptoms are the visible response of a plant against a causative agent over some time. for example, discoloration or coloration of a specific plant part, dead spots in leaves or barks, change in the shape of a fruit or any other part, swelling in the roots or branches.

Signs indicate the population of pathogen or individual pathogens seen on a host plant indicating some observable features such as 1. Fungal spores, 2. Insects on plants.

Range of Symptoms

1. Symptoms of color change
2. Symptoms of death of the cell
3. Hypertrophy and Hyperplasia
4. Hypotrophy and Hypoplasia

1. **Symptoms of color change:** These include three common symptoms in plants:
    - **Chlorosis:** Appearance of yellowish leaves due to loss of chlorophyll causes inadequacy of leaves to produce food, which causes nutrient deficiencies and that leads to permanent toxicity of the fruit.
    - **Etiolation:** Elongation and yellowing of leaves and stems during dark conditions.
    - **Mosaic:** Light to dark and green to yellow mosaic pattern in leaves.

2. **Symptoms of death of cells:** The death of cells usually occurs in two ways: general necrosis and local necrosis.
    - **General Necrosis:** It occurs in the whole part of the plant, for example, soft rot, dry rot, wilt, blight, etc. as already discussed above in the diseases.
    - **Local Necrosis:** It occurs in the limited part of the plant, for example, leaf spots and cankers, etc.

3. **Hyperplastic deformities:** This is the symptom of abnormal outgrowth of/from a specific part of the plant. This includes hypertrophy and hyperplasia.
    - **Hypertrophy:** Excessive and overgrowth of a plant organ or plant size because of the increase in the cellular sizes, for example, gall (localized swellings), that is, crown gall caused by *Agrobacterium tumefaciens,* Root-knot caused by *Meloidogyne* spp. (Nematoda), Witches Brown (clusters of small shoots) and Curls (twisted or curled leaves), etc.
    - **Hyperplasia:** Excessive and outgrowth of the plant organ due to the increase in the number of cells by rapid multiplications.

4. **Hypoplasia and Hypertrophy:**
   - Hypoplasia is a condition of the retarded growth or the dwarfing of a plant due to reducing cell number.
   - Hypertrophy is a condition of retarded growth for the drafting of plants due to reduced cell sizes (Balodi et al., 2017).

Diagnostic Techniques

The principle on which the diagnostic techniques work is to understand the following factors in the diseased plant:

- Cause of the disease in plants.
- Understanding the physiology of the host and its parasitic nature.
- Understanding the genetics of the host with its parasitic nature.
- Epidemiology and plant disease control measures.

Detection of phytopathogen in a host as well as in the environment, such as soil, water, and the air is required to quantify pathogen inoculum and to assess the effectiveness of methods of managing pathogens. There are various technologies and molecular methodologies that are accountable for the detection of bacterial diseases in plants. This is a very informative step to be followed in the system of management of bacterial diseases in plants. The diagnostic techniques are as given below:

1. **Visual inspection:** This technique uses the plant germplasm and selects the healthy material. The method employed include light imaging, hyperspectral imaging, thermal imaging, and chlorophyll fluorescence imaging. This method is influenced by the environment and biological factors of the pathogen, and is responsible for diagnosing the symptoms being visible in diseased plants (Camargo and Smith, 2009).

2. **Sero diagnostics:** This method is used for accurate detection of the pathogen present in a plant. This is a fast and quick technique for detection, and this method is based on antibodies (immunoglobulins (Ig)) and antigen reactions. The most commonly used method in this technique is ELISA, Dot-Blot immune binding assays, etc. (Camargo and Smith, 2009).

3. **Flowcytometry (FCM):** This is a technique based on optical properties, which uses the laser for cell screening, detection of biomarker proteins; However, this technique is not extensively used in plant

pathology, but still, it finds its application in pathogen detection, in the assessment of bacterial genome size (Camargo and Smith, 2009).

4. **Molecular methods:**

A. **PCR (Polymerase Chain Reaction)**

This is a primary directed method. Enzyme reactions in this method influence the exponential replication of DNA. This method is the very sensitive, faster, and ideal method used for the detection of bacteria. This method has certain DNA-based detection techniques, such as fluorescence in situ hybridization (FISH) and nucleic acid sequence-based application (NASBA) (Camargo and Smith, 2009; Schaad and Frederick, 2002).

B. **Real-time PCR**

This is a seed detection assay. This method is more reliable in terms of understanding the epidemiology diagnostics of specific diseases caused by a specific bacterium. This method employs the quantification of the genetic material (DNA or RNA), applications of information computational technology, mapping, and analysis of different DNA and protein sequences. This helps in the formation of a 3D structure of molecules to understand its chemical cause and interference. Some fluorescence methods are based on hybridization of the fluorescently labeled probe, such as fluorescence resonance energy transfer (FRET), molecular beacons. (Miller and Martin, 1988).

C. **Colloidal Gold:**

This is the immune histochemical marker in electron microscopic studies of bacterial pathogenicity. The entire technique is based on an antibody tag for dot-immunobinding (Miller and Martin, 1988).

D. **Immunofluorescence Assay**

Indirect immunofluorescence assay (IFA) is to detect the pathogenicity of the pathogen; pathogen-specific antibodies are conjugated with fluorescent dye molecules (Miller and Martin, 1988; Schaad and Frederick, 2002).

E. **Immuno Reagents**

These are the most crucial components of any immunoassay. Hence, in immuno reagents, polyclonal antibodies are used to contain many antibodies that react with different determinants of the immunogens. Hence, standardized antibody reagents with appropriate specificity and sensitivity are used (Schaad and Frederick, 2002).

F. Other Techniques

Certain nucleic acid-based detection techniques are used in plant pathology along with molecular biology to read and detect the genomic sequences, genotypes, and phenotypes of the specific pathogen to relate the pathogen with the specific type of disease found in a specific plant. Some of the techniques are:

- Dot-Blot assay
- Non-radioactive labels
- Restriction fragments length polymorphism
- Synthetic probes (Schaad and Frederick, 2002)

10.6 DISEASE CONTROL AND TREATMENTS

The main principle of studying the pathogen, their behavior, functioning, and the disease they are causing is to understand how diseases can be controlled. Whereas, the conservation of plant population is the target of the control and management system. Control of the disease is essential to ensure the existence of the particular plant of the crop so that it can be conserved for the ecosystem, and to maintain its yield and economic importance (Tsegaye et al., 2018).

10.6.1 VANDERPLANCK'S EQUIVALENCE THEOREM

1. Effects of the pathogen on host and environment can be translated in terms of rate parameter of an epidemic.
2. Changes in any component create the equivalent effect on the disease.
3. The effects of the disease depend upon the susceptible host, aggressive pathogen, favorable environment, to what extent it can cause disease to a host, and what environmental conditions are supporting the efficiencies of the pathogen.

10.6.2 THE BASIC PRINCIPLE OF DISEASE CONTROL

Based on the effect of the development of resistance to the control measures by the pathogens, the measures of the principles are divided into two main groups:

1. **Eradicated control measures:** This principle works on eliminating the entire pathogen population from a specific plant.
2. **Management control measures:** This principle works on eliminating or destroying a portion of the pathogen population (Tsegaye et al., 2018).

10.6.3 TYPES OF DISEASE CONTROL

1. **BIOLOGICAL CONTROL**

   Biological hosts, entities, antagonists, and microorganisms are manipulated.

   a. **Host resistance:** Distance mechanism based on genetic composition.
      Host resistance is basically of two types described by Vander Plank:

      i. **Vertical resistance:** Resistance is effective against some races of pathogens only and does not entirely reduce the rate of disease development.

      ii. **Horizontal resistance:** Resistance is effective against all races of pathogens and can entirely decrease the effects of diseases.

   b. **Tolerance:** Disease plants do not deal less than a healthy plant.

   c. **Induced resistance:** Any susceptible plant is treated with the antipathogenic stream which gives it resistance to decrease the disease effect.

   d. **Antagonists:** Microbial control for inhibiting the development, growth, and other life processes of the pathogen. In plant pathology, the term applies to the use of microbial antagonists to suppress diseases as well as the use of host-specific pathogens to control weed populations. In both fields, the organism that suppresses the pest or pathogen is referred to as the biological control agent (BCA).

e. **Controlling the insect vectors:** Various insects are responsible for injecting the specific bacteria into the plant, for example, leafhoppers. Hence, control of these insects can reduce the spread of disease.

f. **Antibiotics:** Pathogens can be inhibited through the antibodies produced by the antagonists.

g. **Types of interactions:** Contributing to biological control are mutualism, commensalism, protocooperation, neutralism, antagonism, competition, parasitism, and predation.

h **Induction of host resistance:** Plants can actively respond to various environmental stimuli, including gravity, light, water, nutritional availability. Plants also respond to various chemical stimuli produced by soil and plant-associated motive. This induction of host resistance can be localized and systemic depending upon the amount of stimulus (Pal and McSpadden Gardener, 2006).

10.6.4 CULTURAL CONTROL

Cultural control is called physical control which includes the manipulation of the environment. Some of the cultural control techniques are as follows:

1. Crop rotation
2. Seedling rate and canopy density
3. Selection of the location
4. Selection of the season
5. Irrigation
6. Controlling insects and weeds
7. Sanitation
8. Heating or refrigeration

There are various physical methods for the control of pathogens and diseases caused by them. Some of them are as follows:

- **Hot water treatment:** The most efficient method for the control of seed-borne pathogens, especially bacteria.

- **Hot air treatment:** This method is also employed for the control of seed-borne disease; however, hot air treatment is less injurious to seed and easy to operate. But it is less effective than hot water treatment.
- **Moist hot air treatment:** This is the most effective method used in sugarcane to eliminate grassy shoot disease.
- **Refrigeration:** Low-temperature freezing point checks the growth and activities of all pathogens, this is one of the most interesting methods where fruits and other vegetables are transported and stored in refrigerated vehicles to avoid the contamination of the fruits or vegetables from pathogenic effect (Sbragia, 1975).

10.6.5 CHEMICAL CONTROL

Pesticides are the chemicals that are used in killing pests. These include fungicide bactericide, and nematicides.

- In chemical control, the most commonly used chemical is antibiotics, which are the chemicals produced by microorganisms that destroy living organisms, particularly bacteria.
- Antibiotics were first described by Alexander Fleming in 1929, however, the search for antibiotics was proven to be serendipity.
- *Properties of antibiotics:* Antibiotics are effective in small amounts, they are specific, there are mostly antibacterial but some are also antifungal; their actions are systemic, they may be temperature-sensitive also.
- *Mode of action:* They have direct action inside the plant tissue and on the plant surface. Their action is after transformation within the plants, the direction is directly on the host sometimes.
- **Examples:** Tetracycline, streptomycin, gentamicin, and penicillin bacitracin.
- Streptomycin is an effective medicine against a few fruit pathogens, such as blights and cankers, and cycloheximide can be used to control some `pathogens of crops.
- For the treatment of seeds, the following chemicals are useful:

Seed disinfectants—seeds must be treated with disinfectants to avoid internal and external seed-borne disease.

Seed protectants—used to protect seed from soil-borne pathogens.

10.6.6 LEGISLATIVE AND REGULATORY CONTROL

This is a regulatory method applied for preventing the plant from getting affected by pathogens repeatedly. This method prevents the pathogens entering the affected regions of the plant, this process is also called exclusion.

Legislative management also focuses on reducing the use of pest species. It can be achieved in the following ways:

- Legislation for the prevention from the introduction of foreign pests, disease, and weeds.
- Legislation to prevent the spread of already established pest diseases and pests within the country or a particular state.
- Legislation for notified campaigns of control against pests.

There are two categories of regulatory measures that are operational for the control of pests and diseases and weeds in India:

1. Legislative measures through Plant Quarantine
2. Legislative measures through State Agricultural Pest and Disease Act.

The Insecticides Act of India, 1986

This act was passed by the parliament for regulating the following objectives:

1. To ensure that the farmers are getting quality products.
2. Minimizing the health hazardous pests from contaminating the plant and its nutritional value.
3. To register only safe and efficacious pesticides (Dehnen-Schmutz et al., 2010).

10.7 MANAGEMENT OF BACTERIAL DISEASES IN PLANTS

Management of bacterial disease is a multiplex and relative approach towards the control and treatments of the diseases in various plants and crops. The continuously increasing human population required stable supplements of food with greater yields and good nutritional quality. Hence, management of the bacterial disease is playing a handful role in achieving this target.

There are many methods and approaches made to achieve this goal.

10.7.1 MOLECULAR PATHOLOGY: AN APPROACH TO BACTERIAL DISEASE MANAGEMENT IN PLANTS

The management of bacterial disease remains a formidable grand challenge for plant pathologists, so many plant pathologists have employed various molecular pathological techniques.

1. The implication of horizontal gene transfer in introducing copper resistance gene is the best way of management of bacterial disease with copper and antibiotics. Chromosomal mutation altering the ribosomal protein is targeted by the antibiotics in the bacteria.
2. SAR, induced systemic resistance (ISR), is an alternative technique of using copper and antibiotics as the above technique does not play a very significant role in plant disease management. In SAR, commercially available chemical inducers are used such as acibenzolar S-methyl (ASM). ASM is successful with its application of inducing bacterial canker of tomato, bacterial wilt of tomato, and fire blight.
3. **Antimicrobial peptides** are small peptides synthesized by bacteria, fungi, oomycetes, as a part of the innate immunity system in response to microbes. Mostly, these peptides are cationic and get inserted into the cell membranes and disrupt the nucleic acid in protein synthesis of the microorganism resulting in its death.
4. **Targeting bacterial biofilms,** biofilm formation is responsible for protecting bacteria from the environmental stress, antimicrobial drugs, and host defense mechanisms; it is proven to be fatal for plants. Therefore, the development of inhibitors of biofilm formation is great and attractive research is progressing toward the control of bacterial growth in the modern molecular biological aspects (Sundin et al., 2016).

10.7.2 BIOTECHNOLOGICAL APPROACHES FOR BACTERIAL DISEASE MANAGEMENT IN PLANTS

1. *Modulating quorum sensing and type III secretion symptoms in bacterial plant pathogens for disease management:* Bacteria generally do not live in colony, it lives a unicellular organism; bacteria

rapidly respond to changes in any molecular configuration and any molecular mixtures into their regulatory network. Quorum sensing (QS) is a cell-to-cell communication process in bacteria that helps bacteria to communicate through their neighboring cells. It is beneficial to a large group of bacteria as it is a cell density-dependent process; QS involves secret signaling molecules which are called autoinducers (AIs) to regulate the gene expression collectively. QS relies on the bacterial perception of an external signaling molecule; therefore, this process represents an excellent potential targeting for manipulation, either through signal degradation or overproduction of signaling molecule. So many AIs degrading enzymes are used as QS-controller traits such as quorum-quenching enzymes which were used as anti-QS strategies in plant disease management.

2. Molecular detection strategies for phytopathogenic bacteria: Diseases in plants are caused by a wide member of each logical agent which majorly includes bacteria and fungi. The nature and biology of these agents determine which potential control strategies can be applied to prevent the spread and recurrence of plant disease. Various strategies are employed in it. Some of which are:

   - **In traditional methods:** This technique involves the inspection for symptoms and signs of disease along with certain laboratory tests such as certain serological tests (ELISA, immunofluorescence microscopy).
   - **Assay formats:** This is for amplifying pathogen-specific nucleic acid involves conventional PCR, immunocapture PCR, PCR-ELISA, BIO-PCR, nested PCR, multiplex PCR, competitive PCR, real-time PCR, NASBA.
   - **Molecular-based methods:** These methods are based on extraction and purification of bacterial DNA from infected plant material. The following techniques are employed under this category:

     a. Immunocapture PCR.
     b. Chelex extraction methods involve tissue destruction of bacteria by excessive heat.
     c. Charge switch technology.
     d. Enzyme protein digestion using proteinase (Sundin et al., 2016; Punja et al., 2007; Woo et al., 2006).

3. **Potential disease control strategies revealed by genome sequencing and functional genetics of plant pathogenic bacteria:** Many bacteria have a symbiotic relationship with the plant but many bacteria are the disease-causing agent. Hence, various technologies are employed for disease control in plants. Some of the methods are as follows:

    - Identification of bacteria by high-throughput screens.
    - For turning of bacterial defenses, degradation of the bacteria such as Gram-positive and Gram-negative bacteria is the first and foremost step; the inhibition of structure and function by using bacterial syringes is one of the major steps in molecular and biotechnology approaches.
    - Various bacterial inhibitors such as type III secretion system inhibitors, pat 1 inhibitors, various enzyme inhibitors, and detoxifying toxins are used to inhibit bacterial actions over the plant.
    - There are pigment production inhibitors as both Gram-negative and positive pathogens produce pigments such as carotenoids and lycopene. These pigments help in protecting the cell from ultraviolet radiation and oxidative damages and are important for epiphytic colonization or their pathogenicity.
    - **Ice nucleation and frost damage:** Bacteria not only cause necrotic spots on leaves, fruits, and seeds; but some bacterial pathogens can cause frost damage to plants by nucleating ice on a cell surface protein array. These ice-nucleating strains are responsible for frost damages on crops during very low temperatures. Hence, bacterial antifreeze proteins are used to reduce the size of ice crystals rather than inhibiting freezing, indicating a good approach in biotechnology for the opening of possibilities to control frost damage on crops (Punja et al., 2007).

10.7.3 SUSTAINABLE MANAGEMENT OF BACTERIAL DISEASES IN PLANTS

Disease management practices are ground for the contribution to sustainability by protecting crop yields and by maintaining and improving the profitability of crops. There are various practices for the management of crop diseases that help in sustainable management; these practices involve host plant resistance, biological practices and control, cultural control, and chemical control. Various sustainable practices are used to achieve this goal sustainably without causing harm to the environment.

1. Boosting plant recognition of infection by using engineering resistance strategies.
2. Disarming the host susceptibility genes.
3. Producing antimicrobial compounds to produce the resistance to diverse ranges of pathogens, the major advantage of antimicrobial substances that one can employ in which the techniques of molecular evolution to broaden the range of molecular targets of such antimicrobials.
4. Silencing of the essential gene in pathogens, the gene which is responsible for causing the strains to the plant is suppressed. The double-stranded RNA in the cytoplasm of eukaryotic cells triggers the natural and targeted process of post-transcriptional gene silencing.
5. Modifying host targets of pathogenicity by producing molecules that play a role in virulence by binding to host target molecules. The genetic modification of targets of pathogen virulence factors increases host resistance without introducing an exogenous biochemical pathway into the plant.
6. Detoxification of the toxins produced by the pathogens by disrupting the important biochemical processes. The best example is phytotoxin. Oxalic acid is central to the pathogenicity of *Chryphonectria parasitic* and the cause of catastrophic epidemics of chestnut blight. But significantly, the disease development has decreased by the introduction of the transformed gene coding for the production of the grated enzyme called oxalate oxidase (Vincelli, 2016).

10.8 CONCLUDING REMARKS

Bacteria are the most successful life forms on the earth by their ubiquity; bacteria also define variability and strains are considered to be within a single species if they are only 70% genetically identical. Bacteria are the most imperative microorganisms responsible for a larger number of beneficial activities. But every coin has two faces, there is similarly a wide range of bacteria that can be stated as the bacterial pathogen, disease-causing bacteria. Over 90% of the human and animal diseases are caused by bacteria, where around 40% of the total plant diseases are bacterial disease. Bacteria were responsible for causing innumerable plant diseases, such as blight, spots, gall, Pierce's disease, rots, blotch. Much etiological and epidemiology evidence is found to be most evident and responsibly helpful in producing control against bacterial diseases. Bacterial disease management in plants, with all those techniques that are helping the biological sciences, agricultural sciences, many botanists, and various biotechnologists leads toward the control of drastic diseases caused by bacteria. The management of bacterial diseases is not only essential for the conservation of plants but also essential for increasing the yields of the plant, especially crops, in enhancing the nutritional value and productivity value in crop plants. The purpose of the management of bacterial diseases is to ensure the safety of human beings, as human beings are consumers of the plants, and consuming diseased plants would rather be injurious to their health. Disease management is also dealing with the conservation of the economy of the world. As the disease in various plants and crops leads to worldwide loss of the economy.

Hence, various physical, biological, and chemical methods are employed to control the spread of disease. With this perspective of control and management of the bacterial disease in plants, plant pathology is giving its real-time participation through molecular, biotechnological, serological, genetic, and pathological approaches.

KEYWORDS

- **bacterial plant diseases**
- **bacterial spots**
- **bacterial wilts**
- **leaf blotch**
- **bacterial soft rot**
- **prokaryotic organisms**
- **seed-borne diseases**

REFERENCES

Agyemang, P. A.; Kabir, M. N.; Kersey, C. M.; Dumenyo, C. K. The Bacterial Soft Rot Pathogens, Pectobacterium Carotovorum and P. atrosepticum, Respond to Different Classes of Virulence-Inducing Host Chemical Signals. *Horticulturae* **2020,** *6,* 1–13.

Alvarez, A. M. Integrated Approaches for Detection of Plant Pathogenic Bacteria and Diagnosis of bacterial diseases. *Annu. Rev. Phytopathol.* **2004,** *42,* 339–366.

Balodi, R.; Bist, S.; Ghatak, A.; Rao, K. H. Plant Disease Diagnosis: Technological Advancements and Challenges. *Indian Phytopath.* **2017,** *70* (3), 275–281.

Barton, I. S.; Fuqua, C.; Platt, T. G. Ecological and Evolutionary Dynamics of a Model Facultative Pathogen: Agrobacterium and Crown Gall Disease of Plants. *Environ. Microbiol.* **2018,** *20* (1), 16–29.

Bashan, Y.; Puente, M. E.; de-Bashan, L. E.; Hernandez, J.-P. Environmental Uses of Plant Growth-promoting Bacteria. *Plant Microbe Interac.* **2008,** 69–93.

Batool, A. Y.; Mughal, M.; Khan, M. M.; Jaskani, M. J.; Abbas, M.; Khan, I. A. Citrus Greening Disease—A Major Cause of Citrus Decline in the World—A Review. *Hort. Sci. (Prague)* **2007,** *34* (4), 159–166.

Berg, G.; Mahnert, A.; Echinger, C. M. Beneficial Effects of Plant-associated Microbes on Indoor Microbiomes and Human Health? *Front. Microbiol.* **2014,** *5* (15), 1–5

Brandl, M. T.; Cox, C. E.; Teplitski, M. Salmonella Interactions with Plants and Their Associated Microbiota. *Phytopathol.* **2013,** *103* (4), 316–325.

Braun, A. C. Tumor Inception and Development in the Crown Gall Disease. *Annu. Rev. Plant. Physiol.* **1962,** *13,* 533–558.

Brunvold, L.; Sandaa, R.-A.; Mikkelsen, H.; Welde, E.; Bleie, H.; Bergh, Ø. Characterisation of Bacterial Communities Associated with Early Stages of Intensively Reared Cod (Gadus morhua) Using Denaturing Gradient Gel Electrophoresis (DGGE). *Aquaculture* **2007,** *272,* 319–327.

Burr, T. J.; Caesar, A.; Schrolh, M. N. Beneficial Plant Bacteria. *Crit. Rev. Plant Sci.* **1984,** *2* (1), 1–20.

Caillaud, M.-C.; Dubreuil, G.; Quentin, M.; Perfus-Barbeoch, L.; Lecomte, P.; Engler, J. D. A.; Abad, P.; Rosso, M.-N.; Favery, B. Root-knot Nematodes Manipulate Plant Cell Functions During a Compatible Interaction. *J. Plant Physiol.* **2008,** *165,* 104–113.

Camargo, A.; Smith, J. S. An Image-processing based Algorithm to Automatically Identify Plant Disease Visual Symptoms. *Biosyst. Eng.* **2009,** *102,* 9–21.

Chung, Y. S.; Goeser, N. J.; Cai, X.; Jansky, S. The Effect of Long Term Storage on Bacterial Soft Rot Resistance in Potato. *Am. J. Potato Res.* **2013,** *90,* 351–356.

da Graca, J. V. Citrus Greening Disease. *Annu. Rev. Phytopathol.* **1991,** *29,* 109–136.

Danos, E.; Berger, R. D.; Stall, R. E. Temporal and Spatial Spread of Citrus Canker Within Groves. *Phytopathology* **1984,** *74,* 904–908.

Das, A. K. Citrus Canker—A Review. *J. Appl. Hort.* **2003,** *5* (1), 52–60.

Davis, R. E.; Sinclair, W. A. Phytoplasma Identity and Disease Etiology. *Phytopathology* **1998,** *88* (12), 1372–1376.

Dehnen-Schmutz, K.; MacLeod, A.; Reed, P.; Mills, P. R. *The Role of Regulatory Mechanisms for Control of Plant Diseases and Food Security—Case Studies from Potato Production in Britain.* Springer; 2010.

Dhingra, O. D.; Sinclair, J. B.; *Basic Plant Pathology Methods*; 2nd ed.; CRC Press, 2019, p 448.

Fox, G. E.; Stackebrandt, E.; Hespell, R. B.; Gibson, J.; Maniloff, J.; Dyer, T. A.; Wolfe, R. S.; Balch, W. E.; Tanner, R.; Magrum, L.; Zablen, L. B.; Blakemore, R.; Gupta, R.; Bonen, L.; Lewis, B. J.; Stahl, D. A.; Luehrsen, K. R.; Chen, K. N.; Woese, C. R. *Science* **1980,** *209,* 457–463.

Gelvin, S. B. Crown Gall Disease and Hairy Root Disease: A Sledgehammer and a Tackhammer. *Plant Physiol.* **1990,** *92* (2), 281–285.

Ghosh, D. K.; Motghare, M.; Gowda, S. Citrus Greening: Overview of the Most Severe Disease of Citrus. *Adv. Agric. Res. Technol. J.* **2018,** *2* (1), 83–100.

Glick, B. R. Plant Growth-Promoting Bacteria: Mechanisms and Applications, *Scientifica* **2012,** 1–15.

Gottwald, T. R.; Graham, J. H.; Schubert, T. S. Citrus Canker: The Pathogen and Its Impact. [Online]. *Plant Health Progress*; 2002.

Gupta, R. S. Protein Phylogenies and Signature Sequences: A Reappraisal of Evolutionary Relationships among Archaebacteria, Eubacteria, and Eukaryotes. *Microbiol. Mol. Biol. Rev.* **1998,** *62* (4), 1435–1491.

Haverkort, A. J.; Struik, P. C.; Visser, R. G. F.; Jacobsen, E. Applied Biotechnology to Combat Late Blight in Potato Caused by Phytophthora Infestans. *Potato Res.* **2009,** *52,* 249–264.

Hayat, R.; Ali, S.; Amara, U.; Khalid, R.; Ahmed, I. Soil Beneficial Bacteria and their Role in Plant Growth Promotion: A Review. *Ann. Microbiol.* **2010,** *60,* 579–598.

Hayward, A. C. Biology and Epidemiology of Bacterial Wilt Caused by Pseudomonas Solanacearum. *Annu. Rev. Phytopathol.* **1991,** *29,* 65–87.

Himel, R. M.; Khan, A. A.; Akanda, A. M.; Karim, M. Characterization and Identification of Soft Rot Bacterial Pathogens of Different Fruits in Bangladesh. *IJB* **2016,** *9* (1), 1–9.

Himel, R. M.; Khan, A. A.; Reza, M. E. Effect of Chemicals Against Bacterial Soft Rot of Fruits. *J. Biosci. Agric. Res.* **2017,** *13* (1), 1087–1091.

Hirano, S. S.; Upper, C. D. Ecology and Epidemiology of Foliar Bacterial Plant Pathogens. *Ann. Rev. Phytopathol.* **1983,** *21,* 243–269.

Hogg, S. *Essential Microbiology*; West, S., Ed.; John Wiley and Sons; 2005.

Hopkins, D. L.; Purcell, A. H. Xylella fastidiosa: Cause of Pierce's Disease of Grapevine and other Emergent Disease. *Plant Dis.* **2002,** *86* (10), 1056–1066.

Jarvis, R. M.; Royston, G. Characterisation and Identification of Bacteria Using SERS. *Che. Soc. Rev.* **2008,** *37,* 931.

Kim, B.-S.; French, E.; Caldwell, D.; Harrington, E. J.; Iyer-Pascuzzi, A. S. Bacterial Wilt Disease: Host Resistance and Pathogen Virulence Mechanisms. *Physiol. Mol. Plant Pathol.* **2016,** *95,* 37–43.

Koch, A. L. *The Bacteria: Their Origin, Structure, Function and Antibiosis*; Springer, 2006.

Kumar, A.; Rawal, R.; Kumar, A.; Kumar, N.; Verma, D. K.; Talukdar, D. Xanthomonas Citri: the Pathogen of Citrus Canker Disease and its Management Practices. In *Microbiology for Sustainable Agriculture, Soil Health, and Environmental Protection*; Apple Academic Press: USA, 2019; pp 189–222.

Kyrkou1, I.; Pusa, T.; Ellegaard-Jensen, L.; Sagot, M.-F.; Hansen, L. H. Pierce's Disease of Grapevines: A Review of Control Strategies and an Outline of an Epidemiological Model. *Front. Microbiol.* **2018,** *9,* 1–23.

Margulis, L. Whittaker's Five Kingdoms of Organisms: Minor Revisions Suggested by Considerations of the Origin of Mitosis. *Evolution* **1971,** *25* (1), 242–245.

Mew, T. W.; Alvamz, A. M.; Leach, J. E.; Swlngs, J. Focus on Bacterial Blight of Rice. *Plant Dis.* **1993,** *77* (1), 1–12.

Miller, S. A.; Martin, R. R. Molecular Diagnosis of Plant Disease. *Ann. Rev. Phytopathol.* **1988,** *26,* 409–432.

Nagendran, K.; Karthikeyan, G.; Peeran, M. F.; Raveendran, M.; Prabakar, K.; Raguchander, T. Management of Bacterial Leaf Blight Disease in Rice with Endophytic Bacteria. *World Appl. Sci. J.* **2013,** *28* (12), 2229–2241.

Odelade, K. A.; Babalola, O. O. Bacteria, Fungi and Archaea Domains in Rhizospheric Soil and their Effects in Enhancing Agricultural Productivity. *Int. J. Environ. Res. Public Health* **2019,** *16,* 3873.

Pal, K. K.; McSpadden Gardener, B. *Biological Control of Plant Pathogens*; The Plant Health Instructor; 2006; pp 1–25.

Punja, Z. K.; De Boer, S. H.; Sanfacon, H. *Biotechnology and Plant Disease Management*; Hardback 2007; pp 1–580.

Romantschuk, M. Attachment of Plant Pathogenic Bacteria to Plant Surfaces. *Annu. Rev. Phytopathol.* **1992,** *30,* 225–243.

Rowan, F. E.; Docherty, N. G.; Coffey, J. C.; O'Connell, P. R. Sulphate-Reducing Bacteria and Hydrogen Sulphide in the Aetiology of Ulcerative Colitis. *Br. J. Surg.* **2009,** *96,* 151–158.

Santos, M. L. D.; Berlitz, D. L.; Wiest, S. L. F.; Schunemann, R.; Knaak, N.; Fiuza, L. M. Benefits Associated with the Interaction of Endophytic Bacteria and Plants. *Braz. Arch. Biol. Technol.* **2018,** *61,* 1–11.

Sbragia, R. J. Chemical Control of Plant Diseases: An Exciting Future. *Annu. Rev. Phytopathol.* **1975,** *13,* 257–269.

Schaad, N. W.; Frederick, R. D. Real-time PCR and Its Application for Rapid Plant Disease Diagnostics. *Can. J. Plant Pathol.* **2002,** *24,* 250–258.

Sharma, S. Food Preservatives and their Harmful Effects. *Int. J. Sci. Res. Public.* **2015,** *5* (4), 1–2.

Shoda, M.; Bacterial Control of Plant Diseases. *J. Biosci. Bioeng.* **2000,** *89* (6), 515–521.

Simeon, A. U.; Abubakar, A. Evaluation of Some Plant Extracts for the Control of Bacterial Soft Rot of Tubers. *Am. J. Exp. Agric.* **2014,** *4* (12), 1869–1876.

Singh, V. K.; Pandey, P. Physical Methods in Management of Plant Diseases. In *Eco-friendly Innovative Approaches in Plant Disease Management*; 2012, pp 21–30.

Sobiczewski, P. Bacterial Diseases of Plants: Epidemiology, Diagnostics and Control. *Zemdirbyste Agric.* **2008,** *95* (3), 151–157.

Su, C.-C.; Chang, C. J.; Chang, C-M.; Shih, H.-T.; Tzeng, K.-C.; Jan, F.-J.; Kao1, C.-W.; Deng, W.-L. Pierce's Disease of Grapevines in Taiwan: Isolation, Cultivation and Pathogenicity of Xylella fastidiosa. *J. Phytopathol.* **2013,** *161,* 389–396.

Sucher, J.; Boni, R.; Yang, P.; Rogowsky, P.; Buchner, H.; Kastner, C.; Kumlehn, J.; Krattinger, S. G.; Keller, B. The Durable Wheat Disease Resistance Gene Lr34 Confers Common Rust and Northern Corn Leaf Blight Resistance in Maize. *Plant Biotechnol. J.* **2017,** *15,* 489–496.

Sundin, G. W.; Castiblanco, L. F.; Yuan, X.; Zeng, Q.; Yang, C.-H. Bacterial Disease Management: Challenges, Experience, Innovation and Future Prospects. *Mol. Plant Pathol.* **2016,** *17* (9), 1506–1518.

Tavares, T. D.; Antunes, J. C.; Padrao, J.; Ribeiro, A. I.; Zille, A.; Amorim, M. T. P.; Ferreira, F.; Felgueiras, H. P. Activity of Specialized Biomolecules against Gram-Positive and Gram-Negative Bacteria. *Antibiotics* **2020,** *9,* 314.

Tinzaara, W.; Gold, C. S.; Ssekiwoko, F.; Tushemereirwe, W.; Bandyopadhyay, R.; Abera, A.; Eden-Green, S. J. Role of Insects on Transmiion of Banana Bacterial Wilt. *Afr. Crop Sci. Jo.* **2006,** *14* (2), 105–110.

Tsegaye, Z.; Assefa, F.; Tefera, G.; Alemu, T.; Gizaw, B.; Abatenh, E. Concept, Principle and Application of Biological Control and their Role in Sustainable Plant Diseases Management Strategies. *Int. J. Res. Stud. Biosci.* **2018,** *6* (4), 18–34.

Vincelli, P. Genetic Engineering and Sustainable Crop Disease for Case-by-Case Decision-Making. *Sustainability* **2016,** *8* (495), 1–22.

Wallace, H. R. The Diagnosis of Plant Diseases of Complex Etiology, *Ann. Rev. Phytopathol.* 1978. *16,* 379–402.

Whittaker, R. H. New Concepts of Kingdoms of Orgs. *Science* **1969,** *163* (3863), 150–160.

Woese, C. R.; Kandlert, O.; Wheelis, M. L. Towards a Natural System of Organisms: Proposal for the Domains. *Proc. Natl. Acad. Sci.* **1990,** *87,* 4576–4579.

Woo, S. L.; Scala, F.; Ruocco, M.; Lorito, M. The Molecular Biology of the Interactions Between Trichoderma spp., Phytopathogenic Fungi, and Plants. *Phytopathology* **2006,** *96* (2), 181–185.

Yang, D. C.; Blair, K. M.; Salamaa, N. R. Staying in Shape: the Impact of Cell Shape on Bacterial Survival in Diverse Environments. *Microbiol. Mol. Biol. Rev.* **2016,** *80* (1), 187–203.

Young, K. D. Bacterial Morphology: Why have Different Shapes? *Curr. Opin. Microbiol.* **2007,** *10* (6), 596–600.

Young, K. D. Bacterial shape. *Mol. Microbiol.* **2003,** *49* (3), 571–580.

Zekri, M.; Dewdney, M.; Burrow, J.; Roberts, P. *Field Identification of Citrus Canker Symptoms and Decontamination Procedures*; IFAS Extn: University of Florida, 214.

CHAPTER 11

Molecular Approaches for Rapid Detection and Identification of Phytoplasmas Associated with Crop Diseases

NGUYEN NGOC BAO CHAU[1] and NGUYEN BAO QUOC[2,3]

[1]*Faculty of Biotechnology, Ho Chi Minh City Open University, Ho Chi Minh City, Vietnam*

[2]*Research Institute for Biotechnology and Environment, Nong Lam University, Ho Chi Minh City, Vietnam*

[3]*Faculty of Biological Sciences, Nong Lam University, Ho Chi Minh City, Vietnam*

ABSTRACT

Phytoplasmas are phloem-limited bacterial plant pathogens that lack a cell wall and can be transmitted between plants by insect vectors, causing severe diseases in a wide range of plant species. Due to problems and difficulties in establishing axenic cultures of phytoplasmas belonging to the class Mollicutes living in the phloem of host plants, erratic contribution, low concentration, seasonal fluctuation, and the inhibitory effects of plant polysaccharide and polymorphic compounds, especially in woody perennial plant hosts, their detection and identification are not always easy. Therefore, innovative approaches for rapid, sensitive, accurate, and early diagnosis of phytoplasma diseases are indispensable to reduce their damage to crops. In the present chapter, several serological and molecular techniques for accurate and sensitive detection and identification of phytoplasmas in plants are reviewed.

Plant Diseases and Their Management: A Sustainable Approach.
Deepa Srivastava, Rajarshi Kumar Gaur, and Ajay K. Tiwari (Eds.)

11.1 INTRODUCTION

In 1967, phytoplasmas were discovered by Japanese scientists and considered agents with structural characteristics similar to mycoplasma-like organisms (MLOs). Phytoplasmas belong to the genus *Mollicutes*, which are free-living and self-replicating plant pathogens without a cell wall and surrounded by a single membrane (80–800 nm) (Doi et al., 1967). They have the closest relationship with bacteria such as *Bacillus*, *Clostridium*, and *Streptococcus*. The locations of phytoplasmas known as intracellular parasites are normally observed in the nutrient-rich cytoplasm of phloem cells and hemolymph in infected plants and insect vectors, respectively (Hogenhout and Music, 2010). Phytoplasmas are transmitted by sap-sucking insects, mainly leafhoppers and planthoppers of the order Hemiptera (Bosco and Tedeschi, 2013). Phytoplasma diseases in plants are generally seen as "yellows," a common yellowing phenomenon in plants that scientists initially thought was probably caused by a virus, now called phytoplasma. Phytoplasma interferes with the development of plants through typical symptoms including phyllody (flowers that turn into leaves), virescence (instead of having normal colors, the development of green pigment is observed on flowers), witches' broom (dense growth of shoots from a single point), an anomalous number of hairy roots, inhibition of flowering, stunted plants, proliferation of shoots, reduced size of flowers, and rottenness (Bertaccini and Duduk, 2009; Hogenhout et al., 2008; Chaturvedi et al., 2010; Bertaccini, 2015). Although there is a loss of many essential metabolic pathways in the genome of phytoplasmas, indicating their dependence on a nutrient-rich host, many essential genes encoding DNA replication, transcription, translation, protein translocation, transport systems, immunodominant membrane proteins (IDPs), AMP/IMP plant actin-binding proteins, and repetitive extragenic palindromes (PhREPs) were found and reported in their genomes. These reveal important roles in phytoplasma interaction with the host, vector specificity, motility, and phytoplasma adaptation to variable niches (Hogenhout et al., 2008; Kube et al., 2012; Oshima et al., 2004; Bai et al., 2006; Christensen et al., 2005; Razin, 2007; Kakizawa et al., 2006; Galetto et al., 2011; Bconrod et al., 2012; Jomantiene and Davis, 2006; Jomantiene et al., 2007).

In the 1970s, many studies were focused on the diagnosis through symptoms and observations under an electron microscope to visualize phytoplasma in phloem cells in host plants and the expansion of phytoplasma-associated diseases on crops. At that time, the biology and classification of phytoplasma pathogens were still a mystery. Until now, the nature of phytoplasma has been discovered through the advances of molecular revolutions in the 1980s.

Serological and molecular analyses including enzyme-linked immunosorbent assay (ELISA) (Chang et al., 1995; Loi et al., 2002; Brzin et al.2003), conventional polymerase chain reaction (PCR), heteroduplex mobility assays (HMAs) (Wang and Hiruki, 2001), single-strand conformation polymorphisms (SSCP) (Music and Skorix, 2013), terminal restriction fragment length polymorphism (t-RFLP) (Hodgetts et al., 2006), PCR and real-time PCR (Marzachi et al., 2008; Cordova et al., 2003; Clair et al., 2003; Christensen et al., 2004), loop-mediated isothermal amplification (LAMP) and recombinase polymerase amplification (RPA) assays (Obura et al., 2011; Valasevich and Schneider, 2017), and droplet digital PCR and next-generation sequencing (NGS) (Mehle et al., 2014; Palmano et al., 2012) are becoming popular approaches for rapid detection and quantification of phytoplasmas on crops.

11.2 DETECTION AND IDENTIFICATION OF PHYTOPLASMA USING MICROSCOPIC TECHNIQUES

Various microscopic techniques are used to localize and identify the presence of phytoplasmas in the infected tissues of plants. Light microscopy can be used for phytoplasma diagnosis in herbaceous and woody plants by using Dienes reagent and DAPI (DNA-binding fluorochrome, 4-6 diamidino-2-phenylindole) as essential stains (Seemuller, 1976; Hiruki and De Rocha, 1986; Schuiling and Forstel-Neuhaus, 1992; Sinclair et al., 1992; Malinowski et al., 1996; Thomas and Balasundaran, 1998; Musetti et al., 1992, 2000). Confocal laser scanning microscopy can also be used to detect alive and active phytoplasmas in living plant tissues using vital dyes such as DiOC7 and SYTO13 (Christensen et al., 2004). However, these methods can be used to detect phytoplasmas with high concentrations but not to detect phytoplasma with low concentration in the phloem tissue of infected plants (Deeley et al., 1979). Scanning electron microscopy (SEM) and transmission electron microscopy (TEM) have been reported for the observation of phytoplasma cells in the phloem tissue of affected plants (Haggis and Sinha, 1978; Poghosyan et al., 2004; Lebsky et al., 2011; Doi et al., 1967; Lee et al., 2000; Chapman et al., 2001). Contrary to the SEM technique, which is less frequently used for phytoplasma diagnosis in plants, the TEM method is reliable and efficient not only to detect phytoplasma but also to study plant-phytoplasma interaction (Poghosyaan et al., 2004; Braun and Sinclair, 1976; Chapman et al., 2001; Musetti et al., 2000). However, the drawbacks of these methods are that they are time-consuming, costly, labor-intensive, and only work in well-equipped laboratories but are not suitable for practical applications (Sinclair et al., 1992). To date, at least 32 16Sr phytoplasma groups have been reported based on RFLP, virtual RFLP, and sequence analysis of the 16S rRNA gene (Zhao et al., 2009; Nejat et al., 2013). Therefore, the fluorescence in situ hybridization (FISH) technique can be applied to differentiate and localize 16Sr phytoplasma groups, subgroups, and strains (Bulgari et al., 2011).

11.3 DETECTION OF PHYTOPLASMA USING SEROLOGY-BASED TECHNIQUES

Serological methods have been used for detecting antibodies or antibody-like substances in association with the presence of pathogens such as viruses and phytoplasmas. There are various types of serological tests, of which ELISA was a popular method used for phytoplasma diagnosis at the beginning of the 1980s. By using monoclonal and polyclonal antiserum types, this method showed reliable results in the detection and identification of several phytoplasma strains such as *Primula* yellows (Clark et al., 1989), peanut witches' broom (Hobbs et al., 1987), lime witches' broom (Shahryari et al., 2011), sweet potato witches' broom (Shen and Lin, 1993), faba bean phyllody (Saeed et al., 1993), flavescene doree (Seddas et al., 1993), apple proliferation (Loi et al., 2002; Brzin et al., 2003), sugarcane grassy shoot/ sugarcane white leaf (Sarindu and Clark, 1993; Viswanathan, 1997), rice yellow dwarf (Chang et al., 1995), Stolbur (Fos et al., 1992), and Bermudagrass white leaf (Biabani Khankahdani and Ghasemi, 2011). Although many innovative serological techniques like immunosorbent electron microscopy, immunofluorescence microscopy, immunogold labeling, and western blot have been developed for phytoplasma detection in different host plants until now, these approaches have not been widely employed due to the difficulties encountered in obtaining good-quality antisera (Brzin et al., 2003) or inappropriate specificity and sensitivity, particularly cross-reaction with plant host proteins and nonspecific background reactivity (Sinha and Benhamous, 1983; Musetti et al., 2002; Cousin et al., 1989; Guozhong et al., 1996; Milne et al., 1995; Clark et al., 1989; Saeed et al., 1992; Adams et al., 2001).

11.4 DETECTION OF PHYTOPLASMA USING DNA-BASED TECHNQIUES

Dot blot and southern blot hybridization assays showed more reliable, sensitive, and specific results than the ELISA method for the detection and differentiation of various phytoplasmas such as alfalfa witches' broom, clover yellow edge, X-disease, clover phyllody, and eastern and western aster yellows in a wide range of host plants and insect vectors (Bertolini et al., 2007; Schneider et al., 1997). However, these methods have become obsolete due to the difficulties in generating specific probes and the insufficient sensitivity of the probes. Another approach that can be considered as gold standard in the detection of various 16Sr phytoplasma-associated diseases is nested PCR based on phytoplasma-specific universal PCR primers designed on the basis of the highly conserved 16S rRNA and 23S rRNA gene sequences and the 16S-23S rRNA intergenic spacer (ITS) (Ahrens and Seemuller, 1992; Davis and Lee, 1993; Deng and Hiruki, 1991a, 1991b; Firrao et al., 1993; Gundersen et al., 1994, 1996; Harrison et al., 1996; Jomantiene et al., 1998; Lee et al., 1993, 1994, 1995; Lorenz et al., 1995; Marcone et al., 2000; Namba et al., 1993; Nejat and Vadamalai, 2010; Schneider et al., 1993; Smart et al., 1996). Until now, various target sequences such as ribosomal protein, elongation factor tuf, nitroreductase, gyrase genes, and so on have been designed to detect each phytoplasma-specific group (Bertaccini and Martini, 1999; Gundersen et al., 1994; Jomantiene et al., 1998; Lim and Sears, 1991; Marcone et al., 2000; Schneider et al., 1997; Jarausch et al., 1994, 1998; Chuang and Lin, 2000; Bertaccini et al., 2019). An advantage of nested PCR is that this method can detect phytoplasma occurring in low titer in the phloem tissues of their host plants, especially in woody perennial plants, and localizing scattered and irregular due to seasonal fluctuation required for diagnostic purposes (Andersen et al., 1998; Gundersen and Lee, 1996; Heinrich et al., 2001). Although the drawbacks of nested PCR, such as cross-contamination in routine analysis resulting in false positive results, were previously reported, this technique has considerably proven to be more sensitive and specific than microscopic, serological, and hybridization methods for detecting phytoplasmas in plant and insect hosts. For detection of phytoplasma using other DNA-based techniques, various methods such as HMA, t-RFLP, pulsed-field gene electrophoresis (PFGE), and real-time PCR have been used for distinction and classification of phytoplasmas; assessment of microbial communities without the need for any genomic sequence information; separation of full-length phytoplasma chromosomes from contaminating host plant nucleic acids; and diagnosis of phytoplasmas, respectively (Cousin et al., 1998; Wang and Hiruki, 1999, 2001; Zhong and Hiruki, 1994; Han and Cha, 2002; Osborn et al.2000; Marcone, 2013; Baric and Dalla Via, 2004).

Identification and differentiation of each phytoplasma strain in plants have been conducted using nested PCR, sequencing, phylogenetic, and virtual RFLP analyses. These approaches required more time, labor, and cost and were not suitable for rapid detection of phytoplasma-associated diseases in plants. Therefore, LAMP assays have been recently considered as an alternative technique for in-field detection of phytoplasma-associated diseases (Dickinson, 2015; Hodgetts, 2018). This approach requires the activity of Bst DNA polymerase and a set of four to six special primers designed to recognize six distinct regions on the target gene, and the completion of the LAMP reaction is usually within 30 min at a single temperature between 60 and 65°C (Notomi et al., 2000; Obura et al., 2011). The advantages of LAMP assays over PCR methods are due to cost-effectiveness, analytical specificity and sensitivity, and particularly their applicability for in-field detection (Fig. 11.1). Therefore, LAMP-based assays have been studied and developed as a rapid, simple, and low-cost diagnostic tool for phytoplasmas within 1 h of sampling in the field that is appropriate to the poor and developing countries (Hodgetts et al., 2011; Tomlinson et al., 2010). In fact, the real-time LAMP system has been successfully demonstrated for rapid detection of phytoplasma-associated diseases in plants (Tomlinson et al., 2013; Bekele et al., 2011; Kogovsek et al., 2015, 2017; Nair et al., 2016; Vu et al., 2016; Sugawara et al., 2012, Quoc et al., 2020, 2021a,b). Also, LAMP-based assays have been applied for in-field detection of 16SrXI phytoplasma causing sugarcane white leaf (SWL) and grassy shoot (SGS) diseases in Vietnam (Fig. 11.2). The rapid detection of SWLD/SGSD in sugarcane plantations is very useful for us to control and manage these diseases from nurseries to the field, resulting in the reduction of damages caused by phytoplasmas in the sugarcane industry. Another approach that can be used for in-field detection of phytoplasma-associated diseases is the RPA. With the advantages of this technique as a rapid, sensitive, and cost-effective tool, RPA has been used to detect *Candidatus Phytoplasma oryzae* (Wambua et al., 2017) and *Candidatus Phytoplasma mali* (Valasevich and Schneider 2017) until now.

11.5 CONCLUSIONS

With the development of science and technology, various innovative techniques have been developed for the diagnosis of phytoplasma, understanding

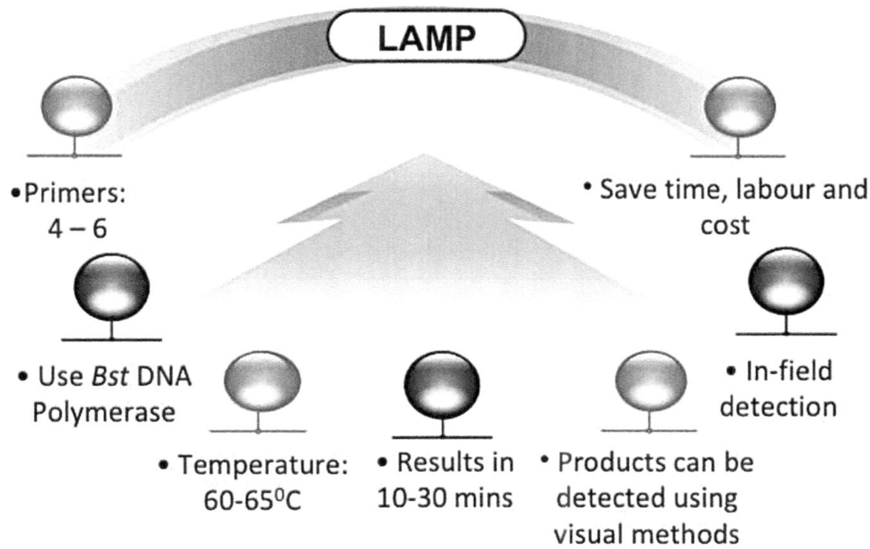

FIGURE 11.1 Advantages of loop-mediated isothermal amplification (LAMP) approach for rapid detectio of phytoplasma-associated diseases in plants.

genetic evolution, genomic information, and molecular mechanisms of interactions between phytoplasmas, plants, and insect vectors. Up to date, in-field detection plays a very important and essential role in the early diagnosis and efficient control of phytoplasmas causing severe damage to crops. Rapid detection techniques like LAMP, microfluidic LAMP, and recombinase polymerase amplification lateral flow assay (RPA-LFA) promise to be alternative approaches over traditional methods for countries with limited resources.

ACKNOWLEDGMENT

This chapter was supported by the FIRST project, Ministry of Science and Technology, Hanoi, Vietnam (No. 21/FIRST/1a/HUAF).

FIGURE 11.2 In-field detection of sugarcane white leaf/grassy shoot diseases was conducted at sugarcane plantation in Nghe An province, Vietnam by using LAMP-based assays with GenieII system (Optigene, the UK).

KEYWORDS

- **phytoplasma disease**
- **nested-PCR**
- **LAMP**
- **RFLP**
- **microscopy**

REFERENCES

Adams, A. N.; Davies, D. L.; Kirby, M. J. Virus and Phyto-plasma Detection in Fruit Trees. *Outlook Agric.* **2001,** *30,* 45–54.

Ahrens, U.; Seemuller, E. Detection of DNA of Plant Pathogenic Mycoplasma Like Organisms by a Polymerase Chain Reaction that Amplifies as Sequence of the 16S rRNA Gene. *Phytopathology* **1992,** *82,* 823–32.

Andersen, M. T.; Beever, R. E.; Gilman, A. C.; Liefting, L. W.; Balmori, E.; Beck, D. L.; Sutherland, P. W.; Bryan, G. T.; Gardner, R. C.; Forster, R. L. S. Detection of Phormium Yellow Leaf Phytoplasma in New Zealand Flax (*Phormium tenax*) Using Nested PCRs. *Plant Pathol.* **1998,** *47,* 188–196.

Bai, X.; Zhang, J.; Ewing, A.; Miller, S. A.; Radek, A. J.; Shevchenko, D. V.; Tsukerman, K.; Walunas, T.; Lapidus, A.; Campbell, J. W.; Hogenhout, S. A. Living with Genome Instability: The Adaptation of Phytoplasmas to Diverse Environments of Their Insect and Plant Hosts. *J. Bacteriol.* **2006,** *188,* 3682–3696.

Baric, S.; Dalla-Via, J. A New Approach to Apple Proliferation Detection: A Highly Sensitive Real-time PCR Assay. *J. Microbiol. Methods* **2004,** *57* (1), 135–45.

Bekele, B.; Hodgetts, J.; Tomlinson, J.; Boonham, N.; Nikolic, P.; Swarbrick, P.; Dickinson. Use of a Real-time LAMP Isothermal Assay for Detecting 16SrII and XII Phytoplasmas in Fruit and Weeds of the Ethiopian Rift Valley. *Plant Pathol.* **2011,** *60,* 345–355.

Bertaccini, A.; Martini, M. Ribosomal and Nonribo-somal Primers for Sensitive Detection and Identification of Phytoplasmas. *Petria* **1999,** *9,* 89–92.

Bertaccini, A.; Duduk, B. Phytoplasma and Phytoplasma Diseases: A Review of Recent Research. *Phytopathol. Mediterr.* **2009,** *48,* 355–378.

Bertaccini, A. Phytoplasma Research Between Past and Future: What Directions? *Phytopathogenic Mollicutes* **2015,** *5,* S1–S4.

Bertaccini A., Fiore, N.; Zamorano, A.; Tiwari, A. K.; Rao, G. P. Molecular and Serological Approaches in Detection of Phytoplasmas in Plants and Insects. In *Phytoplasmas: Plant Pathogenic Bacteria - III Genomics, Host Pathogen Interactions, and Diagnosis*. Springer Nature Singapore: Singapore; 2019, pp 105–136.

Bertolini, E.; Torres, E.; Olmos, A.; Martín, M. P.; Bertaccini, A.; Cambra, M. Co-operational PCR Coupled with Dot Blot Hybridization for Detection and 16SrX Grouping of Phytoplasmas. *Plant Pathol.* **2007,** *56* (4), 677–682.

Biabani Khankahdani, R.; Ghasemi, S. *Serological Aspects of Phytoplasma Associated with Bermudagrass White Leaf (BGWL) Disease*. International Conference on Asia Agriculture and Animal IPCBEE: Singapore, 2011; pp 106–110.

Boonrod, K.; Munteanu, B.; Jarausch, W.; Krczal, G. An Immunodominant Membrane Protein (Imp) of *Candidatus* Phytoplasma Mali Binds to Plant Actin. *Mol. Plant Microbe* **2012,** *25,* 889–895.

Bosco, D.; Tedeschi, R. Insect Vector Transmission Assays. *Methods Mol. Biol.* **2013,** *938,* 73–85.

Braun, E. J.; Sinclair, W. A. Histopathology of Phloem Necrosis in *Ulnus americana*. *Phytopathology* **1976,** *66,* 598–607.

Brzin, J.; Ermacora, P.; Osler, R., et al. Detection of Apple Proliferation Phytoplasma by ELISA and PCR in Growing and Dormant Apple Trees. *J. Plant Dis. Prot.* **2003,** *110,* 476–483.

Bulgari, D.; Casati, P.; Faoro, F. Fluorescence in situ Hybridization for Phytoplasma and Endophytic Bacteria Localization in Plant Tissues. *J. Microbiol. Meth.* **2011,** *87,* 220–223.

Chang, F. L.; Chen, C. C.; Lin, C. P. Monoclonal Antibody for the Detection and Identification of a Phytoplasma Associated with Rice Yellow Dwarf. *Eur. J. Plant Pathol.* **1995,** *101,* 511–518.

Chapman, G. B.; Buerkle, E. J.; Barrows, E. M.; Davis, R. E.; Dally, E. L. A Light and Transmission Electron Microscope Study of a Black Locust Tree, *Robinia pseudoacacia* (Fabaceae), Affected by Witches Broom, and Classification of the Associated Phytoplasma. *J. Phytopathol.* **2001,** *149,* 589–597.

Chaturvedi, Y.; Rao, G. P.; Tiwari, A. K.; Duduk, B.; Bertaccini, A. Phytoplasma on Ornamentals: Detection, Diversity and Management. *Acta Phytopathol. Entomol. Hung.* **2010,** *45,* 31–69.

Christensen, N. M.; Nicolaisen, M.; Hansen, M.; Schulz, A. Distribution of Phytoplasmas in Infected Plants as Revealed by Real-time PCR and Bioimaging. *Mol. Plant Microbe* **2004,** *17,* 1175–1184.

Christensen, N. M.; Axelsen, K. B.; Nicolaisen, M.; Schultz, A. Phytoplasmas and Their Interactions with Their Hosts. *Trends Plant Sci.* **2005,** *10,* 526–535.

Chuang, J. G.; Lin, C. P. Cloning of gyrB and gyrA Genes of Phytoplasma Associated with Peanut Witches' Broom. *Plant Pathol. Bull.* **2000,** *9,* 157–166.

Clair, D.; Larrue, J.; Aubert, G.; Gillet, J.; Cloquemin, G.; Boudon-Padieu, E. A Multiplex Nested-PCR Assay for Sensitive and Simultaneous Detection and Direct Identification of Phytoplasma in the Elm Yellows Group and Stolbur Group and its Use in Survey of Grapevine Yellows in France. *Vitis* **2003,** *42,* 151–7.

Clark, M. F.; Barbara, D. J.; Davies, D. L. Production and Characteristics of Antisera to *Spiroplasma citri* and Clover Phyllody-associated Antigens Derived From Plants. *Ann. Appl. Biol.* **1983,** *103,* 251–259.

Clark, M. F.; Morton, A.; Buss, S. L. Preparation of Myco-plasma Immunogens From Plants and a Comparison of Poly-clonal and Monoclonal Antibodies Made Against Primula Yellows MLO-associated Antigens. *Ann. Appl. Biol.* **1989,** *114,* 111–124.

Cordova, I.; Jones, P.; Harrison, N. A.; Oropeza, C. In situ PCR Detection of Phytoplasma DNA in Embryos From Coconut Palms with Lethal Yellowing Disease. *Mol. Plant Pathol.* **2003,** *4* (2), 99–108.

Cousin, M. T.; Dafalla, G.; Demazeau, E.; Theveu, E.; Grosclaude, J. *In situ* Detection of MLOs for Solanaceae Stolbur and Faba Bean Phyllody by Indirect Immunofluorescence. *J. Phytopathol.* **1989,** *124,* 71–79.

Cousin, M. T.; Roux, J.; Boudon-Padieu, E.; Berges, R.; Seemüller, E.; Hiruki, C. Use of Heteroduplex Mobility Analysis (HMA) for Differentiating Phytoplasma Isolates Causing Witches'-broom Disease on *Populus nigra* cv. Italica and Stolbur or Big Bud Symptoms on Tomato. *J. Phytopathol.* **1998,** *146,* 97–102.

Davis, R. E.; Lee, I. M. Cluster-specific Polymerase Chain Reaction Amplification of 16 s rDNA Sequences for Detection and Identification of Mycoplasma-like Organisms. *Phytopathology* **1993,** *83,* 1008–1011.

Deeley, J. W.; Stevens, A.; Fox, R. T. V. Use of Dienes' Stain to Detect Plant Diseases Induced by Mycoplasma-like Organisms. *Phytopathology* **1979,** *69,* 1169–1171.

Deng, S.; Hiruki, C. Genetic Relatedness Between two Nonculturable Mycoplasma-like Organisms Revealed by Nucleic Acid Hybridization and Polymerase Chain Reaction. *Phytopathology* **1991a,** *81,* 1475–1479.

Deng, S.; Hiruki, C. Amplification of 16S rRNA Genes from Culturable and non Culturable Mollicutes. *J. Microbiol. Meth.* **1991b,** *14,* 53–61.

Dickinson, M. Loop-Mediated Isothermal Amplification (LAMP) for Detection of Phytoplasmas in the Field. *Methods Mol. Biol.* **2015,** *1302,* 99–111.

Doi, Y.; Teranaka, M.; Yora, K.; Asuyama, H. Mycoplasma or P.L.T. Group-like Microorganisms Found in the Phloem Elements of Plants Infected with Mulberry Dwarf, Potato Witches'-Broom, Aster Yellows or Paulownia Witches'-broom. *Ann. Phytopathol. Soc. Jpn.* **1967,** *33,* 259–266.

Firrao, G.; Gobbi, E.; Loci, R. Use of Polymerase Chain Reaction to Produce Oligonucleotide Probes for Myco-plasma-like Organisms. *Phytopathology* **1993,** *83,* 602–606.

Fos, A.; Danet J-L, Zreik, L.; Garnier, M.; Bové, J.-M. Use of a Monoclonal Antibody to Detect the Stolbur Mycoplasmalike Organism in Plants and Insects and to Identity a Vector in France. *Plant Disease* **1992,** *76,* 1092–1096.

Galetto, L.; Bosco, D.; Balestrini, R.; Genre, A.; Fletcher, J.; Marzachí, C. The Major Antigenic Membrane Protein of *Candidatus* Phytoplasma Asteris Selectively Interacts with ATP Synthase and Actin of Leafhopper Vectors. *PLoS One* **2011,** *6,* e22571.

Gundersen, D. E.; Lee, I. M.; Rehner, S. A.; Davis, R. E.; Kingsbury, D. E. Phylogeny of Mycoplasmalike Organisms (Phy-toplasmas): A Basis for Their Classification. *J. Bacteriol.* **1994,** *176,* 5244–5254.

Gundersen, D. E.; Lee, I. M. Ultrasensitive Detection of Phytoplasma by Nested-PCR Assays Using Two Universal Primer Sets. *Phytopathol. Mediterr.* **1996,** *35,* 144–151.

Guozhong, T.; Xijin, Z.; ShuiFang, Z.; Chengliang, Z.; Fei, L.; Wensheng, H. Detection of Mycoplasma-Like Organisms in the Infected Paulownia by Indirect Immunofluores-cence Microscopy. *J. For Res.* **1996,** *9,* 1–6.

Haggis, G. H.; Sinha, R. C. Scanning Electron Microscopy of Mycoplasma-like Organisms After Freeze Fracture of Plant Tissues Affected with Clover Phyllody and Aster Yellows. *Phytopathology* **1978,** *68,* 677–680.

Han, S.; Cha, B. Genetic Differentiation of Phytoplasma Isolates by DNA Heteroduplex Mobility Assay and Single-strand Conformation Polymorphism Analysis. *Plant Pathol. J.* **2002,** *18,* 308–312.

Harrison, N. A.; Richardson, P. S.; Tsai, J. H. PCR Assay for Detection of the Phytoplasma Associated With Maize Bushy Stunt Disease. *Plant Dis.* **1996,** *80,* 263–269.

Heinrich, M.; Botti, S.; Caprara, L.; Arthofer, W.; Strommer, S.; Hanzer, V.; Katinger, H.; Bertaccini, A.; Laimer da Câmara Machado, M. Improved Detection Methods for Fruit Tree Phytoplasmas. *Plant. Mol. Biol. Rep.* **2001,** *19,* 169–179.

Hiruki, C.; Da Rocha, A. Histochemical Diagnosis of Mycoplasma Infections in *Catharanthus roseus* by Means of a Fluorescent DNA-binding Agent, 4', 6-diamidino-2-pheny-lindole-2HC1 (DAPI). *Can. J. Plant Pathol.* **1986,** *8,* 185–188.

Hobbs, H. A.; Reddy, D. V. R.; Reddy, A. S. Detection of a Mycoplasma-like Organism in Peanut Plants with Witches' Broom Using Indirect Enzyme-linked Immunosorbent Assay (ELISA). *Plant Pathol.* **1987,** *36,* 164–167.

Hodgetts, J.; Ball, T.; Boonham, N.; Mumford, R.; Dickinson, M. Use of Terminal Restriction Fragment Length Polymorphism (T-RFLP) for Identification of Phytoplasmas in Plants. *Plant Pathol.* **2006,** *56* (3), 357–360.

Hodgetts, J. Rapid Sample Preparation and LAMP for Phytoplasma Detection. In *Phytoplasmas – Methods and Protocols*; Musetti, R., Pagliari, L., Eds.; Humana Press, 2018; pp 187–201.

Hodgetts, J.; Tomlinson, J.; Boonham, N.; Gonzalez-Martin, I.; Nikolic, P.; Swarbrick, P.; Yankey, E. N.; Dickinson, M. Development of Rapid in-field Loop-mediated Isothermal Amplification (LAMP) Assays for Phytoplasmas. *Bull. Insectol* **2011,** *64* (Supplement), S41–S42.

Hogenhout, S. A.; Oshima, M.; Ammar, E. D.; Kakizawa, S.; Kingdom, H. N.; Namba, S. Phytoplasmas: Bacteria That Manipulate Plants and Insects. *Mol. Plant Pathol.* **2008,** *9,* 403–423.

Hogenhout, S. A.; Music, M. S. Phytoplasma Genomics, From Sequencing to Comparative and Functional Genomics — what Have we Learnt? In *Phytoplasmas — Genomes, Plant Hosts and Vectors*; Weintraub, P. G., Jones, P., Eds.; CABI: Wallingford, 2010; pp 19–36.

Jarausch, W.; Saillard, C.; Dosba, F.; Bové, J. M. Differentiation of Mycoplasma-Like Organisms (MLOs) in European Fruit Trees by PCR Using Specific Primers Derived From Sequence of a Chromosomal Fragment of the Apple Proliferation MLO. *Appl. Environ. Microbiol.* **1994,** *60,* 2916–2923.

Jarausch, W.; Lansac, M.; Saillard, C.; Broquaire, J. M.; Dosba, F. PCR Assay for Specific Detection of European Stone Fruit Yellows Phytoplasmas and its use for Epidemiological Studies in France. *Eur. J. Plant Pathol.* **1998,** *104,* 17–27.

Jomantiene, R.; Davis, R. E.; Maas, Dally, E. L. Classification of new Phytoplasmas Associated with Diseases of Strawberry in Florida, Based on Analysis of 16S rRNA and Ribosomal Protein Gene Operon Sequences. *Int. J. Syst. Bacteriol.* **1998,** *48,* 269–277.

Jomantiene, R.; Davis, R. E. Clusters of Diverse Genes Existing as Multiple, Sequence Variable Mosaics in a Phy-toplasma Genome. *FEMS Microbiol. Lett.* **2006,** *255,* 59–65.

Jomantiene, R.; Zhao, Y.; Davis, R. E. Sequence-variable Mosaics: Composites of Recurrent Transposition Characterizing the Genomes of Phylogenetically Diverse Phyto-plasmas. *DNA Cell Biol.* **2007,** *26,* 557–564.

Kakizawa, S.; Oshima, K.; Namba, S. Diversity and Functional Importance of Phytoplasma Membrane Proteins. *Trends Microbiol.* **2006,** *14,* 254–256.

Kogovsek, P.; Hodgetts, J.; Hall, J.; Prezelj, N.; Nikolic, P.; Mehle, N.; Lenarcic, R.; Rotter, A.; Dickinson, M.; Boonham, N.; Demastia, M.; Ravnikar, M. LAMP Assay and Rapid Sample Preparation Method for on-site Detection of Flvescence Doree Phytoplasma in Grapevine. *Plant Pathol.* **2015,** *64* (2), 286–296.

Kogovsek, P.; Mehle, N.; Pugelj, A.; Jakomin, T.; Schroers, H. J.; Ravnikar, M.; Dermastia, M. Rapid Loop-mediated Isothermal Amplification Assays for Grapevine Yellows Phytoplasmas on Crude Leaf-vein Homogenate has the Same Performance as qPCR. *Eur. J. Plant Pathol.* **2017,** *148* (1), 75–84.

Kube, M.; Mitrovic, J.; Duduk, B.; Rabus, R.; Seemüller, E. Current View on Phytoplasma Genomes and Encoded Metabolism. *Sci. World J.* **2012,** *2012,* 185942.

Lebsky, V.; Hernández-González, J.; Arguello-Astorga, G.; Cardenas-Conejo, Y.; Poghosyan, A. Detection of Phytoplasmas in Mixed Infection with Begomoviruses: A Case Study of Tomato and Pepper in Mexico. *Bull. Insectol.* **2011,** *64* (Suppl), S55–S56.

Lee, I. M.; Hammond, R. W.; Davis, R. E.; Gundersen, D. E. Universal Amplification and Analysis of Pathogen 16S rDNA for Classification and Identification of Mycoplasma-Like Organisms. *Phytopathology* **1993,** *83,* 834–842.

Lee, I. M.; Gundersen, D. E.; Hammond, R. W.; Davis, R. E. Use of Mycoplasma-Like Organism (MLO) Group-specific Oligo-nucleotide Primers for Nested-PCR Assays to Detect Mixed-MLO Infections in a Single Host Plant. *Phytopathology* **1994,** *84,* 559–566.

Lee, I. M.; Bertaccini, A.; Vibio, M.; Gundersen, D. E. Detection of Multiple Phytoplasmas in Perennial Fruit Trees with Decline Symptoms in Italy. *Phytopathology* **1995,** *85,* 728–735.

Lee, I. M.; Davis, R. E.; Gundersen-Rindal, D. E. Phyto-plasma: Phytopathogenic Mollicutes. *Annu. Rev. Microbiol.* **2000,** *54,* 221–255.

Lim, P. O.; Sears, B. B. DNA Sequence of the Ribosomal Protein Genes rp12 and rps19 from a Plant-Pathogenic Mycoplasma-like Organism. *FEMS Microbiol. Lett.* **1991,** *84,* 71–74.

Loi, N.; Ermacora, P.; Carraro, L., et al. Production of Monoclonal Antibodies Against Apple Proliferation Phytoplasma and Their Use in Serological Detection. *Eur. J. Plant Pathol.* **2002,** *108,* 81–86.

Lorenz, K. H.; Schneide, B.; Ahrens, U.; Seemüller, E. Detection of the Apple Proliferation and Pear Decline Phytoplasmas by PCR Amplification of Ribosomal and Non-ribosomal DNA. *Phytopathology* **1995,** *85,* 771–776.

Malinowski, T.; Zandarski, J.; Komorowska, B.; Zawadzka, B. Application of DAPI Staining and PCR Amplification of DNA for the Identification of Pear Decline Phytoplasma in Declining Trees in Poland. *Phytopathol. Polonica* **1996,** *12,* 103–110.

Marzachi, C.; Veratii, F.; Bosco, D. Direct PCR Detection of Phytoplasmas in Experimentally Infected Insects. *Ann. Appl. Biol.* **2008,** *133* (1), 45–54.

Marcone, C.; Lee, I. M.; Davis, R. E.; Ragozzino, A.; Seemüller, E. Classification of Aster Yellows-Group Phytoplasmas Based on Combined Analyses of rRNA and *tuf* Gene Sequences. *Int. J. Syst. Evol. Microbiol.* **2000,** *50,* 1703–1713.

Marcone, C. Pulsed-field Gel Electrophoresis for Isolation of Full-length Phytoplasma Chromosomes From Plants. *Methods Mol. Biol.* **2013,** *938,* 395-403.

Mehle, N.; Dreo, T.; Ravnikar, M. Quantitative Analysis of "Flavescence Doree" Phytoplasma with Droplet Digital PCR. *Phytopathog. Mollicutes* **2014,** *4* (1), 9–15.

Milne, R. G.; Ramasso, G. E.; Lenzi, R.; Masenga, V.; Sarindu, S.; Clark, M. F. Pre- and Post-Embedding Immunogold Labeling and Electron Microscopy in Plant Host Tissues of Three Antigenically Unrelated MLOs: Primula Yellows, Tomato Big bud and Bermudagrass Whiteleaf. *Eur. J. Plant Pathol.* **1995,** *101,* 57–67.

Musetti, R.; Favali, M. A.; Carraro, L.; Osler, R. An Attempt to Differentiate by Microscopic Methods Two Plant Myco-plasma-like Organisms. *Cytobios* **1992,** *72,* 71–82.

Musetti, R.; Favali, M. A.; Pressacco, L. Histopathology and Polyphenol Content in Plants Infected by Phyto-plasmas. *Cytobios* **2000,** *102,* 133–147.

Musetti, R.; Loi, N.; Carraro, L.; Ermacora, P. Application of Immunoelectron Microscopy Techniques in the Diagnosis of Phytoplasma Diseases. *Microsc. Res. Tech.* **2002,** *56,* 462–464.

Music, M. S.; Skorix, D. Single-strand Conformation Polymorphism Analysis for Differentiating Phytoplasma Strains. *Methods Mol. Biol.* **2013,** *938,* 217-22.

Namba, S.; Kato, S.; Iwanami, S.; Oyaizu, H.; Shiozawa, H.; Tsuchizaki, T. Detection and Differentiation of Plant-pathogenic Mycoplasma-like Organisms Using Polymerase Chain Reaction. *Phytopathology* **1993,** *83,* 786–791.

Nair, S.; Manimekalai, R.; Ganga Raj, P.; Hegde, V. Loop Mediatedisothermalamplification (LAMP) Assay for Detection of Coconut Root Wilt Disease and Arecanut Yellow Leaf Disease Phytoplasma. *World J. Microbiol. Biotechnol.* **2016,** *32,* 108.

Nejat, N.; Vadamalai, G. Phytoplasma Detection in Coconut Palm and Other Tropical Crops. *Plant Pathol. J.* **2010,** *9,* 101–110.

Nejat, N.; Vadamalai, G.; Davis, R. E.; Harrison, N. A.; Sijam, K.; Dickinson, M.; Abdullah, S. N. A.; Zhao, Y. '*Candidatus* Phytoplasma Malaysianum', a Novel Taxon Associated with Virescence and Phyllody of Madagascar Periwinkle (*Catha-ranthus Roseus*). *Int. J. Syst. Evol. Microbiol.* **2013,** *63,* 540–548.

Notomi, T.; Okayama, H.; Masubuchi, H.; Yonekawa, T.; Watanabe, K.; Amino, N.; Hase, T. Loop-Mediated Isothermal Amplification of DNA. *Nucleic Acids Res.* **2000,** *28,* E63.

Obura, E.; Massiga, D.; Wachira, F.; Gurja, B.; Khan, Z. R. Detection of Phytoplasma by Loop-mediated Isothermal Amplification of DNA (LAMP). *J. Microbiol Methods* **2011,** *84,* 312–316.

Osborn, A. M.; Moore, E. I. R.; Timmis, K. N. An Evaluation of Terminal-restriction Fragment Length Polymorphism (T-RFLP) Analysis for the Study of Microbial Community Structure and Dynamics. *Environ. Microbiol.* **2000,** *2,* 39–50.

Oshima, K.; Kakizawa, S.; Nishigawa, H.; Jung, H. Y.; Wei, W.; Suzuki, S.; Arashida, R.; Nakata, D.; Miyata, S.; Ugaki, M.; Namba, S. Reductive Evolution Suggested From the Complete Genome Sequence of a Plant-pathogenic Phytoplasma. *Nat. Genet.* **2004,** *36,* 27–29.

Palmano, S.; Saccardo, F.; Martini, M.; Eemacora, P.; Scortichini, M.; Abba, S.; Marzachi, C.; Loi, N.; Firrao, G. Insights Into Phytoplasma Biology Through Next Generation Sequencing. *J. Plant Pathol.* **2012,** *94* (4), S4-50-S4-51.

Poghosyan, A. V.; Lebsky, V. K.; Arce-Montoya, M.; Landa, L. Possible Phytoplasma Disease in Papaya (*Carica papaya* L.) From Baja California Sur: Diagnosis by Scanning Electron Microscopy. *J. Phytopathol.* **2004,** *152,* 376–380.

Quoc, N. B.; Xuan, N. T. T.; Nghiep, N. M.; Phuong, N. D. N.; Linh, T. B.; Chau, N. N. B.; Chuong, N. D. X.; Nien, N. C.; Dickinson, M. Loop-mediated Isothermal Amplification for Detection of Sesame Phyllody Phytoplasmas in Vietnam. *Folia Microbiol.* **2021a,** *66* (2), 273–283.

Quoc, N. B.; Xuan, N. T. T.; Phuong, N. D. N.; Trang, H. T. T.; Chau, N. N. B.; Duong, C. A.; Dickinson, M. Development of Loop Mediated Isothermal Amplification Assays for the Detection of Sugarcane White Leaf Disease. *Physiol. Mol. Plant Pathol.* **2021b,** *113,* 101595.

Quoc, N. B.; Chau, N. N. B.; Duong, C. A. Sugarcane White Leaf and Grassy Shoot Management for Healthy Seed Production in Vietnam. In *Advances in Seed Production and Management*; Tiwari, A. K., Eds.; Springer: Singapore; 2020. https://doi.org/10.1007/978-981-15-4198-8_4.

Razin, S. Molecular Biology and Genomics of Mollicutes. *Bull Insectol.* **2007,** *60,* 101–103.

Sarindu, N.; Clark, M. F. Antibody Production and Identity of MLOs Associated with Sugarcane White Leaf Disease and Bermudagrass White Leaf Disease From Thailand. *Plant Pathol.* **1993,** *42,* 396–402.

Saeed, E.; Rage, P.; Cousin, M. T. Determination of the Antigenic Protein Size Associated with Faba Bean Phyllody MLO by Using (SDS-PAGE) Electrophoresis and Immuno-transfer. *J. Phytopathol.* **1992,** *136,* 1–8.

Saeed, E. M.; Roux, J.; Cousin, M. Studies of Polyclonal Antibodies for the Detection of MLOs Associated with Faba Bean (*Vicia faba* L.) Using Different ELISA Methods and Dot-blot. *J. Phytopathol.* **1993,** *37,* 33–43.

Schneider, B.; Ahrens, U.; Kirkpatrick, B. C.; Seemüller, E. Classification of Plant Pathogenic Mycoplasma-like Organisms Using Restriction-site Analysis of PCR Amplified 16 s rDNA. *J. Gen. Appl. Microbiol.* **1993,** *139,* 19–527.

Schneider, B.; Gibb, K. S.; Seemüller, E. Sequence and RFLP Analysis of the Elongation Factor Tu Gene Used in Differentiation and Classification of Phytoplasmas. *Microbiology* **1997,** *143,* 3381–3389.

Schuiling, M.; Forstel-Neuhaus, A. The use of the Fluorochrome DAPI in the Diagnosis of Lethal Disease of Coconut Palm (*Cocos nucifera*) in Tanzania. *Z. Pflanzenk Pflanzensch.* **1992,** *99,* 614–616.

Seddas, A.; Meignoz, R.; Daire, X.; Boudon-Padieu, E.; Caudwell, A. Purification of Grape-Vine "Flavescence Dorée" MLO (Mycoplasma-like Organism) by Immunoaffinity. *Curr. Microbiol.* **1993,** *27,* 229–236.

Seemuller, E. Fluorescence Optical Demonstration of Mycoplasma-like Organisms in the Pholoem of Trees with Pear Decline or Proliferation Disease. *Phytopathol. Z.* **1976,** *85,* 368–372.

Shen, W. C.; Lin, C. P. Production of Monoclonal Antibodies Against a Mycoplasmalike Organ- ism Associated with Sweet Potato Witches' Broom. *Phytopathology* **1993,** *83,* 671–675.

Sinclair, W. A.; Griffiths, H. M.; Davis, R. E.; Lee, I. M. Detection of ash Yellows Mycoplasmalike Organisms in Different Tree Organs and in Chemically Preserved Specimens by a DNA Probe vs. DAPI. *Plant Dis.* **1992,** *76,* 154–158.

Sinha, R. C.; Benhamou, N. Detection of Mycoplasmalike Organism Antigens From Aster Yellows Diseased Plants by two Serological Procedures. *Phytopathology* **1983,** *73,* 1199–1202.

Smart, C. D.; Schneider, B.; Blomquist, L.; Guerra, J.; Harrison, N. A.; Ahrens, U.; Lorenz, K. H.; Seemüller, E.; Kirkpatrick, B. C. Phytoplasma-specific PCR Primers Based on Sequences of the 16S-23S rRNA Spacer Region. *Appl. Environ. Microbiol.* **1996,** *62,* 2988–2993.

Sugawara, K.; Himeno, M.; Keima, T.; Kitazawa, Y.; Maejima, K.; Oshima, K.; Namba, S. Rapid and Reliable Detection of Phytoplasma by Loop-mediated Isothermal Amplification Targeting a Housekeeping Gene. *J. Gen. Plant Pathol.* **2012,** *78* (6), 389–397.

Shahryari, F.; Safarnajad, M. R.; Shams-Bakhsh, M.; Ataiee, S. Use of a Recombinant Protein for Development of a DAS-ELISA Serological Kit for Sensitive Detection of Witches' Broom Disease of Lime. *Bull. Insectol.* **2011,** *64* (Supplement), S43–S44.

Thomas, S.; Balasundaran, M. *In situ* Detection of Phytoplasma in Spike-disease-affected Sandal Using DAPI Stain. *Curr. Sci.* **1998,** *74,* 989–993.

Tomlinson, J. A.; Boonham, N.; Dickinson, M. Development and Evaluation of a One-hour DNA Extraction and Loop-mediated Isothermal Amplification Assay for Rapid Detection of Phytoplasmas. *Plant Pathol.* **2010,** *59,* 465–471.

Tomlinson, J. A.; Ostoja-starzewska, S.; Adams, I. P.; Miano, D. W.; Abidrabo, P.; Kinyua, Z.; Alicai, T.; Dickinson, M. J.; Peters, D.; Boonham, N.; Smith, J. Loop-mediated Isothermal Amplification for Rapid Detection of the Causal Agents of Cassava Brown Streak Disease. *J. Virol. Methods* **2013,** *191,* 148–54.

Valasevich, N.; Schneider, B. Rapid Detection of "Candidatus Phytoplasma Mali" by Recombinase Polymerase Amplification Assays. *J. Phytopathol.* **2017,** *165* (11–12), 762–770.

Vu, N. T.; Pardo, J. M.; Alvarez, E.; Le, H. H.; Wyckhuys, K.; Nguyen, K. L.; Le, D. T. Extablishment of a Loop-mediated Isothermal Amplification (LAMP Assay for the Detection of Phytoplasma-associated Cassava Witches' Broom Disease. *Appl. Biol. Chem.* **2016,** *59* (2), 151–156.

Viswanathan, R. Detection of Phytoplasmas Associated with Grassy Shoot Disease of Sugarcane by ELISA Techniques. *Z. PflanKrankheit PflanSchutz* **1997,** *104,* 9–16.

Wambua, L.; Schneider, B.; Okwaro, A.; Wanga, J. O.; Imali, O.; Wambua, P. N.; Agutu, L.; Olds, C.; Jones, C. S.; Masiga, D.; Midega, C.; Khan, Z.; Jores, J.; Fischer, A. Development of Field-Applicable Tests for Rapid and Sensitive Detection of *Candidatus* Phytoplasma oryzae. *Mol. Cell. Probes* **2017,** *35,* 44–56.

Wang, K.; Hiruki, C. Rapid Detection of a Single-basepair Mutation in Phytoplasmal Conserved Genes by DNA Heteroduplex Mobility Assay (HMA). *Proc. Jpn. Acad.* **1999,** *75B,* 259–262.

Wang, K.; Hiruki, C. Use of Heteroduplex Mobility Assay for Identification and Differentiation of Phytoplasmas in the Aster Yellows Group and the Clover Proliferation Group. *Phytopathology* **2001,** *91,* 546–552.

Zhong, Q.; Hiruki, C. Genetic Differentiation of Phyto-plasma Isolates Determined by a DNA Heteroduplex Mobility Assay. *Proc. Jpn. Acad.* **1994,** *70B*, 127–131.

Zhao, Y.; Wei, W.; Lee, I. M.; Shao, J.; Suo, X.; Davis, R. E. Construction of an Interactive Online Phytoplasma Classification Tool, *i*PhyClassifier, and its Application in Analysis of the Peach X-disease Phytoplasma Group (16SrIII). *Int. J. Syst. Evol. Microbiol.* **2009,** *59,* 2582–2593.

CHAPTER 12

Viral Disease of Pulses and Their Management

TULIKA MISHRA[1], SHAIL PANDE[2], PALLAVI SHARMA[3], and NEHA KUMARI SONKAR[3]

[1]*Department of Botany, D.D.U Gorakhpur University, Gorakhpur, India*

[2]*Principal DAV Degree College, Gorakhpur, India*

[3]*Department of Botany, D.D.U Gorakhpur University, Gorakhpur, India*

ABSTRACT

In recent eras, poverty, hunger, and malnutrition have been the biggest problems in the world. India has varied agroclimatic regions like temperate, tropical, subtropical, and climatic zones; hence, it has one of the wealthiest biodiversities in the world. The subtropical and tropical climatic conditions are also best suited for the growth of various plant pathogens and pests. Pulses are a great source of proteins, but pulse crops are heavily damaged by various pathogens in agricultural fields. In recent times, several diseases have been caused by viruses of the families *Begomovirus, Badnavirus, Cucumovirus, Potyvirus, and Ilarvirus,* which are common in pulses. Diseases like alfalfa mosaic (AM), mung bean yellow mosaic (MYM), pea seed-borne mosaic (PSM), and bean common mosaic (BCM) are also commonly found in pulses. A variety of viruses are disseminated by the seeds of infected plants over long distances. Sometimes, vectors are responsible for localized spread from weeds or infected crop plants. Precise methods for identifying, detecting, and characterizing viruses are important precautions for disease control measures. Implied management of viral diseases in general demands consolidated measures like disease-free seeds, as well as

Plant Diseases and Their Management: A Sustainable Approach.
Deepa Srivastava, Rajarshi Kumar Gaur, and Ajay K. Tiwari (Eds.)

the use of virus-tested certified seeds, spatial isolation from spotted virus inoculum origins, and checking vector populations, which are also required with the production of virus resistance or disease tolerance via selection and breeding. In this chapter, the management of diseases of economically important pulses caused by many viral pathogens is discussed in detail.

12.1 INTRODUCTION

The worldwide population is anticipated to increase from the present 7.8 billion (2021) to ~ 9 billion in the next 30 years. Therefore, enhancing food production for food and nutritional security may be a challenge. Food security and nutrition seem crucial to understanding the conditions of the world's ecosystem and the ever-growing population. The agricultural sector has a major role in government and the economy. Most developing countries, like India, are heavily involved in agriculture for their livelihood. Therefore, the planting of crops should be well cared for, and if the crop is infected, the first diagnosis of this disease can help farmers, and the necessary steps are often taken on time. These two challenges are also addressed by diversifying and cultivating legume crops. In India, leguminous crops are generally known as pulses. These constitute ~ 750 genera and ~ 18,000 species (Graham and Vance, 2003). The rosid dicot family is made up of essential grains, oilseeds, fodder, and agroforestry varieties. The trace of agricultural legume crops has been spotted since neolithic times. Chickpea (*Cicer arietinum*) was found in the neolithic period (Lev-Yadun et al., 2000). The earliest domesticated leguminous crops include soybeans (*Glycine max*), lentils (*Lens culinaris*), and beans (*Phaseolus vulgaris*) (Kaplan and Lynch, 1999). Pulses have a crucial role in the regular diet, mainly in a few developed nations and some developing nations, and thus legumes are also called the meat of the poor man. Pulses are produced on ~12–15% of the world's arable land, and their contribution to the total human dietary protein nitrogen requirement is ~30% (Graham and Vance, 2003). The most important legumes used as food include chickpeas (gram), beans, lentils, mung beans (green gram), urdbeans (black gram), field peas (matar), pigeon peas (tur), and cowpeas. Soybeans and groundnuts are important legumes that meet about 35% of the world's oil demand (Graham and Vance, 2003). The world's pulses production area and yield in 2013 were ~73 million tons (MT), ~80.8 million ha, and ~904 kg ha^{-1}, respectively (FAO, 2019). Asia and Africa collectively contribute ~49 MT, that is, 67% of worldwide pulse production. In 2019, India's production of various legumes was: pigeon pea (~4.74 MT), chickpea (~13.1 MT), lentils (~1.13 MT), dried peas (0.6 MT), peanut nuts with shell (9.4 MT), and soybean (11.95 MT) (FAO, 2019). Worldwide legume production was enhanced by approximately 1.7 times between 1961 and 2011, and through that exact time, the production of cereals increased by ~3 times. The yield per hectare of pods increased only ~1.4 times as compared to ~3.0 times in cereals. All these factors contribute to the scarcity of legumes worldwide, especially pulses in India. Some leguminous crops are cultivated in several parts of India, of which mung bean, pigeon pea, chickpea, field pea, urban, and lentil are the most common. In 2018–19, India generated 19.27 MT of pulses, and approximately 3.18 MT of pulses costing Rs. 11,038 crores (the US$1.8 billion) were imported from Australia, Turkey, Canada, Syria, Myanmar, etc. during the same year. Results from previous demonstrations have shown that pulses' cultivation is often improved to the required level when appropriate technology transfer efforts are made. India has seen magnificent growth in pulses' production over the past 5 years, with a well-combined growth rate. The expansion rate of pulse production (2.61%) over the past decade was higher than the rate of expansion of rice (1.59%), total cereals (1.88%), and wheat (1.89%). This had an effect on the individual supply of pulses (39.4 g/capita/day from as early as 36 g/capita/day). Among the various pulses, the best rate was observed in the production of chickpea (5.89%), followed by pigeon pea (2.61%). The general production of pulses increased to 786 kg ha^{-1} during 2012–13, in contrast to 577 kg ha^{-1} during 2004–05. The praise goes to the enhanced diversities and manufacture of breeder seed, exposition of pulses production technologies, frontline expositions, policy support, and various government schemes like Rashtriya Krishi Vikas Yojana (RKVY), accelerated pulses' production program (A3P), and National Food Security Mission (NFSM), etc. started by the central government to push pulses production and so on to confirm self-sufficiency. The requirement for pulses within the nation is expected to be 39 million tons by the year 2050, at an annual rate of 2%. This may require a smart change in research and development strategies, besides proper policy support from the government.

Viral infections can lead to lower yields, which in turn can lead to greater economic losses. A variety of methods designed to diagnose diseases in plants involve local specialists performing cellular and serological assays. These methods are time-consuming, expensive, and manual. Large personnel and large facilities are required to detect and identify infections in large agricultural fields. Most farmers cannot afford these types of services and workers. Physical examination of such types of diseases requires many staff and should result in boredom, which can suffer accuracy and be liable to human errors. Therefore, the need for automatic detection and a method of diagnosing phytopathogen infection may be performed in less time and with greater accuracy. Various pathogens affect the visible part of the plant, as the leaves may appear between the yellow and brown spots on the surface. Thus, recognizing those symptoms often results in a faster, more automatic, and more accurate diagnosis of various plant diseases. But it is still a challenge to find such symptoms because of their random nature. Strategies for controlling pathogens often include controlling the number of vectors using pesticides, the use of sterile disinfectants, appropriate cultural practices, and the use of disease-resistant strains. However, each of the above methods has its drawbacks. Legumes can be a plant in the Fabaceae family, the seeds of this plant, or the fruit. When utilized as a dry grain, the seeds are also called pulses. Legumes are farmed for agriculture, mainly for our consumption, forage, and silage, and for soil enrichment as compost. The most popular legumes include peanuts, alfalfa, clover, beans, peas, lentils, lupins, carob, mesquite, soybeans, tamarind, and chicks. Leguminous plants produce a special kind of fruit, a dry fruit that grows from a simple carpel and frequently dehisces on both sides (Siddique et al., 2012). Most of the legumes have nitrogen-fixing bacteria in their roots. Because of it, they have a crucial role in crop rotation. Legumes are a good source of dietary minerals, protein, dietary fiber, and carbohydrates compared to staple grains. Legumes are a reliable source of health-boosting acids such as linoleic, linolenic, palmitic, and oleic acids. In addition, many vegetables are a rich source of carbohydrates (30–60%) (United States Department of Agriculture (USDA), 2013). There are various carbohydrates in leguminous crops, including monosaccharides—ribose, fructose, and glucose; disaccharides—sucrose and maltose; oligosaccharides—ciceritol, verbascose, raffinose, and stachyose; and polysaccharides—starch, hemicelluloses, and cellulose. Higher oligosaccharide content in chickpeas results in higher flatulence (Jaya et al., 1979). Most of the carbon reserve in legume starch is also composed of amylose and amylopectin (Chibbar et al., 2010). The deficiency of protein is one of the most important problems in the world, as almost 1 billion people suffer from a deficiency of protein and other nutrients, and consequently, malnutrition. In the current context, germination plants are important in tackling this malnutrition because they are rich in protein and micronutrients, as well as cheap food sources in destitute and developing nations. Worldwide, 70.41 million pulse crops are produced on 77.5 million ha of land. The average pulse production is 907 kg/ha. Important pulse crops, such as regular beans, peas, green gram, black gram, soybeans, lentils, cowpeas, dried peas, fava beans, and pigeon peas, make up about 90% of pulses. These legumes are a rich source of fiber, and the total dietary fiber content of legumes is ~ 8–27.5% (Guillon and Champ, 2002). Several health benefits are related to enhanced consumption of dietary fiber, which include decreased threat of various diseases (diabetes, cardiovascular disease, corpulency, and cancer) and lowered blood cholesterol levels (Tosh and Yada, 2010; Marlett et al., 2002).

12.2 VIRAL DISEASES

Plant pests are economically important, reducing the yield and quality of many crops. There are more than 700 known plant viruses that will cause serious damage to infected plants. Globally, plant viruses can cause total failure, even with updated measures to control pests and diseases. The viruses that cause the spread of symptoms in infected plants can also be present in plants without showing any symptoms. Common symptoms of infection are local lesions, mosaic patterns, yellowing, leaf rolling or curling, ring spots, necrosis, and complete plant death. Some viruses are seed-borne, others are transmitted mechanically, and most viruses are transmitted by insects. A viral infection is characterized by symptoms. Their research can help identify viral diseases and contribute to understanding why and how the virus harms plants and causes plant losses. Viruses are attached to living cells to replicate. Symptoms, which are the result of impaired cell function or deficiency, can vary greatly not only from the virus but also from the host genotype and environmental conditions. Common (and/or chronic) *alfalfa mosaic virus* (AMV), *cucumber mosaic virus* (CMV), b*ean yellow mosaic virus* (BYMV), and *pea seed-borne mosaic virus* (PSbMV) in pulses, so the main persistent viruses (phloem limited) are *turnip yellows virus* (TuYV), *bean leaf roll virus* (BLRV), and *subterranean clover stunt virus* (SCSV). The first source of nonpersistent viral infection is an infected seed. If the infection is 1%, one in every 100 plants will be infected; therefore, it will be distributed randomly across the sector. The second infection occurs with an aphid. Aphids get the virus from primary infections or weeds and spread it to healthy plants in a brief period of time. Aphids lose infectivity after some probes because these viruses are not kept for long in their mouths and do not infect each other for the rest of their lives. These viruses do not go through a life cycle like most other pathogens. They cannot survive long without a plant and rely solely on aphids for pollination among living plants. These viruses can live for a long time in seeds, but they do not have the stages that allow them to live in the soil as the origin of infection in future plants. Strong viruses are confined to phloem and are only propagated by aphids. Aphids must use plants from hours to days to transmit the virus. The aphid remains infected throughout its life. These terms do not appear to form seeds. The source of the virus is perennial weeds.

12.3 VIRAL DISEASES OF PIGEON PEA

Pigeon pea [*Cajanus cajan* (L.) Millsp.] is a shrub or could be a perennial shrub, with its traditional origins found in India. It is a supreme legume plant widely cultivated in India as a pivotal source of protein. Globally, pigeon pea is grown on nearly 6.22 million ha, and 75% of this land is in India (FAOSTAT, 2014). It is also grown in other parts of the earth, including the local area and the region of Southeast Asia. It grows in desert areas, due to the ability of the crop to withstand drought, and in downgraded areas (Upadhyaya et al., 2014). Crops are planted mainly year-round using cultivars with different ripening periods: short grown within 90 days, medium grown within 120 days, and up to several heights within 180 days. Pods are eaten as a vegetable; the dried ripe grain is used for cooking. Approximately 15 viruses are reported to be naturally present in pigeon pea (Kumar et al., 2008). Sterility mosaic disease (SMD), which gives rise to the pigeon pea sterility mosaic virus (PPSMV), is the most economically deteriorating disease in India, causing an approximate loss of more than US$300 million per year (Reddy et al., 1998). SMD was propounded in 1931 from Pusa (Bihar) (Mitra, 1931) and is commonly found in India, Nepal, Bangladesh, and Myanmar. The disease has also been found in Thailand (Nene, 1995; Newton and Peiris, 1953). Symptoms indistinguishable from those of SMD were noticed in pigeon peas planted in China (G. V. R. Rao, personal communications, ICRISAT, Patancheru, India), but no data were found related to virus confirmation. Conformity is not known to occur outside of Asia. The pathophysiology of SMD remained a riddle for almost seven decades despite intensive efforts, making it one of the most rigorous diseases of the 20th century in India (Nene, 1995). Improvement was achieved in 1999, which finally resulted in the full manifestation of the SMD-charged virus (Kumar et al., 2000).

12.3.1 YELLOW MOSAIC DISEASE

Family: Geminiviridae
Genus: *Begomovirus*

The yellow mosaic disease of pigeon pea is appearing in several agroclimatic areas of India (Biswas et al., 2008). The pigeon pea yellow mosaic reportedly gave rise to the mung bean yellow mosaic virus (MYMV) in the area of whitefly transmission (*Bemisia tabaci*) and symptomatology (Nene, 1972). It is noticed in areas north and south of Delhi, in the province of Andhra Pradesh, and in Karnataka (Muniyappa and Veeresh, 1984; Manjunatha et al., 2015). The virus was found in naturally infected pigeon plants, and the germ particles were 15–18 × 30 nm (Muniyappa et al., 1987). Infection of the mung bean yellow mosaic India virus (MYMIV) within pigeon pea plants through whitefly transmission was found in the mung bean infected with MYMIV as a source plant (Biswas et al., 2008). Based on the sequence of proteins of the *Begomovirus* that causes yellow mosaic disease in Karnataka, pigeon peas can be closely related to the pulse yellow mosaic virus and the legume yellow mosaic virus (Manjunatha et al., 2015). The virus was transmitted to healthy pigeon plants from plants with whiteflies (Raj et al., 2005). The genetic sequence of coat proteins from these plants was found to be closely related to the different types of tomato leaf curl Indian capital virus (ToLCNDV).

Symptoms: Symptoms of the disease are marked as yellow mosaic, mottling, shortening of decayed leaves, and producing only a few pods.

FIGURE 12.1 Yellow mosaic disease of pigeon pea.

Source: Reprinted with permission from Mediga et al., 2023.

12.3.2 RHYNCHOSIA MOSAIC DISEASE

Family: Geminiviridae
Genus: *Begomovirus*

Rhynchosia mosaic (RhM) is a disease of *Rhynchosia minima* (L.) DC., which is an ubiquitous weed. The virus spreads by the Sida biotype of *Bemisia tabaci* (Genn.) and has been named RhM virus-PR (RhMV-PR) (Idris et al., 2002) (same name, *Rhynchosia mosaic virus* [RMV]). The RhM signs in *R. minima* included a foliar yellow mosaic and stunting. The virus has a wide range of tests and infects species of Fabaceae, including *R. minima*, pigeon pea (*Cajanus cajan* (L.) Millsp.), and *Clitoria falcate* (L.) (Idris et al., 2002). However, to date, RhMV has not been identified in naturally infected pigeon peas or *Clitoria falcata. B. tabaci,* as a vector, transferred whitefly from the same source to test plants, *R. minima* and *C. cajan,* confirming that the inoculated hosts were indistinguishable from both isolates.

Symptoms: Symptoms include severe stunting, leaf deformities, and a greenish-yellow mosaic on the leaves.

12.3.3 STERILITY MOSAIC DISEASE

Family: Rhabdoviridae
Genus: *Emaravirus*

Pigeon pea sterility mosaic virus (PPSMV), a genus *Emaravirus*, causes sterility mosaic disease (SMD) in pigeon pea. The disease, called "green disease," as affected plants live in the vegetation area in the absence of flowers, has been reported in many countries in Southeast Asia and India. SMD is estimated to lead to a loss of more than US$300 million in India per year alone. The etiology of SMD, which remained a mystery for more than 70 years, was resolved with the discovery of PPSMV in 2000 and its complete genetic sequence in 2014.

Symptoms: The differential diagnosis of SMD remained a mystery for over 70 years, despite many attempts that did not produce a clear response but suggested the possible participation of a virus based on symptoms and modes of transmission (Nene, 1995). The inconsistent association of the species of eriophyid mite *Aceria cajani* with SMD led to speculation that the disease may be caused by the toxic effect of feeding on eriophyid mites. However, more detailed transmission studies with eriophyid larvae grown in healthy plants showed that mosaic symptoms are not caused by insect feeding, which strongly indicates a causal correlation (Ghanekar et al., 1992). The first evidence of consistent viral transmission was given in 1999, and it was named PPMSV

(Kumar et al., 2000; Kumar et al., 2002; Kumar et al., 2003). A termination of flower production (offspring), overgrowth of plants, stagnation, chlorotic rings or mosaic markings on leaves, and a decrease in their size are the symptoms of plants affected by SMD. However, the character that appears depends on the genotype type, and therefore, the time of infection. New growth from pigeon peas—the method of harvesting plants by leaving the roots and lower parts uncut to allow new growth from the stems—shows more severe symptoms. The disease is often referred to as "green disease" as infected plants are pale green and overgrown and do not bloom well. Under favorable conditions, it spreads throughout the fields by way of disease. In addition, the disease causes the plants to become infected with fungal infections and colonized by spider larvae. Yield losses due to SMD vary and depend on genotype type and stage of infection; infection before flowering (45 days after planting) can result in a yield loss of 26% to 97% (Kannaiyan et al., 1984).

FIGURE 12.2 (A and B) Sterility mosaic disease of pigeon pea.

Source: Reprinted with permission from Sayiprathap et al, 2020. Copyright © 2020, Indian Phytopathological Society

Management: Acaricides and insecticides, often used for chemical treatment of vectors that transmit viral infections (Hoy, 2011; Marcic, 2012; Van Leeuwen et al., 2010), are rarely used to control eriophyid larvae to treat SMD. Biological control using entomopathogenic fungi is also a choice because eriophyid mosquitoes are organisms with no cuticular barriers, and therefore fungi commonly enter the cuticle (McCoy, 1996). However, the best way to minimize disease-related losses is through the use of host-treated plants or the transmission of disease-resistant species. The latest advancement in SMD research has made it easier to select the most productive varieties with strong resistance to SMD. Origins of resistance are recognized within the germplasm group at ICRISAT, Patancheru, India (Kumar et al., 2008; Sharma et al., 2012). However, differentiation of PPMSV occurs in several provinces of India, and widespread resistance to any or all of those strains is rare in genotypes that have been planted.

12.4 VIRAL DISEASES OF MUNG BEAN

The gram of gold [*Vigna radiata* (L.) Wilczek, Syn., *Phaseolus aureus* Roxb., and *Phaseolus radiatus* L.], belonging to the Fabaceae family, is one of the 13 edible crops grown in India and the third most important pulse plant in India. In the region of approximately 3.43 Mha (approximately 15% of the area of the national pulse crop), the production of 1.71 M tons of 498 kg/ha (Anonymous Project Co-ordinator's Report. 2012) of grain production to meet worldwide demand is important to increase the current global production rate (400 kg/ha) in addition to increasing yields in new areas (Nair et al., 2012). Mung bean has many common names, namely, Green gram, Mungo, Mung, Moong, and Golden bean. It is located in the India–Burma region of southeast Asia. In India, *Vigna radiata* is most abundant in the provinces of Rajasthan, Madhya Pradesh, Maharashtra, Karnataka, Odisha, Haryana, Gujarat, Delhi, and Bihar in the times of Kharif and Zaid. Maharashtra ranks first within the area under cultivation of mung bean, followed by Rajasthan and Karnataka, while in terms of productivity, Maharashtra is followed by Andhra Pradesh and Bihar. Being a crop of tropical and subtropical regions, *Vigna radiata* is in a position to tolerate high temperatures as high as 40°C. It is considered "poor men's protein," excluding 26% protein, contains 10% moisture, 51% carbohydrates, 4% minerals, and three vitamins. Mung bean is grown mainly for its high-protein seeds, which are used as human food by cooking, boiling, digesting, or germinating and used to make soups, curries, pieces of bread, sweets, noodles, salads, papad, etc. The mung bean crop is packed with a few biotic and abiotic features. Among the organisms, the *mung bean yellow mosaic virus* (MYMV), urdbean *leaf crinkle virus* (ULCV), *mung bean leaf curl virus* (MLCV), and *mosaic mottle virus* (BCMV) are critical in reducing yields.

12.4.1 *MUNG BEAN YELLOW MOSAIC DISEASE*

The yellow mosaic disease is the most serious cause of mung bean cropping. The emergence of the *mung bean yellow mosaic virus* was first reported in Delhi, India (Narian, 1960). Maharashtra ranks first within the area under cultivation of mung bean, followed by Rajasthan and Karnataka, while in terms of productivity, Maharashtra is followed by Andhra Pradesh and Bihar. Being a crop in several regions, *Vigna radiata* is in a position to tolerate high temperatures as high as 40°C. Yield losses because of this disease differ from 5 to 100% depending on the seriousness of the disease, the tendency of the crop, and the number of whiteflies (Rathi, 2002; Karthikeyan et al., 2014; Singh et al., 1982). Infection not only significantly reduces yields but also slows pod growth (Rishi, 2009) and reduces the number of pods per plant. In mung beans, the occurrence of yellow mosaic disease in farmers' fields may be up to 100 (Singh et al., 1979). Weeds are a source of infection and a source of primary inoculum. The virus first spreads when the monsoon begins in India. It is transmitted via whitefly, poinsettia strain, and grafting under commendatory conditions in a very circular, nondiffused, and interconnected manner. It is not a seed, soil, or liquid that is transmitted.

Symptoms: Initially, yellow spots or small patches appear on the green lamina of leaves. It soon becomes a bright yellow mosaic or a golden yellow mosaic. The yellow color changes gradually, and the green leaves completely turn yellow. Maturity is delayed in diseased plants, and the formation of flowers and pods is greatly reduced. Seeds that shoot up in highly infected plants are small and immature. The pods are short and distorted. The early infection causes the plant to die before the production of seeds.

FIGURE 12.3 Mung bean yellow mosaic disease.

Source: Department of Seed Science and Technology, 2020

12.4.2 LEAF CRINKLE DISEASE

Family: Geminiviridae
Genus: *Begomovirus*

ULCV disease was reported to be a natural disease of the Indian Ocean in India by Singh (Kadian, 1982). ULCV has the potential to cause significant crop losses in mung bean beans. In newly infected plants, seed production may fail. The grain yield losses range from 2–95% in mung beans depending on plant age and symptom onset (Singh, 1980). The virus also reduced crop yields such as plant length, root length, nodules, pods/plant, pod length, seed/pod, and 100 seed weight, but increased the area of leaves and dry weight of roots and shoots (Bharadwaj and Dubey, 1986). The *ULCV* is spread through milk, grafts, insects, and seeds, as reported. Seeds are transferred to 15% of mung bean beans. Many workers have reported its infection by various insects, namely, aphids, *Aphis craccivora* and *A. gossypii* (Dhingra, 1975; Beniwal, 1983); beetle, *Henosepilachna dodecastigma (*Narayansami and Jaganathan, 1974*)*; and whitefly, *Bemisia tabaci* (Singh, 2013). Also, the presence of weed controllers, such as *Aristolochia bracteata* and *Digera arvensis, is seen*. The Kharif season is highly susceptible.

Symptoms: The first symptoms appear on small leaves, such as chlorosis around certain lateral veins and their branches near the line. The leaves show the curvature of the border facing downward. Some leaves show twisting. The veins show a reddish-brown color change in the lower back and extend to the petioles. Symptomatic plants 5 weeks after sowing remain fragile, and most deaths are due to high necrosis within 1 or 2 weeks. Plants infected with

FIGURE 12.4 Leaf crinkle disease of mung bean (Bashir, 2006).

Source: Batzer et al., 2022, https://creativecommons.org/licenses/by/4.0/

this disease at the end of the germinating phase do not show great curvature and twisting of the leaves but do show chlorosis of bright veins anywhere in the lamina of the leaf. Petioles, as well as internodes, are abbreviated. The infected plant gives it a dry and bushy appearance. Flowering is delayed; the inflorescence, when formed, is poorly arranged with small flowers and fails to open. Pollen reproduction and pod formation are greatly reduced in infected plants.

12.4.3 MUNG BEAN LEAF CURL DISEASES

Family: Tospoviridae
Genus: *Orthotospovirus*

ULCV was reported in natural infections of mung bean in India (Makkouk, 2003). ULCV can cause heavy yield losses in the mung bean. In early-infected plants, seed production may fail. The grain yield losses range from 2–95% in mung beans, depending on the age of the plant and symptom onset (Kaiser and Mossahebi, 1974). The virus also reduces yield components like plant height, root length, nodulation, pods/plant, pod length, seeds/pod, and weight of 100 seeds, but it increases the leaf area and dry weight of roots and shoots (Tsuchaizaki et al., 1996). *ULCV* is transmitted through the sap, grafting, insect vectors, and seeds, as reported. It has been seed-transmitted to an extent of 15% in mung bean. Many workers have reported its transmission by various insects, namely, aphids, *Aphis craccivora* and *A. gossypii* (Talens and Talens, 1978; Walter and Mossahebi, 1974); beetle, *Henosepilachna dodecastigma* (Shanmugasundaram, 1988); and whitefly, *Bemisia tabaci* (Dubey and Singh, 2013). The presence of weed hosts like *Aristolochia bracteata* and *Digera arvensis is also seen*. The Kharif season crop is highly susceptible.

Symptoms: The initial symptoms visible on the youngest leaves are chlorosis around some lateral veins and its branches near the tip. The leaves show curling of the margin downward. Some leaves show twisting. The veins show reddish-brown discoloration on the undersurface, which also extends to the petioles. Plants that show symptoms within 5 weeks after sowing invariably remain stunted, and the majority of these deaths are due to top necrosis within 1 or 2 weeks. Plants infected in the late stages of growth do not show severe curling and twisting of the leaves but show conspicuous veinal chlorosis anywhere on the leaf lamina. Petioles, as well as internodes, are shortened. The infected crop has a stunted and bushy appearance. Flowering is delayed; inflorescence, if formed, is malformed with small flower buds and fails to open. Pollen fertility and pod formation are severely reduced in infected plants.

FIGURE 12.5 Leaf curl disease of mung bean.

Image credit: Howard F. Schwartz, Colorado State University, Bugwood.org

12.4.4 MOSAIC MOTTLE DISEASE

Family: Potyviridae
Genus: *Potyvirus*

In 1968, mosaic mottle disease of the mung bean was first reported from Iran, and it is also quite common in India (Ghosh et al., 2009). The virus infects the mung bean and urdbean and is distributed in several countries in southeast Asia, including the Philippines and Thailand (Zeshan et al., 2012). The virus is spread both through seed and sap. It is transmitted by seeds in mung bean at a rate of 8–32% and by aphids (*Aphiscra ceivora* and *A. gossypii*).

Symptoms: The disease initially appears in the form of irregular light green patches, alternating with normal green areas. The size of the leaf is reduced, and the edges show upward rolling. Later, the young leaves show puckers and blisters, the normal green areas appear on both sides of the long veins, and the leaves become rough and brittle. Affected plants are stunted and often show excessive branches. When the infection is severe, the entire inflorescence is transformed into a leafy structure, resulting in complete seed loss.

Management:

- To grow resistant varieties, that is, PantMung-6, Pusa Vishal, Basanti, ML267, ML-337, PDM-54, PDM 139, LGG407, LGG-460, Narendra mung-5, HUM1, TM-99-37, HUM-16, TARM-1, Pusa95-31, Sweta, and Samrat against mung bean yellow mosaic virus, mosaic mottle disease against T-2, Hyb-43, L-51-1 and L-293 29,30, mung bean leaf curl virus against SML 99, SML 100 resistant 31 and Pant Mung-3 moderately resistant.
- Handling sowing dates and plant density.
- Manuring with two rows of maize (60 × 30 cm) or sorghum (45 × 15 cm) in all 15 rows of green gram prevents suitable space for vectors and reduces their movement.
- Removal of associated weeds and eradicating diseased plants.
- Treatment of seeds with hot water at 52°C for 20–30 min to reduce seed infection.
- Thiamethoxam is a seed treatment (4 g/kg seed) that protects a healthy plant from MYMV with its following spray (0.02%).
- Seed treatment with Imidacloprid-70WS @ 4 g/kg seeds and the next two sprays in 21 to 35 days protects the crop up to the stage of pod formation.
- The application of phosphorus, nitrogen, zinc, boron, potash, and naphthalene acetic acid (NAA) effectively reduces ULCV infection.
- Six foliar sprays of 10% aqueous root extract of *Boerhavia diffusa* and eight sprays of leaf extract of *Clerodendrum aculeatum* are very effective and suppress symptomatic severeness by 62–90% under natural conditions.
- The use of neem extract at 2% reduced the risk of ULCV infection.
- Neem seed kernel extract at 5% and neem oil at 2% will work as effectively as monocrotophos 0.05%.
- Foliar spray of Thiomethaxam 0.02% at 15 days after showing successively decreased whitefly population and incidence of leaf crinkle.

12.5 VIRAL DISEASES OF CHICKPEA

Chickpea (*Cicer arietinum* L.) is grown in different regions of the world; commonly, two types of chickpeas are cultivated: "Kabuli" type and "desi" type. Some of the viruses that infect chickpeas are discussed below (Characterization, Diagnosis, & Management of Plant Viruses).

12.5.1 ALFALFA MOSAIC VIRUS

Taxonomy: Family—*Bromoviridae*, Genus—*Alfamovirus*

Symptoms: It can infect more than 600 plant species, including many economically important crops like pea, tomato, potato, tobacco, etc. Its symptoms in chickpeas include less production of pods, premature death of plantlets, dwarfing, proliferation, and chlorosis of axillary shoots.

Transmission: *Alfalfa mosaic virus* is transmitted by aphids via sap and seeds.

Management: As the virus is transmitted nonpersistently by aphids, insecticides have limited use in preventing transmission of the virus. Resistance against the *alfalfa mosaic virus* in legumes has not been reported (Characterization, Diagnosis, & Management of Plant Viruses). The management approaches applied to this virus are:

- Vector (aphid) control
- Usage of virus-free and healthy seeds
- Over-wintering host by spatial separation for control of virus spread

12.5.2 BEAN LEAF ROLL VIRUS

Taxonomy: Family—*Luteoviridae*, Genus—*Polerovirus*

Symptoms: This virus infects only plants of leguminous species. It shows various symptoms in chickpeas, like yellowing or discoloration of leaves, thickening of stems and leaves, and internodes shortening, which leads to stunting and browning of the phloem. This disease is also called "chickpea stunt disease" because in early infection, stunting is easily observed.

Transmission: The *bean leaf roll virus is transmitted by many species of aphids.* This virus is not transmitted via seeds or saps.

Management: There are some chickpea varieties known that show resistance against *the bean leaf roll virus (*Johnstone and Rapley, 1979*)*. The management approaches applied to the virus are:

- Use of aphicides
- Treatment of seeds with Imidacloprid before sowing
- By manipulating the planting date

12.5.3 BEET WESTERN YELLOWS VIRUS

Taxonomy: Family—*Luteoviridae*, Genus—*Polerovirus*

Symptoms: This virus infects plant members of the Brassicaceae, Compositae, Fabaceae, etc. Its symptoms in chickpeas include stunting and yellowing, discoloration of the phloem, and a loss in grain production.

Transmission: *Beet western yellows virus* is transmitted by many species of aphids.

Management: There are no resistant chickpea varieties to the virus (Characterization, Diagnosis, & Management of Plant Viruses). Viral incidence is minimized by cultural practices like:

- Sowing density
- Planting date
- Identifying and removing infected plants early in the season
- Far agricultural site from chickpea irrigation area

12.5.4 CHICKPEA STUNT DISEASE-ASSOCIATED VIRUS

Taxonomy: Family—*Luteoviridae*, Genus—*Polerovirus*

Symptoms: The virus causes symptomless infection in many plant species, but its natural infection is only reported in chickpeas. Its symptoms in chickpeas are similar to *those of the beet western yellows virus* (described earlier in this chapter).

Transmission: *M. persicae* and *A. craccivora* are vectors of this virus that transmit the virus in a propagative and circulative way. Mechanical inoculation is not a means of transmission for this virus (Characterization, Diagnosis, & Management of Plant Viruses).

Management: There is no evidence, but a likely effective approach to management is through cultural practices like:

- Growing chickpeas far from infection sources
- Rouging infected plants
- Removal of weeds
- Adjusting planting dates and sowing densities

12.5.5 CHICKPEA CHLOROTIC DWARF VIRUS

Taxonomy: Family—*Geminiviridae*, Genus—*Mastrevirus*

Symptoms: This virus is not restricted to but is also found in members of the Solanaceae, Chenopodiaceae, and Fabaceae. It shows various symptoms in chickpeas, like reddening of leaves and stunting.

Transmission: *Leafhopper Orosius orientalis (Matsumura) is the vector of the chickpea chlorotic dwarf virus.* Mechanical inoculation does not transmit the virus.

Management: Some local varieties of chickpeas are less prone to the virus. To significantly reduce the viral incidence, it is suggested to hold up for 3–4 weeks from the date of planting with brief irrigation intervals (Hamed and Makkouk, 2002).

12.5.6 CUCUMBER MOSAIC VIRUS

Taxonomy: Family—*Bromoviridae*, Genus—*Cucumovirus*

Symptoms: It infects chickpeas, fava beans, lentils, and peas. The number of flowers and pod production drastically decreases in diseased plants. Bushiness and stunting are the main symptoms of this virus in chickpeas.
Transmission: The virus is transmitted by many species of aphids; it is also seed-transmitted and spread mechanically.

Management: The transmission of the virus is mainly controlled by prevention (Characterization, Diagnosis, & Management of Plant Viruses). Some ways of management are:

- Development of new varieties resistant to the virus by biotechnology or by the breeding method
- Insecticide spray
- Planting barrier crops and covering the ground with aluminum foil boards
- Spraying crops with mineral oil

12.5.7 FABA BEAN NECROTIC YELLOWS VIRUS

Taxonomy: Family—Nanoviridae, Genus—*Nanovirus*

Symptoms: This virus infects only members of the family Fabaceae. This virus shows symptoms like stunting, interveinal chlorosis, and thickening of leaves. Usually, infected plants expire within a few weeks after infection.

Transmission: This virus is transmitted by three aphids: *A. craccivora, A. fabae*, and *A. pisum*, cumulatively (Makkouk et al., 2001).
Management: The most effective management approaches are:

- One-to-two times aphicide spray
- Seed treatment before treatment with Imidacloprid
- Appropriate seeding rate
- Observing and removing infected plants during the early season

KEYWORDS

- **virus**
- **vectors**
- **mosaic**
- **pulses**
- **management practices**

REFERENCES

Anonymous Project Co-ordinator's report AICRP on MULLaRP crops, IIPR, Kanpur. 2012.

Bashir, M. Occurrence and Distribution of Viral Diseases of Mungbean and Mashbean in Punjab, Pakistan. *Pakistan J. Bot.* **2006,** *38.*

Batzer J. C., Singh A, Rairdin A, Chiteri K, Mueller D S,. Mungbean: A Preview of Disease Management Challenges for an Alternative U.S. Cash Crop, Journal of Integrated Pest Management, Volume 13, Issue 1, 2022, 4, https://doi.org/10.1093/jipm/pmab044

Beniwal, S. P. S. *Urdbean Leaf Crinkle Virus*, Viruses of Legumes. In *Descriptions and Keys from VIDE*; Boswell, K. F., Gibbs, A. J., Eds.; Austral. Natl. Univ.: Canberra; 1983, pp 85–86.

Bharadwaj, S. V.; Dubey, G. S. Studies on the Relationship of *Urdbean Leaf Crinkle Virus* and its Vectors, *Aphis Craccivora* and *Acyrtho Siphompisum. J. Phytopath.* **1986,** *83,* 115.

Biswas, K. K., et al. Diagnosis of Symptomless *Yellow Mosaic Begomo Virus* Infection in Pigeonpea by Using Cloned *Mungbean Yellow Mosaic India virus* as Probe. *J. Plant Biochem. Biotechnol.* **2008,** *17,* 09–14.

Characterization, Diagnosis, & Management of Plant Viruses; Rao, G. P., Kumar, P. L., Holgium-Pena, R. J. Eds.; Stadium Press: Texas, USA. Vol. 3: Vegetable and Pulse Crops, pp 213–234.

Chibbar, R. N.; et al. Molecular Diversity in Pulse Seed Starch and Complex Carbohydrates and its Role in Human Nutrition and Health. *Cereal Chem.* **2010,** *87,* 342–352.

Department of Seed Science and Technology, CCS, HAU, Hisar: Online facebook accessed on 10.04.2022: htts://m.facebook.com/2204097723194243/posts/yellow-mosaic.

Dhingra, K. L. Transmission of *Urdbean Leaf Crinkle Virus* by Two Aphid Species. *Indian Phytopath.* **1975,** *28,* 80.

Dubey, S. C.; Singh, B. Integrated Management of Major Diseases of Mungbean by Seed Treatment and Foliar Application of Insecticides, Fungicides and Bioagent. *Crop. Protection.* **2013,** *47,* 55–60.

FAO. *Food and Agriculture Organization of the United Nations*; Rome, Italy; 2019.

FAOSTAT. Food and Agriculture Organization of the United Nations, 2014.

Ghanekar, A. M., et al. Sterility Mosaic of Pigeonpea. In *Plant Diseases of International Importance*, Vol. 1, *Diseases of Cereals and Pulses*; Singh, U. S., Mukhopadhyay, A. N., Kumar, J., Chaube, H. S., Eds.; 1992, pp 415–428.

Ghosh, D.; Laha, S. K.; Biswas, N. K. Effect of Different Pesticides on Incidence of *Mungbean Yellow Mosaic Virus. Int. J. Plant Protect.* **2009,** 2 (1), 67–70.

Graham, P. H.; Vance, C. P. Legume Importance and Constraints to Greater Use. *Plant Physiol.* **2003,** *131,* 872–877.

Guillon, F.; Champ, M. M. Carbohydrate Fractions of Legumes: Uses in Human Nutrition and Potential for Health. *Br. J. Nutr.* **2002,** *3,* 293–306.

Hamed, A. A.; Makkouk, K. M. Occurrence and Management of Chickpea Chlorotic Dwarf Virus in Chickpea Fields in Nouthern Sudan. *Phytopathol. Mediterranea* **2002,** *34,* 193–198.

Hoy, M. A. *Agricultural Acarology: Introduction to Integrated Mite Management*; CRC Press, Taylor & Francis Group: Boca Raton, FL, 2011.

Idris, A. M.; Bird, J.; Rogan, D. M.; Brown, J. K. Molecular Characterization of *Rhynchosia Mosaic Virus*-Puerto Rico Associated with Symptomatic *Rhynchosia minima* and *Cajanus Cajan* in Puerto Rico. *Plant Dis.* **2002,** 86 (5), 558.

Jaya, T. V., et al. Effect of Germinated Legumes on the Rate of in-vitro Gas Production by *Clostridium perfringens*. *Nutr. Rep. Int.* **1979,** *20,* 393–401.

Johnstone, G. R.; Rapley, P. E. L. The Effect of Time of Sowing on the Incidence of Subterranean Clover Red Leaf Virus in Broad Bean (*Vicia faba*). *Ann. Appl. Biol.* **1979,** *91,* 345–351.

Kadian, O. P. Yield Loss in Mungbean and Urdbean Due to Leaf Crinkle Virus. *Indian Phytopath.* **1982,** *35,* 642–644.

Kaiser, W. J.; Mossahebi, G. H. Disease of Pulse Crops (Edible legumes). *Plant Dis. Rep.* **1974,** *52,* 683–691.

Kannaiyan, J.; et al. Prevalence of Pigeonpea Diseases and Associated Crop Losses in Asia, Africa, and the Americas. *Trop. Pest Manag.* **1984,** *30,* 62–71.

Kaplan, L.; Lynch, T. F. *Phaseolus* (Fabaceae) in Archaeology. AMS. *Econ. Bot.* **1999,** *53,* 261–272.

Karthikeyan, A.; et al. *Mungbean Yellow Mosaic Virus* (MYMV): A Threat to Green Gram (*Vigna radiata*) Production in Asia. *Int. J. Pest Manag.* **2014,** *60* (4), 314–324.

Kumar, P. L.; et al. Breakthrough in the Identification of the Causal Agent of Pigeonpea Sterility Mosaic Disease. *J. Mycol. Plant Pathol.* **2000,** *30,* 249.

Kumar, P. L.; Jones, A. T.; Reddy, D. V. R. A Novel Mite-transmitted Virus with a Divided RNA Genome Closely Associated with Pigeonpea Sterility Mosaic Disease. *Phytopathology* **2003,** *93,* 71–81.

Kumar, P. L.; Jones, A. T.; Reddy, D. V. R. Mechanical Transmission of *Pigeonpea Sterility Mosaic Virus*. *J. Mycol. Plant Pathol.* **2002,** *32,* 88–89.

Kumar, P. L.; Jones, A. T.; Waliyar, F. Virus Diseases of Pigeonpea. In *Characterization, Diagnosis, and Management of Plant Viruses*. Vegetable and Pulse Crops 2008; 3: 235–258. Stadium Press: Texas, USA

Kumar, P. L.; Jones, A. T.; Waliyar, F. Virus Diseases of Pigeonpea. In *Characterization, Diagnosis and Management of Plant Viruses*. Vol. 3: *Vegetable and Pulse Crops*; Rao, G. P., Kumar, P. L., Holguin-Pena, R. J., Eds.; 2008, pp 235–258.

Kumar, S.; Subbarao, B.; Hallan, V. Molecular Characterization of Emaraviruses Associated with Pigeonpea Sterility Mosaic Disease. *Sci. Rep.* **2017,** *7,* 11831.

Lev-Yadun, et al. The Cradle of Agriculture. *Science* **2000,** *288,* 1602–1603.

Makkouk, K. M. Other Legumes. In *Virus and Virus Like Disease of Major Crops in Developing Countries*; Leobension, D., Thollapilly, G., Eds.; Klieur Academic Publisher; 2003; pp 447–476.

Makkouk, K. M.; Kumari, S. G.; Sarker, A.; Erskine, W. Registration of Six Lentil Germplasm Lines with Combined Resistance to Viruses. *Crop. Sci.* **2001,** *41,* 931–932.

Manjunatha, N.; et al. Molecular Detection and Characterization of Virus Causing Yellow. *Int. J. Pure App. Biosci.* **2015,** *3,* 258–264.

Marcic, D. Acaricides in Modern Management of Plant-feeding Mites. *J. Pest. Sci.* **2012,** *85,* 395–408.

Marlett, J. A.; McBurney, M. I.; Slavin, J. L. Position of the American Dietetic Association: Health Implications of Dietary Fibre. *J. Am. Diet. Assoc.* **2002,** *102,* 993–1000.

McCoy, C. W. Pathogens of Eriophyoid Mites. In *Eriophyoid Mites—Their Biology, Natural Enemies and Control*; Lindquist, E. E., Sabelis, M. W., Bruin, J.; eds.; 1996, pp 481–490.

Mediga, K.R.; Sunkad, G.; Kulkarni, S.; Chandran, U.S.S.; Ghosh, R.; Kshirsagar, D.; Sonnappa, M.; Katravath, S.; Parthasarathy, A.; Sharma, M. Multi-Environment Testing Based G × E Interactions Reveal Stable Host-Plant Resistance against Sterility

Mosaic Disease in Pigeonpea. Agronomy 2023, 13, 2859. https://doi.org/10.3390/agronomy13122859

Mitra, M. Report of the Imperial Mycologist. *Sci. Rep. Agric. Res. Inst. Pusa* **1931,** *19,* 58–71.

Muniyappa, V.; et al. Isolation and Characterization of Gemini Virus Causing Yellow Mosaic Disease of Horsegram (*Macrotyloma uniflorum* (Lam.) Verdc.) in India. *J. Phytopathol.* **1987,** *119,* 81–87.

Muniyappa, V.; Veeresh, G. K. Plant Virus Diseases Transmitted by Whiteflies in Karnataka. *Proc. Indian Acad. Sci. (Anim. Sci.)* **1984,** *93,* 397–406.

Nair, R. M., et al. Genetic Improvement of Mungbean Productivity. In *Proc. of the12th ABRAO Congress on Plant Breeding towards 2025: Challenges in a Rapidly changing World*; 2012, pp 27–28.

Narayansami, P.; Jaganathan, T. Characterization of *Blackgram Leaf Crinkle Virus*. *Madras Agric. J.* **1974,** *16,* 979.

Narian, T. K. Yellow Mosaic of Mungbean (*Phaseolus aeurcus*). *Indian Phytopath.* **1960,** *13,* 24–29.

Nene, Y. L. A Study of Viral Diseases of Pulse Crops in Uttar Pradesh. Res. Bull., No.4. G. B. Pant. Unvi. Agri. Tech., Pantnagar. 1972; 144.

Nene, Y. L. Sterility Mosaic of Pigeonpea: The Challenge Continues. *Indian J. Mycol. Plant Pathol.* **1995,** *25,* 1–11.

Newton, W.; Peiris, J. W. L. Virus Diseases of Plants in Ceylon. *FAO Plant Protect. Bull.* **1953,** *2,* 17–21.

Raj, S. K.; Khan, M. S.; Singh, R. Natural Occurrence of a Begomovirus on Pigeonpea in India. *New Disease Rep.* **2005,** *11,* 4.

Rathi, Y. P. S. Epidemiology, Yield Losses and Management of Major Diseases of Kharif Pulses in India. In: Plant Pathology and Asian Congress of Mycology and Plant Pathology, Oct 1–4, University of Mysore, Mysore, India. Management technology. *J. Gen. Plant Pathol.* **2002,** *75,* 1–18.

Reddy, M. V.; Raju, T. N.; Lenne, J. M. Diseases of Pigeonpea. In *The Pathology of Food and Pasture Legumes*; Allen, D. J., Lenne, J. M., Eds.; 1998, pp 517–558.

Rishi, N. In *Significant Plant Virus Diseases in India and a Glimpse of Modern Disease*; Presence of Urdbean Leaf Crinkle Virus in Seeds of Mungbean Germplasm. *Indian Phytopath.* 2009, 33; p 360.

Sayiprathap, B. & Patibanda, Anil & Kumari, V. & Jayalalitha, K. & Vatluri, Srinivasa & Sharma, Mamta & Sudini, Hari. (2020). Prevalence of sterility mosaic disease (SMD) and variability in pigeonpea sterility mosaic virus (PPSMV) in southern-India. Indian Phytopathology. 73. 10.1007/s42360-020-00246-7.

Schwartz, F. H. *Colorado State University*, Bugwood.org.

Shanmugasundaram, S. A. *Catalog of Mungbean Cultivars Released Around the World.* Asian Vegetable Research and Development Centre: Shanhua, Taiwan, 1988.

Sharma, M.; et al. New Sources of Resistance to Fusarium Wilt and Sterility Mosaic Disease in a Mini-core Collection of Pigeonpea Germplasm. *Eur. J. Plant Pathol.* **2012,** *133,* 707–714.

Siddique, K.; et al. Innovations in Agronomy for Food Legumes: A Review Agron. *Sustain. Dev.* **2012,** *32,* 45–64.

Singh, B. R.; et al. Yield Loss in Mungbean Due to Yellow Mosaic. *Sci. Cult.* **1982,** *48,* 435–436.

Singh, J. P. Effect of Virus Diseases on Growth Component and Yield of Mungbean (*Vigna radiata*) and Urdbean (*Vigna mungo*). *Indian Phytopath.* **1980,** *8,* 405–08.

Singh, J. P.; Kadian, O. P.; Verma, J. P. Survey of Virus Diseases of Mungbean and Urdbean in Haryana. *Haryana Agric. Univ. J. Res.* **1979,** *9,* 345.

Singh, R. G. Management of Insect Vectors in Mungbean and Urdbean. In *Advances in Pest Management in Legume Crops*; Saini, R. K., Sharma, S. S., Eds.; Campendium of CAFT, CCS Haryana Agriculture University: Hisar; 2013, pp 196–209.

Talens, L. T.; Talens, A. Mungbean Viruses in the Philippines II. Host Legume, Symptomatology and Effect of Mungbean Mottle Virus Growth of Mungbean (*Vigna radiata* (L.) Willczek). *Philippines J. Crop. Sci.* **1978,** 3 (4), 242–246.

Tosh, S. M.; Yada, S. Dietary Fibre in Pulse Seeds and Fractions: Characterization, Functional Attributes, and Applications. *Food Res. Int.* **2010,** *43,* 450–460.

Tsuchaizaki, T.; et al. Bean Common Mosaic Virus Isolated From Mungbean (*Vigna radiata*) in Thailand. *Tech. Bull. Trop. Agric. Res. Centre* **1996,** *21,* 184–188.

United States Department of Agriculture (USDA) National Nutrient Database for Standard Reference; 2013.

Upadhyaya, H.; et al. Enhancement of the use and Impact of Germplasm in Crop Improvement. *Plant Genet. Resour.* **2014,** *12,* 155–159.

Van Leeuwen, T.; et al. The Control of Eriophyoid Mites: State of the Art and Future Challenges. *Exp. Appl. Acarol.* **2010,** *51,* 205–224.

Walter, J. K.; Mossahebi, G. H. Natural Infection of Mungbean by *Bean common Mosaic Virus*. *Phytopathology* **1974,** *64,* 1209–1214.

Zeshan, A. M.; et al. Role of Nutrients and Naphthalene Acetic Acid in the Management of *Urdbean Leaf Crinkle Virus*. *Pak. J. Phytopathol.* **2012,** *24,* 79–81.

CHAPTER 13

Viral Diseases of *Ipomoea batatas* and Their Control and Management Strategies

NEHA RAUNIYAR and DEEPA SRIVASTAVA

Department of Botany, Deen Dayal Upadhyay Gorakhpur University, Gorakhpur, India

ABSTRACT

In the progression of history till date, viral diseases have ranged from simple marginal infection to several severe plagues. Since there is a large variation in the epidemiology and pathogenesis of different viruses, it is hard to deter a single approach for their management and control as each virus, affecting either plant or animal, has its own set of specificity. The most effective evolution in the management of the viruses are the development of vaccines along with vector control and maintenance of sanitation along with some known antiviral drugs which widely help in the control of these viruses. Threat to sustainable agriculture and huge amount of economic loss are the common effects of plant viral diseases. The frequency with which the viruses are evolving is due to the wide range of climate change, the ability of rapid evolution of virus, and, in some aspects, due to international trade. The control of the disease could be easily overcome if at the very initial stage the causal agent is detected or identified as methods could be developed to remove the infection, cross protect the plant, control the natural vectors, etc. The development of pathogen-resistant strategies varies from coat protein-based interference of virus production to RNA-mediated virus gene silencing. This chapter aims to highlight several viruses which are reported in *Ipomoea batatas*, their symptomology, and, also, the method which would help manage these viral diseases and various approaches are discussed in detail.

Plant Diseases and Their Management: A Sustainable Approach.
Deepa Srivastava, Rajarshi Kumar Gaur, and Ajay K. Tiwari (Eds.)

13.1 INTRODUCTION

A dicotyledonous plant *Ipomoea batatas* belonging to the family Convolvulaceae is an important tuberous root crop that is mainly cultivated throughout the warm temperature regions along with tropical regions where the amount of water is sufficient to support the growth of the plant (Austin, 1988; Vincent, 2009). Sweet potato is an Edaphic plant that can tolerate a wide range of climatic conditions and is known as a dryland crop. Because of its high tolerance to cold than other tuberous crops and tropical root, it is easy to grow it at high altitude as high as 2500 m (Luisa and Robert, 2000). A growth season of 4–5 months with an optimum temperature of 20–25°C is most beneficial for sweet potato (Stathers et al., 2013). In addition to the existence of protein and Niacin (vitamin B3), sweet potato is known to provide over 90% of essential nutrients for each calorie consumed; this is the foremost reason why it is a seasonal staple when other foodstuffs are in shortage (Tesfaya et al., 2011). It's starch richness, which is representative of more than 50% of the components of carbohydrate, is higher than any other kind of tubers (Ellong et al., 2014). Starch content is nearly 70% of the total dry weight of *Ipomoea batatas* which helps in providing an important source of nutrition and acts as a major source of alcohol, animal fodder, and starch (Woolfe, 1992).

13.2 IPOMOEA BATATAS—AN OVERVIEW OF THE PLANT

Sweet potato has a tuberous root and is a perennial herbaceous climber, its leaves being egg-, oval-, or heart-shaped to up to 10 cm long, either three-lobed or mostly unlobed with margins being toothed. The petals are generally lavender, pale purple, or white in color with trumpet shape of a flower being up to 7 cm long, inside of the tube being darker. The fruits are dry, dehiscent, and ovoid capsule. The classification of *Ipomoea batatus* is described in Table 13.1.

13.2.1 SYNONYMOUS VARIETY OF IPOMOEA BATATAS

The synonymous variety of *Ipomoea batatas* are *Batatas batatas* (L.) Karst., *Batatas edulis* (Thunb. ex Murray) Choisy, *Batatas glaucifolius* (L.) G. Don, Batatas wallii Morr., *Batatas xanthorhiza* Boj., *Convolvulus acuminatus* Salzm. ex Meisn., *Convolvulus apiculata* Martens and Galeotti,

TABLE 13.1 Demonstrates the Classification of Genus *Ipomoea batatus*

Rank	Name (common/scientific)
Kingdom	Plantae
Division	Magnoliophyta
Class	Magnoliopsida
Order	Solanales
Family	Convolvulaceae (morning-glory family)
Genus	*Ipomoea L.* (morning-glory)
Species	*Batatas*
Origin	America (tropical regions)
Tuber color	Red, purple, brown, yellow-brown, white
Shape of tuber	Fusiform, elongated subterranean.
Major nutrients	Carbohydrate (20.58%)
	Copper (22.33%)
	Vitamin A (134.71%)
	Vitamin B5 (21.28%) Vitamin B6 (21.38%)
Health benefits	Enhances immunity, reduces inflammation, cures asthma, treat bronchitis, manage arthritis
Calories present	114 Kcal/cup
Scientific nameame	*Ipomoea batatas* (L.) Lam. (sweet potato)

Convolvulus attenuata Martens and Galeotti, *Convolvulus batatas* L., *Convolvulus candicans* Sol. ex Sims, *Convolvulus chrysorhizus* Forst. ex Wien., *Convolvulus cordatifolius* Vell., *Convolvulus cordifolius* Salzm. ex Meisn., *Convolvulus edulis* Thunb. ex Murray, *Convolvulus esculentus* Vell., *Convolvulus essequebensis* Spreng., *Convolvulus glaucifolius* (L.) Spreng., *Convolvulus macrorhizus* A. Gray ex Choisy, *Convolvulus panduratus* L., *Convolvulus platanifolius* Vahl, *Convolvulus roseus* Mill., *Convolvulus scaber* (J.F.Gmel.) Colla, *Convolvulus septangularis* Steud., *Convolvulus tuberifer* Steud., *Convolvulus tuberosus* Vell., *Convolvulus varius* Vell., *Ipomoea alba* Garcke, *Ipomoea batatas* f. trifida Moldenke, *Ipomoea batatas* var. dissoluta Kuntze, *Ipomoea batatas* var. edulis (Thunb. ex Murray) Makino, *Ipomoea batatas* var. erythrorrhiza (Voigt) M.R.Almeida, *Ipomoea batatas* var. fastigata (Sweet) Kuntze, *Ipomoea batatas* var. leucorrhiza (Voigt) M.R.Almeida, *Ipomoea batatas* var. leucorrhiza Griseb., *Ipomoea batatas* var. lobata Gagnep. and Courchet, *Ipomoea batatas* var. porphyrorhiza Griseb., *Ipomoea candicans* Sweet, *Ipomoea catesbaei* G.

Mey., *Ipomoea davidsoniae* Standl., *Ipomoea edulis* (Thunb. ex Murray) Makino, *Ipomoea fastigiata* var. ciliata Huber, *Ipomoea mucronata* Schery, *Ipomoea purpusii* House, *Ipomoea scabra* Forsk., *Ipomoea vulsa* House, *Ipomoea wallii* (Morren) Hemsl., *Pharbitis forskoeli* G. Don, and *Tirtalia pandurata* (L.) Rafin.

13.2.2 ORIGIN AND DISTRIBUTION

The domestication of *Ipomoea batatas* was done around 6000 AD in tropical America; in 16th century, the Spanish introduced the crop to the Philippines from where it spread to many other islands and mainlands of Asia; the crop was recorded in South China in 1594 to promote drought mitigation. In India, Africa, Europe, and another parts of South-East Asia, the crop was introduced by Portuguese sea farers (Jia, 2013). According to the recent report of Food and Agriculture Organization, UN, there are over 117 countries currently cultivating sweet potato with 104 million ton of production in 2011, annual production is about 80%. Asia is the world's largest producer region of sweet potato along with Oceania 1%, America 3%, and Africa 16%, respectively, according to FAOSTAT, 2011 (Gruneberg et al., 2017).

13.2.3 NUTRITIONAL EVALUATION OF IPOMOEA BATATAS

Vegetative propagation is the main method for the cultivation of *Ipomoea batatas* by using either the stored root or stem cut method. The stem being cylindrical in shape grows based on the availability of water present in the soil. The industrial part of the sweet potato is its storage root which has high nutritional value (Mohanraj and Sivashankar, 2014).

The smooth skin root tuber color ranges from yellow, orange, red, brown, purple, and beige and its flesh color also ranges from white, pink, violet, and yellow. The sweetness depends on the color of flesh, that is, flesh with yellow or white are less sweet and flesh with red, pink, and orange flesh are sweeter (Loebenstein and Thottappilly 2009). Table 13.2 provides details of the nutrition per 100 gm in sweet potato (National Nutrient Database) (Mohanraj and Sivasankar, 2014).

TABLE 13.2 Demonstration of Nutritional Value Per 100 gm of *Ipomoea batatus*

Nutrition in Ipomoea batatas	Value per 100 gm
Energy	360 kg
Carbohydrates	20.1 g
Starch	12.7 g
Sugar	4.2 g
Dietary fiber	3.0 g
Fat	0.1 g
Protein	1.6 g
Vitamin A equivalent	89%
β-carotene	79%
Lutein and zeaxanthin	0%
Vitamin B1	0.1 mg
Vitamin B2	0.1 mg
Vitamin B3	0.61 mg
Vitamin B5	0.8 mg
Vitamin B6	0.2 mg
Vitamin B9	11 mg
Vitamin C	2.4 mg
Vitamin E	0.26 mg
Calcium	3%
Iron	5%
Magnesium	7%
Phosphorus	7%
Potassium	7%
Sodium	4%
Zinc	3%

The presence of β-carotene is high as observed in orange fleshed sweet potato. In recent literature studies, it has been proved that the potential to raise blood level by Vitamin A, which is very beneficial for children, has been reported in *Ipomoea batatas*. The high content of carbohydrates in *Ipomoea batatas* is a very efficient diet for people doing a strenuous job and is also known to be beneficial in case of patients suffering from stomach cancer (You et al., 1988). The reports also show the presence of certain phytonutrients which help lower the health risk posed by the free radical (Huang et al., 2004). Sucrose, glucose, fructose, and maltose are some simple sugars which are known to be present along with starch (Cereda et al., 2001). The presence of essential nutrient such as calcium, magnesium, manganese, phosphorus, potassium, sodium, iron, and low level of toxicants such as total oxalate, tannins, phytic acid, and cyanide along with rich content of retinol (Vit A) and ascorbic acid (Vit C) makes the leaf of *Ipomoea batatas* more nutritious then the tuber and that is why it is used as essential nutrients to supplement the dietary allowances (Antia et al., 2006).

13.2.4 PHARMACOGNOSY OF IPOMOEA BATATAS

Diseases such as arthritis, diabetes, hypertension, dysentery, constipation, fatigue, rheumatoid disease, hydrocephaly, meningitis, kidney ailment, and inflammations are known to be cured by several species of Ipomoea (Meira et al., 2012). Extracts from the leaves of *Ipomoea batatas* are proved to be an effective astringent, bactericide, demulcent, fungicide, laxative, and are used in several tonics (Ludvik et al., 2004). The powdered form of the tuber is used as flour and is known to be effective in the prevention of ethanol-induced gastric ulcers as it helps in the suppression of Edema formation and help in wound healing (Hermes et al., 2013). Along with being very helpful in treating tumors of mouth and throat, nausea, stomach distress, and whitlows, they are also proved to be good for the cardiovascular system (Duke and Wain, 1981; Yoshimoto et al., 2002).

Presence of a various number of vitamins, minerals, and nutrients along with a high amount of phosphorus and β-carotene has various health benefits such as are essential for cardiac, ocular health, and for maintenance of veins and arteries as described.

A. Immunity enhancement.
B. Reduction of inflammation.
C. Curing of asthma.
D. Treatment of bronchitis.
E. Management of arthritis.
F. Helpful in healthy digestion
G. Cancer treatment.
H. Treats ulcer of the stomach.
I. Diabetes treatment.
J. An essential role in gaining weight.
K. Helpful in balancing weight.

The main characteristic property of *Ipomoea batatas* in pharmacological aspects are:

I. Effects on the cardiovascular system.

II. Anti-cancerous activity
III. Effect on diabetic patients.
IV. Effects on the immune system.
V. Anti-oxidant property.
VI. Safety evaluations.
VII. Anti-ulcer property.

I. **Effects on cardiovascular system:** The vasorelaxation mechanism or mode of action of *Ipomoea batatas* was similar to that of acetylcholine; *Ipomoea batatas* shows 97% of relaxation activity for preparation of endothelium intact aortic ring but is 35% for mesenteric vascular bed (Runnie et al., 2004).

II. **Anti-cancerous activity:** It has been reported by several kinds of research that the chances of breast cancer, colorectal cancer, gallbladder, and kidney cancer could be reduced by regular consumption of sweet potato. It has also been reported over 10 year-case study that regular consumption of *Ipomoea batatas* is associated with a decrease in the risk of disease considering certain factors such as medical history, dietary and lifestyle consideration, anthropometry, etc. (Washio et al., 2005). National Cancer Institute was the first to discover the compound in the sweet potato which was exactly responsible for possessing anti-cancerous property with the help of biochemical biological rationale method and it was 4-Ipomeanol (Lakhanpal et al., 2001). In normal tuber, the presence of 4-Ipomeanol is not reported as it is only known to be present in the infected sweet potato only which shows a significant rate of cytotoxic activity and helped in the discovery of a drug that could treat lung cancer (Hsu et al, 2003).

III. **Effect on the diabetic patient:** When compared to the standard drug for treating diabetes, that is, diabense, higher antidiabetic property of *Ipomoea batatas* are reported. Consumption of high polyphenol in the diet for a week is known to modulate antioxidant status and decrease oxidative damage induced by exercise and pro-inflammatory cytokine secretion due to presence of polyphenol, sweet potato has higher antidiabetic property as compared to diabense (Oke et al., 1999). The entire vegetable, that is, flesh along with peel has a larger amount of protein content and is greatly efficient in lowering the blood glucose in a diabetic patient. Such kind of person having diabetes have lower content of adiponectin, which is a hormone for fat cells, proteinaceous in nature and *Ipomoea batatas* is reported to have the presence of adiponectin which is helpful for a diabetic person The patients with type 2 diabetes have a low glycemic index which could be maintained by the consumption of *Ipomoea batatas* which could be used as a convectional drug with low cost and least side effect making it an efficient antidiabetic agent.

IV. **Immune system effect:** Certain therapeutic agents in combination with *Ipomoea batatas* is known to be a therapy for a skin wound. The response of this agent in combination with *Ipomoea batatas* to wound was it reduced the wound size and its severity.

V. **Anti-oxidant property:** As compared to ascorbic acid (Vit C), the total antioxidant capacity of sweet potato is 42.94% (Pochapski et al., 2011). The presence of total phenolic content in the plant in all parts such as leaves, stem, and roots are a useful indicator of the antioxidant activity of the plant. The antioxidant content also depends on the color of flesh, that is, purple-fleshed has higher antioxidant activity as compared to white-fleshed sweet potato (Teow et al., 2007). 1,1- diphenyl 1-2 picrylhydrazyl (DPPH) stain along with metal ion dependent hydroxyl radical, ferric thiocyanide (FTC) reducing powder, etc., are the method used for the study of antioxidant property of *Ipomoea batatas* (Dong et al.).

VI. **Safety evaluation:** Oxalic acid reported in *Ipomoea batatas* is also a component of some vegetable which are known to be present in the form of oxalate crystals in the urinary tract of some people so the patients with such issues of urinary tract oxalate stone are advised to consume these sweet potatoes and are suggested to intake a large amount of water so that these crystals could be released out through urine.

VII. **Anti-ulcer property:** A specific increase in glutathione, superoxide dismutase, catalase, glutathione peroxidase, glutathione reductase levels, and also known inhibition of mean ulcer score or ulcer index, reduction in the level of lipid peroxidation in a dose-dependent manner. These all proved the gastroprotective activity of sweet potato which was studied while evaluation of methanolic extract of *Ipomoea batatas* on cold stress and aspirin (Panda and Sonkamble, 2012).

Table 13.3 describes about different types of root tubers of *Ipomoea batatus* with different flesh color.

TABLE 13.3 Different Types of *Ipomoea batatas* Root Tubers.

Variety of root tubers	Type of *Ipomoea batatas*	Color of flesh
1- Orange sweet potato variety	A. Beauregaurd	Orange color
	B. Allgold	Medium orange flesh
	C. Apache	Pale orange
	D. Bunch Porto Rico	Light yellow
	A. Centennial	Carrot color
	B. Cherokee	Orange flesh
2- Red flesh sweet potato variety	A. Amish red	Orange flesh red skin
	B. Continental red	Creamy light orange
	C. Carolina ruby	Deep orange
	D. Dianne	Dark orange
	E. Garnet	Orange to yellow
	F. Georgia jet	Deep orange
	G. Cordner's red	Moist orange flesh
3- White flesh sweet potato variety	A. Star leaf	Light orange
	B. White delite	Yellow fleshed
	C. Nancy hall	Yellow fleshed
	D. Sumor	White yellow fleshed
	E. Picadita	White fleshed

13.2.5 PLANT PARTS AND RELATED HEALTH BENEFITS (MOHANRAJ AND SIVASANKAR, 2014)

A. **Root tubers:** The root tubers of *Ipomoea batatas* is a rich source of dietary fibers, antioxidants, vitamins, and minerals and have no evidence of cholesterol or saturated fat. The major source of energy in root tubers is due to the presence of starch, that is, a carbohydrate. The ratio of amylose to amylopectin is higher which raises blood sugar levels as compared to simple sugars so it is known as a healthy food substance and can be also consumed by diabetic patients. The root tubers are known to be a good source of flavonoids which help as a protectant for lung and cavity cancer. The presence of vitamin A makes it a great antioxidant, also helpful in making healthy skin and healthy mucus membrane. The tuber is also known to be a rich source of certain essential minerals such as magnesium, manganese, calcium, iron, and potassium that are essential for certain enzymatic activity and another metabolism of the body (Woolfe, 1992).

B. **Leaves:** The leaf's of *Ipomoea batatas* are known to be an immune booster, helpful in curing certain diseases and infections. The leaves are also known to overcome the problem related to eyes, help to prevent macular degeneration and cataract which are age-related disorders that occur due to the presence of Lutein. Earlier literature shows that leaves are antidiabetic, antimutagenic, anti-cancerous, and antibacterial, and essentially help in promoting good health to human which is basically due to occurrence of polyphenols.

13.2.6 PHYTOCHEMICALS AND ITS STRUCTURE REPORTED IN IPOMOEA BATATAS

Pochapski et al. (2011) have reported that the leaves of *Ipomoea batatas* have the presence of alkaloid, anthraquinones, coumarins, flavonoids, phenolic acid, saponins, steroids, or triterpenes. The terpenoid present in the leaves acts as anticarcinogenic, which helps to lower LDL—cholesterol level in the body, the carotenoids help to reduce mutagenesis in a cell, and has great antioxidant capacity. The ability of leaf extract to fight against cancer is due to the presence of chemical compounds such as quercetin and chlorogenic acid.

I. **Anthocyanin**—The genes of purple anthocyanin, that is, IbMYB1 and IbMYB2, basically known to be rich in purple sweet potato, are a source of anti-inflammatory, antioxidant, and is known to reduce certain health risk (Montilla et al., 2010).

II. **Caffeoylquinic acid derivatives**—Certain derivatives of caffeoylquinic acid has been isolated from the leaves of *Ipomoea batatas* such as chlorogenic acid (Ch A), caffeic acid, and 3,4 di CQ A, 4,5 di CQ A, 3,4,5 tri CQ A compounds are known to prevent cancerous cell growth in humans and are antimutagenic in nature (Kurata et al., 2011).

III. **Coumarin**—They are known to inhibit HIV replication due to their anticoagulation property due to presence of aesculetin, scopoletin umbelliferon, and scopoletin (a phytoalexin) in *Ipomoea batatas*; it possesses hepatoprotective activity, known to inhibit the growth of cell by inducing cell death of human androgen-independent prostate adenocarcinoma cell (PC 3) (Liu et al., 2001; Sukanya et al., 2016).

IV. **Furanoterenoids**—1-Ipomeanol, 4-Ipomeanol, 1,4- Ipomeadiol and Ipomeanine, etc. (Mohanraj and Sivasankar, 2014) are a group of furanoterpenoids isolated from sweet potato infected by *Fusarium solani* as Ipomeanol is only isolated from infected sweet potatoes (Boyd and Wilson, 1972).

V. **Phenolic compounds**—Certain phenolic compounds are reported to be present in *Ipomoea batatas* as isolated by the chromatographic method by extracts of methanol and hydromethanol. The stem of *Ipomoea batatas* has a significant amount of feruloyl quinic acid and very least or minimum amount of 4-caffeoyl- feruloylquinic acids (Zheng, 2008; Gil and wianowska, 2017).

VI. **Triterpenes**—These are obtained from boehmeryl acetate and act as an ovipositional stimulant for the weevil of *Ipomoea batatas*. The compound β-amyrin acetate is also a triterpene and is proved to be an antinociceptive when analyzed on mice (Tan et al., 1995; Krogh et al., 1999).

This chapter aims at describing various and major kinds of viruses, and their methods of management. *Ipomoea batatas* have a great role in the economic system of the world as it is one of the largest produced staple foods and has much of the health benefits along with pharmaceutical use. The infection of viruses on these plant not only reduce the growth and development of plant but also have a devastating effect on the economic system all over the world, there are many kinds of viruses reported till date which is infecting the plant and is known to change the physiology of the plant leading to a decrease in the production of fleshed tubers or malfunctioning in the plants, also these viral infections are transferred to other plants with are economically beneficial to the farmers by causing a decrease in their productivity; the presence of these viruses may lead to a decrease in the content of metabolites or compounds which are pharmaceutically very important leading to the great loss of the pharmaceutical industry in the development of certain medicinal drugs, decreasing the phytoconstituent which would lead to a decrease in the growth of plant and spread a virus in severe terms when studied at the molecular level. There are many different kinds of viruses all over the world reported in genus Ipomoea that infect the plant to date. This chapter would emphasize describing each kind of virus infecting genus *Ipomoea* in detail, certain kinds of methods that could be applicable which would reduce the infection of viruses helping in the development of healthy and infection-free plants.

However, it is not the only virus which decreases the sweet potato's productivity but certain other biotic and the abiotic factor is also responsible for infection in the plant but the most common among these are Weevil damage and viral infections of any kind (Tairo et al., 2005; Valverde et al., 2007). Several literature surveys have provided the evidence of nearly 30 different kinds of viruses that severely infect the sweet potato; among these the three major kinds are sweet potato leaf curl virus (SPLCV), SPCSV, and SPFMV (Christopher et al., 2012). These viruses are mainly described to be of different families such as Begomovirus of family Geminiviridae and Badnavirus, Cavemovirus of family Caulimoviridae (Christopher et al., 2012; Clark et al., 2012; Cuellar et al., 2015). The different kinds of propagation method such as vegetative propagation, grafting, etc., lead to the accumulation of certain pathogens and viruses; sweet potato is highly susceptible to certain kind of viruses and their co-infections, so it acts as the reservoir of many different kinds of viruses known (Souto et al., 2003; Rey et al., 2012; Loebenstein, 2015).

13.3 GENETIC MULTIPLICITY OF SPCSV, SPFMV, AND SPMMV

For a better understanding of epidemiology of virus at the molecular level, a good resolution of the population genetic variability and global distribution of strain in SPCSC, SPFMV, and SPMMV is required. The different multiplicity of the virus at different genetic level affects the serological; the intraspecific viral diversity knowledge is essential for the determination of virus-specific antibodies; PCR primers and nucleic acid probes could be available to detect different virus isolated of different geographical origin. Each virus resistance engineering mechanism are strain-specific these make them prone to overcome unless the local virus strains are characterized and others are introduced. To import on the plant, the material the different viral strain is avoided by following certain protocols—effective quarantine procedures and sequencing of reliable virus (Tairo et al., 2005).

13.3.1 CHARACTERISTIC OF VIRUSES SO FAR EVOLVED AND REPORTED IN IPOMOEA BATATAS

The viruses so far reported in genus Ipomoea is around 30 in number belonging to nine different families, among which the highest number of the virus is reported from the family Geminiviridae (15 virus), followed by potyviridae (9 viruses), Caulimoviridae (3 viruses), Closteroviridae (1 virus), Comoviridae (1 Virus), Bunyaviridae (1 virus), and Bromoviridae (1 Virus) (Christopher et al., 2012). The infection reported as the result of these viruses could be either symptomatic or most often non-symptomatic. Some species of *Ipomoea* are known to be the mediator of severe virus, that is, SPCSV. The recent research reports that the DNA virus infecting sweet potato is more diverse as compared to RNA virus when their sequencing is done i.e. the molecular characterization of these viruses could be obtained by rolling circle amplification (RCA) (Mardis, 2011). The past 15–20 years have shown major evolution in the methods of molecular characterization showing how these viruses evolved, the complex of consumption of sweer potato virus, the effect of these viral infections on the production of sweet potato and related varieties, how the plant and virus interact, and several management approaches of these viral diseases (Valverde et al., 2007).

13.4 SOME VIRUSES OF IPOMOEA BATATAS IN DETAIL

13.4.1 FEW UNCOMMON VIRUSES OF IPOMOEA BATATAS

1. **Ipomoea crinkle leaf curl virus (ICLCV)**—This virus belongs to the genus Begomovirus of the family Geminiviridae, the main cause of transmission of this genus is whitefly, The virus is distinct from the SPLCV. The ICLCV is mainly transmitted by grafting the mechanism by the whitefly *Bemisia argentifolii* but do not infect mechanically, the virus transmitted by these *Bemisia* is associated to be a geminate particle so belongs to family Geminiviridae and it induces symptoms that are distinctly spread on different species of genus *Ipomoea* but these symptoms are not observed on *Ipomoea batatas* cv. Georgia Jet; also the transmission of this virus on another genus except *Ipomoea* is not reported. The hybridization of bean yellow mosaic virus riboprobe, the extracts of infected *Ipomoea* plant, and tomato yellow leaf curl virus riboprobe showed that the Ipomoea crinkle leaf curl virus is very different, the name was proposed in 1997.
2. **Ipomoea yellow vein virus (IYVV)**—Common to be known as SPLCV belonging to the genus Begomovirus of family Geminiviridae, *Ipomoea purpurea* is the main host plant for Ipomoea yellow vein virus (IYVV) (Geetanjali et al., 2013). The *Bemisia tabaci,* whitefly, is the main vector through which the virus is transmitted in a circulative non-propagative manner through grafting but not transmitted by plant contact or mechanical sap inoculation. The symptom by the virus, that is, leaf curl and yellow vein are the major symptoms of the virus which infect the genus *Ipomoea*, that is, morning glory in common. In general, SPLCV association with beta satellite is not known but in India. Croton yellow vein mosaic beta satellite and papaya leaf curl beta satellite are two different beta

satellite through which SPLCV is associated in yellow vein and leaf curl the disease of *Ipomoea purpurea* (Geetanjali et al., 2013).

3. **Cucumber mosaic virus (CMV)**—Sweet potato is only infected by CMV only if the first infection is known to be caused by SPCSV making it a helper virus for CMV (Cohen et al., 1997); however, some time CMV is known to cause infection without the help of SPCSV, that is, an isolate from *Arracacia xanthorrhiza* shows that CMV could infect plant without SPCSV (Ishak, 2002; Untiveros et al., 2007). This shows that SPCSV epidemiologically enhances the chance of infection of CMV so act synergistically.
4. **Sweet potato ringspot virus (SPRSV)**—This virus infects mainly some genus of *Ipomoea, Chenopodium and nicotiana.* Though not much of the history of this virus is known neither much of its serological sequencing is done but still it is known to be the only sweet potato virus that shares some characters of Nepovirus, that is, its physiochemical properties (Brown, 1988).

Supplementary viruses—Sweet potato leaf specking virus (SPLSV) rarely detected belongs to family Luteoviridae and could be detected by several methods such as PCR of coat protein which is targeting the genome for this virus. Bunyaviridae which have been known to be present in few isolates of C-3 virus which could be mainly known to cause infection in *Ipomoea setosa, Ipomoea nil,* and *Nicotiana benthamiana* (Loebenstein, 2009).

1. **Family—Caulimovirids**

   In recent sequencing of Badnavirus, some genetic distinction is reported when collected from all over the world and this sanction leads to the origin of Sweet potato pakakuy virus along with its significant name. Sweet potato collusive virus is known to belong to Cavemovirus when the sequence of virus and its characterization is studied which was done through PCR and partial viral purification methods (De Souza and Cuellar, 2011). The sweet potato Badnavirus has been showing mixed infection, sometimes non-symptomatic infection when its characteristic feature is studied (Kreuze et al., 2000, 2008, 2009). Sweet potato vein clearing virus (SPVCV) also belongs to caulimoviridae being different in the genomic organization than that of SPCV which now fall into Solevidovirus, a newly accepted genus. The natural infection of SPCSV is not so well known but is identified when it acts in synergistic relation with SPCV and SPVCV. For this family of a virus, no insect vector has been so far reported and this virus is not also known to be transmitted mechanically through grafting (Cuellar et al., 2011). The effect on growth and yield of these viruses on the plant is not so well known. The infection of these viruses has been shown as integrated with *Ipomoea setosa* as a sequence of these caulimovirids has not been clearly detected (Valkonen et al., CIP unpublished).

2. Family Potyviruses

To this family basically nine types of viruses have been reported which belongs to distinct genus, some of these are as follows:

a. Sweet potato feathery mottle virus (SPFMV)
b. Sweet potato latent virus (SPLV)
c. Sweet potato mild mottle virus (SPMMV)
d. Sweet potato mild speckling virus (SPMSV)
e. Sweet potato virus 2 (SPV 2)
f. Sweet potato virus C (SPV C)
g. Sweet potato virus G (SPV G)
h. Sweet potato yellow dwarf virus (SPYDV)

P1-N domain provides specific benefits to Potyviruses when they are reported to infect the Ipomoea species. A region of high genetic diversity shows known similarity in the N terminal of P1 protein in SPMMV and SPFMV. The homologous region of SPMMV shows great similarity with N terminal of P1 protein of SPFMV whereas P1 region of many different potyviruses shows similarity with C terminal of P1 protein of SPFMV (Tugume, 2010; Untiveros et al., 2010). There is interval recombination observed between sweet potato virus C (SPV C) and sweet potato feathery mottle virus (SPFMV) when the viral genome sequencing of these viruses are done, not much of the reports have been recognized about the SPV C but is known to mainly affect genus Ipomoea ranging to several species.

When different isolates of SPFMV was extracted from the cultivated sweet potato the recombinant quality of these viruses could be studied but no reports of interspecific recombination are observed, these viruses are also helpful to study the interhost transmission which could be similarly observed in SPMMV, that is, sweet potato mild mottle virus (Tugume et al., 2010; Virus taxonomy, 2012). SPMMV and SPFMV group of potyvirus had a unique P1-N domain which could be confirmed by sweet potato latent virus (SPLV) is also a member of the Potyvirus family (Chung et al., 2008).

3. **Family Carlaviruses**

   Sweet potato C-6 virus—A weak antiserum against sweet potato C-6 virus suggest its relation with Carlavirus (Loebenstein et al., 2009).

   Sweet potato chlorotic fleck virus (SPCFV)—The isolates of these viruses when analyzed phylogenetically showed geographically differentiation, the virus was basically isolated from CIP germplasm collection and has characteristic feature similar to the Carlavirus (Mukasa, 2003; Aritua, 2007, 2009).

4. **Family Crinivirus**

   Sweet potato chlorotic stunt virus (SPCSV)—The great loss in the yield of sweet potato is mainly due to mixed infection of SPCSV, that is, it is one of the main kinds of virus causing a great loss to the economy of sweet potato cultivation (Tairo et al., 2005; Valverde et al., 2007). The SPCSV belongs to the family Closteroviridae and is bipartite in nature. The spread of this virus is phloem limited (Loebenstein et al., 2015). The SPCSV is also known to cause synergistic relation with another genus of RNA virus, that is, Potyvirus, Ipomovirus, Carlavirus, and Cucumovirus whereas some are DNA viruses like Begomovirus, Cavemovirus, and Solendovirus, which causes symptoms when their synergistic association is reported but when they are singly present, they are mainly symptomless (Cuellar et al., 2015). These coinfections due to synergistic effect make it very difficult for separation of purification and genomic sequencing of these SPCSV virus (Qin et al., 2014). This virus is mainly reported in the areas where the vegetation of sweet potato is very common (Tairo et al., 2009). Sweet potato is more susceptible to SPCSV in a very wide range and this susceptibility make it a more common type of virus infecting the genus Ipomoea especially *Ipomoea batatas* (Cuellar et al., 2015).

   Till date, only two strain of SPCSV has been identified and studied, that is, SPCSV–EA (East Asia) and SPCSV–WA (West Asia). A small sample of orange-fleshed sweet potato from Burkina Faso shows the presence of coinfection of all the three most commonly known viruses, that is, SPLCV, SPCSV, and SPFMV (Tibiri et al., 2020).

   An estimated loss of approximately 70–95% have been reported due to infective cause of sweet potato viral diseases (SPVD) which is the result of co-infection of SPCSV and SPFMV. Further molecular analysis has shown that the RNA viral titers of SPFMV have been increased to about 600-fold and the titer of SPCSV is equal to the single infection when associated with SPCSV (Mukasa et al., 2006; Loebenstein, 2015).

5. **Family Geminiviridae**

Genus Begomovirus is a member of the family Geminiviridae which is monopartite in nature commonly known to be a sweet potato virus (Trenado et al., 2011). The occurrence of these viruses has shown major systems in different parts of the world. The Rep gene, which is almost an incomplete genome copy is helpful to diagnose different isolated and also helpful in the characterization of different biological isolates (Paprotka et al., 2010). The transmission of these viruses is mainly due to *Bemisia tabaci* which could be transmitted to *Ipomoea setosa* through grafting. The isolated from *Ipomoea setosa* and also from sweet potato through sequencing could confirm the presence of the same kind of Begomovirus. To render the infectious clones, sweet potato virus have been recalcitrant when genome sequencing was done, unlike the other Gemini viruses.

Mainly three novel species of Begomoviruses to date are reported:

1. Sweet potato leaf curl canary virus (SPLCCaV).
2. Sweet potato leaf curl Lanzarote virus (SPLCLaV).
3. Sweet potato leaf curl Spain virus (SPLCSV).

Phylogenetic analysis of sweet potato infecting Begomovirus showed a wide range of distinction between new world and old world Begomovirus (Wasswa et al., 2011).

Ipomoea indica shows the infection of Ipomoea yellow vein virus which is known to adapt some of the specific host range which proved the frequent recombination of Begomovirus to various other species (Lozano et al., 2009; Zhang and Ling, 2011; Christopher et al., 2012; Tibiri et al., 2020). As the infection and co-infection of Begomovirus is mainly symptomless in both cases. So, the infection of Sweepovirus is not easily detected in new and different areas which could attain high titers (Trenado et al., 2011; Tibiri et al., 2020). The clones isolated from the binary vector showed complete tandem dimeric genome which could be confirmed by PCR, dot blot, and southern blot hybridization (Trenado et al., 2011).

13.5 TRANSMISSION OF VIRUSES IN IPOMOEA BATATAS THROUGH GRAFTING

Tibiri et al. (2020) stated that the selection of viruses to study the transmission of the virus in sweet potato is done with the help of NCM- ELISA, PCR, and RT-PCR. Some infections which could not be confirmed by ELISA were made sure by RT-PCR. Sweet potato is a natural reservoir of several viruses which has been proved by the isolation of several viruses such as SPFMV, SPLCV, SPCS, and SPCFV. These infections could be easily studied within a time period of two months when the actual infection occurs (Christopher et al., 2012). These infections could decrease the production and yield of sweet potato. Few species of *Ipomoea* are known to be a natural reservoir of several infecting viruses, these species are *Ipomoea wrightii, Ipomoea hederacea, Ipomoea hederifolia, Ipomoea trichocarpa, Ipomoea hederacea, Ipomoea lacunose, Ipomoea setosa,* and *Ipomoea alba* could be differentiated by studying the mechanism of host grafting which could transmit the virus. These different viruses are phylogenetically very distinct from each other and have been studied with the help of different scientists.

In many cases of the viral infections, the field symptoms are very common but the decrease in the yield of crop production is not so much known. The symptoms typically are spots of chlorosis, ringspot, and banding of vein (Hoy et al., 2006; Valverde, 2007). Some of the important symptoms of the disease could be characterized by isolation of internal cork. The systemic analysis of the viruses is done from time to time utilizing the pervious DNA or RNA sequence analysis (Valverde et al., 2007; Christopher et al., 2012; Tibiri et al., 2020) The prevalence of these viruses are most common being sweet potato feathery mottle virus (SPFMV 24%), followed by SPLCV (18%) then the sweet potato chlorotic stunt virus (SPCSV 2%) (Valverde et al., 2007).

13.5.1 DETECTION METHODS AND DIAGNOSIS TECHNIQUES OF SWEET POTATO VIRUS

To rapidly control the viral infection, it is very important to learn the methods for the detection of viral infection and its identification as precise and early detection of the virus would lead to help in the early management of disease (Valverde et al., 2007; Christopher et al., 2012; Tibiri et al., 2020). The source resistance, control methods, and several indexing procedures could be developed easily once the kind of viruses has been detected. In sweet potato, viral detection is a major challenging task due to low viral task as well as the high concentration of inhibitors which could interfere with the serology and as well as PCR-based viral detection methods. So, in the developing years it has been seen significant progress in various techniques which could help in easy detection of sweet potato virus (Mukasa et al., 2006; Tairo et al., 2006).

Earlier the methods used for the detection of viruses were grafting the plant mechanically. Basically, *Ipomoea setosa* was used for it as it is used as an indicator plant for viral detection but later advances have shown the presence of great technological development in the field of nucleic acid-based assays and serological methods which could help in easy detection of the causative agent of the disease and infections but if we look back it was reported that the infection caused by SPFMV often masked the infection of other viruses especially Potyvirus which hinders the isolation of Potyvirus and other viruses (Valverde et al., 2012; Tibiri et al., 2020).

ELISA carried out on the microtiter plate or that on Nitrocellulose membrane is one of the serological methods used to detect viruses. For serological detection of most of the RNA viruses of Sweet potato diagnostic kit is to be provided by CIP, PCR based approaches are also for the detection with fast technologies. Different technologies have been described by many researchers all over the world which are as follows—

1. ELISA, molecular hybridization, PCR (Valverde et al., 2007).
2. PCR-based approaches; RT-PCR, generic novel molecular detection, that is, rolling circle amplification (RCA); deep sequencing of SiRNA (Christopher et al., 2012).
3. Nitrocellulose membrane ELISA for virus diagnosis, reverse transcription PCR and PCR for amplification of CP genes; sequencing and sequence analysis of RT-PCR (Tibiri et al., 2020).

13.5.1.1 MOLECULAR HYBRIDIZATION

Sweet potato viruses, that is, SPLCV and SPFMV could be easily diagnosed with the help of nucleic acid hybridization. DNA or RNA viral probes consisted of radioactive and non-radioactive labels. An in vitro transcribes RNA probes was developed for SPFMV virus detection, the development of this probe was done by using CDNA which consisted of 3' terminal of capsid protein cistron region. When compared to immunological assays, the riboprobes were found to be more sensitive as it was able to overcome host factors interference through which immunodiagnostic assays reliability could be compromised. Some species of *Ipomoea* specifically *Ipomoea batatas* were used for the detection of strong signals using DNA for which SPLCV specific probe with direct labels and horseradish peroxidase, this proved that SPLCV infected *Ipomoea batatas* and several species showed strong signals to it (Moyer et al., 1992; Valverde et al., 2004a; Valverde et al., 2007; Christopher et al., 2012).

13.5.1.2 ROLLING CIRCLE AMPLIFICATION (RCA) AND SiRNA

Due to the great range of evolution in viruses till date, it was difficult to recognize viral strains with the help of PCR, so RCA in combination with restriction fragment length polymorphism (RFLP) as it served as a great emerging tool in the diagnosis of different type of sweet potato, the strains, novel species, variants of different sweet potato was identified by using the combined mechanism of RCA with RFLP along with sequencing (Haible et al., 2006; Albuquerque et al., 2011; Paprotka et al., 2010; Zhang and Ling, 2011; Chirstopher et al., 2012).

PCR amplification and Sanger sequencing of Badnavirus and Mastrevirus confirmed the presence of these viral strains. The siRNA sequence of assembly help to determine primers of these viruses so, the new PCR-based detection tool of viruses was developed using this SiRNA deep sequencing approaches and this siRNA sequencing utilized to design newly primers for some viruses such as SPSMV-1 and SPPV

Several kinds of these viruses are determined by these techniques such as SPVG, SPV2 SPLV, SPFMV-RC, and some of the newly detected virus such as new strains of SPFMV, Cavemovirus, Solendovirus, Badnavirus, and other new sweet potatoes. At present day, it has been observed that siRNA deep sequencing is becoming more affordable in price which was a challenging to new world viruses as previous the equipment was time-consuming and was not able to easily identify much of the viruses (Christopher et al., 2012; Malabisana et al., 2018; Maina et al., 2018).

13.5.1.3 POLYMERASE CHAIN REACTION (PCR)

The sensitivity and reliability of virus detection by PCR is mainly affected by inhibitory compounds found in sweet potato, PCR methods help in its detection. For the rapid analysis, detection and confirmation of the different type of viruses of Sweet potato including Potyviruses. Several primary approaches were included such as RFLP analysis, Broad spectrum PCR and sequence amplification, etc., several kits are used for DNA extraction which helps to detect the infecting viruses for PCR-based detection techniques.

The cheap and reliable alternative over expensive commercial PCR primer kit is LiCl, a semiautomatic homogenizer; Cetyl trimethyl ammonium bromide (CTAB), etc. The cheapest being the NaOH Tris protocol of DNA extraction for several plant tissues. PCR analysis is done using Immune capture reverse transcription PCR the method which does not require much labor work for the isolation of nucleic acid (Fenby et al., 1998; Valverde et al., 2007; Christopher et al., 2012; Mohammed et al., 2017).

Fifteen partial sequences out of which 13 of SPFMV, 1 of SPCSV, and 1 of SPLCV were obtained from PCR and RT-PCR as a result of de novo assembly in Genes (Tibiri et al., 2020). The PCR techniques have the great ability to detect virus even if they are unevenly distributed or even if they are present at a very low level in the diseased plants. Through this PCR technique, it was easy to determine the taxonomic position of each of identified viruses (Colinet et al., 1998).

The infection of the sweet potato plant as a complex could be easily sorted out using PCR based approaches; Oligo T25 Primer along with reverse transcription method was used to distinct two strains of SPFMV and SPV 2 virus several other procedures could be developed to utilize the efficiency of PCR and discover a new technique for the detection of viruses (Tairo et al., 2006; Valverde et al., 2007; Christopher et al., 2012; Mbanzibwa et al., 2014; Cuellar et al., 2015).

13.5.1.4 REAL TIME-POLYMERASE CHAIN REACTION (RT-PCR)

Most sensitive and accurate method for the detection of plant viruses is qualitative RT-PCR. Several genes in *Ipomoea batatas* have been used as internal controls in qPCR as the amplification of mRNA or coding sequence of host gene requires internal control for the detection of DNA and RNA viruses, some of these genes include 18s and 26s RNA, plant mitochondrial NADH dehydrogenase (NAD 5) and sometimes Cytochrome oxidase gene (CoX), which help to differentiate between RNA and DNA concentration of the sample from the infected plants. qPCR help in the easy and efficient detection of several viruses such as SPLCUV, SPLCV, SPFMV, etc., Several other techniques could be used to detect the viruses by mixing certain primers and using distinctly labeled fluorophores probe. Although much of the benefits of qPCR is known, that is, it saves time, is cost-effective, and least labor required high detection sensitivity. However, it also has some of the demerits such as the sequence specificity of the primers in the condition of routine virus detection is limited; expensive reagents and TaqMan probes are required, TaqMan probes are used for detection of Sweet potato viruses for which the sequences are available (Simmons et al., 2009; Ling et al., 2010; Wassawa et al., 2011; Christopher et al., 2012; Tibiri et al., 2020).

13.5.1.5 ENZYME-LINKED IMMUNOSORBENT ASSAY (ELISA)

The coat protein of virus and viral particles' serological studies are done to study and identify the plant viruses of different kind. The ELISA is a simple, accurate, sensitive and adaptable tool that is the best-preferred test for the detection of viruses but sometimes the irregular virus distribution and symptomology make it difficult for ELISA to detect the kind of viruses. For documentation of several viruses new or old is easier to be identified by the ELISA technique. The ELISA test for detection of several viruses have shown to be positive when detected in different areas of world and sometimes negative where the infection is not found for example in Burkina Faso the sample which tested negative by NCM-ELISA tested positive by RT-PCR describes the multiple assays and highlighting the associated technique of virus detection (Tibiri et al., 2020).

Christopher et al. (2012) has described that for viral disease detection using microtiter plate or nitrocellulose membrane with the help of ELISA in sweet potato. These are the several methods which are used for the early detection of the viral infection in the plant tissues of sweet potato through which we can easily manage these kinds of viruses infecting several species of sweet potatoes which had reduced the yield of sweet potato and also its metabolites. By these methods, managing viruses become easy as it helps in early detection of virus and reduce its effectiveness (Phylogenetic tree of Begomovirus infecting *Ipomoea*).

13.5.2 MANAGEMENT AND CONTROL STRATEGIES OF VIRAL DISEASES OF IPOMOEA BATATAS

The epidemiology of the disease is very difficult to study as we know that viruses infecting their host and its machinery completely and that's the reason it is difficult to understand the mechanism of the virus infecting the plants. So, to study the kind of pathogen infecting the plant. We study their symptomology that appears in different parts of the plant. These symptomologies help the researchers to understand the pathogen and then help in managing the infecting pathogen so that its effect could be reduced or minimized and the plant could be managed healthy. By the clear-cut study of these pathogens, it helps in developing methods that could be essentially used to control the infection and to develop several other methods, techniques that could combat the effect of these pathogens and viruses are also developing day by day in a much complex mechanism of their behavior so with the evolution in time and pathogen it has become a very important to develop comprehensive management strategies for the viruses and pathogen which could effectively help to reduce the infection. The first and foremost role in line to manage viral diseases is to study the epidemiology of the kind of virus infecting the plant. Earlier it was difficult to identify kind of viruses due to lack in the tool and techniques or the expensive nature of tools and also the detection method of these viruses is very sensitive. In *Ipomoea batatas* earlier, it was ELISA and PCR which helped in the detection and confirmation of the virus infecting it (Aritua et al., 2007; Valverde et al., 2007). If we consider the different range of viruses in different geographical areas which were known to infect the sweet potato, it helped the researchers understand a different kind of viruses infecting *Ipomoea batatas*. When the single infection of the viruses like SPFMV, SPCSV, SPLCV was studied. It was observed that it caused rapidly or nearly 100% infection in the sweet potato fields of different area whereas the infection of these viruses were seen to be much reduced when co-infection or mixed infection was reported (Valverde et al., 2007)

Some of the known methods to detect the viruses are described as follows:

13.5.2.1 SEED MANAGEMENT PROGRAM

The first and foremost step that could be taken ahead to reduce the ill effect of viruses is to develop disease-resistant seed which could not only help to overcome the symptoms of disease but could also help to increase the production and yield of *Ipomoea batatas* in the field. The size, shape, and development system of the seed may vary according to the type of the infection being caused, due to different type of climate, farming system, etc., so, it is an essential urge to maintain the seed to provide effective control to certain kind of diseases arisen from different kind of viruses such as SPCSV, SPVD, SPFMV, SPLCV, etc.

Christopher et al. (2012) described that for the development of a healthy and sustainable seed by the farmers. Certain requirements should be provided to them which are:

- Providing adequate supply and quality of plant material in the proper time of cultivation to the farmers.
- Affordable plant and plant product range.
- Providing an efficient means for controlling the quality of the material of plants.
- For the dissemination of new cultivars channels must be provided.

It has been reported that with the practice of selecting healthy-appearing plant material the effectiveness of controlling or eliminating SPVD could be reduced (Valverde et al., 2007; Christopher et al., 2012; Tibiri et al., 2020). The consideration of important factors along with the economic factor helped in providing the necessity for seed programs to manage SPVD and provided with success full results (Fugile et al., 1999). As reported in few surveys that the symptoms caused by SPVD are very dramatic. So, it is readily possible to identify the affected plant in the field unlike other common viral diseases of sweet potato. So, to control these viral infections in plants from the infected seeds. We need to evaluate the rate of mixed infection with different viruses, how environmental condition varies, what is host resistance with special consideration of the systemic movement of viral infection towards the shoot tip (Valverde et al., 2007; Kim et al., 2017). So, in near future is important to develop long-term economically beneficially and sustainable seed systems with the help of farmers and several private sectors who could help in the development of certain tissue culture techniques for the development of proper diagnostic techniques and also major therapies for viruses.

13.5.2.2 QUARANTINE APPROACHES IN VIRAL DISEASE MANAGEMENT

International Board for Plant Genetic Resource (IBPGR) and Food Agricultural Organization provide technical guidance for the safe sweet potato germplasm movement and general quarantine approaches. Not all seeds and plants are allowed to travel or move to different countries without problem, quarantine guidelines and various other guidelines of different countries help in protecting the agriculture of different countries and in different geographical areas, its crops from different pathogen or viruses. Only such seeds are allowed to travel which has no evidence of disease by the process called 'Seed movement' but some viruses such as Nepovirus are known to be transmitted only through seed infection. So certain programs of tissue culture to develop small plantlets invitro is the most effective method for vegetative propagation of *Ipomoea batatas*. During vegetative propagation, *Ipomoea setosa* as an indicator plant for several viral diseases in *Ipomoea batatas* with the new methods of viral detection along with quarantine is PCR, ELISA, RCA, etc. (Christopher et al., 2012; Anonymous, 2007; Chen et al., 2008; Jackson and Simmons, 2014).

13.5.2.3 APPROACHES TO DEVELOP RESISTANCE TO A DIFFERENT TYPE OF VIRUSES

There are several viral diseases which are symptomless and provide a challenge for its management which could be detected from the indicator plant *Ipomoea setosa*. Sometimes NCM-ELISA and RT-PCR this virus is particularly SPFMV. In a report by Tibiri et al. (2020), it shows that physiologically the sweet potato showed no symptoms of up to 95%, that is, it was negative to viral infection until it was detected by NCM-ELISA and among the most prevalent was SPFMV.

Management of SPLCV—These viruses are mainly found to be symptomless and are known to reduce yield in sweet potato with the approximation of about

20–80% (Ling et al., 2010; Tibiri et al., 2020). For management of SPLCV, the process is the screening of germplasm or SPLCV titers, management of whitefly population since it is the main insect to transfer the virus. The control of these two methods helps to properly manage these diseases (Qin et al., 2014).

Management of sweet potato feathery motile virus (SPFMV)—The infection of these viruses is serious and common as it is the most widespread sweet potato viruses around the world (Maina et al., 2018; Tibiri et al., 2020). These viruses are managed by reducing the inoculum of the virus by using the seed from the limited generation that would be virus-free. Introduction of non-pathogen tested plant material increase disease in the plant. The process of RNA silencing, which could be done only to a certain variety of sweet potato viruses, can reduce the chance of infection in healthy plants (Mwanga et al., 2003; Mulabisana et al., 2018).

Management of sweet potato curl Spain virus (SPCSV)—In tropics, this is one of the major kinds of virus infecting *Ipomoea batatas* which cause a reduction in both foliage and root yield of plants (Christopher et al., 2012). The infection of SPCSV is in general an autoinfection caused in collaboration with SPFMV, SPCSV an important resistant variety develop which was high yielding and SPVD resistant is NASPOT-1 (Mwanga et al., 2003).

Development of SPVD disease-resistant plant—The most effective measure to control SPVD is to develop plant cuttings taken from symptomless parent plant which is known to exclude SPVD. Virus-free cuttings could be developed by using apical cuttings from healthy-looking plant vines which provided resistance to SPCSV in East Africa (Gibbon et al., 1997). Management of whitefly aphids which is known to be a short distant carrier of SPCSV virus. Rogueing of the affected plant controls the disease effectively when done within the first month of planting sweet potato.

It has been reported that SPCSV, SPFMV, and SPVD are not found to be infected in wild species of *Ipomoea*.

Using the expression of coat protein by different scientist, great resistance to SPFMV could be obtained. These could help develop plants resistant to SPFMV (Okada et al., 2001; Christopher et al., 2012; Maina et al., 2018; Tibiri et al., 2020) and also through the development of rice cystine proteinase inhibitor-mediated resistance to potyviruses.

13.6 DISCUSSION

These are some of the effective and well-known methods which are discussed above being used over several years to get control over certain kind of viruses infecting the *Ipomoea batatas* which are becoming a great challenge day by day as with the passage of time the nature, complexity, and kind of viruses are also evolving. Through this chapter, it could be concluded that the infection of this sweet potato virus is unevenly distributed in the world which has to be managed time to time by the development of new methods and techniques as *Ipomoea batatas* are highly nutritious tuberous root with much of the potentiality to overcome a certain kind of irregularity and diseases of the bottle, being rich in content of vit A, β-carotene, and certain other phytochemical has made this tuber an essential important compound in the diet of people all over the world. When these plant of *Ipomoea batatas* is infected by certain viruses, it not only reduces its market value and its production but it also reduces the content of the phytochemical present in the plant which could be effective in one or the other way to the human being so it has become prime importance to protect the plant from these kinds of viral diseases. The proper knowledge of host-vector mechanism, vectors biology, transmission mechanism, etc., is an essential part to manage the disease. At present, it has become essential to develop such a technique that is cost-effective and less time-consuming to detect the viruses with and without symptoms by just studying their epidemiological studies, proper methods of disease-free plants, and seed development could be implemented as soon as possible, the use of tissue culture should be effectively brought in use to overcome these viral infections, the eradication and reduction of these viruses require proper management and planning of certain people who are known to be the real cultivars of these sweet potatoes in collaboration with the researchers who are being a supporting hand in reducing or combating these effects.

KEYWORDS

- **pharmacognosy**
- **nutritive value**
- **viruses**
- **transmission of viruses**
- **diagnosis of viruses**

REFERENCES

Albuquerque, L. C.; Inoue-Nagata, A. K.; Pinheiro, B.; Rubeiro, S. D.; Resende, R. O.; Moriones, E.; Navas-Castillo, J. A Novel Monopartite Begomovirus Infecting Sweet Potato in Brazil. *Arch. Virol.* 2011. doi: 10.1007/s00705-011-1016-x

Anonymous. Ipomoea (Sweet potato/Kumara) Post-Entry Quarantine Testing Manual. Biosecurity New Zealand, Ministry of Agriculture and Forestry [Online], 2007 Aukland. http://www.biosecurity.govt.nz/files/regs/imports/plants/high-value-crops/ipomoea-testing-manual.pdf

Antia, B. S.; Akpan, E. J.; Okon, P. A.; Umoren, I. U. Nutritive and Antinutritive evaluation of Sweet potato (*Ipomoea batatas*) Leaves. *Pak. J. Nutr.* **2006,** *5,* 166–168.

Boyd, M. R.; Wilson, B. J. Isolation and Characterization of 4-Ipomeanol, a Lung-toxic Furanoterpenoid Produced by Sweet Potatoes (Ipomoea batatas). *J. Agric. Food Chem.* **1972,** *20,* 428–430.

Boyd, M. R.; Wilson, B. J. Isolation and Characterization of 4-Ipomeanol, A Toxic Furanoterpenoid Produced by Sweet Potato. *J. Agric. Food Chem.* **1972,** *20,* 428–430.

Cao, M.; Lan, P.; Abad, F. J.; et al. Genome Characterization of Sweet Potato Symptomless Virus 1: A Mastrevirus with an Unusual Non-nucleotide Sequence, 2017.

Cereda, M.; Franco, C. M. L.; Daiuto, E. R.; Demiate, J. M.; Carvalho, I. J. C. B.; Leonel, M.; et al. *Propriedades Gerais do Amido*; Fundaco Cargil: Csmpinas, Brazil, 2001; pp 21.

Chen, L.; Du, Z.; Hamaguchi, T.; Sugita, T.; Nagata, R.; Terao, H.; Tsuzuki, E. Clonal Propagation and Quick Detection of Virus-free Plants of Sweetpotato, *Ipomoea batatas. Bull. Minamikyushu Univ.* **2008,** *38A,* 1–5.

Christopher, A.; Abad, A.; Settumba, B.; Sweet Potato Viruses: 15 Years of Progress on Understanding and Managing Complex Diseases. *Plant Dis.* **2012,** 96 (2), 168–185.

Chung, B. Y.-W.; Miller, W. A.; Atkins, J.F.; Firth, A. E. An Overlapping Essential Gene in the Potyviridae. *Proc. Natl. Acad. Sci. U. S. A.* **2008,** *105,* 5897–5902.

Cohen, J.; Milgram, M.; Antignus, Y.; Pearlsman, M.; Lachman, O.; Loebenstein, G. Ipomoea Crinkle Leaf Curl Caused by Whitefly Transmitted Gemini-like Virus. *Ann. Appl. Biol.* **1997,** *131,* 273–282.

Colinet, D.; Nguyen, M.; Kummert, J.; Lepoivre, P. Differentiation among Potyviruses Infecting Sweet Potato Based on Genus- and Virus-Specific Reverse Transcription-polymerase Chain Reaction. *Plant Dis.* **1998,** *82,* 223–229.

Cuellar, W. J.; De Souza, J.; Barrantes, I.; Fuentes, S.; Kreuze, J. F. Distinct Cavemoviruses Interact Synergistically with Sweet Potato Chlorotic Stunt Virus (genus Crinivirus) in Cultivated Sweet Potato. *J. Gen. Virol.* **2011,** *92,* 1233–1243.

Cuellar, W. J.; Galvez, M.; Fuentes, S.; Tugume, J.; Kreuze, J. Synergistic Interactions of Begomoviruses with Sweet Potato Chlorotic Stunt Virus (Genus Crinivirus) in Sweet Potato (*Ipomoea batatas* L.) *Mol. Plant Pathol.* **2015,** *16,* 459–471.

De Souza, J.; Cuellar, W. J. Sequence Analysis of the Replicase Gene of 'Sweet Potato Caulimo-like virus' Suggests that this Virus is a Distinct Member of the Genus Cavemovirus. *Arch. Viro.* **2011,** *153,* 535–537.

Dina, L.; Valverde, A.; Murai, N. Sweet Potato Leaf Curl Virus: Coat Protein Gene Expression in E. coli and Product Identification by Mass Spectrometry. *Am. J. Plant Sci.* **2015,** *6,* 3013–3024.

Duke, J. A.; Wain, K. K. Medicinal Plants of the World. Computer Index with More than 85,000 Entries. 3 Vols, 1981.

Ellong, E. N.; Billard, C.; Adenet, S. *Food Nutr. Sci.* **2014,** *5,* 196–211.

Family Potyviridae. *Virus Taxonomy*; 2012; pp 1069–1089.

Fenby, N.; Seal, S.; Gibson, R.; Foster, G. RNA Extraction from a Virus-diseased Sweet Potato for Reverse Transcriptase-polymerase Chain Reaction Analysis. *Mol. Biotechnol.* **1998,** *10,* 187–190.

Fugile, K. O.; Zhang, L.; Salazar, L. F.; Walker, T. S. Economic Impact of Virus-Free Sweet Potato Planting Material in Shandong Province, China. *International Potato Center Program Report for 1997–98*; 1999; pp 249–254.

Geetanjali, A. S.; Shipi, S.; Mandal, B. Natural Association of Two Different Beta Satellite with Sweet Potato Leaf Curl Virus in Wild Morning Glory (Ipomoea purpurea) in India. *Virus Gene* **2013,** *47,* 184–188.

Gibson, R.; Mwanga, R. O. M.; Namanda, S.; Jeremiah, S. C.; Barker, I. Apparent Absence of Viruses in Most Symptomless Field-grown Sweet Potato in Uganda. *Ann. Appl. Biol.* **1997,** *130,* 481–490.

Hahn, S. K. Effects of Viruses (SPVD) on Growth and Yield of Sweet Potato. *Exp. Agric.* **1979,** *15,* 252–256.

Haible, D.; Kober, S.; Jeske, H. Rolling Circle Amplification Revolutionizes Diagnosis and Genomics of Geminiviruses. *J. Virol. Methods* **2006,** *135,* 9–16.

Hermes, D.; Dudek, D. N.; Maria, M.; Horta, L. P.; Lima, E. N.; Fatima, A.; et al. In Vivo Wound Healing and Antiulcer Properties of White Sweet Potato (*Ipomoea batatas*). *J. Adv Res.* **2013,** *4,* 411–415.

Hsu, H.; Rainov, N. G.; Quinones, A.; Eling, D. J.; Sakamoto, K. M.; Spear, M. A. Combined Radiation and Cytochrome CYP4B1/ 4-Ipomeanol Gene Therapy Using the EGRI Promoter. *Anticancer Res.* **2003,** *23,* 2723–2728.

http:// www.healthbenefitstimes.com/ Sweet potato

Huaman, Z. *Systematic Botany and Morphology of the Sweet Potato Plant*; Lima Peru: International Potato Center (CIP), 1992; pp 5–11.

Huang, D. J.; Lin, C. D.; Chen, H. J.; Lin, Y. H. Antioxidant and Antiproliferative Activities of Sweet Potato (*Ipomoea batatas* [L.] Lam; "Tainong 57" Constituent, *Bot. Bull. Acad. Sin.* **2004,** *45,* 179–186.

Ishak, J. A. Pathological Studies on Some Virus Diseases of Sweet Potato. M.Sc. Thesis, Ain Shams University, Cairo, Egypt, 2002.

Jackson, L.; Simmons, H. Management of Sweet potato Leaf Curl Virus in Sweet Potato Using Insecticides. *J. Agric. Urban Entomol.* **2014,** *30,* 82–95.

Jan, F K.; Ana, P.; Milton, U.; et al. Complete Viral Genome Sequence and Discovery of Novel Viruses by Deep Sequencing of Small RNAs: A Generic Method for Diagnosis and Sequencing of Viruses. *Virology* **2009,** *388,* 1–7.

Kim, J.; Yang, J.; Kwak, H. R.; Kim, M. K.; Seo, J. K.; Chung, M. N.; et al. Virus Incidence of Sweet Potato in Korea from 2011 to 2014. *Plant Pathol. J.* **2017,** *33,* 467–477.

Kurata, R.; Yahara, S.; Yamakawa, O.; Yoshimoto, M. Simple High Yield Purification of 3,4,5-Tri-O-Caffeoylquinic Acid from Sweet potato (*Ipomoea batatas* L.) Leaf and Its inhibitory Effects on Aldose Reductase. *Food Sci. Technol. Res.* **2011,** *17* (2), 87–92.

Lakhanpal S.; Donehower, R. C.; Rowinsky, E. K. Phase II Study of 4-Ipomeanol, a Naturally Occurring Alkylating Furan, in Patients with Advanced Hepatocellular Carcinoma. *Invest. New Drugs* **2001,** *19,* 69–76.

Lin, S. S. M.; Peet, C. C.; Chen, D. M.; LOHF. Sweet Potato Production and Utilization in Asia and the Pacific. In *Sweet Potato Products a Natural Resource for the Tropics*; CRC Press: Boca Raton, Florida; 1985; pp 139–148.

Ling, K. S.; Jackson, D. M.; Harrison, H.; Simmons, A. M.; Pesic-VanEsbroeck, Z. Field Evaluation of Yield Effects on USA Heirloom Sweet Potato Cultivars Infected by Sweet Potato Leaf Curl Virus. *Crop Prot.* **2010,** *29,* 757–765.

Loebenstein, G.; Thottappilly, G. *The Sweet Potato*; Springer Verlag; 2009, pp 391–425.

Loebenstein, G.; Thottappilly, G.; eds. *Virus and Phytoplasma Disease*; pp 2009; pp 105–134.

Lozano, G.; Trenado, H. P.; Valverde, R. A.; Navas-Castillo, J. Novel Begomovirus Species of Recombinant Nature in Sweet Potato (*Ipomoea batatas*) and Ipomoea Indica: Taxonomic and Phylogenetic Implication. *J. Gen. Virol.* **2009,** *90,* 2550–2562.

Ludvik, B.; Neuffer, B.; Pacini, G. Efficacy of *Ipomoea batatas* (Caiapo) on Diabetes Control in Type 2 Diabetic Subjects Treated with Diet. *Diabetes Care,* **2004,** *27,* 400–436.

Maina, S.; Barbetti, M. J.; Edwards, O. R.; de Almeida, L.; Ximenes, A.; Jones, R. A. C. Sweet Potato Feathery Mottle Virus and Sweet Potato Virus C from East Timorese and Australian Sweet Potato: Biological and Molecular Properties and Biosecurity Implication. *Plant Dis.* **2018,** *103,* 589–599.

Mardis, E. R. A Decade's Perspective on DNA Sequencing Technology. *Nature* **2011,** *470,* 198–203.

Gil, M. Wianowska, D. Chlorogenic Acid Their Properties. *Occur. Anal.* **2017,** *72* (1), 61–104.

Mbanzibwa, D. R.; Tugume, A. K.; Chiunga, E.; Mark, D., Tairo, F. D. Small RNA Deep Sequencing-based Detection and Further Evidence of DNA Viruses Infecting Sweet Potato Plants in Tanzania. *Ann. Appl. Biol.* **2014,** *165,* 329–339.

Meira, M.; Pereira da silva, E.; David, J. M.; David, J. P. Review of Genus Ipomoea: Traditional Uses, Chemistry and Biological Activities. *Rev. Bras Farmacogn.* **2012,** *22,* 682–713.

Mohammed, H. S.; El Siddig, M. A.; El Hussein, A. A.; Ibrahim, F. A.; Navas-Castillo, J.; Fiallo-Olive, E. First Report of Sweet Potato Leaf Curl Virus Infecting Sweet Potato in Sudan. *Plant Dis.* **2017,** *101,* 8–9.

Mohanraj, R.; Sivasankar, S. Sweet Potato (*Ipomoea batatas* [L]. Lam); A Valuable Medicinal Food: A Review. *J. Med. Food* **2014,** *7* (17), 733–741.

Montilla, E. E.; Hillebrand, S.; Winterhalter, P. Anthocyanins in Purple Sweet Potato (*Ipomoea batatas* L.) Varieties. *Fruit Veg. Cereal Sci. Biotechnol.* **2010,** *5* (2), 19–24.

Mulabisana, M. J.; Cloete, M.; Mabasa, K. G.; Laurie, S. M.; Oelofse, D.; Esterhuizen, L. L.; et al. Surveys in the Gauteng, Limpopo and Mpumalanga Provinces of S. *Afr. J. Bot.* **2018,** *114,* 280–294.

Mwanga, R. O. M.; Odongo, B.; Turyamureeba, G.; Alajo, A.; Yencho, G. C.; Gibson, R. W.; Smit, N. E. J. M.; Carey, E. E. Release of Six Sweet Potato Cultivars ('NASPOT 1' to 'NASPOT 6') in Uganda. *HortScience* **2003,** *38,* 475–476.

Paprotka, T.; Boiteux, L. S.; Fonseca, M. E. N.; Resende, R. O.; Jeske, H.; Faria, J. C.; et al. Genomic Diversity of Sweet Potato Chlorotic Stunt Virus. *J. Plant Pathol.* **2010,** *92,* 363–366.

Simmons, A. M.; Ling, K.; Harrison, H. F. Jr.; Jackson, D. M. Sweet Potato Leaf Curl Virus: Efficiency of Acquisition, Retention and Transmission by Bemisia tabaci (Hemiptera: Aleyrodidae). *Crop Prot.* **2009,** *28,* 1007–1011.

Stather, T.; Carey, E.; Mwanga, R.; Njoku, J.; Malinga, J.; Njoku, A.; Gibson, R.; Namanda, S. *Everything You Ever Wanted to Know about Sweet Potato Reaching Agents of Change TOT Training Manual*; 2013, vol 4.

Sukanya, D.; Parndaeng, K.; Sakul, C. W. Determination of Phytochemical Compounds and Tyrosinase Inhibitory and Antimicrobial Activities of Bioactive Compounds from *Streblus ilicifolius* (S Vidal) Cornor. *Trop. J. Pharm. Res.* **2016,** *15* (3), 497–506.

Tairo F, Mukasa SB, Jones RA, et al. Unravelling the Genetic Diversity of the Three Main Viruses Involved in Sweet Potato Virus Disease and Its Practical Implication. *Mol. Plant Pathol.* **2005,** *6* (2), 199–211.

Teow, C. C.; Truong, V. D.; et al. Antioxidant Activities, Phenolic and β. Carotene Content of Sweet Potato Genotypes with Varying Flesh Color. *Food Chem. 103,* **2007,** 829–838.

Tesfaya, T.; Feyissa, T.; Abraham, A. Survey and Serological Detection of Sweet Potato (*Ipomoea batatas* (L). Lam) Viruses in Ethiopia. *J. Appl. Biosci.* **2011,** *41,* 2746–2756.

Tibiri, B.; Pita, S.; Tiendrebeoge, B.; et al.; Characterization of Virus Species Associated With Sweet Potato Virus Disease in Burkina Faso. *Plant Pathol.* **2020,** *69,* 1003–1017.

Tibiri, E. B.; Pita, S.; Tiendrebeogo, F.; Bangratz, M.; et al. Characterization of Virus Species Associated with Sweet Potato Virus Disease in Burkina Faso. Wiley. *Plant Patho.* **2020,** *69,* 1003–1017.

Tugume, A. K.; Mukasa, S. B.; Kalkkinen, N.; Valkonen, J. P. T. Recmbination and Selection Pressure in the Ipomovirus Sweet Potato Mild Mottle Virus (Potyviridae) in Wild Species and Cultivated Sweet Potato in the Center of Evolution in East Africa. *J. Gen. Virol.* **2010,** *91,* 1092–1108.

Untiveros, M.; Fuentes, S.; Salazar, L. F. Synergistic Interaction of Sweet Potato Chlorotic Stunt Virus (Crinivirus) with Carla-, Cucumo-, Ipomo-, and Potyviruses Infecting Sweet Potato. *Plant Dis.* **2007,** *91,* 669–676.

Untiveros, M.; Quispe, D.; Kreuze, J. Analysis of Complete Genomic Sequences of Isolates of the Sweet Potato Feathery Mottle Virus Strains C and EA: Molecular Evidence for Two Distinct Potyvirus Species and Two P1 Protein Domains. *Arch. Virol.* **2010,** *155,* 2059–2063.

Valverde, R. A.; Clark, C. A.; Valkonen, J. P. T. Viruses and Virus Disease Complexes of Sweet Potato. *Plant Viruses* **2007,** *1,* 116–126.

Gruneberg, W. J.; et al. Advances in Sweet Potato Breeding from 1992-2012; Genetic Improvement of Tropical Crops, **2017**. doi: 10.1007/978-3-319-59819-2-6.

Washio, M.; Mori, M.; Ozasa, K.; et al. A Risk Factor for Kidney Cancer in a Japanese Population, Findings from the JACC Study. *J. Epidericol.* **2005,** *15* (Supp 2), S203-S211.

Wasswa, P.; Otto, B.; Maruthi, M. N.; Mukasa, S. B.; Monger, W.; Gibson, R. W. First Identification of a Sweet Potato Begomovirus (sweepovirus) in Uganda: Characterization, *Detection and Distribution Plant Pathol*; 2011.

Woolfe, J A. *Sweet Potato; An Untapped Food Resources*; Cambridge University Press: Cambridge, UK, 1992.

Yoshimoto, M.; Yahara, S.; Okuno, S.; Islam, M. S.; Ishiguro, K.; Ya-Makawa. Anti Mutagenicity of Mono-, di- and tri Caffeoylquinic acid Derivatives Isolated from Sweet Potato (*Ipomoea batatas* [L].) Leaf. *Biosci. Biotechnol. Biochem.* **2002,** *66,* 2236–2341.

You, W. C.; Blot, W. J.; Chand, Y. S.; Ershow, A. G.; Yang, Z. T.; An, Q.; et al. Diet and High Risk of Stomach Cancer in Shandong, *China Cancer Res.* **1988,** *48,* 3518–3523.

Zhang, S.-C.; Ling, K.-S. Genetic Diversity of Sweet Potato Begomoviruses in the United States and Identification of a Natural Recombinant between Sweet Potato Leaf Curl Virus and Sweet Potato Leaf Curl Georgia Virus. *Arch. Virol.* **2011,** *156,* 955–968.

CHAPTER 14

Management of Temperate Fruit Viruses

RASHMI SUMAN, NARAYAN RISHI, and VANITA CHANDEL

Amity Institute of Virology & Immunology, Amity University, Noida, Uttar Pradesh, India

ABSTRACT

Fruits are important source of various vitamins and other nutrients and complete the diet of humans. Globally, there are various tropical and temperate fruits grown based on the climatic conditions. Temperate fruits are grown in colder parts of the world. Fruits like apple, peach, plum, cherry, apricot, nectarine, pear, all comes under the umbrella of temperate fruits. Quality and quantity of these fruits are affected by virus incidence on these fruits. A number of viruses have been reported on these fruits. Management of viral diseases is not easy due to unavailability of chemical control. For efficient management, diagnosis plays a very important role. Viruses can be managed efficiently using conventional/cultural and biotechnological approaches. Emerging management approaches include nanophytovirology where nanoparticles can be helpful in studying plant–virus interactions.

14.1 INTRODUCTION

Temperate fruits are well suited at middle latitudes and adapted to cold climates. They require low chilling temperatures of around 2–3°C and high relative humidity of 90–95%. Climatic conditions for different species may vary but almost all of them requires low temperatures for growth and breaking dormancy. However, some of them may survive even at –20°C from freezing point and prefer to be at dormant stage. Upon getting favorable

Plant Diseases and Their Management: A Sustainable Approach.
Deepa Srivastava, Rajarshi Kumar Gaur, and Ajay K. Tiwari (Eds.)

growth conditions, that is, at low temperature, it starts growing its leaves and fruits. Fruits in *Rosaseae* family such as pome fruits (apple, pear, and quince) and stone fruits (peach, plum, apricot, almond, and cherry) are considered temperate fruits. These are highly nutritive as well as delicious. They are rich in vitamins and fibers and significantly reduces the chances of cancer and cardiovascular disease after daily consumption.

Temperate fruits are distributed across Asia, Europe, America, North Africa, Australia, and New Zealand. More than 50% of the total volume of temperate fruits in the world are produced by China, USA, Brazil, Italy, and Spain. China is the leading producer of apples with production of 41 million MT in 2020, followed by European Unions and United States. China also accounts largest producer of peach, plum, and pear. USA is the largest producer of almond followed by Spain. Turkey ranks first in apricot and sweet cherry production while Iran comes second in this category (FAO, 2019).

In India, the temperate fruits are grown in the Himalayan states like Jammu and Kashmir, Himachal Pradesh, Uttarakhand, and some parts of North Eastern states. These regions have a suitable chilling temperature from September to March. However, after March, flowering and fruiting starts as it breaks the dormancy period. India is the second largest producer of fruits but the productivity is low as compared to developed countries. Majority of temperate fruits are introduced from other countries but walnuts and apple are indigenous and accounts for nearly 5.6% and 79% in production among temperate fruits (Verma et al., 2018; Chandel et al., 2012).

Change in climate condition might influence the pathogens like viruses and viroids on temperate fruits and its closely related vector species. Some of the viral diseases remain unnoticed due to asymptomatic nature; however, they reduce the fruit size and quality, alter fruit composition, decrease growth and life span of trees, and many such things affects the overall production of temperate fruits. Few viruses cause serious disease to the temperate fruits which can be observed directly like mosaic pattern on leaves, chlorotic spots, distortion, and stunting. Fruit loss may be upto 30% in case of mixed infections. These viruses are transmitted through either infected seeds or grafts, cutting, and budding. Hence, proper disease management is required to control the spread of viruses. This requires virus-free planting material for raising healthy orchards and increasing the production of great qualities of temperate fruits. Before management, proper diagnosis is also very important. Techniques such as ELIA and PCR are commonly used for virus diagnosis. Several techniques have been developed to control viruses which includes virus eradication program, proper grafting techniques, and transgenic methods. Combined approaches where both conventional and transgenic/modern biotechnological approaches are used for the management of temperate fruit viruses are more effective in controlling viruses. Integrated disease management approach is being used for management of many plant pathogens. Since for plant viruses no chemical control is available, we still rely on other approaches such as use of certified planting material, effective quarantine measures, use of resistant/tolerant varieties, and use of techniques such as meristem tip culture, thermotherapy, chemotherapy, etc.

14.2 VIRUSES AND VIROIDS INFECTING TEMPERATE FRUITS

Viruses infecting temperate fruits are mainly latent showing either no or low symptoms but when infected on susceptible cultivars of fruit, they might cause severe symptoms like distortion, rolling, mottling, mosaic and chlorotic spots, necrotic spots on leaves, grooving and pitting on bark or stems, deep crinkle, ring spots and line pattern on fruits with reduced size, quality, and productivity. All these factors associated with viral disease on pome and stone fruits leads to severe damage to fruit industry, export and import of nation, and overall economy of the world. Some of the viruses infecting temperate fruits are listed below with taxonomic details, diagnostic methods, and distribution worldwide (Table 14.1).

14.2.1 APPLE STEM PITTING VIRUS

It is a type member of genus *Foveavirus* and family *Betaflexiviridae*. It has a wide host range as it has been detected in pear, apple, and crabapple. It is asymptomatic but sometimes when infected, scion is grafted to rootstock, it is often hypersensitive which may cause stem pitting, curling of leaves, and weakening of rootstock. Infection rate reported by ASPV is upto 25–100%. It is transmitted by infected grafting material. No vector transmission has been reported.

14.2.2 APPLE CHLOROTIC LEAF SPOT VIRUS (ACLSV)

It belongs to the genus *Trichovirus* and family *Betaflexiviridae*. It infects most of the temperate fruits like apple, pear, plum, almond, sweet cherry,

TABLE 14.1 Major Temperate Fruit Viruses their Detection and Distribution.

No.	Virus	Genus	Family	Diagnostics	Distribution	References
1	*Apple mosaic virus* (ApMV)	*Ilarvirus*	*Bromoviridae*	RT-PCR, DAS-ELISA, multiplex PCR, LAMP	China, Turkey, Spain, USA, India, Italy, Poland	Kapoor et al., 2018; Laxmi et al., 2007
2	*Apple chlorotic leaf spot virus* (ACLSV)	*Trichovirus*	*Betaflexiviridae*	ELISA, RT-PCR, multiplex-PCR, LAMP	Germany, Greece, Poland, Canada, Europe, China, India, Iran, Japan, Turkey and Africa	Kundu et al., 2010) Rana et al., 2007, 2012)
3	*Apple stem grooving virus* (ASGV)	*Capillovirus*	*Betaflexiviridae*	ELISA, RT-PCR, multiplex- PCR, LAMP	Germany, Greece, Poland, Canada, Europe, China, India, Iran, Japan, Turkey and Africa	(Kundu et al., 2010)
4	*Apple stem pitting virus* (ASPV)	*Foveavirus*	*Betaflexiviridae*	ELISA, RT-PCR, Multiplex- PCR, LAMP	Germany, Greece, Poland, Canada, Europe, China, India, Iran, Japan, Turkey and Africa	(Kundu et al., 2010) Dhir et al., 2010
5	*Apple green crinkle associated virus*(AGCaV)	*Foveavirus*	*Betaflexiviridae*	RT-PCR, ELISA	Japan, Italy, Canada and New Zealand	(James et al., 2013)
6	*Apple rubbery wood virus-2*	*Unclassified Ribiviria*	*Phenuiviridae*	RT-PCR	China, Korea	(Hu et al., 2021)
7	*Apple scar skin viroid* (ASSVd)	*Apscaviroid*	*Pospiviroidae*	RT-PCR, Indexing, Northern blotting	Japan, Italy, China, UK,	(Desvignes et al., 1999)
8	*Little cherry virus A* (CVA)	*Ampelovirus*	*Closteroviridae*	RT-PCR, ELISA	Columbia, Canada, China, Poland, Germany, Great Britain	(Komorowska and Cieślińska, 2004), Noorani et al., 2010
9	*Cherry leaf roll virus* (CLRV)	*Nepovirus*	*Secoviridae*	RT-PCR	Europe, North America, China, Japan, Australia, New Zealand	(Woo, Clover and Pearson, 2012)

TABLE 14.1 *(Continued)*

No.	Virus	Genus	Family	Diagnostics	Distribution	References
10	*Cherry mottle leaf virus* (CMLV)	*Trichovirus*	*Betaflexiviridae*	RT-PCR, Dot blot hybridization	North America, China	(Su, Bhagwat and Bernardy, 2016)
11	*Cherry rasp leaf virus* (CRLV)	*Cheravirus*	*Secoviridae*	RT-PCR, Dot blot hybridization, Real time RT-qPCR	China, Canada, United States, Columbia	(James, Howell and Mink, 2001)
12	*Cherry twisted leaf associated virus* (CTLV)	*Robigovirus*	*Betaflexiviridae*	RT-PCR, ELISA	Canada and British Columbia	(James, Varga and Lye, 2014)
13	*Cherry necrotic rusty mottlevirus* (CNRMv)	*Robigovirus*	*Betaflexiviridae*	RT-PCR, ELISA, multiplex RT-PCR	North America, Europe, Africa, Asia, New Zealand and Chile	(Osman, Rwahnih and Rowhani, 2017), Noorani et al., 2013
14	*Prunus necrotic ringspot virus* (PNRSV)	*Ilarvirus*	*Bromoviridae*	RT-PCR, DAS-ELISA, Northern Blotting	China, Jordan, Egypt, Turkey, Brazil, India, Croatia, Czech Republic,	(Kapoor et al., 2018) Chandel et al., 2008
15	*Prune dwarf virus* (PDV)	*Ilarvirus*	*Bromoviridae*	RT-PCR, ELISA	China, USA, India,	(Kapoor et al., 2018)
16	*Peach rosette mosaic virus* (PRMV)	*Nepovirus*	*Secoviridae*	RT-PCR, ELISA, Immuno-capture RT-PCR	North America	(Ho et al., 2018)
17	*Peach latent mosaic viroid* (PLMVd*)*	*Pelamoviroid*	*Avsunviroidae*	Dot blot hybridization, RT-PCR, Microarray	Europe, North America, South America, Asia	(Flores et al., 2006)
18	*Pear blister canker viroid* (PBCVd)	*Apscaviroid*	*Pospiviroidae*	RT-PCR, Indexing, Dot blot hybridization	France, Argentina, Canada, Australia, China, Japan, Turkey	(Desvignes et al., 1999)
19	*Plum pox virus* (PPV)	*Potyvirus*	*Potyviridae*	RT-PCR, ELISA	Germany, Greece, Poland, Canada, Europe, China, India, Iran, Japan, Turkey and Africa	(Levy et al., 2011)
20	*Apricot latent virus (ALV)*	*Foveavirus*	*Betaflexiviridae*	ELISA, RT-PCR	France, Italy, Turkey, Iran, Egypt, Spain, Lebanon and Czech Republic	(Grimová and Ryšánek, 2012)

apricot, and quince. It has large host range and infection rate is upto 80–100%. ACLSV is latent virus but susceptible cultivars are severely affected. Symptoms specific to this virus are chlorotic spots, ring pattern and mosaic, distortion, and stunting of leaves. It is also associated with graft incompatibility, bud necrosis, fruit deformation, and severe yield reduction. ACLSV is graft transmissible but not seed or vector transmissible (Rana et al., 2007, 2010).

14.2.3 APPLE STEM GROOVING VIRUS (ASGV)

ASGV is a type member of genus *Capillovirus* and family *Betaflexiviridae*. It has a wide host range like apple, pear, and quince and it is also graft transmissible. ASGV is not transmitted by any vector or seeds. The virus is latent but susceptible cultivars may have severe symptoms like grooving and brown line on stems and graft union incompatibility causing top working disease in many cultivars. Infection rate of ASGV is upto 100% (Dhir et al., 2010).

14.2.4 APPLE MOSAIC VIRUS (APMV)

ApMV belongs to genus *Ilarvirus* and family B*romoviridae*. It has a wide host ranging from apple, pear, apricot, peach, plum, cherry, and almond. ApMV incidence on temperate fruits are often symptomless but on severe cases it may show symptoms like yellow mosaic, blotches, ring or line pattern on leaves, irregular or band pattern on veins, and premature leaf drop. Transmission is reported by infected grafting material, budding, and pollen (Laxmi et al., 2011).

14.2.5 PRUNUS NECROTIC RING SPOT VIRUS (PNRSV)

PNRSV also belongs to genus *Ilarvirus* and family *Bromovidae*. Host range of PNRSV is mostly prunus species like cherry, peach, almond, and pear. It is transmitted by budding and grafting, pollens, and infected propagative material. Occurrence of this virus is worldwide including India, Turkey, Brazil, Egypt, Croatia, Czech Republic, and Jordan. It has been reported in peach, plum, cherry, apple, and pear with characteristic brown line, rings, leaf curling, yellow mottling, blotching, necrosis, and leaf deformity. PNRSV is responsible for significant yield loss upto 70% (Chandel et al., 2007, 2008).

14.2.6 PRUNE DWARF VIRUS (PDV)

PDV belongs to the same group of *Ilarvirus* and it causes stunting, short internodes, and leaf malformation in plum. In prune, it causes chlorosis and leaf reduction, while in cherry it causes mottling of leaves, chlorotic spots, and rings. Host range for PDV is wide including almond, cherry, peach, and plum with different infection rate causing significant yield reduction in fruit by 60%. Transmission occurs by seeds and pollen.

14.2.7 PLUM POX VIRUS (PPV)

PPV belongs to genus *Potyvirus* and family *Potyviridae*. PPV induces sharka disease which was first reported in Bulgaria in 1900s. It is one the most devastating disease causing large economic losses. It has been detected in peach, plum, cherry apricot, and almonds showing characteristic symptoms like pale yellow spots, rings, leaf mottling, flower discoloration, deformed fruits with poxes, and irregular lines or rings on surface. PPV is transmitted by vegetative propagation, grafting, and budding as well as by vector aphids such as *Aphis spiraecola* and *Myzus persicae* in a nonpersistent manner.

14.2.8 LITTLE CHERRY VIRUS (LCV)

LCV is a type member of genus *Ampelovirus* and family *Closteroviridae*. Little cherry disease is a serious disease reported in Columbia and Canada during 1930s. It is distributed worldwide in cherry plants. It has two variants LCV-1 and -2 each with distinct symptoms but graft transmissible. However, LCV-2 is transmitted by vector mealybug like *Pseudococcus maritimus* and *Phenacoccus aceris*. Symptoms characterized by these viruses are reduction in size, color, and taste of fruit.

14.2.9 APPLE RUBBERY WOOD VIRUS 2 (ARWV-2)

ARWV-2 is negative sense ssRNA having three genome segments placed in the family *Phenuiviridae* and unclassified genus *Ribiviria*. It was first reported in China infecting pear and then in apple showing rusty skin symptoms (Hu et al., 2021).

14.2.10 APPLE SCAR SKIN VIROID (ASSVD)

ASSVd is a member of genus *Apscaviroid*, and family *Pospiviroidae*. It was first reported from China in 1930s and also observed in other countries such as Japan, Korea, India, Europe, and North America. ASSVd is naked ssRNA without capsid protein infecting mainly woody stone and pome fruits. It is transmitted by vegetative propagation and grafting. One of the report has mentioned the active transmission of ASSVd by whitefly *Trialeurodes vaporariorum* (Walia et al., 2015). Cucumis sativus Phloem protein 2 (CsPP2), a plant protein, facilitates translocation of viroid RNAs as evident by the presence of the complex PP2/ASSVd in cucumber plant. Symptoms observed in apple due to ASSVd are scar skin and dapple and on pear are rusty skin disease and fruit crinkle disease. One of the study has reported that ASSVd infection was observed only in the "Niitaka" cultivar of pear and the infection rate was around 13.8% (Kim et al., 2019) which is higher than the infection rate (1.7%) observed in 2010.

14.2.11 PEACH LATENT MOSAIC VIROID (PLMV)

PLMVd is categorized under genus *Pelamoviroid* and family *Avsunviroidae* having a characteristic feature of chloroplastic viroid with hammerhead ribozymes and 33-351nt circular RNA adopting branched conformation stabilized by pseudoknot between two kissing loops (Flores et al., 2006). The viroid is transmitted horizontally by infected propagative material and pruning tools but not by seeds and pollen. Symptoms are conspicuous but certain isolates induces leaf mosaic and blotches, an extreme albinism (peach calico), delayed flowering and ripening, deformed fruits, stem, pitting, bud necrosis, and premature ageing of trees (Malfitano et al., 2003). It infects peach, plum, pear, and apricot. Incidence of this viroid is reported upto 45% in peach cultivar in Egypt, 40% in Syria, and 34% in Lebanon (El-Dougdoug et al., 2017). PLMVd was first identified in France in 1976 and later distributed all over the world including Europe, Asia, North America, and South America. Detection of PLMVd was done through dot blot hybridization, RT-PCR, microarrays, and NGS (Serra et al., 2017).

14.2.12 PEAR BLISTER CANKER VIROID (PBCVD)

PBCVd was detected in pear trees in Australia using PAGE and dot blot hybridization technique as well as RT-PCR (Joyce et al., 2006). It has single circular RNA as genome of upto 315 nt assuming branched conformation. It is a member of genus *Apscaviroid* and family *Pospiviroidae* (Hernfindez et al., 1992). It does not produce any symptoms unless susceptible cultivar is infected which displays pustule cankers on bark. PBCVd is transmitted by infected pruning and propagative materials. It has been reported in several European and Asian countries, Japan, Australia, America, and China.

14.3 SYMPTOMS

Plant viruses in the field can be identified on the basis of their symptoms. In the nomenclature of the virus, symptom is always mentioned. For example, ApMV—in this, mosaic is the symptom shown by the virus; ACLSV—in this, chlorotic leaf spot is the symptom; in ASPV, pitting is major symptom; ASGV—in this, stem grooving is major symptom. So, it is not very difficult to identify virus symptoms in temperate fruit trees.

14.4 DIAGNOSIS OF TEMPERATE FRUIT VIRUSES

Management of any pathogen depends upon sensitive and reliable diagnostic methods. Thus, diagnosis is very important part of management of any plant virus. Detection of viruses early in planting material or fruit trees can manage viruses and hence can reduce losses caused by these viruses. Without proper identification of the viruses or wrong identification means both loss in money and time. Plant viruses are detected by three main techniques—biological, serological, and molecular.

Biological diagnosis is done by identification of symptoms on the tree/plant or mechanically transmitting virus to indicator plants or also by grafting in case of fruit trees.

Serological detection is based on antigen antibody reaction by ELISA. This method by used by Clark and Adams (1977) for the very first time for the detection of PPV. Commercial kits are available for all temperate fruit viruses are widely used for detection of viruses. ACLSV has been detected on apple tree using ELISA kits as first method of detection of virus (Rana et al., 2007, 2012). Other viruses such as ASGV, ApMV, ASPV, and PNRSV are also detected using antibodies against the viruses (Dhir et al., 2010; Laxmi et al., 2011; Chandel et al., 2007, 2008). Serological methods are more economic as compared to molecular methods and when large number of samples are to be tested preliminary testing by ELISA is always beneficial. ELISA can also be used for identification of different virus isolates or even serotypes using monoclonal antibodies. In PNRSV, 34 serotypes and PPV serotypes were identified by this technique (Myrta et al., 2001; Cambra et al., 2011).

Molecular methods for the diagnosis of temperate fruit viruses are more sensitive and specific and are based on nucleic acid. These methods are used mainly in disease survey, plant quarantines and in certification programs. Most widely used molecular method for temperate fruit virus detection is polymerase chain reaction which amplify target genes. Since most of temperate fruit viruses have RNA genome, so, first step before PCR is always reverse transcription of RNA to cDNA which is done by using reverse transcriptase enzymes (Chandel et al., 2008, 2011, 2012).

Multiplex PCR is also very commonly used for virus detection in temperate fruit trees. Its benefit is that in one reaction two to five viruses can be identified which saves time and money both. Four cherry viruses were identified using multiplex PCR—cherry virus A (CVA), cherry necrotic rusty mottle virus (CNRMV), little cherry virus 1 (LChV-1), and prunus necrotic ringspot virus (PNRSV) (Noorani et al., 2013). Real-time PCR, which is a quantitative method for virus detection, has also been used in the detection of many temperate fruit viruses. ASGV and apple green crinkle-associated virus (AGCaV) were detected using real-time PCR assay in a study conducted by E. Beaver-Kanuya et al. (2019).

Microarray is another important method for the plant virus detection. Microarray has an advantage that it can detect many viruses in one time if probes are available but it can be done for only known viruses. It has been used for the identification of temperate fruit viruses (Barba and Hadidi, 2011).

Next-generation sequencing (NGS) is the latest technique for virus detection and whole genome sequencing of temperate fruit viruses (Barba et al., 2014; Hadidi and Barba 2012; Barba et al., 2014).

14.5 DISEASE MANAGEMENT

Intensive agriculture practices to meet the demands of growing population and favorable climatic conditions has increased the viral disease incidence with huge losses upto 100%. Hence, there is need to develop integrated management practices as there is no viricides developed yet. Basically, there are two strategies behind controlling the viral disease, which are (1) decreasing the source of infection, and (2) minimizing the rate of spread.

14.5.1 *CONVENTIONAL METHODS*

Earlier, the viral disease was controlled by use of proper agricultural practices. Use of certified virus free planting material or seedlings were used in fields after being tested in nurseries. Meristem tip culture was promoted in laboratories to avoid viruses in planting material.

14.5.2 *EXCLUSION OF PATHOGENS BY QUARANTINE AND CERTIFICATION*

Trade is important for the economy of any country. With trade movement of viruses also takes place with planting material, seeds, or other crops. In order to restrict the movement of viruses from one place/country to another government has imposed certain regulations and guidelines. Plant quarantine regulatory measures are operative through the "Destructive Insects & Pests Act, 1914 (Act 2 of 1914)" in India. Phytosanitary activities at the international level are governed on the basis of agreement on the application of sanitary and phytosanitary measures under World Trade Organization general agreements of tariffs and trade.

Import and export regulations also depend on individual countries. As in European Union, it is mandatory that fruit tree materials or fruits imported should be free of pathogens if they are not reported in EU. In this way, they prevent entry of new pathogens to the country (Reed and Foster, 2011).

Another way by which fruit viruses spread is by use of infected planting material such as root stocks, buds, grafted material, etc. In order to prevent entry of viruses using infected material, there is certification which is required for such planting material stating that the material is free from any kind of viruses. Certification is provided by many referral laboratories which are established by the Government of India. Usually it's done by checking the planting material for viruses using serological methods such as ELISA or molecular methods such as RT-PCR. In this way, growers will know that the planting material that they will be using is free from viruses. This helps in the management of viruses as in case of plant viruses once virus established in the field it's not possible to remove it, so better way in plant virus management is always preventing virus from establishing in the field. This technique is though very well suited for smaller plants and crops where it can be easily done. Whereas for development of fruit orchards it takes many years and those years of efforts by the growers. Therefore, it is only recommended in highly infected trees or small trees which are not very old. In European countries, PPV is major problem in temperate fruits, and in order to prevent its spread to new loci many times, trees were eradicated which were infected with PPV (Barba et al., 2015).

14.5.3 VECTOR CONTROL BY CHEMICAL MEANS

Viral vectors play very important role in the spread of the disease in the field. In management of plant viruses, focus is on managing viral vectors along with other management techniques. In temperate fruits PPV is major nuisance and it is transmitted by various species of aphids. Use of insecticide is recommended to control aphids though it leads to reduction in vector population but it does not affect the affect the transmission of viruses by aphids and the possible reason for this is that there are number of aphid species which transmit the virus PPV. In some reports, use of oil application in the nurseries to reduce the spread of PPV has been reported (Barba et al., 2014; Carraro et al., 2011; Migliori et al., 1998).

Several researchers have proven the use of pyrethroids effective against thrips vectors as well as aphids and leafhoppers (Sastry and Zitter, 2014). The efficacy of controlling viruses by chemical means depends on the persistent of insecticides on plant surface and their application frequency.

14.5.4 ELIMINATION OF VIRUSES USING MERISTEM TIP CULTURE/ THERMOTHERAPY/CRYOTHERAPY

Viruses can be eliminated from the plants using certain techniques such as tissue culture, thermotherapy, and cryotherapy (Facolli et al., 1998). Meristem tip culture is one important technique in plant tissue culture which has been widely used to generate pathogen free plants. In a study done by Vivek and Modgil (2018), apple plants were made free of ACLSV, ASPV, and ASV using thermotherapy and meristem tip culture.

In some plants, cryogenic methods has been useful in eradication of viruses from planting material (Brisson et al., 1997). Hu et al. (2021) reported thermotherapy for apple virus eradication in which they heat-treated infected apple shoots at 38°C for 30 days and excised the shoot tip for regeneration. The total survival frequency of ASGV, ACLSV, ApMV, and ASPV was 30% (Hu et al., 2017).

Vitrification technique has successfully eliminated PPV from *prunus* species with eradication rate upto 45–60%, irrespective of size of shoot tip.

Shoot tip cryotherapy has been used to eradicate viruses from apple (Jaime A. Teixeira da Silva 2019). Romadanova et al. has used the excised shoot tips that were treated at 22°C for 8 h and –1°C for 16 h and subjected to vitrification cryotherapy. It resulted into virus elimination of ACLSV, ASGV, and ASPV by 100%, 67%, and 50%, respectively (Romadanova et al., 2016). Shoot-tip cryotherapy was failed completely in eradication of ASGV and Raspberry bushy dwarf virus (RBDV) but combination of thermotherapy and shoot-tip cryotherapy has proven its efficacy upto 33% in controlling RBDV (Wang et al., 2009) and 100% in case of ASGV (Zhao et al., 2018).

Hot water treatment can also inactivate viruses proven by study of Pandey et al. (1972) where budwood of the apple which was infected by ApMV, almond infected by almond mosaic virus and plum infected by plum line patten virus, all were treated in water at 50°C for maximum 14 min and observed that it could inactivate viruses.

Chemotherapy technique applies antiviral drugs like ribavirin and 2-thiouracil along with combining thermotherapy and meristem culture to eradicate viruses. Paprstein et al. reported chemotherapy of infected apple shoots by treating with medium containing ribavirin at 20 mg/l for four weeks in first cycle and 100 mg/l in next cycle for next four weeks. Shoot tips containing 2–3 leaf primordial was excised and regenerated. It was observed that plantlets were free of ASGV by 65% and ASPV by 35% in first cycle while ACLSV by 76% in the second cycle (Paprstein and Sedlák, 2015). This technique of chemotherapy combining with thermotherapy was applied successfully in eradicating apple viruses such as ACLSV, ASGV, and ASPV from infected apple in vitro shoots (Hu et al., 2015). They treated the infected apple shoots with ribavirin at 25 mg/L for 40 days followed by heat treatment at 36°C for 20 days and observed for total virus elimination frequency of ASGV, ASPV, and ACLSV was almost 94–100% in axillary shoot tip.

Micrografting is a technique which implies attachment of small section of microshoot and meristem onto the rootstock cultured in vitro. This technique of micrografting in combination with thermotherapy is applied in case of woody plants such as apple (Hu et al., 2017), pear, and citrus (Kim et al., 2005) as shoot culture is difficult in these case.

14.5.5 USE OF RESISTANT/TOLERANT PLANTING MATERIAL

Plant breeding for the development of new and resistant varieties is a century-old technique being used specially for perennial plants/trees. Use of virus resistant or tolerant rootstock has been used in temperate fruits such as apple and stone fruits in conventional plant breeding programs. The only drawback of this strategy is that it takes very long time, many years to develop such variety and also only at the end one can know whether the breeding was successful or not (Garcia et al., 2013). This technique has been used in PPV in stone fruits and cherry leaf roll virus in walnut (Decroocq et al., 2011). Use of conventional breeding has not been done in case of other temperate fruits so widely. Another technique that is being used in the management of temperate fruits is the selection of resistant/tolerant germplasm. In some wild plants, resistance has been identified such as in wild species of *Prunus davidiana* resistance to PPV (Decroocq et al., 2005).

14.5.6 TRANSGENIC APPROACHES FOR VIRUS MANAGEMENT

14.5.6.1 TARGETING VIRUS GENOME

To overcome the drawbacks of conventional plant breeding techniques, transgenic methods were introduced. Gene revolution in the form of genetic engineering techniques was considered as new and effective technique for management of viruses. Cross protection where resistance is induced in the tree or plant by use of mild strain of the virus, has been used widely since ages as technique for management of plant viruses. It was successfully done in case of ApMV and it demonstrated that mild strain of ApMV virus was able to provide protection against infection from severe strain of the same virus (Chamberlain et al., 1964). Same technique has also been used for PNRSV and PPV in plum (Laimer, 2012). Mechanism of cross protection was always supported by several mechanism involved. The mechanisms involved protein mediated protection, such as CP-mediated resistance, RNA-mediated resistance, and RNA silencing have been suggested as mechanisms of cross protection. Pathogen-derived resistance has been widely used from early 1990s and the main idea behind this was to reduce or stop the virus multiplication using virus genes/proteins; hence the name pathogen-derived resistance (Stanford and Johnstein, 1985). This technique was very commonly used in case of PPV. Later, when post-transcriptional gene silencing (PTGS) was discovered, it became clear that this resistance was based on RNA-mediated mechanism/RNA silencing (Simon matio-Garcia, 2011).

Pandolfini et al. (2003) transformed the *N. benthamiana* by PPV-D intron spliced hairpin RNA (ihpRNA) under the control of promoter *rol*C. This promoted resistance against systemic infection rather than local (Pandolfini et al., 2003). Another report has mentioned the usage of four ihp RNA constructs having 5'UTR, P1, P3, and HC pro of PPV-M ISPaVe44 in *N. benthamiana* and conferred resistance in different strains of PPV efficiently (Di Nicola-Negri et al., 2005). Interestingly they found that ihpRNA UTR/P1 plants were highly efficient than other construct against other viruses as well. Hence, these studies on model plant concluded that h-UTR/P1 construct was the best candidate for transforming *prunus* plants. In another report, CP gene was incorporated in ihpRNA construct along with CaMV35S promoter or peach Cab promoters were introduced to *N. benthamiana* and it was successfully resistant to PPV (Hily et al., 2007), so it was further employed to transform plum. Transgenic plums after were subjected to different isolates of PPV-D, Rec, EA, and M strains for over multiple cycle of vegetative growth and plants showed good progress over fourth and final bud-break cycle after initial failure (Ravelonandro et al., 2014). Liu et al. reported the production of pTRAP6i (ihpRNA construct), covering gene fragments from six stone fruit infecting viruses like PPV, PDV, PNRSV, peach mosaic virus (PMV), tobacco ringspot virus (TRSV), and *American plum line pattern virus.* But unfortunately none of the transgenic cherry leaf explants showed resistant and so this construct failed due to longer length of around 2500 bp covering 400–500 bp fragment from each virus genome (Liu et al., 2007).

Advantages of using ihpRNA construct in developing transgenic plants are—firstly; the double-stranded structure and very low steady state transcripts reduces the possibility of recombination between incoming virus and ihpRNA construct; secondly, there is no expression of transgenic protein which might complement with virus or cause toxicity and/or allergy to the plants (Lemgo et al., 2013); thirdly; it induces consistent resistance to high percentage of stone fruits plants.

14.5.6.2 TARGETING HOST

Recent availability of molecular mechanism involved in plant host–virus interaction have paved the way for development of novel strategy in control of virus infection in plants. There are some host factors which play important role virus in replication and infection and one of such is eukaryotic translation initiation factor eIF(iso)4E. It was seen that upon disruption of eIF(iso)4E gene by transposon in *A.thaliana*, transgenic plants was able to confer resistance against five different isolates of PPV (Decroocq et al., 2006). When this host factor was silenced by ihpRNA construct in plum lines, 80% of the transgenic plants were resistant to PPV (Wang et al., 2013).

Engineered Nanoparticles (ENPs)—With size ranging 1–100 nm, large surface area-to-volume ratio and high reactivity, such as nanometals, carbon nanoparticles, metalloids, metal oxides, liposomes, quantum dots, and dendrimers, have all the required features to be used as suitable agrochemicals for sustainable disease management. They have tremendous use in various aspects like nano-fertilizers, nano-carriers, bio-stimulants, and anti-microbial agent. Carbon-based nanoparticles (C60 and CNTs) at 200 mg/L (Adeel et al., 2021), SiO_2-NPs, and ZnO-NPs (Cai et al., 2019) have inhibited the replication of TMV by directly interacting with their capsid protein and disrupting the viral genome in *N.benthamiana*. Foliar spray of Silver based NPs at 50 mg/L on tomato plants have reduced the infectivity of ToMV. Several studies have demonstrated that NPs triggers the defensive activity by elevating the level of phytohormones such as jasmonic acid, abscisic acid, salicylic acid, zeatin riboside, and brassino steroids, increasing the production of anti-oxidants and reactive oxygen species (ROS). It also increases photosynthesis in plants and regulates several gene expression for early immune response. Mitter and group in their report has mentioned the use of nanocarriers that is nontoxic and degradable-layered double hydroxide bioclay nanosheets (BioClay) for delivering dsRNA in plants which would effectively trigger RNA interference (RNAi) (Mitter et al., 2017). This technique can also be used in the temperate fruit viruses in future.

CRISPR Cas system is the latest technology being used now-a-days to improve resistance in plants from pathogens specially viruses. It can be used to introduce targeted mutations in host plant genes which affects viral replication in their vicinity. Host factor eIF4E interacts with 5'UTR of virus through VPg protein and this interaction could be impaired with disrupting eIF4E gene with the gene editing technique like CRISPR cas9 system. Genome editing in apple using a single gRNA and Cas9 in first-generation transgenic apple. This will be a way forward in plant breeding programs in apple (Chikako Nishitani et al., 2016).

14.6 CONCLUSION

Plant viruses cause significant economic losses to temperate fruits and are real threat for sustainable agriculture. The emergence of new viral disease is result of rapid evolution in viruses, climatic changes, and international trade. Disease management depends upon fast and specific diagnosis of causal agent which requires detailed knowledge about viruses, their genetic relationship with same, and different groups. Apart from this, we also need to practice proper quality control measures before plantation of seedlings, certification of propagative materials in nurseries, border control to prevent entry of new viruses in the country, and most important dispersal of viruses through vectors in the field. The aim of integrated disease management (IDM) is to protect the plants from diseases as well as increase the yield of fruits to its maximum. This is possible only when preventive measures is taken before plantation along with curing the plants suffering from diseases. Preventive measure includes basic agricultural practices with proper sanitation, pruning, crop rotation, use of chemical and biological pesticides, nanoparticles, micrografting, and advanced gene editing technologies that have successfully resulted in controlling plant viruses. Despite the advanced technology and breeding programs for developing virus resistant cultivars, there are still certain restrictions in few countries about the ecological impact of transgenic plants on human beings as well as animals and environment. The challenge for plant virologist is to apply the innovative and new technologies along with previous knowledge to its greatest effect for the disease management of temperate fruits.

KEYWORDS

- **temperate fruit viruses**
- **viroid's**
- **diagnosis**
- **diseases management**
- **vectors**
- **elimination**

REFERENCES

Adeel, M.; Farooq, T.; White, J. C.; Hao, Y.; He, Z.; Rui, Y. Carbon-based Nanomaterials Suppress Tobacco Mosaic Virus (TMV) Infection and Induce Resistance in Nicotiana Benthamiana. *J. Hazard Mater.* **2021,** *404* (Pt A), 124167.

Beaver-Kanuya, E.; Szostek, S. A.; Harper, S. J. Development of Real-time RT-PCR Assays for Two Viruses Infecting Pome Fruit. *J. Virol. Methods* **2019,** *266,* 25–29.

Barba, M.; Hadidi, A. DNA Microarrays and Other Future Trends in Detection and Typing of Viruses, Viroids, and Phytoplasmas. In *Virus and Virus-like Diseases of Pome and Stone Fruits*; Hadidi, A., Barba, M., Candresse, T., Jelkmann, W., Eds.; APS Press: T. Paul, M. N., 2011; pp 361–372.

Barba, M.; Czosnek, H.; Hadidi, A. Historical Perspective, Development and Applications of Next-generation Sequencing in Plant Virology. *Viruses*. **2014,** *6* (1), 106–136.

Barba, M.; Ilardi, V.; Pasquini, G. Control of Pome and Stone Fruit Virus Diseases. *Adv. Virus Res.* **2015,** *91,* 47–83.

Brison, M.; de Boucaud, M. Y.; Pierronet, A.; Dosba, F.. Effect of Cryopreservation on the Sanitary State of a cv *Prunus* Rootstock Experimentally Contaminated with Plum Pox Potyvirus. *Plant Sci.* **1997,** *123,* 189–196.

Cai, L.; Liu, C.; Fan, G.; Liu, C.; Sun, X. Preventing Viral Disease by ZnONPs Through Directly Deactivating TMV and Activating Plant Immunity in Nicotiana Benthamiana. *Environ. Sci.: Nano.* **2019,** *6,* 3653–3669.

Cambra, M.; Boscia, D.; Gil, M.; Bertolini, M.; Olmos, E. Immunology and Immunological Assays Applied to the Detection, Diagnosis and Control of Fruit Tree Viruses. In Hadidi, A.; Barba, M.; Candresse, T.; Jelkmann, W., Eds.; *Virus and Virus-Like Diseases of Pome and Stone Fruits*; T. Paul MN: APS Press. 2011; pp 303–314.

Carraro, L.; Labonne, G.; Dallot, S. Vector Control of Fruit Tree Viruses and Phytoplasmas. In Hadidi, A.; Barba, M.; Candresse, T.; Jelkmann, W. (Eds.), *Virus and Virus-like Diseases of Pome and Stone Fruits*; T. Paul, M. N.: APS Press. 2011; pp 379–382.

Chamberlain, E. E.; Atkinson, J. D.; Hunter, J. A. Cross-protection between Strains of Apple Mosaic Virus, *N. Z. J. Agric. Res.* **1964,** *7* (4), 480–490.

Chandel, V.; Rana, T.; Hallan, V.; Zaidi, A. A. Wild Himalayan Cherry (Prunus cerasoides) as Natural Host of Prunus Necrotic Ring Spot Virus in India. *Plant Dis.* **2007,** *91* (12), 1686.

Chandel, V.; Rana, T.; Handa, A.; Thakur, P. D.; Hallan, V.; Zaidi, A. A. Incidence of *Prunus Necrotic Ring Spot Virus* on *Malus Domestica* in India. *J. Phyotopathol.* **2008,** *156,* 382–384.

Chandel, V.; Rana, T.; Hallan, V.; Zaidi, A. A. Detection of Prunus Necrotic Ring Spot Virus in Plum, Cherry and Almond by Serological and Molecular Techniques from India. *Arch. Pytopathol. Plant Protec.* **2011,** *44* (18), 1779–1784.

Chandel, V.; Rana, T.; Kumar, Y.; Hallan, V.; Zaidi, A. A. Stone Fruit Viruses: Economic Importance and Biotechnological Approaches towards Management Strategies. In *Recent Trends in Plant Virology*; Rao, G. P., Baranwal, V. K., Mandal, B., Rishi, N., Eds.; Studium Press LLC: USA, 2012; pp 301–332.

Clark, M. F.; Adams, A. N. Characteristics of the Microplate Method of Enzyme-linked Immunosorbent Assay for the Detection of Plant Viruses. *J. Gen. Virol.* **1977,** *34* (3), 475–483.

Decroocq, V.; Foulongne, M.; Lambert, P.; Gall, O. L.; Mantin, C.; Pascal, T.; Schurdi-Levraud, V.; Kervella, J. Analogues of Virus Resistance Genes Map to QTLs for Resistance to Sharka Disease in Prunus Davidiana. *Mol. Genet. Genom.* **2005,** *272* (6), 680–689.

Decroocq, V.; Sicard, O.; Alamillo, J. M.; Lansac, M.; Eyquard, J. P.; García, J. A.; Candresse, T.; Le Gall, O.; Revers, F. Multiple Resistance Traits Control Plum Pox Virus Infection in Arabidopsis Thaliana. *Mol. Plant Microbe. Interact.* **2006,** *19* (5), 541–549.

Decroocq, V.; Badenes, M. L.; Neumuller, M. B. R. Plum Pox Virus. In Hadidi, A.; Barba, M., Candresse, T., Jelkmann, W., Eds.; *Virus and Virus-like Diseases of Pome and Stone Fruits*; T. Paul MN: APS Press, 2011; pp 403–408.

Desvignes, J. C.; Grasseau, N.; Boyé, R.; Cornaggia, D.; Aparicio, F.; Di Serio, F.; Flores, R. Biological Properties of Apple Scar Skin Viroid: Isolates, Host Range, Different Sensitivity of Apple Cultivars, Elimination, and Natural Transmission. *Am. Phytopathol. Soc.* **1999,** *83* (8), 768–772.

Dhir, S.; Tomar, M.; Thakur, P. D.; Ram, R.; Hallan, V.; Zaidi, A. A. Molecular Evidence for *Apple Stem Pitting Virus* Infection in India. *Plant Pathol.* **2010,** *59,* 393.

Di Nicola-Negri, E.; Brunetti, A.; Tavazza, M.; Ilardi, V. Hairpin RNA-mediated Silencing of Plum Pox Virus P1 and HC-Pro Genes for Efficient and Predictable Resistance to the Virus. *Transgen. Res.* **2005,** *14* (6), 989–994.

El-Dougdoug, K. A.; Çağlayan, K.; Elleuch, A.; Al-Tuwariqi, H. Z.; Gyamera, E. A.; Hadidi, A. Geographical Distribution of Viroids in Africa and the Middle East. In *Viroids and Satellites*; Hadidi, A.; Flores, R.; Randles, J. W.; Palukaitis, P. B. T.-V.; S., Eds.; Academic Press: Boston, 2017; pp 485–496.

Faccioli, V. C.; Marani, F. Virus Elimination by Meristem Tip Culture and Tip Micrografting. In *Plant Virus Disease Control*; Hadidi, A., Khetarpal, R. K., Koganezawa, H., Eds.; APS Press: St. Paul, 1998; pp 346–380.

FAOSTAT 2019. http://www.fao.org/faostat/en/#data/QC.

Flores, R.; Delgado, S.; Rodio, M.-E.; Ambrós, S.; Hernández, C.; Serio, F. D. I. Peach Latent Mosaic Viroid: Not so Latent. *Mol. Plant Pathol.* **2006,** *7* (4), 209–221.

Grimová, L.; Ryšánek, P. Apricot Latent Virus—Review. *Hortic. Sci. (Prague).* **2012,** *39,* 144–148.

Garcia, J. A.; Glasa, M.; Cambra, M.; Candresse, T. *Plum Pox Virus* and Sharka: A Model Potyvirus and a Major Disease. *Mol. Plant Pathol.* **2013,** *15,* 226–241.

Hadidi, A.; Barba, M. Next Generation Sequencing: Historical Perspective and Current Applications in Plant Virology. *Petria.* **2012,** *22,* 262–277.

Hernfindez, C.; Elena, S. F.; Moya, A.; Flores, R.; Cientificas, S. D. I.; Roig, C. J.; et al. Pear Blister Canker Viroid is a Member of the Apple Scar Skin Subgroup (apscaviroids) and Also has Sequence Homology with Viroids from Other Subgroups. *J. Gen. Virol.* **1992,** *73,* 2503–2507.

Hily, J.; Petri, C.; Ravelonandro, M.; Bassett, C.; Damsteegt, V.; Liu, Z.; Scorza, R. Plum Pox Virus Coat Protein Gene Intron-hairpin-RNA (ihpRNA) Constructs Provide Resistance to Plum Pox Virus in *Nicotiana benthamiana and Prunus domestica.* **2007,** *132* (6), 850–858.

Ho, T.; Harris, A.; Katsiani, A.; Khadgi, A.; Schilder, A.; Tzanetakis, I. E. Genome Sequence and Detection of Peach Rosette Mosaic Virus. *J. Virol. Methods* **2018,** *254,* 8–12.

Hu, G.; Dong, Y.; Zhang, Z.; Fan, X.; Ren, F.; Zhou, J. Virus Elimination from In Vitro Apple by Thermotherapy Combined with Chemotherapy. *Plant Cell Tissue Organ. Cult.* **2015,** *121* (2): 435–443.

Hu, G. J.; Dong, Y. F.; Zhang, Z. P.; Fan, X. D.; Ren, F.; Li, Z. N. Efficacy of Virus Elimination from Apple by Thermotherapy Coupled with In Vivo Shoot-Tip Grafting and In Vitro Meristem Culture. *J. Phytopathol.* **2017,** *165* (10), 701–706.

Hu, G. J.; Dond, Y. F.; Zhang, Z. P.; Fan, X. D.; Ran, F.; Lu, X. K. First Report of Apple Rubbery Wood Virus 2 Infections of Apple in China. *Plant Dis.* **2021,** *105,* 519.

James, D.; Howell, W. E.; Mink, G. I. Molecular Evidence of the Relationship between a Virus Associated with Flat Apple Disease and Cherry Rasp Leaf Virus as Determined by RT-PCR. *Plant Dis.* **2001,** *85* (1), 47–52.

James, D.; Varga, A.; Jesperson, G. D.; Navratil, M.; Safarova, D.; Constable, F., et al. Identification and Complete Genome Analysis of a Virus Variant or Putative New Foveavirus Associated with Apple Green Crinkle Disease. *Arch. Virol.* **2013,** *158* (9), 1877–1887.

James, D.; Varga, A.; Lye, D. Analysis of the Complete Genome of a Virus Associated with Twisted Leaf Disease of Cherry Reveals Evidence of a Close Relationship to Unassigned Viruses in the Family Betaflexiviridae. *Arch. Virol.* **2014,** *159,* 2463–2468.

Joyce, P. A.; Constable, F. E.; Crosslin, J.; Eastwell, K.; Howell, W. E.; Rodoni, B. C. Characterisation of Pear Blister Canker Viroid Isolates from Australian Pome Fruit Orchards. *Australas Plant Pathol.* **2006,** *35,* 465–471.

Kapoor, S.; Sharma, A.; Shylla, B.; Handa, A. Ilarviruses and the Importance of Certified Elite Planting Material in Apple Production System—An Overview. *Int. J. Curr. Microbiol. Appl. Sci.* **2018,** *7* (04), 2444–2462.

Kim, D.; Shim, H.; Kwon, H.; Lee, S.; Lee, J.; Kim, K.; Hyun, J.. Production of Virus-Free Stocks from Citrus Plant by the Shoot-Tip Grafting and Heat Treatment. *J. Plant. Biotechnol.* **2005,** *32* (1), 45–50.

Kim, N. Y.; Lee, H. J.; Kim, N. K.; Oh, J.; Lee, S. H.; Kim, H.; Moon, J. S.; Jeong, R. D. Occurrence Pattern of Viral Infection on Pear in Korea and Genetic Characterization of Apple Scar Skin Viroid Isolates. *Hortic. Sci. Technol.* **2019,** 767–778.

Komorowska, B.; Cieślińska, M. First Report of Cherry virus A and Little cherry virus-1 in Poland. *Plant Dis.* **2004,** *88* (8), 909.

Kundu, J. K.; Paunovic, S.; Komorowska, B.; Gospodaryk, A. Genetic Diversity of Flexiviruses Infecting Pome Fruit Trees. *J. Plant Pathal.* **2010,** *92* (3), 685–691.

Lakshmi, V.; Hallan, V.; Ram, R.; Ahmed, N.; Zaidi, A. A.; Varma, A. Diversity of *Apple mosaic virus* Isolates in India Based on Coat Protein and Movement Protein Genes. *Indian J. Virol.* **2011,** *22* (1), 44–49.

Laimer, M. In *Natural and Induced Pathogen Resistance in Temperate Fruit Trees and Grapevines*. Proc. XXVIIIth IHC—IS on the Challenge for a Sustainable Production, Protection and Consumption of Mediterranean Fruits and Nuts; D'Onghia, A. M., et al., Ed.; Acta Hort. ISHS, 2012; p 940.

Lemgo, G. N. Y.; Sabbadini, S.; Pandolfini, T.; Mezzetti, B. Biosafety Considerations of RNAi-mediated Virus Resistance in Fruit-tree Cultivars and in Rootstock. *Transgen. Res.* **2013,** *22,* 1073–1088.

Levy, L.; Damsteegt, V.; Scorza, R.; Kolber, M. Plum Pox Potyvirus Diseases of Stone Fruits. *Am. Phytopathal. Soc.* **2011,** 1–8.

Liu, Z.; Scorza, R.; Hily, J.; Scott, S. W.; James, D.. Engineering Resistance to Multiple Prunus Fruit Viruses Through Expression of Chimeric Hairpins. *Am. Soc. Hortic. Sci.* **2007,** *132* (3), 407–414.

Malfitano, M.; Di Serio, F.; Covelli, L.; Ragozzino, A.; Hernández, C.; Flores, R. Peach Latent Mosaic Viroid Variants Inducing Peach Calico (Extreme Chlorosis) Contain a Characteristic Insertion that is Responsible for this Symptomatology. *Virology* **2003,** *313* (2), 492–501.

Mitter, N.; Worrall, E. A.; Robinson, K. E.; Li, P.; Jain, R. G.; Taochy, C.; Carroll, B. J.; Fletcher, S. J.; Lu, G. Q.; Xu, Z. P. Clay Nanosheets for Topical Delivery of RNAi for Sustained Protection Against Plant Viruses. *Nat. Plants* **2017,** *3,* 16207.

Migliori, A.; Quiot, J. B.; Labonne, G.; Boudon, J. P.; Lauriaut, F.; Freydier, M., et al. Mineral Oil, a Means of Preventive Control Against the Agent of Sharka Spread by Aphids in Nurseries. *Phytoma* **1998,** *51,* 32–35.

Myrta, A.; Terlizzi, B. Di.; Boscia, D.; Choueiri, E.; Gatt, M.; Gavriel, I.; Çaglayan, K.; Varveri, C.; Zeramdini, H.; Aparicio, F.; Pallàs, V.; Savino, V. Serological Characterisation of Mediterranean Prunus Necrotic Ringspot Virus Isolates. *J. Plant Pathol.* **2001,** *83* (1), 45–49.

Nishitani, C.; Hirai, N.; Komori, S.; et al. Efficient Genome Editing in Apple Using a CRISPR/Cas9 system. *Sci. Rep.* **2016,** *6,* 31481.

Noorani, M. S.; Awasthi , P.; Singh, R. M.; Ram, R.; Sharma, M. P.; Singh, S. R.; Ahmed, N.; Hallan, V.; Zaidi, A. A. 2010 Complete Nucleotide Sequence of Cherry Virus A (CVA) Infecting Sweet Cherry in India. *Arch. Virol.* **2010,** *155* (12), 2079–2082.

Noorani, M. S.; Awasthi, P.; Sharma, M. P.; Ram, R.; Zaidi, A. A.; Hallan, V. Simultaneous Detection and Identification of Four Cherry Viruses by Two Step Multiplex RT-PCR with an Internal Control of Plant nad5 mRNA. *J. Virol. Methods* **2013,** *193* (1), 103–107.

Osman, F.; Al Rwahnih, M.; Rowhani, A. Real-time RT-qPCR Detection of Cherry Rasp Leaf Virus, Cherry Green Ring Mottle Virus, Cherry Necrotic Rusty Mottle Virus, Cherry Virus A and Apple Chlorotic Leaf Spot Virus in Stone Fruits. *J. Plant Pathol.* **2017,** *99* (1), 279–285.

Pandolfini, T.; Molesini, B.; Avesani, L.; Spena, A. Expression of Self-complementary Hairpin RNA Under the Control of the rolC Promoter Confers Systemic Disease Resistance to Plum Pox Virus Without Preventing Local Infection. **2003,** *15,* 1–15.

Paprstein, F.; Sedlák, J. Micropropagation of Czech Apple Cultivars. *Acta Hortic.* **2015,** *1083,* 267–271.

Pandey, B. N.; Sindhan, G. S.; Singh, B. E. Note on the Effect of Heat Treatment on the Inactivation of Some Temperate-Fruit Viruses (Hot Water Treatment). *Progr. Hortic.* **1972,** *4,* 53–58.

Rana, T.; Chandel, V.; Handa, A.; Thakur, P. D.; Hallan, V.; Zaidi, A. A. Molecular Evidence of *Apple chlorotic leaf spot virus* in Himachal Pradesh. *Indian J. Virol.* **2007,** *18* (2), 70–74.

Rana, T.; Negi, A.; Dhir, S.; Thockchom, T.; Chandel, V.; Walia, Y.; Singh, R. M.; Ram, R.; Hallan, V.; Zaidi, A. A. Molecular Diagnosis of Apple Virus and Viroid Pathogens from India. *Arch. Pytopathol. Plant Protec.* **2012,** *44,* 505–512.

Reed, P. J.; Foster, A. Exclusion of Pome and Stone Fruit Viruses, Viroids, and Phytoplasmas by Certification and Quarantine. In *Virus and Virus-like Diseases of Pome and Stone Fruits*; Hadidi, A., Barba, M., Candresse, T., Jelkmann, W., Eds.; T. Paul MN: APS Press, 2011; pp 391–396.

Ravelonandro, M.; Scorza, R.; Michel, H. J.; Briard, P. The Efficiency of RNA Interference for Conferring Stable Resistance to Plum Pox Virus. *Plant Cell Tissue Organ Cult.* **2014,** *118,* 347–356.

Romadanova, N. V.; Mishustina, S. A.; Gritsenko D ilyara, A.; Omasheva, M. Y.; Galiakparov, N. N.; Reed, B. M.; et al. Cryotherapy as a Method for Reducing the Virus Infection of Apples (*Malus sp.*). *Cryo Lett.* **2016,** *37* (1), 1–9.

Sastry, K. S.; Zitter, T. A. *Plant Virus and Viroid Diseases in the Tropics*; Dordrecht: Springer Netherlands, 2014.

Serra, P.; Bertolini, E.; Martínez, M. C.; Cambra, M.; Flores, R. Interference between Variants of Peach Latent Mosaic Viroid Reveals Novel Features of its Fitness Landscape: Implications for Detection. *Nature* **2017,** 1–10.

Sanford, J. C.; Johnston, S. A. The Concept of Parasite-Derived Resistance—Deriving Resistance Genes from the Parasite's Own Genome. *J. Theor. Biol.* **1985,** *113,* 395–405.

Simón-Mateo, C.; García, J. A. MicroRNA-guided Processing Impairs Plum Pox Virus Replication, but the Virus Readily Evolves to Escape this Silencing Mechanism. *J. Virol.* **2006,** *80* (5), 2429–2436.

Su, L.; Bhagwat, B.; Bernardy, M.; Wiersma, P. A.; Cheng, Z.; Xiang, Y. Molecular Analysis of Three New Cherry Mottle Leaf Virus Isolates Reveals Intra-Species Genetic Diversity and Inter-species Gene Transfer. *Phytoparasitica* **2016,** *44,* 571–574.

Teixeira da Silva, J. A.; Gulyás, A.; Magyar-Tábori, K.; Wang, M. R.; Wang, Q. C.; Dobránszki, J. In vitro Tissue Culture of Apple and Other *Malus* Species: Recent Advances and Applications. *Planta.* **2019,** *249,* 975–1006.

Verma, M. K.; Mir, J. I.; Lal, S.; Sahoo, T. Recent Advances in Temperate Fruit Crop Improvement. *Int. J. Sci. Technol. Soc.* **2018,** *4,* 1–2.

Vivek, M.; Modgil, M. Elimination of Viruses Through Thermotherapy and Meristem Culture in Apple Cultivar 'Oregon Spur-II.' *Virusdisease* **2018,** *29* (1), 75–82.

Walia, Y., Dhir, S.; Zaidi, A. A.; Hallan, V. Apple Scar Skin Viroid Naked RNA is Actively Transmitted by the Whitefly Trialeurodes Vaporariorum. *RNA Biol.* **2015,** *12* (10), 1131–1138.

Wang, Q. C.; Panis, B.; Engelmann, F.; Lambardi, M.; Valkonen, J. P. T. Cryotherapy of Shoot Tips: A Technique for Pathogen Eradication to Produce Healthy Planting Materials and Prepare Healthy Plant Genetic Resources for Cryopreservation. *Ann. Appl. Biol.* **2009,** *154* (3), 351–363.

Wang, X.; Kohalmi, S. E.; Svircev, A.; Wang, A.; Sanfacon, H.; Tian, L. Silencing of the Host Factor eIF(iso)4E Gene Confers Plum Pox Virus Resistance in Plum. *PLoS One.* **2013,** *8* (1), e50627.

Woo, E. N. Y.; Clover, G. R. G.; Pearson, M. N. First Report of Cherry Leaf Roll Virus (CLRV) in *Malus domestica. Australas Plant Dis. Notes.* **2012,** *7* (1), 151–156.

Zhao, L., Wang, M. R., Cui, Z. H., Chen, L., Volk, G. M., Wang, Q. C. Combining Thermotherapy with Cryotherapy for Efficient Eradication of Apple Stem Grooving Virus from Infected In-vitro-Cultured Apple Shoots. *Plant Dis.* **2018,** *102* (8), 1574–1580.

CHAPTER 15

Vegetable Production in West Africa: The Challenges of Viral Diseases and Their Management

MICHAEL KWABENA OSEI[1], ALLEN OPPONG[1], JOSEPH ADJEBENG-DANQUAH[2], HERBAUD PHANUEL FORTUNAT ZOHOUNGBOGBO[3], RUTH PREMPEH[1], LEANDER DEDE MELOMEY[4], JOSEPH ADOMAKO[1], RICHARD YAW AGYARE[2], JACINTA ADOMA OPOKU[1], and EMMANUEL ASAMOAH ADJEI[1]

[1]*CSIR-Crops Research Institute, Kumasi, Ghana*

[2]*CSIR-Savannah Agricultural Research Institute, Nyankpala, Ghana*

[3]*World Vegetable Center West and Central Africa (WCA)—Coastal and Humid Regions IITA-Benin Campus, Cotonou, Benin*

[4]*West Africa Centre for Crop Improvement (WACCI), University of Ghana, Accra, Ghana*

ABSTRACT

Vegetables constitute a very important component of the human diet, supplying essential nutrients, micronutrients, and minerals necessary for good health and vitality. In West Africa, vegetable crops are mostly clustered as fruit, root, and leafy vegetables, depending on the part utilized. Vegetables may be specifically cultivated or obtained from cultivated crops or parts of indigenous plant species such as baobab, roselle, etc. Vegetable cultivation is, however, faced with a number of challenges encompassing biotic and abiotic stresses. This chapter enumerates some important vegetable crops in West

Plant Diseases and Their Management: A Sustainable Approach.
Deepa Srivastava, Rajarshi Kumar Gaur, and Ajay K. Tiwari (Eds.)

Africa and how they are utilized in terms of food and medicine. Challenges associated with variety, agronomic practices, pests, and diseases, including viruses, as well as postharvest handling challenges, are highlighted. Some common viruses and their economic importance in vegetable cultivation in West Africa are identified. Furthermore, some common viral diseases and modes of transmission, which can be persistent or nonpersistent, are mentioned. The chapter then identifies some common symptoms of viral diseases as well as methods of identifying viruses, which can be conventional, serological, or molecular. Some management measures against viruses in West Africa using cultural and phytosanitary practices, vector management, host plant resistance, and integrated approaches are indicated. Finally, progress made in the development of virus-resistant varieties and the prospects of breeding virus-resistant vegetable crops in West Africa are highlighted. The availability of resistant varieties and the promotion of such varieties will minimize the challenges associated with the cultivation of vegetables and boost vegetable cultivation in West Africa.

15.1 INTRODUCTION

Vegetables form a fundamental constituent of the diets of humans and have been recognized globally as a source of micronutrients, vitamins, minerals, and other essential nutrients, resulting in the nutrition security of consumers (Fadairo et al., 2020). According to the FAO, a daily intake of 200 g of vegetables in the diets of humans is enough to ensure an adequate supply of essential nutrients to the body (ARC, 2013). This implies that vegetables are an essential addition to every meal. In West Africa, the majority of countries have a vegetable sector employing the majority of small-scale farmers from rural and urban regions engaged in farming (Schreinemachers et al., 2021). The production and marketing of vegetables aid or offer a crucial role in income generation for a considerable proportion of the populace, including smallholder farmers and traders. Vegetables used by West Africans can be classified based on the part used for food. These include root, fruit, and leafy vegetables. Some indigenous plant species, such as baobab, roselle, etc., are also either purposely cultivated or harvested from the wild and used as vegetables (Levasseur et al., 2007). In West Africa, vegetables are used in a wide range of dishes, either as the main ingredient in soups and stews or as spices to give flavor to dishes.

Most vegetables grown in West Africa are of short duration, which requires small pieces of land. Vegetable cultivation is faced with numerous challenges that range from biotic, abiotic, and socioeconomic issues (Adeoye, 2020; Fadairo et al., 2020). Among the biotic stresses, viral diseases are highly important economically. West African countries continue to lose vital revenue to the scourge of viruses, and consequently, livelihoods. The impact of viruses and their effect on the quality of vegetable produce in West Africa is overwhelming. This effect is likely to exacerbate due to the rapid advent of new strains of viruses that are virulent. Virus disease pandemics and epidemics over the years have seriously reduced the yields and quality of vegetables (Osei et al., 2010). This has serious economic consequences on the rural and peri-urban populace, which rely mainly on vegetables for their livelihood (Schippers, 2000). Unlike other food crops, vegetables are not recognized as food or cash crops in some West African countries. Farmers continue to cultivate vegetables on small, part-time basis, often as stop-gap means of livelihood coupled with staples like cereals and legumes (Fadairo et al., 2020). In West Africa, viral diseases are a major threat to the production of vegetables. The key issues influencing the new rise of viruses are: i) monocropping system characterized by high planting densities, which exposes to pests and other pathogens; ii) germplasm exchange of seeds and/or planting materials to other environments; iii) climate change and its effects on host plants, vectors, and their relationships; and iv) the evolution of viruses to acclimate to the changing environment (Anderson et al., 2004; Jones, 2009; Elena et al., 2014). At the moment, there are no remedies to cure plants with virus infection. Even though fungal and bacterial infections can be treated with antifungal and antibacterial agents, respectively, this is not so with viruses. As such, the management of viral disease hinges on using resistant plants to viral infection, integrated management that will use multiple strategies involving each virus, host, and environment. To evaluate strategies for disease management, diagnostics and identification are vital using specific tools. It is, however, important to also know the mode of transmission, as it helps in viral diagnosing and averting the emergence of diseases. This chapter focuses mainly on the challenges and management of viral diseases in vegetables, with particular reference to West Africa. Information on the production of vegetables in West Africa, demonstrating important or common vegetables in the region, as well as their utilization, is reviewed. Others include the economic importance of viruses in vegetable cultivation, the mode of transmission of viruses in vegetable production, the main or common viral diseases of vegetables in West Africa, the symptoms of common viral diseases in vegetables, and the identification of the main viruses affecting vegetables in West Africa. Lastly, disease management strategies are described, indicating both cultural and phytosanitary practices, vector management with insecticides, host plant resistance against viruses, and integrated pest management approaches. This chapter provides a realistic approach intended to coach or guide scientists to plan and use a more precise method of viral detection of vegetables in West Africa.

15.2 VEGETABLE PRODUCTION IN WEST AFRICA

15.2.1 IMPORTANT/COMMON VEGETABLES IN WEST AFRICA

Vegetables form a basic component of the diets of humans and have been recognized globally as a source of micronutrients, vitamins, minerals, and other essential nutrients needed for the nutrition security of consumers (Fadairo et al., 2020). According to the FAO, a daily intake of 200 g of vegetables in human diets is enough to ensure an adequate supply of essential nutrients to the body (ARC, 2013). This means that vegetables are considered a very important addition to every meal. Vegetables used by West Africans can be classified into leafy vegetables, fruit vegetables, and root vegetables, depending on the part used for food. Some indigenous plant species, such as baobab, roselle, etc., are also either intentionally cultivated or harvested from the wild and used as vegetables (Levasseur et al., 2007). The main vegetables grown in West Africa range from exotic species like *Capsicum* spp., *Cucumis sativus*, *Celosia argentea*, *Beta vulgaris*, *Latuca sativa*, *Lycopersicon esculentum*, *Alium* spp., *Moringa oleifera*, and *Amaranthus* to indigenous species such as *Vernonia amygdalina, Telfairia occidentalis, Corchorus olitorius*, *Luffa aegyptiaca, Momordica charantia, Pterocarpus soyauxi, Tamarindus indica, Talinum triangulare*, *Ocimum gratissimum*, and *Zanthoxylum zanthoxyloides* (Ogwu et al., 2016). Catarino et al. (2019) reported that about 24 indigenous plant species with edible leaves have been identified for consumption as leafy vegetables in Guinea-Bissau. Some of these species include *Bombax costatum, Adansonia digitate,* and *Sesamum radiatum*. The young leaves of *Hibiscus sabdariffa* and *Amaranthus hybridus,* or other domesticated wild plants with edible foliage that are widely distributed across the region, are also consumed. In some communities, the leaves of even staple food crops like cowpea, cassava, sweet potato, okra, *Xanthosoma*, and taro are harvested periodically and utilized as vegetables. Cowpea leaves are also preferred by consumers in parts of northern Ghana. *Talinum triangulare,* a crop considered a weed in some communities, is used as a vegetable in soups in parts of Ghana. This therefore implies that the classification of a crop as a food crop, weed, or vegetable may be due to the location and its role in the food preferences of the consumers predominant in that area. The mature or sometimes immature fruits of crops such as tomato, eggplant, cucumber, pepper, okra, etc., together with seeds are harvested and utilized as vegetables in food preparations. Other species such the cucurbits are cultivated mainly for their seeds, which are harvested, dried, and utilized in food preparations. Young immature cowpea pods are also harvested and utilized as vegetables in some parts of Ghana. In the case of other vegetable crops, the storage roots from the economic part are harvested and utilized as vegetables. Examples include carrot, radish, beetroot, and other crops whose storage roots are harvested and utilized as vegetables.

15.2.2 UTILIZATION OF VEGETABLES IN WEST AFRICA

Vegetables are used in a wide range of dishes, either as the main ingredient in soups and stews or as spices to give flavor to dishes. They provide minerals, vitamins, and other essential nutrients. Vegetables are utilized in two main ways: as a source of food and as medicine. Vegetables such as tomato, eggplant, cucumber, pepper, and okra are eaten raw or with little cooking. In Nigeria and parts of Ghana, eggplants are eaten raw as a snack or with groundnut paste. Some, such as chili pepper, are used in small quantities in stews as spices, while eggplant may constitute the main ingredient in stew or soup in Ghana. Egusi (melon seeds) is used in addition to cocoyam leaves or other leafy vegetables to prepare a popular stew known as *palaver sauce* in Ghana and egusi soup in Nigeria. In addition to being utilized as food, vegetable crops that have medicinal values are utilized for their therapeutic roles in the treatment of several ailments (Kimywe et al., 2007). Indigenous African leafy vegetables have healing properties in addition to the nutritive qualities they possess and are currently attracting research attention. Ayodele (2005) reported 28 different medicinally important vegetables in Nigeria that treat various ailments ranging from tapeworm infestation, low sperm count, malaria, stomach ache, respiratory disorders, etc.

15.2.3 MAJOR CHALLENGES ASSOCIATED WITH VEGETABLE CULTIVATION IN WEST AFRICA

Vegetable cultivation is faced with several challenges, which range from biotic, abiotic, and socioeconomic issues like changes in demand and supply with the associated price fluctuations and disparities (Adeoye, 2020; Fadairo et al., 2020). Climatic variability, with its accompanying fluctuations in temperature and rainfall, has aggravated most of the biotic and abiotic constraints associated with vegetable cultivation. In some parts of West Africa, the vegetable subsector is not as organized as the staple food crops subsector. Farmers cultivate vegetables on a small, part-time basis, often as stop-gap means of livelihood, in addition to cultivating staple food crops like cereals and legumes (Fadairo et al., 2020).

15.2.3.1 LACK OF IMPROVED VARIETIES AND SEEDS

In West Africa, most farmers rely on indigenous varieties, mostly landraces of common crops such as pepper, tomatoes, eggplants, onions, and okra. Some farmers also depend on imported seeds of these crops, which are normally expensive (Osei et al., 2021). The landraces have low yield potential and are susceptible to pests and diseases predominant in the growing areas. Since quality seed is a prerequisite for any vibrant crop production system to flourish, the lack of quality seeds for most vegetables, and in some cases, limited access has been a serious impediment to vegetable producers and consumers in rural areas. In the absence of a formal local vegetable seed system, most vegetable farmers in rural areas depend on inter- or intra-community farmer seed transfer or rely on their own saved seeds from the fruits of vegetable crops they obtain from other sources. Some of these seeds sometimes have questionable viability and yield potential, and the minimum quality standards of a good seed may not be met. In Ghana, however, there exists a traditional seed system for vegetables that is not formally regulated by law (Etwire et al., 2013) but operates in most communities through inter-farmer seed exchange or farmer-saved seeds obtained through extraction from fruits purchased by the farmer. Common vegetables treated this way include tomatoes, okra, peppers, and eggplants (Addo-Quaye and Djokoto, 2013). Such seeds normally have an unknown pedigree and are of poor quality, thereby posing serious marketing challenges for such seeds.

15.2.3.2 POOR AGRONOMIC PRACTICES

Most vegetable growers do not follow good agricultural practices related to the cultivation of a particular vegetable and the safe application of agro-chemicals. Quite a number rely on a calendar spraying regimen with often inappropriate chemicals, which result in large deposits on the harvested produce. Such contamination has resulted in the failure of such commodities to meet export market requirements. The use of inappropriate chemicals for the control of pests has resulted in the contamination of the produce and has resulted in casualties among consumers if they are not properly washed before utilization. Unlike staple crops that have several recommended practices, most vegetable growers, especially in rural areas, do not have access to such information. Production guides are often not available to such farmers to aid them in the production of wholesome produce.

15.2.3.3 EXTREMES OF WEATHER/ERRATIC RAINFALL ASSOCIATED WITH CLIMATE CHANGE

Vegetables require a lot of water during the cultivation cycle. Therefore, an adequate supply of water during the production process is crucial to attaining the optimum yield and profitability of the production system. Dams have been constructed in certain parts of West Africa to supplement rainfed cultivation and for off-season vegetable cultivation (Fadairo et al., 2020; Kpéra et al., 2017). This is, however, not the case for all countries. For instance, few of such dams exist in other places, resulting in severe land pressure and overcultivation on the same pieces of land for several years, thereby resulting in the buildup of pests and disease-causing organisms (Fadairo et al., 2020). Vegetable cultivation has been affected by the changing climate, which has resulted in low productivity. Prolonged dry spells have resulted in farmers having to resort to using unwholesome water, particularly in urban and peri-urban vegetable cultivation. This has resulted in contamination with pathogens and fecal matter, especially on leafy vegetables (Datta, 2013; Deuter, 2014). Extremes of temperature affect the yield and nutritional quality of vegetables (Datta, 2013; Nelson et al., 2010). This has been compounded by the absence of greenhouses and infrastructure for the cultivation of vegetables in controlled environments to provide the optimum conditions of temperature, moisture, and humidity necessary for vegetable cultivation.

15.2.3.4 INCIDENCE OF PESTS AND DISEASES

Vegetables are prone to attacks by several pests, which range from viruses, bacteria, fungi, nematodes, and insect pests, that directly impact the yield and quality of the harvested produce or predispose them to other disease-causing pathogens (Mohajar et al., 2015; Sharma et al., 2009; Barkai-Golan, 2005). The changing climate has also impacted negatively on vegetable cultivation due to the associated variations in the incidence and distribution of disease pathogens, pests, and different weed species (Deuter, 2014). A wide range of insect-transmitted geminiviruses, including tomato leaf curl virus, okra mottle virus, tomato golden vein virus, and tomato yellow vein streak virus, have been reported to affect tomatoes, okra, and pepper, respectively, thereby resulting in poor performance (inoue-Nagata et al., 2016; Albuquerque et al., 2013).

15.2.3.5 HIGH POSTHARVEST LOSSES AND MARKETING CHALLENGES

Vegetables are highly perishable and must be consumed or processed immediately after harvesting unless they are kept in appropriate structures. Most vegetables (especially fruit vegetables) pass through a lot of handling during harvesting, which results in bruises that render them unwholesome or a decline in the quality of such produce. Due to the seasonal nature of vegetable cultivation, most farmers plant and harvest their crops at the same time. This results in periods of gluts at the time of harvesting and scarcity during the lean season. As a result, most farmers are not able to attract good prices for their produce. This situation has been made worse by middlemen and women, who often inflate farmgate prices at the expense of consumers (Wongnaa et al., 2019). One major challenge that has reduced the profitability of vegetable cultivation in West Africa is the importation of vegetables from other parts of Africa and Europe into West Africa. This creates an imbalance in the pricing of these vegetables to the disadvantage of the local producers, who are unable to recoup their investments in most cases. Additionally, there is limited inclusion of fresh vegetables in the diets of most urban dwellers, thereby resulting in a limited market for such large supplies.

15.3 ECONOMIC IMPORTANCE OF VIRUSES IN VEGETABLE CULTIVATION

In most West African countries, the vegetable industry employs the vast majority of small-scale farmers from both rural and urban areas (Schreinemachers et al., 2021). For a large number of smallholder farmers and traders, vegetable production and sale provide an essential source of income and employment. For example, in Ghana, where agriculture accounts for around 22% of GDP, smallholder farmers make up about half of the workforce (Gonzalez et al., 2016). They earn a decent income from the cultivation of these vegetables because an appreciable yield could be obtained on a small area of land within a short time (Abdulai et al., 2016).

Some of these vegetables are also exported to many other African countries as well as countries outside of Africa, thus contributing to the revenue mobilization of the exporting countries. For example, between 2015 and 2019, revenue obtained from food exportation from Ghana to Burkina was estimated to be GHC 868.13 million. Notwithstanding, Burkina is ranked first in food importation with 92,319 tons (Fresh Plaza, 2021). Tomato exportation from Burkina Faso to Ghana is estimated to earn the former about US$99.5 million annually (Ghana Business News, May 26, 2018). Cote d'Ivoire is ranked second with 14,222 tons, followed by Mali (1180) (Ghanaweb.com, March 2020).

Production of vegetables such as tomatoes and onions in West Africa has steadily risen over the years (Fig. 15.1) (FAOSTAT, 2019), though there has been relative stagnation in production of other vegetables including lettuce, dry and green chili, eggplant, cucumber, and okra.

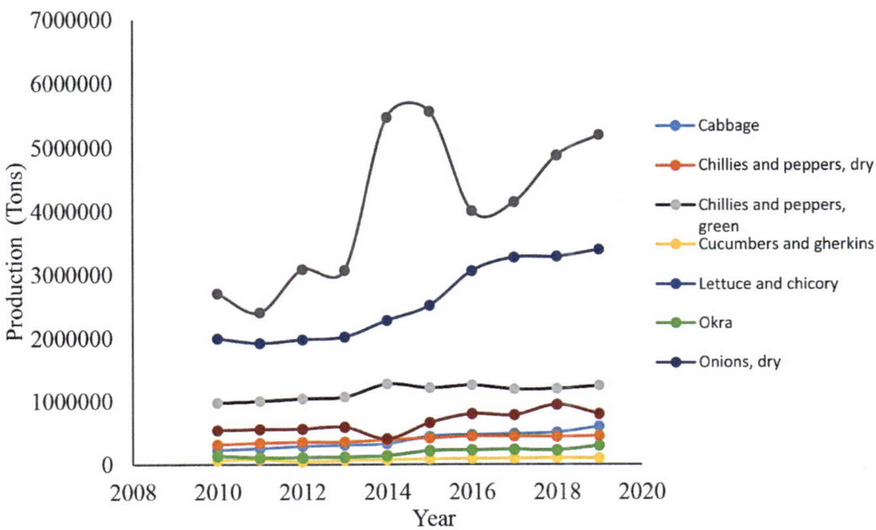

FIGURE 15.1 Ten years (2010–2019) production trends of some vegetables in West Africa.
Source: FAOSTAT, 2019.

Despite the many benefits of vegetables, there are various setbacks in their production around the world. Some of them have been connected to abiotic or biotic stressors, with virus infections being a major contributor. Virus disease pandemics and epidemics over the years have seriously reduced the yields and quality of vegetables (Osei et al., 2010). This has serious economic consequences on the rural and peri-urban populace, which rely mainly on vegetables for their livelihood (Schippers, 2000). Hence, leading to food and nutrient insecurity, they are required for a balanced diet and good human health. The demand for food and nutrients for the growing population [411,024,235 (United Nation, 2021)] and the difficulty in effectively controlling viral diseases amidst climate change are making vegetable production unattractive.

Several viruses affect plants, especially vegetables. However, the underlisted viral diseases are among the most important viruses ranked by 250 plant virologists for causing serious diseases to plants (Rybicki, 2014).

1. Tobacco mosaic tobamovirus (TMV)
2. Tomato spotted wilt tospovirus (TSWV)
3. Tomato yellow leaf curl begomovirus (TYLCV)
4. Cucumber mosaic cucumovirus (CMV)
5. Potato virus Y potyvirus (PVY)
6. Cauliflower mosaic caulimovirus (CaMV)
7. African cassava mosaic begomovirus (ACMV)
8. Plum pox potyvirus (PPV)
9. Brome mosaic bromovirus (BMV)
10. Potato virus X potexvirus (PVX)
11. Pepper veinal mottle virus (PVMV)
12. Tobacco etch virus (TEV)

Among the viruses that affect vegetable crops, geminiviruses, potyviruses, and tenuiviruses are the most important. For the purpose of this chapter, the economic importance of some specific viruses that affect important vegetables cultivated in West Africa will be discussed in brief.

15.3.1 CUCUMOVIRUSES

These belong to the family Bromoviridae and infect over 1200 plant species from about 100 families (Edwardson and Christie, 1991). The viruses are transmitted by aphids' vectors in a nonpersistent manner (Tomlinson, 1987). Cucumoviruses are economically important viruses due to the wide range of cultivated plants and weedy forms that serve as reservoirs for subsequent crop infections. Losses mounting up to 80% of total vegetable yield have been reported, though this varies with the stage of plant growth at which infection occurs (Avilla et al., 1997b).

15.3.2 PEPPER VEINAL MOTTLE VIRUS (PVMV)

Pepper veinal mottle virus (PVMV) is commonly found in many West African countries, affecting many crops belonging to the Solanaceae family (Thottappilly, 1992). Alebgejo and Abo (2002) and Fajinmi (2011) reported between 54.5% and 64.3% economic losses in chili and sweet pepper manifested as diminished yield and fruit quality.

15.3.3 PEPPER YELLOW LEAF CURL VIRUS (PYLCV) AND PEPPER YELLOWS VIRUS (PYV)

These viruses of pepper have been identified in Africa as well as in other parts of the world, including North America and Asia. These viruses were also being identified on pepper in Europe (Dombrovsky et al., 2010; Alamabi et al., 2015). New races of this virus have emerged, infecting several crops globally, especially pepper, due to the highly significant variations in their genetics (Velásquez, et al., 2018). Hence, their infestation can lead to total crop loss (100%) (Dogimont et al., 1996).

15.3.4 POTYVIRUSES

Potyviruses belong to the family Potyviridae and are the major plant viruses with a wide host range worldwide. They are mostly found on solanaceous crops and are vectored by aphids in a nonpersistent way. Sometimes, they are transmitted by direct contact between plants and also via infected seeds (Bujarski, 2021). Yield loss of up to 100% may occur depending on the age of the crop at which infection occurred (Dogimont et al., 1996).

15.3.5 ALFAMOVIRUSES

They belong to the family Bromoviridae, which is distributed worldwide and is vectored by over 20 aphid species (Marchoux et al., 2008). They can infect about 250 plant species, including herbs, within 48 families. Economically, they deform fruits and subsequently infest their seeds (Edwardson and Christie, 1997) when the infection occurs at the seedling stage of the plant.

15.3.6 BEGOMOVIRUSES

These viruses are spread by the whitefly (*Bemisia tabaci*), usually in a persistent manner. They belong to the family Geminiviridae. Amongst them are the most devastating and studied: the tomato yellow leaf curl virus (TYLCV), cotton leaf curls Gezira virus, okra leaf curl disease, and okra yellow crinkle virus.

15.3.7 TOMATO YELLOW LEAF CURL VIRUS (TYLCV)

This virus is a host of tomatoes, sweet peppers, and peppers worldwide. TYLCV was first reported on pepper (*C. annuum*) plants in Europe (Reina et al., 1999). The virus was later reported in America (Salati et al., 2002), North Africa (Gorsane et al., 2004), Asia (Tahir and Haider, 2005), and West Africa (Osei et al., 2010). TYLCV may cause economic losses of up to 100% in tomato in many tropical and subtropical regions (Rochester et al., 1994). Severe economic damage occurs when infection starts at the onset of flowering. Some of the symptoms of the virus include reduced fruit size and weight with low market value (low quality fruits). The economic impact of the virus makes the pathogen a serious threat to tomato production. Thus, TYLCV has turned out to be one of the major factors limiting tomato production in many areas (Ioannou, 1985b).

15.3.8 OKRA YELLOW CRINKLE VIRUS, OKRA LEAF CURL DISEASE, AND COTTON LEAF CURL GEZIRA VIRUS

These begomoviruses infest mostly crops in the Malvaceae, like okra (Shih et al., 2007). They can cause 100% yield losses, especially when plants are infected at the early seedling stage (Idris and Brown, 2002). These viruses are rampant in many African countries (Seka, et al., 2016).

15.3.9 PEPPER MILD MOTTLE VIRUS (PMMOV)

This virus is considered the most serious tobamovirus of pepper. The virus spreads easily on the field, causing severe yield losses and reduced fruit quality. Symptoms of the virus include leaf chlorosis, leaf distortion, and irregular fruit sizes with varying shapes and colors. Furthermore, necrosis on leaves and fruits has been reported on pepper (Moury and Verdi012; Kenyon et al., 2014a). Pepper mild mottle virus spreads by contact during field operations and sometimes through infected seeds. The virus is highly stable and can survive on plant debris awaiting the succeeding crop (Johansen et al., 1994; Genda et al., 2011).

15.3.10 CRINIVIRUSES-IPOMOVIRUSES-POLEROVIRUSES-CARLAVIRUS

Many vegetable crops belonging to the Cucurbitaceae family are hosts of many viral species. The major symptoms of these viruses include the yellowing of leaves. The viruses outlined below are of economic importance to vegetable crops.

a. Squash vein yellowing virus (SqVYV) and cucumber vein yellowing virus (CVYV) belong to the genus *Ipomovirus* of the Potyviridae family;
b. Cucumber yellow stunting disorder virus (CYSDV) and beet pseudo-yellows virus (BPYV) belong to the family Closteroviridae and genus *Crinivirus*; and
c. Cucurbit aphid-borne yellows virus (CABYV) belongs to the Luteo-viridae family and genus *Polerovirus*.

These viruses of cucurbits have positive-sense single-stranded RNA (Kassem, et al., 2007; Abrahamian and AbouJawdah, 2014). Some of these emergent viruses of cucurbits are transmitted by the whitefly (*Bemisia tabaci*) in a semipersistent way. Nevertheless, the cucurbit aphid-borne yellow virus, CABYV, is vectored by aphids in a persistent manner (Kassem et al., 2007). The occurrence of these viruses of cucurbits worldwide is an indication of the prevalent nature of the vector, *Bemisia tabaci* (De Barro et al., 2011; Navas-Castillo et al., 2011).

15.3.11 TOBACCO MOSAIC VIRUS

Tobacco mosaic virus (TMV) belongs to the genus *Tobamovirus* and family Virgaviridae. It is one of the economically important viruses in West Africa, attacking several solanaceous crops. The virus causes severe yield losses mostly in solanaceous crops such as pepper, tomato, and tobacco, with reported yield losses as high as 90% in bell pepper (Chitra et al., 2002).

Though predominantly found to infect pepper in eastern (Uganda and Tanzania) and southern (Zimbabwe, Sudan, and Zambia) parts of Africa, TMV has been found in (Waweru et al., 2019) West African countries such as Ghana and Nigeria (Appiah et al., 2014; Olawale et al., 2015). Diseased pepper plants are characterized by stunting, leaf curling, and leaf chlorosis, with deformed leaf blades showing mosaic patterns (Pazarlar et al., 2013; Kumar et al., 2011).

15.4 MODES OF TRANSMISSION OF VIRAL DISEASES IN VEGETABLE CROPS

The mode of transmission of any disease in plants is the process by which a healthy plant becomes infected with a particular pathogen. It is important to know the mode of transmission because it helps in diagnosing the diseases. It also helps in the prevention of disease occurrence. For example, knowing the insects that spread viruses helps in choosing the appropriate management strategy. Additionally, it helps in controlling the disease (Schreinemachers et al., 2015). For instance, the speed of transmission of tomato spotted wilt virus (TSWV) has resulted in limited effectiveness in the application of chemicals to control the vector (Funderburk, 2009). The mode of viral transmission plays a significant role in understanding virus effects on host–vector interactions (Mauck et al., 2012). Viruses are immobile by nature (Roossinck, 2013), and there are many ways by which they spread from one plant to another by a living organism (vector) or carrier. However, insects are the main pathway of transmitting viral diseases, of which aphids are the major vectors (Pinheiro, et al., 2019; Hull, 2014; Zitter and Murphy, 2009). There are variations in the transmission efficiency of vectors depending on the vector, prevailing environmental conditions, and period (Hamelin et al., 2016; Zitter and Murphy, 2009). Disease transmission can be continuous or discontinuous (Zhang and Holt, 2001). Continuous transmission occurs naturally when a pathogen exists in an area or host and starts to grow, multiply, and spread from host to host when conditions are favorable. On the other hand, discontinuous transmission occurs in a new area where the pathogen is introduced by agents such as the dissemination of infected seeds or any other planting materials, farm machinery or tools, and animals or humans. The mode of feeding of the vector involves piercing and sucking, which allow the insects to collect sap from the plant and transfer it to another. Additionally, transmission can also be described as horizontal or vertical. Horizontal transmission involves vectors, farm tools, and other direct, external contamination. Vertical transmission occurs when a plant is infected by its parent plant, for instance, through vegetative propagation or sexual reproduction via infected seeds.

Based on control measures, disease transmission is grouped into direct and indirect. Direct transmission occurs when the pathogen is conceded outside or inside the propagating material. An example of direct viral transmission is potato mosaic and leaf roll carried inside infected tubers. Indirect transmission occurs when the pathogen spreads itself through agents such as wind, water, animals, humans, insects, mites, nematodes, and birds.

Viruses are classified based on the mode of transmission as persistent, semipersistent, and nonpersistent. In the persistent mode, the insect acquires the virions and becomes active for a longer period of time before transmitting them to a healthy plant. Characteristics of persistent transmission include:

i. Insects feed for a longer period (several hours) before the virus is acquired and transmitted (Chen et al., 2018).
ii. The virus passes through the salivary glands of the insect before transmission.
iii. A latent period occurs during which transmission cannot occur.
iv. Persistently transmitted viruses decrease some fitness features and contend with the second virus in the host plants (Chen et al., 2018).
v. Persistently transmitted viruses improve host plant quality for insect vectors, hence increasing the rate of transmission (Chen et al., 2018; Mauck et al., 2012).
vi. They remain with their plant hosts forever and are transmitted vertically at very high rates, almost 100% (Roossinck, 2013).

According to Mauck et al. (2012), there are two types of persistent transmission.

1. Nonpropagative, where the virus circulates through the insect's body and is retained within the vector salivary glands without multiplying, and the vectors are mainly aphids.
2. Propagative, where the virus needs to grow within the cells of the insect vector and stay longer.

For nonpersistent virus transmission, the insect gets and spreads the virus quickly (a few seconds or minutes) through simple probing. Characteristics include:

i. Making the plant hosts unattractive to insect vectors.

ii. Have no or reduced effect on host plant quality (Mauck et al., 2012).
iii. The virus does not go through the gut of the insect, and therefore no latent period involved (Pinheiro et al, 2019).
iv. Virus particles remain on insect mouthparts for a short period of time.
v. The insect needs to refeed on another infected plant if further transmission is to occur.

Semipersistent transmission viruses are noncirculative and attach to the insect's foregut.

i. They can be acquired within a short period of time.
ii. Transmissions can also be mechanical. Mechanical transmission involves the introduction of an infectious virus or biologically active virus into a suitable site in the living cells through wounds on abrasions in the plant surface when plants come into contact with other plants, leaves rub together, and by the action of humans (Sharma, http://www.hillagric.ac.in › *edu* › *coa* › *ppath* › *lect*).

Three main pathways of virus transmission have been identified, namely pollen, seed, and vector (Hamelin et al., 2016). Pollen transmission involves the movement of virion from the infected parent to the offspring. Seed transmission involves the movement of infected seeds from one location to another, which can be a short or long distance, providing an initial local source of virion for insect vectors. Viruses transmitted through seeds are termed seed-borne.

15.5 SYMPTOMS OF COMMON VIRAL DISEASES

15.5.1 TOMATO YELLOW LEAF CURL DISEASE (TYLCD) SYMPTOMS

Symptoms of TYLCD are observed after weeks of infection. The typical symptoms of TYLCD observed on the leaves include yellowing, curling, upward cupping, shrinkage, leaf margin chlorosis, and mottling. The whole plant may become stunted whether it is infected early or late (Kil et al., 2016; Choe et al., 2021), and the top of the infected plant appears as the head of broccoli. There will be flower abscission and the production of fewer fruits (Glick et al., 2009). Infected tissues appear harder and/or with anthocyanin production. Early infection during the developmental phase results in a bushy appearance of the plant as a result of shorter internodes and reduced comma-shaped leaflets (Blancard, 2012).

15.5.2 PEPPER YELLOW LEAF CURL VIRUS (PEPLCV) SYMPTOMS

The symptoms of the pepper yellow leaf curl virus are very severe when the infection occurs at an early stage of growth. Leaves curl upward, flower buds do not develop to full size, and pollen development is hindered, leading to no fruit set or deformed fruits that are unmarketable (Kumar et al., 2006).

15.5.3 SYMPTOMS OF TOBACCO MOSAIC VIRUS (TMV)

To some extent, the symptoms of TMV depend on the host plant. Observed symptoms include mosaic, necrosis, mottling, leaf curling, yellowing of plant tissues, and severe stunted growth of the entire plant. During periods of low light, leaflets may appear filiform (Blancard, 2012). The plant produces fewer and distorted fruit as well as delayed and nonuniform fruit color (Scholthof, 2000). Collapsed leaves and plants are also observed in plants infected with TMV (Prakash et al., 2019). Plants infected with TMV were reported to have flower abscission, dwarfism, leaf distortion, produce fruits with intense yellow spots covering the fruit surface, and have low yields (Hernández-Santiago et al., 2021).

15.5.4 TOMATO SPOTTED WILT TOSPOVIRUS (TSWV) SYMPTOMS

TSWV can induce different symptoms on the same host species depending on the cultivar, age, nutrition, and environmental conditions of the plant. On tomato leaves, TSWV symptoms include bronzing, small brown flecks, curling, necrosis streaks and spots, dieback of growing tips, and the presence of dark-brown streaks on the leaf petiole, stems, and growing tips. Ring spots appear on green fruit and turn yellow on mature fruits. Severe necrosis on plants can lead to total death. On pepper, the leaves may show chlorotic mosaic or yellow flecking, or chlorotic line patterns with necrotic spots. Necrotic streaks may appear on stems that extend from the terminal buds. There has been a report of yellow spots with concentric rings or necrotic streaks on fruits. The symptoms of TSWV in lettuce appear as chlorotic rings with brown batches on one side of the outer leaves. The infection then progresses to the inner leaves, and then growth on that side of the plant ceases (EPPO Data Sheets, n.d.; Turina et al., 2012).

15.5.5 TOMATO MOSAIC VIRUS (TOMV) SYMPTOMS

The severity of ToMV depends on the strains, host, stage of infection, temperature, light intensity, soil nitrogen, and boron content of the soil. The leaves show alternating yellow, green, and/or white patches. Leaves appear embossed and wrinkled, curved, reduced in size, and deformed in a fern-like or filiform shape, with pale leaf veins followed by marked mottling and white patches of mosaic. Infected plants will exhibit flower abortion, mottling, or discolored fruits, with the occasional presence of chlorotic rings on green or mature fruits. In some cases, fruits will show internal necrosis of vascular tissues. ToMV leads to reduced yield and plant growth, especially when plants are attacked early (Blancard, 2012).

15.5.6 EGGPLANT MOTTLED DWARF VIRUS (EMDV) SYMPTOMS

The symptoms of eggplant mottled dwarf virus on leaves include distinct mottling and crumpling of leaves and puckered with chlorotic to yellow discoloration of the veins and the adjacent tissues, which generally turn into a general chlorotic mottling. No observable symptoms of EMDV are expressed on flowers, but fruits may be deformed, small, and roughened by suberized cracks. The entire plant will appear mildly to severely stunted and reduced to no fruit production (Martelli, 1969; Ladipo, 1977).

15.6 IDENTIFICATION OF THE MAIN VIRUSES AFFECTING VEGETABLES IN WEST AFRICA

Viruses have a great economic impact on vegetable production, leading to unappreciable yields and fruit quality. They can be diagnosed by identifying and detecting the presence of the pathogen, which will subsequently result in the development of an effective management system. Methods currently used for virus diagnosis in vegetables include conventional, serological, and molecular techniques. The conventional method involves the visual observation of symptom development, while the serological and molecular methods are based on the coat protein and/or the nucleic acid (Narayanaswamy, 2011), respectively.

15.6.1 CONVENTIONAL APPROACH

This approach uses symptomatology and transmission tests to identify the virus(es) causing the disease(s). Symptomatology is the visual inspection for plant viral symptoms on different parts of the plant. Viruses have unique symptoms, and a characteristic symptom(s) indicates the particular virus being studied. Transmission tests currently available to identify viruses in vegetables are grafting and budding or mechanical inoculation (Narayanaswamy, 2011; Jones, 1993).

One of the most widely used transmission tests for identifying viruses is the grafting method. This method involves the joining of tissues from two plants, so they grow together into a single plant. Two types of grafting methods are used: approach grafting and cleft/wedge grafting. With the approach grafting, the stems of virus-infected and virus-free plants are cut lengthwise so the cambium is exposed and both surfaces are put together. Subsequently, viral symptoms will be observed in the healthy plant. Cleft/wedge grafting is where the scion of the healthy plant is tightly fixed into the slit of the diseased plant, and the joint is wrapped with plastic tape. Viral symptoms subsequently appear in some parts of the healthy plant. Cleft/wedge grafting is a method of investigating the movement of viruses in vegetables (Spanò et al., 2020).

Mechanical inoculation involves the application of an inoculum (virus extract) on an artificially wounded leaf surface, after which the virus spreads systematically within the host. Subsequently, symptoms appear as either local or systematic lesions. Viruses such as tomato mosaic virus (ToMV), tomato spotted wilt virus (TSWV) (Nono-Womdim, 2003), potato virus Y (PVY) (Ibaba & Gubba, 2011), and chili veinal mottle virus (ChiVMV) (Nono-Womdim et al., 2001) have been identified in vegetables using this method.

Although visual inspection is relatively easy, many factors, such as viral strain, variety, time of infestation, and location, can contribute to the extent of symptoms expressed (Matthews, 1980). In addition, plants can also exhibit virus-like symptoms as a response to other biotic and abiotic stresses. This makes results derived from conventional identification often inconclusive. The conventional method may also not be sensitive enough, and as such, novel methods for detecting and identifying viruses have been developed. Serological and nucleic acid-based assays have been preferred because of their distinct advantages over conventional methods.

15.6.2 SEROLOGICAL APPROACH

This approach is based on the antigen–antibody reaction (Clark and Adams, 1977). Serological tests are important in identifying viruses in vegetables, and the commonly used methods are the agar gel double diffusion test, enzyme-linked immunosorbent assay (ELISA), and immunosorbent electron microscopy (ISEM).

The agar gel double diffusion test is a type of serological test where the antigen from the virus binds with antibodies to form a visible precipitin in an agar gel. This method has been used to detect viruses such as tomato ringspot virus (ToRSV) and cucumber mosaic cucumovirus (CMV) (Gumedzoe, 1992).

The ELISA test is very useful for identifying viruses in large samples, such as those collected from surveys, and screening for resistance in large populations. The most commonly used ELISA test is the double antibody sandwich method (DAS ELISA). This has been used to detect pepper veinal mottle virus (PVMV), cucumber mosaic cucumovirus (CMV), and potato virus Y (PVY) in tomatoes and peppers (Ayo-John et al., 2017). The immunosorbent electron microscopy method is a combination of both serology and electron microscopy, where the coat protein of the virus particle bound to specific antibodies can be seen under an electron microscope.

The different methods described above are tests used in identifying virus(es) in crude sap and purified preparations; however, this requires training and special equipment, which may not be commonly available in some laboratories. Previously, begomoviruses infecting vegetables in Africa were detected using serological and hybridization methods in okra leaf curl disease (OLCD) and tomato leaf curl disease (ToLCD) (Czosnek and Laterrot, 1997; Konate et al., 1995), but with the development of more sophisticated technologies, existing and new bogomoviruses have been identified.

15.6.3 MOLECULAR APPROACH

This approach is based on the amplification of target nucleic acid using specified primers and high-throughput sequencing of targeted viral regions. It provides precise, reliable, and reproducible results rapidly, facilitating early disease management decisions. There are two types of nucleic acid-based tests, namely, polymerase chain reaction (PCR) amplification and isothermal amplification.

PCR makes millions of DNA copies of a specified region of the virus genome, which is subsequently visualized under ultraviolet rays or by the use of fluorescent probes. It is the most important and sensitive molecular technique currently used in identifying viruses in vegetables. The technique is suitable for both DNA and RNA viruses; however, for viruses with an

RNA-based genome, this has to be converted into complementary DNA (cDNA) by reverse transcription (RT) before PCR is conducted. There are many types of PCR, and this includes conventional PCR, real-time quantitative PCR (qPCR), reverse transcription (RT) PCR, and multiplex PCR.

Conventional PCR is a qualitative test that determines the presence or absence of the viral nucleic acid. qPCR, on the other hand, quantifies the nucleic acid and monitors the reaction progress in real time. Cucumber mosaic cucumovirus (CMV) and cucumber green mottle mosaic tobamovirus (CGMMV) were detected from the seeds of *Capsicum annuum* and cucumber by reverse transcription-polymerase chain reaction (RT-PCR). Multiplex PCR is a type of PCR where there is simultaneous detection of different nucleic acids in a single reaction to identify more than one virus at a time (Nassuth et al., 2000). For the identification of CMV, a multiplex-nested RT-PCR has been developed (Bertolini et al., 2003).

Identification of viruses by PCR, rolling cycle amplification (RCA), cloning, and sequencing in Burkina Faso showed the presence of four known African monopartite begomoviruses (CLCuGV, PepYVMV, ToLMLV, and ToLCGHV) (Ouattara et al., 2016).

Isothermal amplification is a rapid technique that efficiently amplifies nucleic acid at a constant temperature. There are two types of techniques for isothermal amplification. These are recombinase polymerase amplification (RPA) and loop-mediated isothermal amplification (LAMP).

High-throughput sequencing, also known as next-generation sequencing, is an advanced technology for virus diagnosis (Villamor et al., 2019; Maree et al., 2018) and does not require much knowledge of viral sequences. It can sequence millions to billions of DNA molecules, detecting all viruses in the plant genome, including both known and unknown viruses (Roossinck, 2015). Through Sanger sequencing, the following viruses have been identified: tomato chlorosis crinivirus (ToCV), tomato torrado virus (ToTV), and tomato curly stunt begomovirus (ToCSV) (Moodley et al., 2019). With the advent of these molecular techniques, new strains of begomovirus/satellite complexes infecting vegetables have been identified (Leke et al., 2015).

In addition, sequencing of the viral genomes of okra has resulted in the identification of two previously known begomovirus species (cotton leaf curl Gezira virus and okra yellow crinkle virus), as well as a new recombinant begomovirus species (okra leaf curl Cameroon virus), a betasatellite (cotton leaf curl Gezira betasatellite), and new alphasatellites (Leke et al., 2015). Out of all these approaches, there is a widespread application of molecular techniques for identifying viruses in vegetables.

15.7 MANAGEMENT OF VIRUSES AFFECTING VEGETABLES IN WEST AFRICA

Viral diseases affecting vegetables are of major economic importance in terms of yield losses in the West African subregion and in the tropics (Varma and Malathi, 2003). The genetic makeup, symptom expression, and mode of transmission vary from one virus to the other. Mutation, recombination, and exchange of chromosomal parts in viruses can lead to diverse viral strains, which can affect their pathogenicity (García-Arenal et al., 2001). Unfortunately, most farmers find it difficult to identify these viral diseases on their favorite vegetable crops, let alone to contemplate any control or management strategy. However, the management of viral diseases in vegetables is not much different from managing any other viral diseases in plants. Abiotic stresses or biological consequences of certain plant processes can cause symptoms such as distortion, streaking, stunting, vein clearing, mosaic, etc. As a result, farmers need to be educated on viral symptoms as a first step of management. Thus, the management of viral diseases of vegetables is a complicated one, and it involves a combination of different management strategies or principles. These strategies can be used alone or in combination with others. These include:

i. Cultural and phytosanitary practices
ii. Vector management using insecticides
iii. Host plant resistance against viruses
iv. Integrated pest management approaches

15.7.1 CULTURAL AND PHYTOSANITARY PRACTICES

Cultural and phytosanitary practices are among the many practices used to manage viral disease in vegetables. Some of the cultural practices used are:

- land preparation (time of tilling),
- use of healthy seeds,
- planting at the right time,
- establishment of a healthy nursery,
- sanitation,
- appropriate fertilizer application, and
- land preparation.

Good land preparation results in good germination and crop establishment. The method of land preparation has an impact on viral disease control.

Poorly prepared lands are reservoirs of viruses, which can easily move on to infect your crop, especially if the viral disease is vector-borne or mechanically transmitted. Also, since most vegetables are nursed in nursery beds before transplanting, poorly prepared nursery beds could be harbingers of viral diseases attacking your crop at the latter stages of plant growth.

15.7.1.1 SELECTION OF HEALTHY SEEDS

Improved and healthy seeds are important for agricultural productivity. Pure seeds with good germination help to establish a uniform and vigorous crop for increased productivity. Vigorous crops can also withstand some common biotic and abiotic stresses. Good seeds are highly productive, yielding more than poor quality seeds. Seeds should always be tested to ascertain their viral status prior to planting. Some viruses are seed-borne and testing them before planting is a good practice to manage viral infections.

15.7.1.2 TIMELY PLANTING

Planting fields on time is very important for the management of viral diseases. Some viral infections are seasonal, and therefore, planting at the right time provides plants with the opportunity to escape infections caused by viruses.

15.7.1.3 RAISING A HEALTHY NURSERY

Some vegetable seeds are sown in nurseries before being transplanted for field establishment. Nursery fields should be clean and free from viral vectors and alternate hosts of viruses. Nurseries should be carefully managed to ensure vigorous plant growth prior to transplanting.

a. Transplanting

   Transplanting gives proper plant spacing, which results in an optimum crop stand for getting maximum yield. It also makes field operations such as weeding, spraying, and fertilizer application easier. Transplanting at the right time also gives a head start over weeds and consequently viral infections.

b. Water Management

   Water is essential for vegetable production. Water management always improves the availability of moisture for crop development; it can lead to better weed control and reduce the risk of water stress, which can result in viral infections and poor yields. When plants are stressed, they become easily prone to viral disease infections. A sufficient amount of water should be applied to ensure the effectiveness of chemicals during herbicide application.

15.7.1.4 BALANCED USE OF FERTILIZERS AS PER RECOMMENDATIONS

The fertilizer requirement of vegetable production is met with the use of inorganic and organic fertilizers. For vegetable production, NPK fertilizers are used as basal and topdressed with either ammonium sulfate or urea. The principle behind this is that a vigorous plant is able to resist or suppress viral disease development, which can result in good yields.

- Removal of weeds from the field

Control of weeds is vital in the management of viruses in vegetable production. Regularly weed farms to avoid the transfer of viruses from alternate hosts into crops of interest.

15.7.2 VECTOR MANAGEMENT WITH INSECTICIDES

Application of insecticides is need-based, making sure to minimize its effects on other unintended targets. The use of synthetic pesticides is recommended only at specific times in the life cycle of target vectors (aphids, whiteflies, and other insects) and always at levels with the least impact on the environment and other beneficial insects. Botanicals are the most preferred chemicals in the control of vectors affecting vegetables.

15.7.2.1 BIOLOGICAL CONTROL PRACTICES

The main approach to biological control is to promote beneficial insects that parasitize target viral vectors. Some biological control agents may not be destroyed because of their beneficial roles: coccinellids, spiders, damsel flies, and dragonflies. Chlorpyriphos could be used for the root dip treatment of vegetable seedlings before transplanting.

15.7.2.2 BEHAVIORAL CONTROL

The use of natural enemies of pests in most cases is the preferred option because of their less detrimental effect on the environment and human health. Naturally occurring biological processes are environmentally friendly and cheaper.

15.7.3 HOST PLANT RESISTANCE AGAINST VIRUSES

Host plant resistance is one of the best ways to fight against virus infections in vegetable crop production. In the establishment of a vegetable production field, the first thing to do is to look for improved varieties that are also resistant to common viruses identified in the location of planting. The utilization of resistant cultivars is the most cost-effective and environmentally safe practice to manage virus infections in a field. Also, contact your seed agent or research institutions that have access to improved seeds with resistance to common viruses for planting.

15.7.4 INTEGRATED PEST MANAGEMENT APPROACHES

Integrated pest management (IPM), or integrated pest control (IPC), is an approach that considers a wide range of management practices in the control of pests in a cost-effective and environmentally friendly manner. The purpose of IPM is to drastically reduce pest populations below the economic injury level (EIL). The UN's Food and Agriculture Organization (FAO) encourages minimal use of pesticides to ensure human health and environmental safety. This management strategy lays emphasis on good plant health, encourages the use of biological agents, and minimizes disruption of agroecological systems. IPM strategies have been advocated for by entomologists and ecologists since the 1970s. Previously, IPM was mainly for managing pests; however, it is currently being used to manage diseases caused by viruses, weeds, and other pests for vegetable production. In West Africa, IPM principles were recommended several years ago (since the 1980s) for the safer management of pest and disease outbreaks. The principles of IPM involve all the principles that have been enumerated above, and they can be used based on a commonsense approach to achieve effective results with the least damage to the environment and human health.

15.8 PROGRESS IN THE DEVELOPMENT OF VIRUS RESISTANCE VARIETY IN VEGETABLES

The most grown and consumed vegetables in West Africa include tomato, chili pepper, okra, onion, and traditional African vegetables. Viral diseases are reported to be the most important cause of reduced productivity and quality of the major vegetables in the region, especially tomato, pepper, and okra. The development and production of resistant varieties of vegetables is an effective strategy to lessen the impact of viral diseases. Many virus resistance genes have been mapped in tomato and pepper, but less in okra, and molecular markers linked to these genes are available. In this section, we will focus on the major vegetable viral diseases in the region, their resistance genes, and genomic approaches to breeding for resistance to virus.

15.8.1 VIRUS RESISTANCE GENES AND THE IMPORTANCE OF BREEDING FOR TOMATO RESISTANCE VARIETIES

Multiple studies have revealed the implication of up to 146 viruses belonging to 33 genera in tomato yield losses around the world. In West Africa, tomato is the main produced and consumed vegetable crop but is affected mainly by begomovirus and tospovirus. These viruses were transmitted, respectively, by whiteflies and thrips. The *Begomovirus* genus is known to be the largest of the family Geminiviridae and contains more than 200 species (Fauquet et al., 2008). In recent years, tomato yellow leaf curl virus (TYLCV) and African cassava mosaic virus (ACMV) have been listed among the 10 most scientifically important and economically damaging crop viruses (Rybick015). TYLCV is known to be the most devastating virus in tomato, and it is also the case in West Africa, especially in the dry season where there is a high presence of whiteflies. Riley et al. (2011) reported that thrips are the main vectors of tospoviruses. At least 14 thrips species are concerned, of which eight are *Frankliniella*, three are *Scirtothrips*, and one species of Scirtothrips, Dictyothrips, and Ceratoyhripoides, respectively. Breeding tomatoes with resistance to begomovirus and tospovirus must be the main challenge for breeders in the region.

Several wild relatives of tomatoes show susceptibility to TYLCD in different ecosystems around the world. Research has identified six resistance genes to the tomato yellow leaf curl disease (Ty-1/Ty-3, Ty-2, Ty-4, Ty-5, and Ty-6) in cultivated tomatoes (Ji et al., 2007; Scott et al., 2015; Verlaan et al., 2013). Among these resistance genes, Ty-1 and Ty-3 are the most used in tomato breeding programs nowadays. Ty-2 is also exploited commercially either alone or in combination with other Ty-genes (i.e., Ty-1, Ty-3, or Ty-5) (Yan et al., 2021). The begomoviruses that infect commercial tomatoes include monopartite and bipartite viruses, and there is high genetic diversity among the first group in West Africa (and sub-Saharan Africa at large), where there is a prevalence of tomato leaf curl Mali virus, tomato leaf curl Burkina

Faso virus, and tomato leaf curl Togo virus. The resistance genes were found in wild tomato germplasm, such as *S. pimpinellifolium, S. peruvianum, S. chilense, S. habrochaites*, and *S. cheesmaniae*.

The tomato spotted wilt virus TSWV (genus *Orthotospovirus* and family Tospoviridae) causes necrotic yellow and necrosis diseases in tomatoes and green peppers. It has been reported that the use of resistant cultivars and insecticides are effective control methods for TSWV. Yet other issues are raised by new strains of TSWV overcoming the resistance genes and developing resistance to common insecticides. (Tomitak019). *S. pimpinellifolium* is the source of the first resistance genes to TSWV reported from wild tomatoes. Progressively, seven TSWV resistance loci have been identified, designated as the dominant and allelic Sw-1a and Sw-1b; three recessive genes: sw-2, sw-3, and sw-4; and three dominant genes: Sw-5, Sw-6, and Sw-77-10. Sw-5, previously introgressed in the cultivar "Stevens," is currently the primary source of TSWV resistance in commercial tomato varieties worldwide.

15.8.2 VIRUS RESISTANCE GENES AND THE IMPORTANCE OF BREEDING FOR RESISTANCE VARIETIES IN PEPPER

A 10-year report of the viruses of pepper in West Africa clearly showed some major viruses in the region identified in the majority of the West African countries. These major viruses include aphid-transmitted viruses [pepper veinal mottle virus (PVMV), cucumber mosaic virus (CMV), chili veinal mottle virus (ChiVMV), potato virus Y (pvy), tobacco etch virus (TEV), alfalfa mosaic virus (AMV), and *Polerovirus*], thrips transmitted viruses [tomato spotted wilt virus (TSWV)], and whitefly transmitted viruses [pepper leaf curl virus (PepLCV) and tomato yellow leaf curl virus (TYLCV)], and viruses not transmitted by invertebrate vectors [tobacco mosaic virus (TMV), tomato mosaic virus (ToMV), and pepper mild mottle virus (PMMoV)] (Afouda et al., 2017; Afouda et al.,2013; Arogundade et al., 2020; Arogundade et al., 2012; Seka et al., 2017; Tiendrébéogo et al., 2011; Tsai et al., 2010; Zohoungbogbo et al., 2022). Many of these viruses are from the *Potyvirus* genus. The genus Potyvirus (family Potyviridae) is among the largest groups of plant viruses, causing diseases in a broad range of plant species. Potyviruses can be spread by aphid vectors, mechanically, or through grafting (Kenyon et al., 2014). PVMV and ChiVMV are found in multiple locations across West Africa, according to the report by Tsai et al. (2010), and represent the main viruses prevalent in most countries in

West Africa. Cucumber mosaic virus or CMV (genus *Cucumovirus*) is also widespread across West Africa.

Several potyvirus resistance (pvr) genes have been reported in *Capsicum* species (Kyle et al., 1997). Most of the pvr genes (pvr1, pvr3, pvr5, pvr6, and pvr8) are associated with a recessive phenotype. In pepper, pvr1 is located on chromosome 3 and confers resistance to several strains of tobacco etch virus (TEV) and potato virus Y (PVY) (Kang et al., 2005), whereas pvr2 was confirmed to be identical to pvr1, and thus these alleles are referred to as pvr1^{1} and pvr1^{2} (Yeam et al., 2005). Moreover, pvr5 is guessed to be another allele of pvr1, namely the eIF4E homolog (Kang et al., 2005). The gene pvr3 is responsible for the resistance to pepper mottle virus (PepMoV) and operates by blocking long-distance movement of the virus (Parrella et al., 2002); however, the genomic position of pvr3 is not yet known. The combination of the pvr6 gene with pvr2^{1} or pvr2^{2} alleles has been identified to confer high levels of resistance to some potyviruses, while individually neither pvr6 nor pvr2^{1} or pvr2^{2} can confer resistance (Caranta et al., 1996). This is aligned with findings reporting that resistance to PVMV and ChiVMV is conferred by pvr6 together with pvr1^{1} or pvr1^{2} (Hwang et al., 2009; Ruffel et al., 2006).

15.8.3 VIRUS RESISTANCE GENES AND THE IMPORTANCE OF BREEDING FOR RESISTANCE VARIETIES IN OKRA

Two main viruses are affecting okra in West Africa, yellow vein mosaic virus (YVMV) and enation leaf curl virus (ELCV). They are both transmitted through whitefly insect vectors. Between them, YVMV is the most serious disease of okra and can infect 100% of plants in a field, with yield losses ranging from 50% to 94% depending on the stage of crop growth at which infection occurs (Sastry et al., 1974). Some wild species (*Abelmoschus crinitus, A. ficulneus, A. angulosus, and A. manihot*) are known to be sources of resistance to ELCV, as are *A. crinitus, A. angulosus, A. manihot, A. pungens, and A. tetraphyllus* for YVMV. Despite huge variation due to the vast number of genotypes within *Abelmoschus esculentus* and the occurrence of more wild relatives, improvement in okra is slow because of various barriers, such as no compatible good source of resistance against diseases like YVMV, ELCV, and fruit borer; polyploidy; difficulties in DNA isolation; and the nonavailability of proper genetic markers (Dhankhar and Koundinya,et al., 2020). Thus, okra breeding is progressing in West Africa to enable farmers to have access to varieties with resistance to YVMV and ELCV.

15.8.4 IDENTIFICATION OF RESISTANCE SOURCES, IMPORTANCE OF STRAINS, AND GENOMIC APPROACHES USED IN DEVELOPMENT OF RESISTANCE VARIETY

Table 15.1 shows the resistance genes of the major viral diseases and some sources of resistance that can be explored to breed for virus resistance in West Africa. Most of the resistance sources were found in wild relatives of the vegetable. They are really important in the identification of resistance genes. Many disease resistance genes have been mapped, and molecular markers associated with these genes are known and can be used for marker-assisted selection (MAS). The use of these molecular markers in resistance breeding can accelerate conventional vegetable breeding schemes. Screening using molecular markers is a quick and accurate method (Watanabe et al., 1995). Marker-assisted selection can be useful and accelerate breeding where phenotypic selection is difficult or where it is not possible (for environmental protection or other issues) to use the virus for direct screening. It can also be useful for backcross breeding, for the introgression of resistance genes from wild species while selecting against the undesirable characteristics of the wild parent (Young et al., 1989). The markers and maps have been used to locate and tag genes, or QTLs, for disease resistance and many other horticultural characteristics. Such information can be used for various purposes, including marker-assisted selection and map-based cloning of desirable genes, or QTLs.

Breeding for resistance with different resistance genes may not be enough when information on the strains present in the target area is not known. Strains and phylotypes of the virus vary among locations and vectors. A cultivar resistant to the virus in one location may be susceptible to virus strains in other locations, and this situation showed the necessity sometimes to do the breeding in the target environment. Especially in West Africa, there is a need to characterize the strains of the major viruses of vegetables to make breeding more efficient in the region.

15.9 PROSPECTS OF BREEDING VIRUS-RESISTANT VEGETABLE VARIETIES IN WEST AFRICA

Vegetables, a major source of vitamins, minerals, dietary fibers, and phytochemicals, are essential for well-balanced diets globally. Generally, vegetable consumption is known to reduce the incidence of diseases such as diabetes

TABLE 15.1 Presenting Major Viral Diseases, Vectors, Sources of Resistance, and Resistance Genes Associated.

Main vegetable	Main virus	Genus	Vectors	Resistance genes	Source of resistance	References
Tomato	Tomato yellow leaf curl virus (TYLCV)	Begomovirus	Whitefly	Six tomato yellow leaf curl disease resistance genes (Ty-1/Ty-3, Ty-2, Ty-4, Ty-5, and Ty-6)	*S. pimpinellifolium, S. peruvianum, S. chilense, S. habrochaites,* and *S. cheesmaniae*	(Ji et al., 2007; Scott et al., 2015; Gill et al., 2019)
	Tomato spotted wilt tospovirus (TSWV)	Tospovirus	Thrips	Swa1, Swb1, Sw2, Sw3, Sw4 Sw5, and Sw6	*S. pimpinellifolium* cultivar "Stevens"	(Stevens et al., 1995; Dhaliwal et al., 2016)
Pepper	Pepper veinal mottle virus (PVMV)	Potyvirus	Aphids	pvr2 and pvr 6	Perennial	(Caranta et al., 1996)
	Chili veinal mottle virus (ChiVMV)	Potyvirus	Aphids	Single dominant resistance source (Cvr1), recessive resistance (cvr4), and oligogenic resistance (Cvr2-1 and Cvr2-2)	Perennial and Yolow wonder	(Caranta et al., 1996; Lee et al., 2017).
	Cucumber mosaic virus (CMV)	Cucumovirus	Aphids	Cucumber mosaic resistant 1 (Cmr1) and CMV resistance gene 2 (cmr2)	*C. annum*: "Perennial"; "Vania"; "Sapporo-oonaga" and "Nanbu oonaga" *C. frutescens*: "BG2814-6" and "LS1839-2-4," and *C. baccatum* "PI439381-1-3"	(Caranta et al., 1996; Grube et al., 2000; Caranta et al., 2002; Suzuki et al., 2003; Kang et al., 2012; Choi et al., 2018)
Okra	Enation leaf curl virus (ELCV)	Begomovirus	Whitefly	-	Some wild species (*Abelmoschus crinitus, A. ficulneus, A. angulosus,* and *A. manihot*)	(Yadav et al., 2018)
	Yellow vein mosaic virus (YVMV)	Begomovirus	Whitefly	-	*A.crinitus, A. angulosus, A. manihot, A. pungens,* and *A. tetraphyllus*	(Shetty et al., 2013)

as well as heart disease, stroke, and different kinds of cancer (Keatinge et al., 2010). Production of vegetables is bedeviled with high disease pressure, compelling farmers to depend largely on pesticides to overcome the challenge of biotic stress. The loss in efficacy of some pesticides due to pathogen evolution and strict regulation in usage in several countries due to increased concerns about their negative effects on the environment, nontarget organisms, coupled with high demand for safe foods and vegetables devoid of harmful pesticide residues will mean a shift in the management of biotic stresses in vegetable production.

A major contribution affecting vegetable production in West Africa has been the issue of plant viruses. Nearly all vegetables are affected by multiple viral diseases that reduce the productivity of the crop, thereby contributing to food insecurity in the region. According to Leke et al. (2015), viruses are prevalent in tropical and subtropical regions due to the abundance of vectors in these regions. Dias and Ortiz (2012) report the lack of virus-resistant germplasm for important vegetable crops as a limitation to the fight against viral diseases in the region. With vegetable intensification and the limited knowledge and expertise available to farmers to mitigate its effect, the problem of viral diseases will continue to pose a threat to sustainable vegetable production. This will further be exacerbated by the noncirculative and nonpersistent transmission modes of vectors (Gonsalves, 1998), like aphid, which require short time frames to acquire and release viral particles from unhealthy and healthy plants, respectively. This requires enormous effort and research to overcome the looming challenge.

Host-plant resistance provides an alternative to pesticides because of the latter's negative effects on humans, animals, and their environment. Resistant germplasm approaches in biotic stress management are environmentally friendly as they reduce the use of hazardous pesticides in the environment, are beneficial to the farmer as less input and action are required, and are highly compatible with other disease management practices.

Notwithstanding the benefits of resistant germplasm, the high diversity, potential of forming new genotypes through recombination, and occurrence of DNA satellites (Leke et al., 2015) complicate the development of virus-resistant vegetable accessions but open up opportunities for virus-resistant vegetable breeding in the region. There will be a need to promote the development and sharing of resistant germplasm through the formulation, harmonization, and adoption of protocols to govern germplasm collection, modification, and exchange of genetic materials among countries, as well as the provision of funds and logistics to support breeding activities.

Continuous screening exercises across the region to identify and select vegetable materials with resistant genes to different viral strains need to be encouraged. However, there is a need for partnerships across countries to conduct multinational screening trials to identify germplasm tolerant to wide strains of viruses of interest due to their high population diversity. For instance, Camara et al. (2013) reported that 12 tomato lines identified to be resistant to tomato yellow leaf curl disease in Senegal showed susceptibility in other West African countries. Unregulated transfer of planting materials and germplasm can perpetuate the spread of diseases and derail any effort in the breeding and delivery of disease-free germplasm. In view of this, plant quarantine services in the region would need to be strengthened to prevent the introduction of exotic strains from other countries through seed and planting material. Due to the diverse strains of viruses, breeding virus-resistant germplasm would need to look at the possibility of employing techniques such as gene pyramiding to stack multiple genes into single varieties to minimize infection. The potential to attack multiple species using transgenic strategies will promote the efficient and timely spread of innovative sources of resistance (Atkinson et al., 2015). Modern biotechnological tools should be investigated and utilized for the development of virus-resistant vegetable varieties, since, according to Tripathi et al. (2015), biotechnological tools facilitate the use of varieties preferred by farmers. Genetic modification and genome editing, for example, need to be explored to expand the genetic tools available to breeders in the region. The success in breeding virus-resistant vegetables will surely mean the adoption of modern biotechnological tools coupled with the utilization of simple and effective mechanisms. Creating public awareness and understanding the importance of using virus-resistant vegetable germplasm will ultimately promote their use.

15.10 CONCLUSION

This chapter has looked at the various aspects of how viruses affect vegetable production in West Africa and their management challenges. The authors have delved deep into literature and coupled it with their own experiences to explicitly unravel how viruses are contributing to practically thwarting farmers' efforts in their farming practices. This chapter also looked at the various virus families, genus, species, and strains that affect various vegetables grown in West Africa. The mode of transmission and the economic impact of viral diseases on vegetable production and how these affect food and nutrition security are all highlighted. Finally, the management strategies that can be employed to address viral diseases in vegetable production, including the use of resistant varieties and other IPM strategies, have been elaborated. It is anticipated that readers, including vegetable growers, extension workers, and the public interested in vegetable production, will find this chapter a useful companion to help manage viral diseases of vegetable crops and raise the productivity of vegetable production in West Africa.

KEYWORDS

- **vegetables**
- **virus**
- **resistant**
- **varieties**
- **symptoms**

REFERENCES

Abdulai, J.; Nimoh, F.; Darko-Koomson, S.; Fallah, K.; Kassoh, S. Performance of Vegetable Production and Marketing in Peri-Urban Kumasi, Ghana. *J. Agric. Sci.* **2016,** *9* (3); 2017.

Addo-Quaye, A.; Djokoto J. Report on Vegetable Seed Systems and Policy Integrated Agricultural Research Systems for the Humid Tropics: A CGIAR Research Program. **2013**.

Adeoye, I. B. *Factors Affecting Efficiency of Vegetable Production in Nigeria: A Review* [Online First], IntechOpen, 2020. doi: 10.5772/intechopen.92702.

Afouda, L. A.; Kotchofa, R.; Sare, R.; Zinsou, V.; Winter, S. Occurrence and Distribution of Viruses Infecting Tomato and Pepper in Alibori in Northern Benin. *Phytoparasitica* **2013**.

Afouda, L.; Kone, D.; Zinsou, V.; Dossou, L.; Kenyon, L.; Winter, S.; Knierim, D. Virus Surveys of Capsicum spp. in the Republic of Benin Reveal the Prevalence of Pepper Vein Yellows Virus and the Identification of a Previously Uncharacterised Polerovirus Species. *Arch. Virol.* **2017,** *162,* 1599–1607.

Agricultural Research Council (ARC). *Production Guidelines for Summer Vegetables*; Vegetables and Ornamental Plants Institute: Pretoria. ISBN 978-1-86949-635-7, 2013; vol. 41, pp 271–276.

Alabi, O. J. A. L.; Rwahnih, M.; Jifon, J. L.; Gregg, L.; Crosby, K. M.; Mirkov, T. E. First Report of Pepper Vein Yellows Virus Infecting Pepper (Capsicum spp.) in the United States. *Plant Dis.* **2015,** *99,* 1656.

Albuquerque, L. C.; Aranha, S. A.; Fernandes, F. R.; Inoue-Nagata, A. K. Further Evidence Reveals that Okra Mottle Virus Arose from a Double Recombination Event. *Arch. Virol.* **2013,** *158,* 181–186.

Alegbejo, M. D.; Abo, M. E. Ecology, Epidemiology and Control of Pepper Veinal Mottle Virus (PVMV), Genus Potyvirus, in West Africa. *J. Sustain Agric.* **2002,** *20,* 5–16.

Alice, K.; Inoue-Nagata, A. K.; Lima, M. F.; Gilbertson, R. L. A Review of Geminivirus Diseases in Vegetables and Other Crops in Brazil: Current Status and Approaches for Management. *Hortic. Bras.* **2016,** *34* (1).

Appiah, A. S.; Quartey, E. K.; Amoatey, H. M.; Nunekpeku, W.; Owusu A. M.; Ofori, S. Response of Nine Cultivars of Pepper (*Capsicum* spp.) to Infection by Four Viruses Under Natural Field Conditions in the Coastal Savanna Zone of Ghana. *Res. J. Appl. Sci. Eng. Technol.* **2014,** *7* (5), 903–907.

Arogundade, O.; Balogun, O. S.; Kareem, K. T. Occurrence and Distribution of Pepper Veinal Mottle Virus and Cucumber Mosaic Virus in Pepper in Ibadan, Nigeria. *Virol. J.* **2012,** *9,* 14.

Arogundade, O.; Balogun, O. S.; Kareem, K. T. Occurrence and Distribution of Pepper Veinal Mottle Virus and Cucumber Mosaic Virus in Pepper in Ibadan, Nigeria. *Virology J.* **2012,** *9* (79). doi:10.1186/1743-422X-9-79

Arogundade, O.; Aderonmu, O. I.; Matthew, J. O.; Ayo-Johh, E. I. First Report of Tomato Mosaic Virus Isolated from Solanum Macrocarpon in Nigeria. *Plant Dis.* **2018,** *102* (2): 458. doi: 10.1094/PDIS-05-17-0727-PDN

Arogundade O.; Balogun O.S. and Kumar L.P. Seed Transmissibility of Cucumber Mosaic Virus in *Capsicum* spp. *Int. J. Veg. Sci.* **2019,** *25* (2):146–153.

Arogundade, O.; Ajose, T.; Osijo, I.; Onyeanusi, H.; Matthew, J.; Aliyu, T. H. Management of Viruses and Viral Diseases of Pepper (Capsicum spp.) in Africa. In *Capsicum*; IntechOpen, 2020.

Arogundade, O.; Ajose, T.; Osijo, T.; Onyeanusi, O.; Matthew, J.; Taye, H.; Aliyu, T. H. Management of Viruses and Viral Diseases of Pepper (*Capsicum* spp.) in Africa, In *Capsicum*; Aman D. A., Ed.; IntechOpen. 2020; doi: 10.5772/intechopen.92266. https://www.intechopen.com/books/capsicum/management-of-viruses-and-viral-diseases-of-pepper-em-capsicum-em-spp-in-africa

Atkinson, H. J.; Roderick, H.; Tripathi, L. Africa Needs Streamlined Regulation to Support the Deployment of GM Crops. *Trends Biotechnol.* **2015,** *33,* 433–435.

Avilla, C.; Collar, J. L.; Duque, M.; Fereres, A. Yield of Bell Pepper (*Capsicum annuum*) Inoculated with CMV and/or PVY at Different Time Intervals. Z. Pflanzenk. Pflanzen. *J. Plant Dis. Protect.* **1997b,** *104,* 1–8.

Ayodele, A. E. The Medicinally Important Leafy Vegetables of South-Western Nigeria Ethnobotanical Leaflets [Online]. 2005; core.ac.uk (accessed June 15, 2021).

Ayo-John, E. I.; Odedara, O. O. Serological Detection of Viruses Infecting Tomato and Pepper in Southwest Nigeria and Their Distribution. *Niger. J. Biotechnol.* **2017,** *33,* 78–82.

Bertolini, E.; Olmos, A.; López, M. M.; Cambra, M. Multiplex Nested Reverse-Transcription Polymerase Chain Reaction in a Single Tube for Sensitive and Simultaneous Detection of Four RNA viruses and Pseudomonas Savastanoi pv. Savastanoi in Olive Trees. *Phytopathology* **2003,** *93,* 286–292.

Blancard, D.; Ed. *Diagnosis of Parasitic and Nonparasitic Diseases*; Academic Press, 2012. doi: 10.1016/B978-0-12-387737-6.50002-9

Bujarski, J. J. Bromoviruses (*Bromoviridae*). In *Encyclopedia of Virology*; 2021; pp 260–267.

Camara, M.; Mbaye, A. A.; Noba, K.; Samb, P. I.; Diao, S.; Cilas, C. Field Screening of Tomato Genotypes for Resistance to Tomato Yellow Leaf Curl virus (TYLCV) Disease in Senegal. *Crop Protect.* **2013,** *44,* 59–65.

Caranta, C.; Palloix, A. Both Common and Specific Genetic Factors are Involved in Polygenic Resistance of Pepper to Several Potyviruses. *Theor. Appl. Genet.* **1996,** *92,* 15–20.

Caranta, C.; Pflieger, S.; Lefebvre, V.; Daubeze, A.; Thabuis, A.; Palloix, A. QTLs Involved in the Restriction of Cucumber Mosaic Virus (CMV) Long-distance Movement in Pepper. *Theor. Appl. Genet.* **2002,** *104,* 586–591.

Catarino, L.; Romeiras, M. M.; Bancessi, Q.; Duarte. D.; Faria, D.; Monteiro, F.; Moldao, M. Edible Leafy Vegetables from West Africa (Guinea-Bissau): Consumption, Trade and Food Potential. *Foods* **2019,** *8* (10), 493. doi: 10.3390/foods8100493

Chen, G.; Su, Q.; Shi, X.; Pan, H.; Jiao, X.; Zhang, Y. Persistently Transmitted Viruses Restrict the Transmission of Other Viruses by Affecting Their Vectors. *Front. Physiol.* **2018,** *9,* 1261. doi: 10.3389/fphys.2018.01261.

Choe, S.; Choi, B.; Kang, J.; Seo, J. *Tolerance to Tomato Yellow Leaf Curl Virus in Transgenic Tomato Overexpressing a Cellulose Synthase-like Gene*; 2021; pp 657–659. doi: 10.1111/pbi.13539

Choi, S.; Lee, J.-H.; Kang, W.-H.; Kim, J.; Huy, H.N.; Park, S.-W.; Son, E.-H.; Kwon, J.-K.; Kang, B.-C. Identification of Cucumber Mosaic Resistance 2 (cmr2) That Confers Resistance to a New Cucumber Mosaic Virus Isolate P1 (CMV-P1) in Pepper (Capsicum spp.). *Front. Plant Sci.* **2018,** *9,* 1106.

Clark, M. F.; Adams, A. N. Characteristics of the Microplate Method of Enzyme-Linked Immunosorbent Assay for the Detection of Plant Viruses. *J. Gen. Virol.* **1977,** *34,* 475–483.

Cohen, S.; Antignus, Y. Tomato Yellow Leaf Curl Virus, a Whitefly-Born Geminivirus of Tomatoes. *Adv. Dis. Vector Res.* **1994,** *10,* 259–288.

Crescenzy, A. Cucumber Mosaic Cucumovirus Populations in Italy Under Natural Epidemic Conditions and after a Satellite-Mediated Protection Test. *Plant Dis.* **1993,** *77,* 28–33. doi: 10.1094/PD-77-0028es.

Czosnek, H.; Laterrot, H. A Worldwide Survey of Tomato Yellow Leaf Curl Viruses. *Arch. Virol.* **1997,** *142,* 1391–1406.

Czosnek, H.; Laterrot H. A Worldwide Survey of Tomato Yellow Leaf Curl Viruses. *Arch. Virol.* **1997,** *142,* 1391–406.

Dafalla, G. A. In *Situation of Tomato and Pepper Viruses in Africa*, Proceedings of International Institute of Tropical Agriculture Conference on Plant Virology in Sub Saharan Africa; Hughes, J. D. A., Odu, B. O., Eds.; 4–8 June 2001, Ibadan, Nigeria, pp 18–24.

Data Sheets, E. *Tomato Spotted Wilt Tospovirus* (n.d.).

Datta, S. Impact of Climate Change in Indian horticulture—A Review. *Int. J. Environ. Sci. Technol.* **2013,** *2* (4), 661–671.

De Barro, P. J.; Liu, S. S.; Boykin, L. M.; Dinsdale, A. B. *Bemisia tabaci*: A Statement of Species Statu. 2011.

Deuter, P. Defining the Impacts of Climate Change on Horticulture in Australia. *Reports Commissioned by the Garnaut Climate Change Review*; 2014.

Dhaliwal, M.; Sharma, A. Breeding for Resistance to Virus Diseases in Vegetable Crops. *Innovations in Horticultural Sciences*; New India Publishing Agency: New Delhi, 2016; pp 303–327.

Dhankhar, S.; Koundinya, A. Accelerated Breeding in Okra. In *Accelerated Plant Breeding*; Springer, 2020; Vol. 2, pp 337–353.

Dias, J. S.; Rodomiro Ortiz, R. Transgenic Vegetable Crops: Progress, Potentials, and Prospects. In *Plant Breeding Reviews*; Jules J.; Eds.; John Wiley & Sons, Inc., 2012.

Dogimont, C.; Palloix, A.; Daubèze, A. M.; Marchoux, G.; Gébré-Sélassié, K.; Pochard, E. Genetic Analysis of Broad Spectrum Resistance to Potyviruses Using Doubled Haploid Lines of Pepper (*Capsicum annuum* L.). *Euphytica* **1996,** *88,* 231–239.

Dombrovsky, A.; Glanz, E.; Pearlsman, M.; Lachman, O.; Antignus, Y. Characterization of Pepper Yellow Leaf Curl Virus, A Tentative New Polerovirus Species Causing a Yellowing Disease of Pepper. *Phytoparasitica* **2010,** *38,* 477–486.

Edwardson, J. R. *CRC Handbook of Viruses Infecting Legumes*, 1st ed.; CRC Press, 1991. doi: 10.1201/9781351071192

Edwardson, J. R.; Christie, R. G. *Viruses Infecting Peppers and other Solanaceous Crops*; University of Florida, 1997; vol 1, p 336.

Etwire, P. M.; Atople, I. D. P.; Buah, S. S. J.; Abdulai, A. L.; Karikari, A. S.; Asungre, P. Analysis of the Seed System in Ghana. *Int. J. Adv. Agric. Res.* **2013,** *1,* 7–13.

Fadairo, O.; Williams, P. A.; Nalwanga, F. S. Perceived Livelihood Impacts and Adaptation of Vegetable Farmers to Climate Variability and Change in Selected Sites from Ghana, Uganda and Nigeria Environment. *Dev. Sustain.* **2020,** *22,* 6831–6849.

Fajinmi, A. A. Agro-ecological Incidence and Severity of Pepper Veinal Mottle Virus, Genus Potyvirus, Family Potyviridae, on Cultivated Pepper (*Capsicum annuum* L.) in Nigeria. *Arch. Phytopathol. Plant Protect.* **2011,** *44,* 307–319.

FAOSTAT. *Food and Agricultural Organization of the United Nations* [Online]. On-line and Multilingual Database, 2019, http://faostat.fao.org

Fauquet, C.; Briddon, R.; Brown, J. K.; Moriones, E.; Stanley, J.; Zerbini, M.; Zhou, X. Geminivirus Strain Demarcation and Nomenclature. *Arch. Virol.* **2008,** *153,* 783–821.

FreshPlaza, 2021. https://www.freshplaza.com/article/9198968/ghana-stop-foodexportdestination-in-west-africa-is-burkina-faso/

Funderburk, J. Management of the Western Flower Thrips (Thysanoptera:Thripidae) in Fruiting Vegetables. *Fla. Entomol.* **2009,** *92,* 1–6.

García-Arenal, F.; Fraile, A.; Malpica, J. M. Variability and Genetic Structure of Plant Virus Populations. *Annu. Rev. Phytopathol.* **2001,** *39,* 157–86. doi: 10.1146/annurev.phyto.39.1.157. PMID: 11701863.

Gaur, A.; Kumar, A.; Kiran, R.; Kumari, P. Importance of Seed-borne Diseases of Agricultural Crops: Economic Losses and Impact on Society. In *Seed-Borne Diseases of Agricultural Crops: Detection, Diagnosis and Management*, 1st ed.; Kumar, R., Gupta, A., Eds.; Springer: Singapore; Karnal, India, 2020; pp 3–23.

Genda, Y.; Sato, K.; Nunomura, O.; Hirabayashi, T.; Tsuda, S. Immunolocalization of Pepper Mild Mottle Virus in Developing Seeds and Seedlings of *Capsicum annuum. J. Gen. Plant Pathol.* **2011,** *77,* 201–208.

Ghanaweb.com. March 2020. Published on 13th March 2020.

Ghanem, G. A. M. Okra Leaf Curl Virus: A Monopartite Begomovirus Infecting Okra Crop in Saudi Arabia. *Arab J. Biotechnol.* **2003,** *6* (1), 139–152.

Gill, U.; Scott, J. W.; Shekasteband, R.; Ogundiwin, E.; Schuit, C.; Francis, D. M.; Sim, S.-C.; Smith, H.; Hutton, S. F. Ty-6, A Major Begomovirus Resistance Gene on Chromosome 10, is Effective Against Tomato Yellow Leaf Curl Virus and Tomato Mottle Virus. *Theor. Appl. Genet.* **2019,** *132,* 1543–1554.

Glick, E.; Levy, Y.; Gafni, Y.; Dagan, B.; Mina, T. *The Viral Etiology of Tomato Yellow Leaf Curl Disease—A Review*. **2009,** *45* (3), 81–97.

Gonsalves, D. Control of Papaya Ringspot Virus in Papaya: A Case Study. *Annu. Rev. Phytopat.* **1998,** *36,* 415–437.

Gonzalez, Y. S.; Dijkxhoorn, Y.; Koomen, I.; Maden, E.; Herms, S.; Joosten, F.; Mensah, S. A. Vegetable Business Opportunities in Ghana. *GhanaVeg Sector Reports*, 2016.

Gorsane, F.; Fekih-Hassen, I.; Gharsallah-Chouchene, S.; Nakhla, M. K.; Maxwell, D. P.; Marrakchi, M.; Fakhfakh, H. In *Typing of Tomato Yellow Leaf Curl Virus Spreading on Pepper in Tunisia*, Proceedings of the XIIth EUCARPIA Meeting on Genetics and Breeding of Capsicum and Eggplant, Noordwijkerhout, Netherlands, 17–19 May, 2004; p 182.

Goyal, G.; Gill, H. K.; McSorley, R. Common Weed Hosts of Insect-Transmitted Viruses of Florida Vegetable Crops ENY-863, UF/IFAS; 2018. website at http://edis.ifas.ufl.edu.

Grube, R. C.; Zhang, Y.; Murphy, J. F.; Loaiza-Figueroa, F.; Lackney, V. K.; Provvidenti, R.; Jahn, M. K. New Source of Resistance to Cucumber Mosaic Virus in Capsicum Frutescens. *Plant Disease* **2000,** *84,* 885–891.

Gumedzoe, M. Y. *Major Virus Diseases of Medicinal Plants in West Africa*, in WOCMAP I-Medicinal and Aromatic Plants Conference: part 3 of 4 331; 1992; pp 307–310.

Hamelin, F. M.; Allen, L. J. S.; Prendeville, H. R.; Hajimorad, M. R.; Jeger, M. J. The Evolution of Plant Virus Transmission Pathways. *J. Theor. Biol.* **2016,** *396* (2016), 75–89 doi: 10.1016/j.jtbi.2016.02.017

Hernández-Santiago, R.; Vargas-Hernández, M.; Zamora-Macorra, E. J. Evaluation of TMV Resistance Inducers in Tomato. *J. Rev. Mex. Cienc. Agríc.* **2021,** *11* (2), Texcoco. doi: 10.29312/remexca.v11i2.2072

Hull, R. *PlantVirology*, 5th ed.; Elsevier: NewYork, 2014.

Ibaba, J. D.; Gubba, A. Diversity of Potato virus Y Isolates Infecting Solanaceous Vegetables in the Province of KwaZulu-Natal in the Republic of South Africa. *Crop Prot.* **2011,** *30* (11), 1404–1408.

Ibaba J. D.; Gubba A. Phylogenetic Studies of Selected Isolates of Potato virus Y (PVY) Infecting Vegetable Crops in KwaZulu-Natal, Republic of South Africa. *S. Afr. J. Plant Soil* **2012,** *29* (2), 117–120.

Idris, A. M.; Brown, J. K. Molecular Analysis of Cotton Leaf Curl virus—Sudan Reveals an Evolutionary History of Recombination. *Virus Genes* **2002,** *24,* 249–256.

Ioannou, N. Yellow Leaf Curl and other Virus Diseases of Tomato in Cyprus. *Plant Pathol.* **1985,** *34* (3), 428–434.

Jalender, P.; Bhat, B. N.; Anitha, K. Studies on Host Range of Cucumber Mosaic virus in Tomato (*Solanum lycopersicum* L.). *Ecol. Environ. Conserv.* **2015,** *21* (Suppl.), S435–S438.

Jalender, P.; Bhat, B. N.; Anitha, K.; Vijayalakshmi, K. Studies on Transmission of Cucumber Mosaic Virus (CMV) by Sap Inoculation in Tomato. *Int. J. Pure Appl. Biosci.* **2017,** *5* (4), 1908–1912.

Ji, Y.; Schuster, D. J.; Scott, J. W. Ty-3, a Begomovirus Resistance Locus Near the Tomato Yellow Leaf Curl virus Resistance Locus Ty-1 on Chromosome 6 of Tomato. *Mol. Breed.* **2007,** *20,* 271–284.

Johansen, E.; Edwards, M. C.; Hampton, R. O. Seed Transmission of Viruses: Current Perspectives. *Annu. Rev. Phytopathol.* **1994,** *32,* 363–386.

Jones, A. T. Experimental Transmission of Viruses in Diagnosis. In *Diagnosis of Plant Virus Diseases*; Matthews, R. E. F., Ed.; CRC Press: Boca Raton, Florida, USA, 1993; pp 49–72.

Kang, B. C.; Yeam, I.; Frantz, J. D.; Murphy, J. F.; Jahn, M. M. The pvr1 Locus in Capsicum Encodes a Translation Initiation Factor eIF4E that Interacts with Tobacco etch Virus VPg. *Plant J.* **2005,** *42,* 392–405.

Kang, W.-H.; Seo, J.-K.; Chung, B. N.; Kim, K.-H.; Kang, B.-C. Helicase Domain Encoded by Cucumber Mosaic Virus RNA1 Determines Systemic Infection of Cmr1 in Pepper. *PLoS One* **2012,** *7,* e43136.

Kassem, M. A.; Sempere, R. N.; Juarez, M.; Aranda, M. A.; Truniger, V. Cucurbit Aphid-borne Yellows Virus is Prevalent in Field-grown Cucurbit Crops of Southeastern Spain. *Plant Dis.* **2007,** *91,* 232–238.

Keatinge, J. D. H.; Waliyar, F.; Jammadass, R. H.; Moustafa, A.; Andrade, M.; Drechsel, P.; d'A, J.; Kardivel, H. P.; Luther, K. Re-learning Old Lessons for the Future of Food: By Bread Alone No Longer—Diversifying Diets with Fruit and Vegetables. *Crop Sci.* **2010,** *60.*

Kenyon, L.; Kumar, S.; Tsai, W.-S.; Hughes, J. d. A. Virus Diseases of Peppers (Capsicum spp.) and Their Control. *Adv. Virus Res.* **2014,** *90,* 297–354.

Kil, E.; Kim, S.; Lee, Y.; Byun, H.; Park, J.; Seo, H.; Kim, C.; Shim, J.; Lee, J.; Kim, J.; Lee, K.; Choi, H.; Lee, S. *Tomato Yellow Leaf Curl Virus Geminivirus in Tomatoes*; Nature Publishing Group, January, 2016; pp 1–10. doi:10.1038/srep19013

Kimiywe, J.; Waudo, J. N.; Mbithe, D.; Maundu, P. Utilization and Medicinal Value of Indigenous Leafy Vegetables Consumed in Urban and Peri-urban Nairobi. *Afr. J. Food Agric. Nutr. Dev.* **2007,** *7* (4).

Konate, G.; Barro, N.; Fargette, D.; Swanson, M. M., Harrison, B. D. Occurrence of Whitefly-Transmitted Geminiviruses in Crops in Burkina-Faso, and Their Serological Detection and Differentiation. *Ann. Appl. Biol.* **1995,** *126,* 121–948.

Kpéra, G. N.; Segnon, A. C.; Saïdou, A.; Mensah, G. A.; Aarts, N.; van der Zijpp, A. Towards Sustainable Vegetable Production Around Agro-pastoral Dams in Northern Benin: Current Situation, Challenges and Research Avenues for Sustainable Production and Integrated Dam Management. *Agric. Food Secur.* **2017,** *6,* 67.

Kumar, S.; Singh, M.; Kumar, A.; Rai, M. Identification of Host Plant Resistance to Pepper Leaf Curl Virus in Chilli (*Capsicum species*). 2006. *110,* 359–361. doi: 10.1016/j.scienta.2006.07.030

Kumar, S.; Udaya Shankar, A. C.; Nayaka, S. C.; Lund, O. S.; Prakash, H. S. Detection of Tobacco Mosaic Virus and Tomato Mosaic Virus in Pepper and Tomato by Multiplex RT-PCR. *Lett. Appl. Microbiol.* **2011,** *53,* 359–363.

Krause-Sakate, R.; Redondo, E.; Richard-Forget, F.; Jadao, A. S.; Houvenaghel, M. C.; German-Retana, S.; Le Gall, O. Molecular Mapping of the Viral Determinants of Systemic Wilting Induced by a Lettuce Mosaic Virus (LMV) Isolate in Some Lettuce Cultivars. *Virus Res.* **2005,** *109,* 175–180.

Kumar, R.; Esakky, R.; Palicherla, S. R. The Emergence of Okra Enation Leaf Curl Virus—an Important Begomovirus Infecting Okra in States Across in India. *Arch. Phytopathol. Plant Prot.* doi: 10.1080/03235408.2019.1606383

Kyle, M.; Palloix, A. Proposed Revision of Nomenclature For Potyvirusresistance Genes in Capsicum. *Euphytica* **1997,** *97,* 183–188.

Ladipo, J. L. A Yellow-Vein Viruslike Disease of Tomato in Nigeria Plant Disease Reporter. **1977,** *61* (11), 958–960.

Lapidot, M.; Polston, J. E. Resistance to Tomato Yellow Leaf Curl Virus in Tomato. In *Natural Resistance Mechanisms of Plants to Viruses*; Lobensteine, G.; Carr, J. P., Eds.; Springer: the Netherlands, 2006; pp 503–520.

Lee, J.-H.; An, J.-T.; Siddique, M. I.; Han, K.; Choi, S.; Kwon, J.-K.; Kang, B.-C. Identification and Molecular Genetic Mapping of Chili Veinal Mottle Virus (ChiVMV) Resistance Genes in Pepper (Capsicum annuum). *Mol. Breed.* **2017,** *37,* 1–10.

Leke, W. N.; Mignouna, D. B.; Brown, J. K.; Kvarnheden, A. Begomovirus Disease Complex: Emerging Threat to Vegetable Production Systems of West and Central Africa. *Agric. Food Sec.* **2015,** *4,* 1. doi: 10.1186/s40066-014-0020-2

Levasseur, V.; Pasquini, M. W.; Kouamé, C.; Temple, L. *A Review of Urban and Peri-Urban Vegetable Production in West Africa*, in *Horticultural Plants in Urban and Peri-Urban Life*, Proceedings of the XXVII International Horticulture Conference. Acta Horticulturae; Lumpkin, T. A.; Warrington, I. J., Eds.; 2007, vol. 762, pp 245–252.

Marchoux, G.; Gognalons, P.; Gébré-Sélassié, K. *Virus des Solanacées Du Génome Viral à la Protection des Cultures*. Editions Quae: Versailles, 2008; p 858.

Maree, H. J.; Fox, A.; Al Rwahnih, M.; Boonham, N.; Candresse, T. Application of HTS for Routine Plant Virus Diagnostics: State of the Art and Chadllenges. *Front. Plant Sci.* **2018,** *9,* 1082. doi: 10.3389/fpls.2018.01082

Martelli, G. P. Bacilliform Particles Associated with Mottled Dwarf of Eggplant (*Solanum melongena L.*). *J. Gen. Virol.* **1969,** *5* (2), 319–320.

Matthews, R. E. F. Host Plant Responses to Virus Infection. In *Comprehensive Virology, Vol. 16, Virus-Host Interaction, Viral Invasion, Persistence, and Diagnosis*; Fraenkel-Conrat, H., Wagner, R. R., Eds.; Plenum Press: New York, USA, 1980, pp. 297–359.

Mishra, G. P.; Singh, B.; Seth, T.; Singh, A. K.; Halder, J.; Krishnan, N.; Tiwari, S. K.; Singh, P. M. Biotechnological Advancements and Begomovirus Management in Okra (*Abelmoschus esculentus L.*): Status and Perspectives. *Front. Plant Sci.* **2017,** *8,* 360.

Moodley, V.; Gubba, A.; Mafongoya, P. L. A Survey of Whitefly-transmitted Viruses on Tomato Crops in South Africa. *Crop Prot.* **2019,** *123,* 21–29.

Moriones, E.; Navas-Castillo, J. Tomato Yellow Leaf Curl Virus, an Emerging Virus Complex Causing Epidemics Worldwide. *Virus Res.* **2000,** *71*, 123–134.

Moury, B.; Verdin, E. Viruses of Pepper Crops in the Mediterranean Basin: A Remarkable Stasis. *Adv. Virus Res.* **2012,** *84,* 127–162.

Narayanasamy, P.; *Microbial Plant Pathogens-Detection and Disease Diagnosis Viral and Viroid Pathogens*; Springer Dordrecht Heidelberg: London, 2011; Vol. 3, p 343.

Nassuth, A.; Pollari, E.; Helmeczy, K.; Stewart, S.; Kofalvi S. A. Improved RNA Extraction and One-Tube RT-PCR Assay for Simultaneous Detection of Control Plant RNA Plus Several Viruses in Plant Extracts. *J. Virol. Meth.* **2000,** *90,* 37–49.

Navas-Castillo, J.; Fiallo-Olivé, E.; Sanchez-Campos, S. Emerging Virus Diseases Transmitted by Whiteflies. *Annu. Rev. Phytopathol.* **2011,** *49,* 219–248.

Nchandi, S.; Benddine, F.; Moumen, A.; Kheddam, M. Status of Tomato Viruses and Control Strategy in Algeria. (Etad des maladies virales de al tomate et strategie de lutte en Algerie). *Bull. OEPP* **2002,** *32* (1), 21–24.

Nono-Womdim, R.; Swai, I. S.; Chadha, M. L.; Gebre-Selassie, K.; Marchoux, G. Occurrence of Chilli Veinal Mottle Virus in Solanum Aethiopicum in Tanzania. *Plant Dis.* **2001,** *85* (7), 801–801.

Nono-Womdim, R. In *An Overview of Major Virus Diseases of Vegetable Crops in Africa and Some Aspects of Their Control*, Plant Virology in Sub-Saharan Africa, Proc. Conf. Organized by IITA. J. d'A. Hughes and J. Odu; International Institute of Tropical Agriculture: Ibadan, Nigeria, pp 213–232.

Ogwu, M. C.; Osawaru, M. E.; Aiwansoba, R. O.; Iroh, R. N. In *Status and Prospects of Vegetables in Africa*, Proceedings of NTBA/NSCB Joint Biodiversity Conference, Unilorin. 2016; pp 47–57.

Olawale, A.; Samuel, B. O.; Solomon, A. S. O.; Kumar, P. L. Surveys of Virus Diseases on Pepper (Capsicum spp.) in South-west Nigeria. *Af. J. Biotechnol.* **2015,** *14* (48), 3198–3205.

Osei, M. K.; Akromah, R. Shih, S. L.; Green, S. K. Evaluation of Some Tomato Germplasm for Resistance to Tomato Yellow Leafcurl Virus (TYLCV) Disease in Ghana. *Asp. Appl. Biol.* **2010,** 96.

Osei, M. K.; Adjebeng-Danquah, J.; Egbadzor, K. F.; Osuman, A. S.; Asamoah, E. A.; Abebrese, S. O.; Melomey, L. D.; Agyare, R. Y. Vegetable Seed Production: Prospects and Challenges: The Case of Ghana. In Tiwari, A. K., Ed.; *Advances in Seed Production and Management*; Springer Nature Singapore Pte Ltd., 2020. doi:10.1007/978-981-15-4198-8_5

Ouattara, A.; Tiendrébéogo, F.; Lefeuvre, P.; Thébaud, G.; Chiroleu, F.; Hoareau, M.; ... Lett, J. M. *Molecular Epidemiology of Emerging Begomovirus Diseases on Vegetable Crops in Burkina Faso*, in *13 International Plant Virus Epidemiology Symposium;* 2016, June; p 165.

Palukaitis, P.; Roossinck, M. J.; Dietzgen, R. G.; Francki, R. I. B. Cucumber Mosaic Virus. *Adv. Virus Res.* **1992,** *41,* 281–349.

Pazarlar, S.; Gumus, M.; Oztekin, G. B. The Effects of Tobacco Mosaic Virus Infection on Growth and Physiological Parameters in Some Pepper Varieties (*Capsicum annuum* L.). *Not. Bot. Horti Agrobot. Cluj Napoca* **2013,** *41* (2), 427–433.

Pinheiro, P. V.; Wilson, J. R.; Xu, Y.; Zheng, Y.; Rebelo, A. R.; Fattah-Hosseini, S.; Kruse, A.; Dos Silva, R. S.; Xu, Y.; Kramer, M.; Giovannoni, J.; Fei, Z.; Gray, S.; Heck, M. Plant Viruses Transmitted in Two Different Modes Produce Differing Effects on Small RNA-Mediated Processes in their Aphid Vector. *Phytobiom. J.* **2019,** *3,* 71–81. doi:10.1094/PBIOMES-10-18-0045-R

Ming, G.; Lin, Q.; Wong, S. iTRAQ-based Analysis of Leaf Proteome Identi fi es Important Proteins in Secondary Metabolite Biosynthesis and Defence Pathways Crucial to Cross-Protection Against TMV. *J. Proteom.* **2019,** *196* (January), 42–56. doi:10.1016/j.jprot.2019.02.002

Riley, D. G.; Joseph, S. V.; Srinivasan, R.; Diffie, S. Thrips Vectors of Tospoviruses. *J. Integr. Pest Manag.* **2011,** *2,* I1–I10.

Roberts, P. D.; Adkins, S.; Pernezny, K.; Jonnes, J. B. Diseases of Pepper and their Management. In *Diseases of Fruits and Vegetables*; Naqvi, S. A. M. H., Ed.; Kluwer Academic; 2004; Vol. 11, pp 333–387.

Rochester, D. E.; DePaulo, J. J.; Fauquet, C. M.; Beachy, R. N. Complete Nucleotide Sequence of the Geminivirus Tomato Yellow Leaf Curl Virus, Thailand Isolate. *J. Gen. Virol.* **1994,** *75* (3).

Roossinck, M. J. Plant Virus Ecology. *PLoS Pathog.* **2013,** *9* (5), e1003304. doi:10.1371/journal.ppat.1003304

Roossinck, M. J. Metagenomics of Plant and Fungal Viruses Reveals an Abundance of Persistent Lifestyles. *Front. Microbiol.* **2015,** *5,* 767. doi: 10.3389/ fmicb.2014.00767

Rybicki, E. P. A Top Ten List for Economically Important Plant Viruses; 2014.

Rybicki, E. P. A Top Ten List for Economically Important Plant Viruses. *Arch. Virol.* **2015,** *160,* 17–20.

Salati, R.; Nahkla, M. K.; Rojas, M. R.; Guzman, P.; Jaquez, J.; Maxwell, D.; Gilbertson, R. L. Tomato Yellow Leaf Curl Virus in the Dominican Republic: Characterization of an Infectious Clone, Virus Monitoring in Whiteflies and Identification of Reservoir Hosts. *Phytopathology* **2002,** *92,* 487–496.

Sastry, K.; Singh, S. Assessment of Losses in Tomato by Tamato Leaf Curl Virus. *Ind. J. Mycol. Plant Pathol.* 1974.

Sastry, K. S. *Seed-Borne Plant Virus Diseases*; Springer: New York, 2013.

Savary, S.; Willocquet, L.; Pethybridge, S. J.; Esker, P.; McRoberts, N.; Nelson A. The Global Burden of Pathogens and Pests on Major Food Crops. *Nat. Ecol. Evol.* **2019,** *3,* 430– 439.

Schippers, R. R. *African Indigenous Vegetables. An Overview of the Cultural Species*; Natural Resources Institute/ACP-EU Technical Centre for Agricultural and Rural Cooperation: Chatham, UK; p 214.

Scholthof, K.-B. G. Tobacco Mosaic Virus. *The Plant Health Instructor*; Texas A & M University: 2000. doi: 10.1094/PHI-I-2000-1010-01

Scholthof, K. B. Tobacco Mosaic Virus: A Model System for Plant Biology. *Annu. Rev. Phytopathol.* **2004,** *42,* 13–34. doi: 10.1146/annurev.phyto.42.040803.140322. PMID: 15283658

Schreinemachers, P.; Balasubramaniam, S.; Boopathi, N. M.; Ha, C. V.; Kenyon, L.; Praneetvatakul, S.; Sirijinda, A.; Le, N. T.; Srinivasan, R.; Wu, M.-H. Farmers' Perceptions and Management of Plant Viruses in Vegetables and Legumes in Tropical and SUBTROPICAL ASIA. *Crop Protec.* **2015,** *75* (2015), 115–123. www.elsevier.com/locate/cropro

Schreinemachers, P.; Howard, J.; Turner, Schreinemachers, P.; Turner, M.; Simon, N.; Dubey G. B.; Mwadzingeni, L.; Chagomoka, T.; Ngugi, M.; Afari-Sefa, V.; Hanson, P.; Wopereis, M. C. S. Africa's Evolving Vegetable Seed Sector: Status, Policy Options and Lessons from Asia. *Food Secur.* 2021.

Scott, J. W.; Hutton, S. F.; Freeman, J. H. Fla. 8638B and Fla. 8624 Tomato Breeding Lines with Begomovirus Resistance Genes ty-5 plus Ty-6 and Ty-6, Respectively. *HortScience* **2015,** *50,* 1405–1407.

Seka, K.; Quattara, A.; Assiri, K. P.; Kra, K. D.; Hoareau, M.; Lefeuvre, P.; Diallo, H. A.; Lett, J. M. (2016). First Reports of Cotton Leaf Curl Gezira Virus and Okra Yellow Crinkle Virus Associated with Okra Leaf Curl Disease in Cote d'Ivoire. *New Dis. Rep.* 2015; 2044–0588.

Seka, K.; Ouattara, A.; Assiri, K.; Kra, K.; Hoareau, M.; Lefeuvre, P.; Atta Diallo, H.; Lett, J.-M. First Report of Pepper Yellow Vein Mali Virus Associated with Pepper Yellow Vein Disease in Cote d'Ivoire, 2017.

Sharma, P. N. (Lecture notes). **Pl. Path. 502 (Cr. Hrs. 2+1)** Department of Plant Pathology, CSK HPKV, Palampur (H.P.) Pl Path 502 Plant Virus disease Management. http://www.hillagric.ac.in › *edu* › *coa* › *ppath* › *lect.*

Shetty, A. A.; Singh, J.;Singh, D. Resistance to Yellow Vein Mosaic Virus in Okra: A Review. *Biol. Agric. Hortic.* **2013,** *29,* 159–164.

Shih, S. L.; Green, S. K.; Tsai, W. S.; Lee, L. M.; Levasseur, V. First Report of a Distinct Begomovirus Associated with Okra Yellow Crinkle Disease in Mali. *Plant Pathol.* **2007,** *56,* 718.

Skelton, A.; Uzayisenga, B.; Fowkes, A.; Adams, I.; Buxton-Kirk, A.; Harju, V.; Forde, S.; Ward, R.; Ritchie, B.; Rutherford, M.; Offord, L.; Umulisa, C.; Waweru, B.; Kagiraneza, B.; Karangwa, P.; Fox, A. First Report of Pepper Veinal Mottle Virus, Pepper Yellows Virus and a Novel Enamovirus in Chilli Pepper (*Capsicum* sp.) in Rwanda. *New Dis. Rep.* **2018,** *37,* 5.

Spanò, R.; Ferrara, M.; Gallitelli, D.; Mascia, T. The Role of Grafting in the Resistance of Tomato to Viruses. *Plants* **2020,** *9* (8), 1042.

Stevens, M.; Lamb, E.; Rhoads, D. Mapping the Sw-5 Locus for Tomato Spotted wilt Virus Resistance in Tomatoes Using RAPD and RFLP Analyses. *Theor. Appl. Genet. 90,* 451–456.

Suzuki, K.; Kuroda, T.; Miura, Y.; Murai, J. Screening and Field Trials of Virus Resistant Sources in Capsicum spp. *Plant Dis.* **2003,** *87,* 779–783.

Tahir, M.; Haider, M. S. First Report of Tomato Leaf Curl New Delhi Virus Infecting Bitter Gourd in Pakistan. *Plant Pathol.* **2005,** *54,* 807.

Thottappilly, G. Plant Virus Diseases of Importance to African Agriculture. *J. Phytopathol.* **1992,** *134,* 265–268.

Tiendrébéogo, F.; Lefeuvre, P.; Hoareau, M.; Traoré, V.S. E.; Barro, N.; Péréfarres, F.; Reynaud, B.; Traoré, A. S.; Konaté, G.; Lett, J.-M. Molecular and Biological Characterization of Pepper Yellow Vein Mali Virus (PepYVMV) Isolates Associated with Pepper Yellow Vein Disease in Burkina Faso. *Arch. virol.* **2011,** *156,* 483–487.

Tomitaka, Y. Studies on the Interaction Between Tomato Spotted Wilt Tospovirus and Thrips. *J. Gen. Plant Pathol.* **2019,** *85,* 465–467.

Tomlinson, J. A. Epidemiology and Control of Virus Diseases of Vegetables. *Ann. Appl. Biol.* **1987**.

Tripathi, L.; Babirye, A.; Roderick, H.; Tripathi, J. N.; Changa, C.; et al. Field Resistance of Transgenic Plantain to Nematodes has Potential for Future African Food Security. *Sci. Rep.* **2015,** *5,* 8127.

Tsai, W.-S.; Abdourhamane, I. K.; Kenyon, L. First Report of Pepper Veinal Mottle Virus Associated with Mosaic and Mottle Diseases of Tomato and Pepper in Mali. *Plant Dis.* **2010,** *94* (3):378–378. doi: 10.1094/PDIS-94-3-0378B

Tsedaley, B. A Review Paper on Potato Virus Y (PVY) Biology, Economic Importance and Its Managements. *J. Biol. Agric. Healthc.* **2015,** *5* (9), 110–126.

Turina, M.; Tavella, L.; Ciuffo, M. Tospoviruses in the Meditterranean Area. In *Viruses and Virus Diseases of the Vegetables in the Mediterranean Basin*; 1st ed.; Elsevier Inc., 2012, Vol. 84. doi:10.1016/B978-0-12-394314-9.00012-9

United Nation. Worldometer (www.worldometers.info). *Elaboration of Data by United Nations*; Department of Economic and Social Affairs, Population Division. World Population Prospects: Live Reporting of Western Africa population, 2021.

Varma, A.; Malathi, V. G. Emerging Geminivirus Problems: A Serious threat to Crop Production. *Ann. Appl. Biol.* **2003,** *142*, 145–164.

Velásquez, A. C.; Castroverde, C. D. M.; He, S. Y. Plant and Pathogen Warfare under Changing Climate Conditions. *Curr. Biol.* **2018,** *28* (10).

Verlaan, M. G.; Hutton, S. F.; Ibrahem, R. M.; Kormelink, R.; Visser, R. G.; Scott, J. W.; Edwards, J. D.; Bai, Y. The Tomato Yellow Leaf Curl Virus Resistance Genes Ty-1 and Ty-3 are Allelic and Code for DFDGD-class RNA–dependent 2013.

Villamor, D. E. V.; Ho, T.; Al Rwahnih, M.; Martin, R. R.; Tzanetakis, I. E. High Throughput Sequencing for Plant Virus Detection and Discovery. *Phytopathology* **2019,** *109,* 716–725. doi: 10.1094/PHYTO-07-18-0257-RVW

RNA polymerases. *PLoS Genet. 9,* e1003399.

Watanabe, K. N.; Orrillo, M.; Golmirzaie, A. M. Potato Germplasm Enhancement for Resistance to Biotic Stresses at CIP. Conventional and Biotechnology-assisted Approaches Using a Wide Range of Solanum Species. *Euphytica* **1995,** *85,* 457–464.

Waweru, B. W.; Kilalo, D. C.; Miano, D. W.; Kimenju, J. W.; Rukundo, P. Diversity and Economic Importance of Viral Diseases of Pepper (*Capsicum* spp.) in Eastern Africa. *J. Appl. Hortic.* **2019,** *21* (1): 70–76

Wongnaa, C. A.; Akuriba, M. A.; Ebenezer, A.; Danquah, K. S.; Ofosu, D. A. Profitability and Constraints to Urban Exotic Vegetable Production Systems in the Kumasi Metropolis of

Ghana: A Recipe for Job Creation. *J. Glob. Entrep. Res.* **2019,** *9* (33), 1–19. doi: 10.1186/s40497-019-0158-2

Yadav, Y.; Maurya, P. K.; Devi, A. P.; Jamir, I.; Bhattacharjee, T.; Banerjee, S.; Dutta, S.; Debnath, D.; Mandal, A. K.; Dutta, S. Enation Leaf Curl Virus (ELCV): A Real Threat in Major Okra Production Belts of India: A Review. *J. Pharmacogn. Phytochem.* **2018,** *7,* 3795–3802.

Yadav, Y.; Maurya, P. K.; Devi, A. P.; Jamir, I.; Bhattacharjee, T.; Banerjee, S.; Dutta, S.; Debnath D.; Mandal, A. K.; Dutta, S.; Chattopadhyay, A. Enation Leaf Curl Virus (ELCV): A Real Threat in Major Okra Production Belts of India: A Review. *J. Pharmacogn. Phytochem.* **2018,** *7* (2): 3795–3802.

Yan, Z.; Wolters, A.-M.A.; Navas-Castillo, J.; Bai, Y. The Global Dimension of Tomato Yellow Leaf Curl Disease: Current Status and Breeding Perspectives. *Microorganisms* **2021,** *9,* 740.

Yeam, I.; Kang, B.-C.; Lindeman, W.; Frantz, J. D.; Faber, N.; Jahn, M. M. Allele-specific CAPS Markers Based on Point Mutations in Resistance Alleles at the pvr1 Locus Encoding eIF4E in Capsicum. *Theor. Appl. Genet.* **2005,** *112,* 178–186.

Young, N.; Tanksley, S. RFLP Analysis of the Size of Chromosomal Segments Retained Around the Tm-2 Locus of Tomato During Backcross Breeding. *Theor. Appl. Genet.* **1989,** *77,* 353–359.

Zhang, X.-S.; Holt, J. Mathematical Models of Cross Protection in the Epidemiology of Plant-Virus Diseases. *Phytopathology* **2001,** *91,* 924–934.

Zitter, T. A.; Provvidenti, R. Vegetable Crops: Virus Diseases of Leafy Vegetables and Celery. Vegetable MD Online, Coop. Ext. Fact Sheet 737.00. Department of Plant Pathology, Cornell University: Ithaca, NY. 1984, http://vegetablemdonline ppath.cornell.edu/factsheets/Viruses_LeafyVege.htm

Zitter, T. A.; Murphy, J. F. Cucumber Mosaic. *The Plant Health Instructor*; 2009. doi: 10.1094/PHI-I-2009-0518-01

Zohoungbogbo, H. P.; Achigan-Dako, E. G.; Honfoga, J.; Lin, S. W.; Lin, T. H.; Wang, Y. W., ... Barchenger, D. W. Incidence and Severity of Aphid-transmitted Viruses and Horticultural Performance of Habanero Pepper (*Capsicum chinense Jacq.*) Breeding Lines in Benin. *HortScience* **2022,** *57* (5), 606–612.

CHAPTER 16

Biology, Epidemiology, and Management of Charcoal Rot of Soybean Caused by *Macrophomina phaseolina*

NAWALE RUTUJA YADAORAO[1], MEENAKSHI RANA[1], SHWETA MESHRAM[1], A.K. TIWARI[2], RAGHAVENDRA REDDY MANDA[3], ADARSH SHARMA[1], VINIT PRATAP SINGH[4] and SEWETA SRIVASTAVA[1]

[1]*School of Agriculture, Lovely Professional University, Punjab, India*

[2]*Sugarcane Research and Seed Multiplication Center, Gola, Uttar Pradesh, India*

[3]*Wageningen University and Research, Wageningen, the Netherlands*

[4]*College of Agriculture, Azamgarh Campus, Acharya Narendra Dev University of Agriculture & Technology, Faizabad, Uttar Pradesh, India*

ABSTRACT

Macrophomina phaseolina is a significant pathogen of Soybean crop. Disease development in any crop requires a susceptible host, a virulent infection, and a suitable environment. *M. phaseolina* is a diverse species and infects over 500 different plant species by causing stem rot, root rot, seedling blight, and charcoal rot. Disease control is difficult due to *M. phaseolina's* broad host range and high existence in soil as microsclerotia. As a saprophyte, it can live in the soil for up to 15 years. *M. phseolina* is a soil-borne pathogen that affects crop quality and quantity. This pathogen is more dangerous in areas where the climate is relatively dry and warm during the growing season; that is why it is also known as drought stress pathogen. During the rainy season,

Plant Diseases and Their Management: A Sustainable Approach.
Deepa Srivastava, Rajarshi Kumar Gaur, and Ajay K. Tiwari (Eds.)

M. phaseolina mycellium or sclerotia infest the soil in field plots in India. Soil with low C: N ratio, high bulk density, and high moisture content are the important factors for the survival of sclerotia. Cultural practices, seed treatment, soil amendments, and host resistance can aid in the management of this disease. Significant yield losses have resulted from this pathogen. Understanding the pathogen's pathogenicity mechanisms and interactions with host crop is therefore critical for managing the disease.

16.1 INTRODUCTION

In the 2020/2021 crop year, soybeans were the leading type of oilseed in the world. That year, some 362 million metric tons of soybeans were produced worldwide (Statista, 2021). Soybean is known as the "Golden Bean," "Miracle Bean," or "Wonder Crop" (Deo et al., 2020). *Macrophomina phaseolina* causes soybean charcoal rot having sclerotial stage—*Rhizoctonia bataticola* also called as *Sclerotium bataticola*, while *Macrophomina phaseolina* is the imperfect pycnidial stage of the same (Young, 1944; Arya et al., 2004). Stem rot, root rotting, charcoal rot, and blight of seedling are the important diseases caused due to *Macrophomina* (Dhingra and Sinclair, 1978; Ghosh et al., 2018). The fungus has a diverse host range as well as a large geographic distribution (Grau et al., 2004). In both conventional and no-till farming systems, Almeida et al. (2001) investigated the prevalence of *M. phaseolina*-caused charcoal rot on soybeans in relation to precipitation. Ghosh et al. (2018) investigated the genetic diversity of *M. phaseolina* species worldwide. The fungus is found all over the world. *Macrophomina phaseolina* causes charcoal rot (Goidanish, 1947; Luna et al., 2017). Uppal (1931) discovered *Macrophomina phaseolina* on Jowar for the first time in India. Veena et al. (2014) identified the pathogen using mycelial and sclerotial characteristics, and pathogenecity was demonstrated using the soil inoculation method. *M. phaseolina* has been identified through morphological and genetic analysis. This pathogen is responsible for root rot in the soybean crop (Dhingra and Sinclair, 1977; Dhingra and Sinclair, 1978; Crous et al., 2006). Every year, 30–50% of the soybean crop is lost (Sharifi et al., 2005; Ramezani et al., 2007; Senthilkumar et al., 2009) Soybean yield losses of 70% in India are due to charcoal rot (Wyllie, 1993). It is most common in Madhya Pradesh, Maharashtra, Delhi, and Rajasthan in India (Gupta and Chauhan, 2005). *M. phaseolina* infected 2.2–15.7% of the seeds, resulting in a 10.8% reduction in grain yield and a 12.3% reduction in seed protein content (Kaushik et al. 1987). When temperatures are high (30–35°C) and soil moisture is low (less than 60%), it creates significant yield losses in soybean crops (Kaur et al. 2012). All the plant parts get infected by these pathogens. The seed coat was the primary site of *macrophomina phaseolina* infection (Sharda and Shetty, 1987). According to Pearson et al. (1984), disease severity rises with increase in air and soil temperatures from 28 to 35°C and lower soil moisture. *Macrophomina phaseolina* is a fungus that causes the diseases charcoal rot, root rot, dry weather wilt, ashy stem blight, and seedling blight in more than 500 plant and weed species, including soybean, sorghum, maize, cotton, peanut, sesame, cowpea, chickpea, and clusterbean (Su et al., 2001; Dhingra and Sinclair, 1977; Diourte et al., 1995). *M. phaseolina* can live in soil for 2–15 years (Meyer et al., 1974). It is a rot-inhibiting fungus that produces black sclerotia (Smith, 1969). These sclerotia are the mainstay of perennation (Smith, 1969; Mirza, 1984; Kaisar et al., 1988). Disease can be controlled through cultural, biological, and chemical means (Bristow and Wyllie, 1975; Gupta, 2004). The most effective strategy for controlling charcoal rot in soybeans is fungicide treatment seeds (Hewidy et al., 2003). Alternative control methods used include biological control, plant extracts and soil solarization (El-Barougy et al., 2009, Dubey et al., 2009), chemical inducers (El-Baz, 2007), resistant cultivars, planting dates, and crop rotation (Franci et al., 1988; Todd, 1993; Mengistu et al., 2007).

16.2 TAXONOMY AND NOMENCLATURE

Macrophomina phaseolina is a fungus that lives in the soil and causes charcoal rot of soybean. The genus *Macrophomina* (subdivision Deuteromycota, form-class Coelomycetes) contains only one species (Mihail, 1992). In both host tissues and culture medium, the pathogen produces two anamorphic forms, i.e., microsclerotia and pycnidiospores. Mihail (1992) found an unverified record of teleomorphic condition of *M. phaseolina* called *Orbilia obscura* reporded by Ghosh and coworkers in 1964, but the record of teleomorph states has not been made since then. *M. phaseolina* has been given several different names in the past due to the pathogen's biological properties and the difficulty of taxonomic categorization (Holliday et al., 1970).

Halsted described *M. phaseolina* in 1890, after infecting *Ipomoea batatas* plant and detecting sclerotia on the roots. *Rhizoctonia bataticola* was the name given to *M. phaseolina* at the time. Tassi (1901) named the fungus *Macrophoma phaseolina* after describing the pycnidial stage. *Macrophomina phaseoli* was renamed *Macrophomina phaseolina* by Goidanich

(1947). *M. phaseolina's* sclerotial condition was previously referred to as *Rhizoctonia* and *Sclerotium,* whereas the pycnidial stage was previously referred to as *Macrophomina, Dothiorella* and *Macrophoma. Macrophomina phaseolina* is the current binomial for the pycnidia and sclerotial anamorph phases (Mihail, 1992). Despite the fact that the teleomorph in this pathogen is unknown (Kaur et al., 2012), *M. phaseolina* is a member of the Botryosphaeriaceae family (Crous et al., 2006).

16.3 HOST RANGE

M. phaseolina is a multiphagous pathogen (Maia et al., 2004) as mentioned in Table 16.1. The pathogen *M. phaseolina* is economically important in subtropical and tropical countries with semiarid climates (Singh et al., 1990; Wrather et al., 1997). It is most common in Madhya Pradesh, Maharashtra, Rajasthan, and Delhi in India (Gupta and Chauhan, 2005). This pathogen affects over 700 plant species, including soybean cotton, maize, sorghum, cassava, peanut, cowpea, and chickpea (Aris et al., 2011; Csondes, 2011; Gupta et al., 2012). *M. phaseolina* has one of the most diverse host ranges, with over 500 different monocotyledonous and dicotyledonous plant species (Wyllie et al., 1989). It infects most of the cereals, pulses, oilseeds, vegetables, and ornamental crops and produces charcoal rot and ashy stem blight symptoms (Small, 1925). The main cultivated hosts of the fungus are sorghum, groundnut, cabbage, pepper, chickpea, soybean, sunflower (Yang and Owen, 1982; Khan, 2007), sweet potato, alfalfa, potato, sesame, bean and corn (Singh et al., 1990), softwood forest trees (Mc cain and Scharf, 1989), citrus and coffee (fruit trees), species of weed (Songa and Hollicks, 1996), and tobacco.

TABLE 16.1 Distribution of *M. phaseolina* Infection on Different Host Plants.

Crop	Disease	Reference
Alfalfa and white clover	Crown rot	Pratt et al. (1998)
Blueberry	Charcoal rot	de los Santos et al. (2019)
Cactus plants	Ashy stem blight	Abdel-Kader et al. (2010)
Canola	Charcoal rot	Khangura and Aberra, 2009)
Cassava	Stem rot	Msikita et al. (1997)
Cherry	Charcoal rot	Nouri et al. (2020)
Chickpea	Dry root rot	Živanov et al. (2019)
Coleus	Root rot	Kamalakannan et al. (2006)

TABLE 16.1 *(Continued)*

Crop	Disease	Reference
Common bean	Charcoal rot/Ashy stem blight	Mayek-Pérez et al. (2003)
Coral hibiscus	Collar rot	Santhakumari et al. (2002)
Cotton	Ashy stem blight	Abd-Elsalam (2010)
Cowpea	Charcoal rot	Sarr et al. (2014)
Cucumber	Charcoal rot	Egel et al. (2020)
Grape	Trunk and Cordon Canker	Nouri et al. (2018)
	Crown and root rot	Nouri et al. (2020)
Groundnut	Charcoal rot	Martínez-Salgado et al. (2021)
Guava	Wilt	Dwivedi (1990)
Faba bean	Charcoal rot	Sun et al. (2019)
Jatropha curcas	Root rot	Sharma and Kumar (2009)
Maize	Charcoal rot	Saleh et al. (2010)
Mungbean	Charcoal rot	Zhang et al. (2011); Basandrai et al. (2021)
Nattrassia mangiferae	Root rot	Msikita et al. (1997)
Pigeonpea	Stem canker	Swamy et al. (2018)
Pistachio	Charcoal rot	Nouri et al. (2020)
Potato	Charcoal rot/black rot	Takooree et al. (2021)
Safflower	Charcoal rot	Razavi and Pahlavani (2004)
Sesame	Charcoal rot	Ibrahim and Abdel-Azeem (2015)
Soda apple	Leaf and stem blight	Iriarte et al. (2007)
Sorghum	Charcoal rot	Purushotham et al. (2020)
Soybean	Charcoal rot	Young (1949); ElAraby and Kurle (2003)
Strawberry	Crown and root rot	Baino et al. (2011); Gerin et al. (2018); Qamar et al. (2019)
	Charcoal rot	Hajlaoui et al. (2015)
Sugar beet	Charcoal rot	Karadimos et al. (2002)
Sunflower	Charcoal rot	Mahmoud and Budak (2011); Qamar et al. (2018)
Tomato	Charcoal rot	Hyder et al. (2018)
Urdbean	Dry root rot/charcoal rot	Sarr et al. (2014); Basandrai et al. (2021)
Watermelon	Charcoal rot	Masafumi et al. (2002)
White spider lilies	Leaf blight	Huda-Shakirah et al. (2019)

16.4 GENETIC DIVERSITY

Macrophomina phaseolina causes a variety of crop diseases, including seedling blight, root and stem rot, wilt, and pre- to post-emergent damping off, all cause stem height, girth, root and head weight loss, as well as death of affected plants (Raut, 1983). The genetic diversity of *M. phaseolina* isolates has been extensively studied, with the majority of the research relying on molecular markers and cluster analysis. In *M. phaseolina* populations, comparative genomic studies from diverse countries showed random amplified polymorphic DNA (RAPD), restriction fragment length polymorphism (RFLP), amplified fragment length polymorphism (AFLP), and rDNA as the most effective techniques (Fuhlbohm, 1997; Mayék-Perez et al., 2001; Su et al., 2001; Almeida et al., 2003; Jana et al., 2005; Babu et al., 2010; Jimenez, 2011; Khan et al., 2017). Several recent research studies have been concentrated on the genetic and pathogenic variations of *M. phaseolina*. Despite the fact that a single root can be infected by multiple genotypes, Almeida et al. (2003) discovered that population diversity was limited and that the pathogen lacked a sexual cycle in their study of Brazilian populations. Several studies have concentrated on the development of sets of microsatellite markers to investigate population variation in *M. phaesolina* (Baird et al., 2009, 2010; Arias et al., 2011).

Despite the fact that *M. phaseolina* does not reproduce sexually, the research demonstrated that isolates of *M. phaseolina* have a lot of genetic diversity. Parasexualism is most likely caused by the union of cells from different hyphae, and as a result, heterokaryons form, which give rise to the diversity, as observed by Almeida et al. (2003). According to Jana et al. (2005), Babu et al. (2010), variability in genetic level has been linked to the origin of host or geographical distribution, whereas Mahdizadeh et al. (2011), Reznikov et al. (2018) found that data clustering was unable to distinguish isolates according to the morphology, pathogenicity characteristics, host or geographical location. *M. phaseolina* genotype is unaffected by sampling site or crop (Khan et al., 2017; Tanci civanov et al., 2019). Furthermore, genetic diversity among *M. phaseolina* isolates from Brazil revealed that a single root could harbor multiple haplotypes (Almeida et al., 2003). Reznikov et al. (2019) believed that studying genotype interactions between plant cultivars and *M. phaseolina* isolates will aid to develop resistant varieties.

16.5 SYMPTOMATOLOGY

M. phaseolina can be found on seedlings as well as on adult plants of soybean. In late July, Yang and Navi (2005) observed infected plants, and by late

August, patches of infected plants wilted and died prematurely. Seedlings can be infected from the moment they emerge through the early vegetative stages, but symptoms do not usually appear until the R5 (starting seed) to R7 (beginning maturity) growth stages (Fehr et al., 1971; Meyer et al., 1974). The taproots and lower stems of infected plants had brown discoloration in the pith. Symptoms could be seen up to the fourth or fifth node. Early symptoms include yellow leaves, decreased vigor, brown to red staining on roots and stems, and an overall withering appearance (Short et al., 1978; Gupta and Chauhan, 2005). Charcoal rot is a gray or silver discoloration caused by the growth of microsclerotia on the lower regions of the stem and roots (Wrather et al., 2008; Gupta et al., 2005; Gupta et al., 2012). Torn open the stems of symptomatic plants, patches of gray dark discoloration in the stem cortex were discovered (El-Araby et al., 2003). Seedling symptoms such as damping off or lesion formation on cotyledons or hypocotyls are common. Depending on plant development stage symptoms also vary which includes root rot, wilting, interveinal chlorosis, gray discoloration of lower stem tissue, and necrosis of leaves, and premature plant mortality usually appears one to four weeks before maturity. Soybean plants field developed charcoal rot symptoms in patches and died prematurely (Bradely and Rio, 2003). The leaves of infected plants remain smaller than normal and eventually turn brown. Microsclerotia are commonly found in the vascular tissue of the stem and roots. Microsclerotia, which form in vascular tissue and restrict water flow, play an important role in the progression of charcoal rot disease. Black microsclerotia were discovered in the vascular tissues of dead palnts' lower stems and roots, giving them a silvery-gray appearance. The leaves are yellow prior to wilting (Gupta and Chauhan, 2005). *M. phaseolina* can be spread through contaminated seed (Hartman et al., 1999; Gupta et al., 2012). Infected seed can also spread *M. phaseolina*, which is usually undetectable because the fungus is hidden by the seed coat (Mengistu et al., 2015) Plants die as a result of charcoal rot, but the leaves remain attached to the petiole as shown in the figure (Fig. 16.1) (Mengistu et al., 2007).

16.6 MORPHOLOGY AND BIOLOGY OF MACROPHOMINA PHASEOLINA

The morphological polymorphism of fungi has been linked to the fungus respone to environmental changes or changes in their host species isolates of *M. phaseolina* (Tok, 2019; Pandey et al., 2020). Reichert and Hellinger

FIGURE 16.1 (A) Healthy plant of soybean; (B) infected plant of soybean; (C) early symptoms of leaf yellowing due to *M. phaseolina.*

(1947) observed the colonies color was grayish to white, having short mycelium and points in the direction of growth with very little aerial mycelium. Hartman et al. (1999) observed hyphal branching at right angle to the main hyphae under the microscope, but subsequent hyphae incline and grow almost parallel to the main hyphae. Hyphae can be found in a variety of dark colors ranging from pale brown to grayish. Septa ranges in width from 2 to 11 meters, and the cells are at a distance of 46 meters long. However, in terms of taxonomy, microsclerotia production size and composition are the most important characteristics (Reichert and Hellinger, 1947).

Depending on the climatic conditions and the microsclerotia's relationship with the host residue, microsclerotia can live for 2–15 years (Meyer et al., 1974; Short et al., 1980; Baird et al., 2003; Gupta et al., 2012). Microsclerotia germinating formed germ tubes, which they used to colonize vascular tissues through natural apertures (Wyllie and Scott, 1988; Bresano et al., 2010). *M. phaseolina* forms black-colored masses called microsclerotia which comprised 50–200 hyphal cells held together by a melanin-like binding agent (Gangopadhyay and Wyllie, 1974; Short and Wyllie, 1978). The size (60–200 m) and number of microsclerotia vary according to the availability of nutrition in the culture media or a particular host. Furthermore, the count of germ tubes and cells is proportional to the microsclerotia size (Dhingra and Sinclair, 1977) The nutritional content of the culture media or the host influences the size (60–200 m) and number of microsclerotia. Furthermore, the number of cells and germ tubes is proportional to the size of the microsclerotia (Dhingra and Sinclair, 1977). Microsclerotia formation has been observed in isolates from a variety of hosts, with barrel-shaped cell fusion being the most common. It was also possible to see the fusion of developed microsclerotia (Reichert and Hellinger, 1947). In two other methods, Wyllie and Brown (1970) went into greater detail about how microsclerotia form. As the hypha expands and interwaves, increase in mass, giving it a circular shape. The hyphae continue to grow, the inner cells of the microsclerotium enlarge, and microsclerotia grow to a size of 50–100 m. When the microsclerotia form as a result of main hyphal thread fusion, a number of bud-like forms emerge over the surface, and the formation of microsclerotia takes place. During microsclerotia development, coating of mucilaginous matrix, and the interior cells turn black.

Pratt (2006) described the morphology of *M. phaseolina* microsclerotia germ tubes and hyphal growth on diluted cornmeal agar (CMA). After 18–24 hours in this environment, an average of four germ tubes, ranging in length from a few micrometers to several hundred, are produced per microsclerotium. Pratt (2006) also observed that germ tubes could be unbranched and curled. Pycnidia of *M. phaseolina* are produced by cultivating the fungus in fresh plant tissue and changing the light routine with 12 hours of darkness and 12 hours of light. In several bean tissues, the pycnidial stage has been observed, including garden and jute beans, as well as soybeans. Ma et al. (2010) described a culture medium that produces optimal conidia and pycnidia: peanut butter extract-saturated filter paper over soynut butter extract agar (PESEA). The size of pycnidia is greater than microsclerotia, it is brownish to blackish in color, rough, globular, or uneven in shape, with beak and ostiole, and are rarely encountered in nature (Lakhran et al., 2018). Pycnidia are globose to fusiform in shape and appear black or grayish (Fig. 16.2). During the initial stages of development, they are immersed in host tissue, but as they mature, they become erumpent. The small truncate ostiole may be imperceptible or have a clear or hyaline aperture (Mihail, 1992; Hartman et al., 1999). Pycnidiospores have a length-to-width ratio of 3:1, are single-celled, hyaline, and ellipsoid to ovoid in shape (Dhingra and Sinclair, 1978).

16.7 MACROPHOMINA PHASEOLINA SOIL POPULATIONS ARE AFFECTED BY SOIL WATER CONTENT

Environmental factors such as temperature, meteorological humidity, and soil water potential influence *M. phaseolina* viability and inoculum potential (Khan, 2007). *M. phaseolina* can form microsclerotia in low-water environments, so the inoculum's survival is effected by soil matric water potentials.

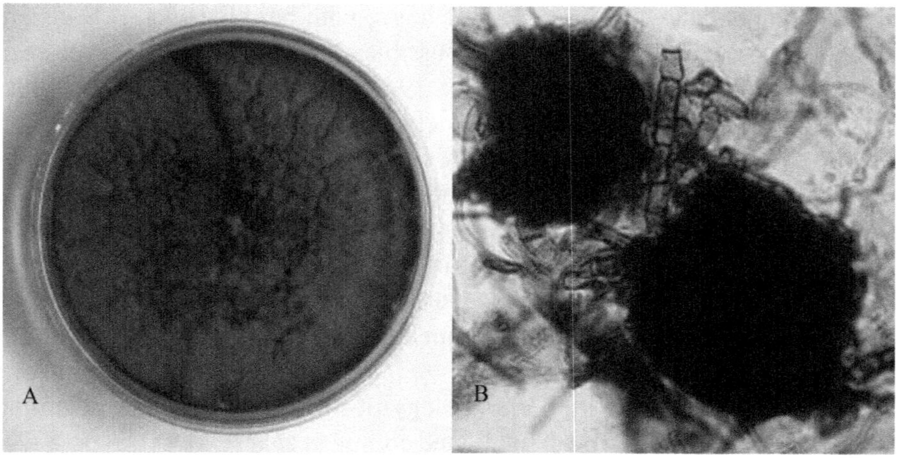

FIGURE 16.2 (A) Colony morphology of *Macrophomina phaseolina* on PDA; (B) microscopic image of pycnidia.

Microsclerotia suvival in a sandy loam soil was significantly less at high water potentials (–30 J/kg, field capacity), but completely unaffected at low water potentials (–1.500 kg/J, permanent wilting threshold) (Olaya et al., 1996).

The amount of water in the soil alters the gaseous conditions, which can reduce microsclerotia survival due to a lack of oxygen. Alcohols, volatiles, and higher CO_2 levels can all be harmful to the inoculum if discovered in flooded soils (Wyllie et al., 1984; Olaya et al., 1996). Germination of microsclerotia was inhibited in artificial environments. Atmospheres with less than 16% oxygen (O_2) content were found in soil column systems, signifying that the loss of viability was not caused by a nutritional deficiency (Wyllie et al., 1984). In anoxic-flooded soils, this information has been utilized to bring down densities of propagule in the soil not only for *M. phaseolina*, but also for *Sclerotinia sclerotiorum* and *Sclerotium cepivorum* (Abawi et al., 1985; Banks and Edgington, 1989).

Goudarzi et al. (2008) described *in vitro M. phaseolina* fungal development with the help of polyethylene glycol (PEG 6000) and sodium chloride to create different matrices and osmotic potentials. Microsclerotia germination and radial growth increase as osmotic and water potentials decrease. However, there is an optimum of –0.6 MPa osmotic potential and –1.2 MPa matric potential, implying that *M. phaseolina's* hypha maintains a positive turgor during growth and that the pathogen's adaptation to survive under water stress.

Soil populations of Microsclerotia were reduced in soils treated with glucose, sucrose, starch, cellulose, and NaNO3 that had a water retention capacity of 60–100%. When soil moisture levels were high, soil bacterial population increased and parasitized *M. phaseolina,* destroying its cell walls. Amino acids, sugars, and organic acids have all been discovered to promote the germination of microsclerotia in soil (Dhingra and Sinclair, 1978).

16.8 SURVIVAL OF *M. PHASEOLINA* ON PLANT DEBRIS, AND SOYBEAN ROOTS IN SOIL

Infected crop residues are a major source of *M. phaseolina* inoculum and inoculum dissemination in the field, as well as a pathway for long-term survival in the soil. Several factors, including soil moisture and temperature, influence plant waste degradation and *M. phaseolina's* relative longevity (Baird et al., 2003). In general, the number of *M. phaseolina* in soybean root debris or soil residue appears to increase over time. The population of *M. phaseolina* at the soil surface grows faster than buried residue; that is why population fluctuation of this pathogen is inversely proportional to the depth of the soil (Short et al., 1980).

The depth of the soil influences tissue deterioration, which has a significant impact on *M. phaseolina* survival. Microorganisms involved in maceration of plant tissue are isolated more frequently when *M. phaseolina*-infected soybean root tissues are buried deeper than surface trash. According to Baird et al. (2003), cellulose degrading fungi *viz., Trichoderma* spp. and basidiomycetes are most vigorous at soil depths where temperature and moisture are favorable for them, which may be part of the reason for reduction in *M. phaseolina* survival on those degrading tissues.

Soil moisture affects the functioning of microsclerotia present in agricultural residue. Microsclerotia embedded in diseased roots can survive during dry and hot weather for longer periods of time. According to Dhingra and Sinclair (1978), microsclerotia recovery from soybean root tissues in soils with less than 40–100% moisture-holding capacity is only 0–12% after four weeks. After eight weeks, the percentage of recovery of microsclerotia from soybean root tissues was 100% in the soil having moisture retention capacity of only 20%. *M. phaseolina* populations tend to increase in soils year after year when soybeans or vulnerable hosts are planted. Cook et al. (1973) discovered a three-year increase in microsclerotia in a soybean field, from 80, 120, and 149 microsclerotia per gramme of soil, respectively.

16.9 DISEASE CYCLE OF MACROPHOMINA PHASEOLINA

M. phaseolina lives in soil in the form of microsclerotia (resting structure) for 15 years or longer duration of time immersed in root debris (Baird et al., 2003; Gupta et al., 2012). Hartman et al. (1999) studied that microsclerotia are brown to black, circular structures that form in plant tissue and are excreted into the soil, allowing the pathogen to thrive even in harsh environments. The globose cells that make up microsclerotium have the ability to germinate on their own.

During the growing season, Microsclerotia germinate in soil at temperatures ranging from 28 to 32°C. Bressano et al. (2010) found that microsclerotia germinate on the root surface or near a susceptible host, resulting in the formation of many germ tubes. Hyphopodia are swelling of mature hyphae that pierce natural opening or epidermal are the most common source of penetration. Infection can occur at any stage of development, including emergence, cotyledon (VC), and early vegetative stages, and can occur up to 2–3 weeks after the seed has been planted (planting (V1–V2). Hartman et al. (1999), Wyllie and Calvert (1969) investigated that the infection does not become active until adverse environmental conditions coincide with plant reproductive stages (R1–R7), allowing the fungus to invade the damaged plant's root and stem tissues. Infected plants lose vitality during the reproductive stage and as they approach maturity, and the stems bear chlorotic and necrotic leaves. When *M. phaseolina* load in the stem and root parts are high enough, a grayish or silvery colored discoloration appears in the epidermal and subepidermal tissues of root and stem (Ammon et al., 1974). Invasive *M. phaseolina* has been linked to poor seed quality and yield reductions. Microsclerotia-laden roots begin to die after harvesting, releasing microsclerotia in the soil (Olaya et al., 1996). Collins et al. (1991) found that the survival of inoculum is affected by moisture in soil, temperature, and the community of microbes.

16.10 HOST-PATHOGEN INTERACTION

Numerous functional genomic methods, comprising transcriptomics and proteomics, have been used to examine the interactions between *M. phaseolina* and many cultivars of diverse host plants to enhance the understanding of the underlying resistance methods. Genes involved in host defense and proteins that are expressed at the time of infection by soil-borne pathogens have been identified, and massive database has been gathered (Table 16.2).

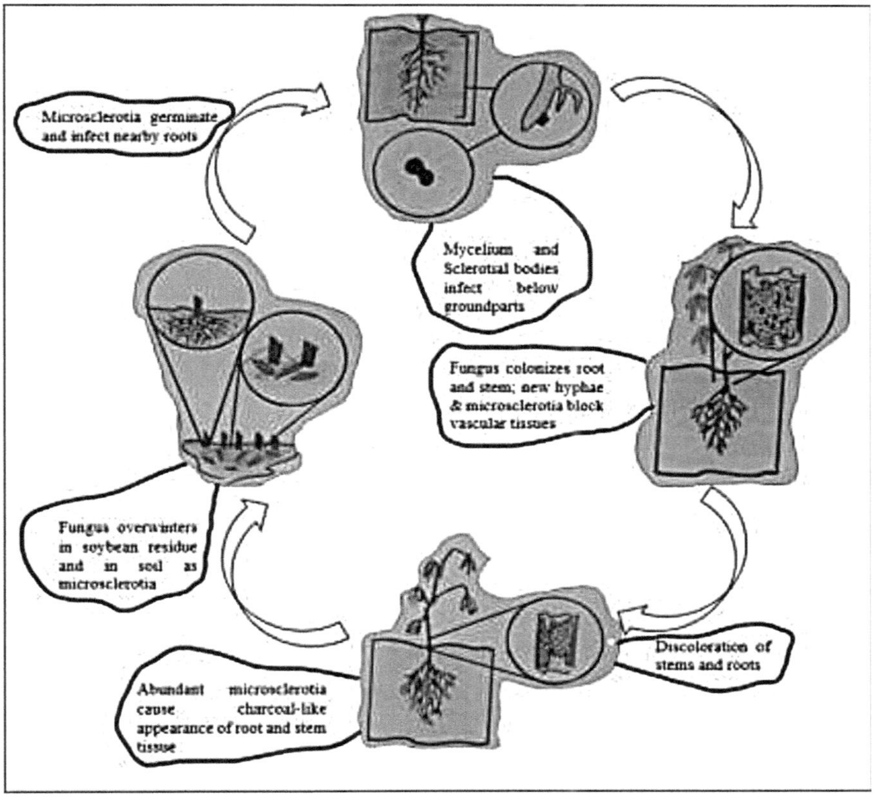

FIGURE 16.3 Disease cycle of charcoal rot of soybean.

Source: Adapted from Luna et al., 2017

Soybean is one of the most important agronomic crops with growing production regions in many parts of the globe. *M. phaseolina*, causing charcoal rot, is a major commercially significant soybean disease (Wrather et al., 2010). As a result, soybean and *M. phaseolina* interaction is likely to be the most researched among crops.

During the early 1980s, the pioneer to look for the resistant varieties of soybean was Pearson et al. (1987). Coser et al. (2017), Mengistu et al. (2018), Reznikov et al. (2018) found despite the fact that the strategy has not been successful yet, numerous researches have now been dedicated toward uncovering additional resistance sources or even a greater grasp of resistance mechanism is needed which will aid future breeding programs. *M. phaseolina* infection is significantly associated with the environmental factors, de Sousa Linhares et al. (2020) proposed that different temperatures be used to better characterize levels of resistance, letting us to select plant cultivars which are best suited to various weather conditions. Mengistu et al. (2018) investigated that *M. phaseolina* severity observed in putative drought-tolerant cultivars, as well as yield loss due to charcoal rot in irrigated as well as nonirrigated conditions. Despite finding just a minor link between the severity of drought tolerance and charcoal rot disease, they observed that in regions where moisture stress in soil and charcoal rot are common, selection for resistance to both the traits is important. The impact of charcoal disease rot on total phenolic compound, lignin, and isoflavone levels in seeds contents in genotypes of soybean with varied levels of resistance to disease and drought circumstances was investigated. Charcoal rot resistance has been linked to more amounts of phenolic chemicals, boron and sugar in seeds, according to the findings (Bellaloui et al., 2012). Furthermore, although there was no evidence of gene or marker overlap in the field as well as greenhouse tests, genome-wide association analyses (GWAS) gave important knowledge for knowing the resistance of genetic underpinnings as well as in the progress of breeding programmes (Coser et al., 2017).

Marquez et al. (2018) observed that *M. phaseolina* infection had a considerable impact on gene expression of soybean, comprising many defence genes of plant associated with signaling compounds, PR proteins, disease-resistant proteins, transcription factors, and secondary metabolism genes, according to the transcriptome profile. Phenylpropanoids such as phytoalexins were found to play a role in plant pathogen interactions which could be poisonous or suppressive to the pathogens. The effect of *M. phaseolina* colonization was measured using transgenic soybean lines with a glyceollin-suppressing gene biosynthesis, glyceollin is collectively known as soybean phytoalexins. The ability of these altered soybeans to manufacture glyceollin was significantly reduced, and they were more susceptible to infection of pathogen, indicates that accumulation of glyceollin plays a critical role in innate soybean defence response (Lygin et al., 2013).

16.11 TOXIN PRODUCTION BY MACROPHOMINA

Siddiqui et al. (1979) observed that the phytotoxin was found in *M. phaseolina* culture filtrates for the first time. Dhar et al. (1982) identified the phytotoxin as an eremophilane sesquiterpenoid, specifically an epoxidized counterpart of phomenone, later. It was made semisynthetically by Kitahara et al. (1991) from (+)-sporogen-AO. Even at low concentrations of 0.60–2.1

TABLE 16.2 Interactions Between Soybean and *Macrophomina phaseolina.*

Study	Tools for study	Results	References
Analysis of susceptible (S)/ moderately resistant (MR) genotypes under irrigated and nonirrigated and under fungal infected and noninfected conditions.	Total phenolics analysis, lignins, and isoflavones in seed.	Under irrigated and nonirrigated conditions, as well as under experimental *M. phaseolina* infested and noninfested conditions, MR genotype seeds had significantly more level of seed coat lignin, isoflavones, sugars, phenolics, and total boron than S genotype seeds, indicates a potential relationship of all these substances resistant to toxin-mediated infection.	Bellaloui (2012)
Genetic architecture of resistance and identification of causal	Genome-wide association studies (GWAS)	For field and greenhouse screening, five and eight loci were found to be related with candidate genes that influence the plant defensive response, respectively. There was no overlapping of genes or markers in the field and greenhouse testings.	Coser et al. (2017)
Defence response under *in vitro* conditions	Transcriptomic profile	Genes induction involved in secondary metabolism, hormone metabolism, signaling, and stress	Marquez et al. (2018)
Transgenic soybean with suppressed synthesis of isoflavones	Molecular and biochemical variability	Increased vulnerability to pathogen infection and reduced root capability to generate glyceollin.	Lygin et al. (2013)

pg/g wet tissue, the toxin decreases seed germination by 50% (Bhattacharya et al., 1994). The amount of toxin has been connected to the percentage of seed germination (Dhar et al., 1982; Bhattacharya et al., 1987).

The exotoxin that is produced by *M. phaseolina* in damaged plant parts was isolated and identified using an enzyme-linked immunosorbent assay (ELISA) (Bhattacharya et al., 1992). Phytopathogenic fungi produce one or more phytotoxins that aid in their penetration, invasion, and colonization of their hosts. This is one of the most common methods for causing plant illness. Mycotoxin biosynthetic genes that regulate mycotoxin production appear as unit in Streptomycetes, same as biosynthetic gene package for antibiotics (Shier et al., 2007). Bhattacharya et al. (1992) investigated that *M. phaseolina* produces a number of phytotoxic chemicals that have been associated to isolate pathogenicity. Isoasperlin, Asperlin, phaseolinic acid, phomenon, phomalactone, and phaseolinone are some of the mycotoxins generated by *M. phaseolina* (Dhar et al., 1982; Mahato et al., 1987).

At a concentration of 25 g/mL, phaseolinone, a non-host specific heat-resistant exotoxin, reduces seed germination in black gramme (*Phaseolus mungo*) (Bhattacharya, 1987). It also causes seedling wilting and necrotic lesions on leaves, which are symptoms similar to those seen in a number of hosts when the pathogen is present (Bilgrami et al., 1979). Toxins, on the other hand, have no known involvement in disease development.

Mycotoxins could suppress or activate the plant response (Berestetskiy, 2008). UV-mutated non-toxigenic, avirulent mutants of *M. phaseolina* and *Aspergillus fumigatus*, a human and animal pathogen, could only infect *Phaseolus mungo* seedlings when phaseolinone was present (Sett et al., 2000). This investigation established the presence of phaseolinone, a substantial phytotoxic compound released by *M. phaseolina* during the start of disease. Phaseolinone toxin appears to have a nonspecific effect on seed immunity. However, it is unknown what role phaseolinone has in the initiation of *M. phaseolina* root infections.

16.12 MANAGEMENT STRATEGIES AGAINST CHARCOAL ROT OF SOYBEAN

Since every management practice entails interruption, modification, or change of at least one of these elements or interaction with least harm to the natural and environment resources, effective charcoal rot control necessitates an integrated approach (Marquez et al., 2021). Farmers with fields with a history of charcoal rot or who live in places where the disease is common should choose cultivars with the best commercial resistance. Charcoal rot damage can be reduced by using cultural management strategies that reduce plant stress throughout reproductive growth stages. Planting at recommended plant populations, employing appropriate fertilization and tillage procedures, suppressing weeds, and irrigating where possible can all help soybeans cope with stress and, as a result, lessen the incidence of charcoal rot. Despite numerous studies aimed at improving disease control tactics for charcoal rot, farmers still have few alternatives for reducing charcoal rot damage (Luna et al., 2017).

TABLE 16.3 A Brief Compilation of Various Management Tactics Used Against Soybean Charcoal Rot.

Management strategies	Experimental condition	Type	References
Genetic resistance	Pot/field experiment	Genotypic analysis, histopathology, QTL mapping	Reznikov et al. (2018); Hemmati et al. (2018); da Silva et al. (2019)
Cultivar selection	*In vivo*	DG3905, Manokin DT97-4290 DT99-16864 DT99-17483 DT98-7553 DT99-17554	Paris et al. (2006); Mengistu et al. (2007); Mengistu et al. (2011); Gillen et al. (2016)
Cultural practices	Field experiment	Crop rotation	Franci et al. (1988)
	Field experiment	Irrigation	Jordaan et al. (2019); Kendig et al. (2000);
	Pot experiment	Fertilization	Spagnoletti et al. (2018); Spagnoletti et al. (2020)
	Field experiment	Soil amendment	Lodha et al. (1997)
	In vivo	Planting at a lower population rate can help in the reduction of charcoal rot and microsclerotia	Gupta et al. (2012)
Fertility and plant nutrients	*In vivo*	Organic amendments such as farmyard manure, neem, and mustard cake can aid in the reduction of charcoal rot disease	Rathore (2000)
Biological control (Fungi)	*In vitro*/pot experiment	*Arbiscular mycorrhiza fungi* (AMF)	Spagnoletti et al. (2017); Marquez et al. (2018); Spagnoletti et al. (2020)
	In vitro	*M.phaseolina* is inhibited by *Trichoderma, Penicillium, Gliocladium, Aspergillus* and *Rhizopus* spp.	Gacitua et al. (2009)
	In vitro	Irrigation combined with *Trichoderma viride* seed treatment at the onset of moisture stress can reduce disease severity by up to 50%	Ansari (2010)
Bacteria	*In vitro*/pot experiment	*Pantoea* sp. *Pseudomonas* sp. *Bacillus* spp.	Vasebi et al. (2013); Simonetti et al. (2015)

TABLE 16.3 *(Continued)*

Management strategies	Experimental condition	Type	References
	In vitro	*Bacillus* sp.	Torres et al. (2016); Hussain and Khan (2020)
	In vitro	In seed treated with *Rhizobacterium* strain FPT721, a plant growth-promoting rhizobacterium, disease incidence was reduced	Senthilkumar et al. (2009); Choudhary (2011)
Plant metabolites	Pot experiment	*Rosmarinus* sp.	Lorenzetti et al. (2018)
	In vitro	*Prosopis* sp. *Anacardium* sp.	Elaigwu et al. (2018)
	In vitro	*Nigella* sp.	Aftab et al. (2019)
	In vitro	*Mentha* sp. *Eucalyptus* spp. *Copaifera* sp. *Lippia* sp.	Ugulino et al. (2018)
Elicitors of plant defense	Pot/field experiment	Benzothiadiazole, chitosan, phenylalanine, salicylic acid	Pawlowski et al. (2016)
Fungicide	*In vivo*/*in vitro*/pot experiment	*Macrophomina phaseolina* is resistant to carbendazim, benomyl, and thiram	Sinha and Khare (1977)
	In vitro	Mancozeb and bavistin inhibited *M. phaseolina* radial growth by 86.3% and 86.4 %, respectively	Dubey (2003)
	In vitro	Seed treatments with carbendazim 50 WP (2.0 g kg seed) and thiophanate-methyl (1.0 g kg seed) were successful in eradicating the pathogen from infected soybean	Kumar and Singh (2000)
	In vitro	In order of decreasing efficacy, benomyl, thiophanatemethyl, thiram, thiobendazole, triforine, and captan reduced the viability of microsclerotia in soil and soybean stem pieces	Ilyas et al. (1976)
	In vivo	Fungicide seed treatments delay infection and reduce overall colonization of other soybean fungal diseases	Bradley (2008); Schulz and Thelan (2008); Ellis et al. (2011)

16.13 CONCLUSION

The focus of this review is on the known yield-limiting diseases affecting Indian soybean production, i.e. *Macrophomina phaseolina*, causing charcoal rot. Charcoal rot management is critical because the importance of their soil-borne fungal pathogen is found not only in the losses it cause, but also in terms of the environmental consequences of nonsustainable management practices. In the coming years, research on charcoal rot management will advance due to advances in understanding the role of resistance genes in cultivar resistance expression, molecular interactions between the host and pathogen, and a better understanding of the role of the environment in disease development. This knowledge will help us screen for resistant genotypes in maturity groups that are appropriate for each location of the United States. We can reduce the impact of charcoal rot on soybean output by improving our understanding of the disease.

KEYWORDS

- **charcoal rot**
- **host-pathogen interaction**
- **macrophomina phaseolina**
- **management**
- **soybean**

REFERENCES

Abawi, G. S.; Grogan, R. G.; Duniway, J. M. Effect of Water Potential on Survival of Sclerotia of *Sclerotinia Minor* in 2 California Soils. *Phytopathology* **1985,** *75,* 217–221.

Abdel-Kader, M. M.; El-Mougy, N. S.; Aly, M. D. E. H.; Lashin, S. M. First Report of Ashy Stem Blight Caused by *Macrophomina phaseolina* on *Aeonium canariense* in Egypt. *J. Plant Pathol. Microbiol.* **2010,** *1,* 101.

Abd-Elsalam, K. A. Genetical and Biological Control of Cotton Ashy Stem Caused by *Macrophomina phaseolina* in Outdoor Pot Experiment. *Saudi. J. Biol. Sci.* **2010,** *17* (2), 147–152.

Aftab, A.; Yousaf, Z.; Javaid, A.; Riaz, N.; Younas, A.; Rashid, M.; Shamsheer, H. B.; Chahel, A. A. Antifungal Activity of Vegetative Methanolic Extracts of Nigella Sativa Against

Fusarium oxysporum and Macrophomina Phaseolina and its Phytochemical Profiling by GC-MS Analysis. *Intl. J. Agric. Biol.* **2019,** *21,* 569–576.

Almeida, Á. M. R.; Abdelnoor, R. V.; Arrabal Arias, C. A.; Carvalho, V. P.; Jaccoud Filho, D. S.; Marin, S. R. R.; Benato, L. C.; Pinto, M. C.; Carvalho, C. G. P. Genotypic Diversity Among Brazilian Isolates of *Macrophomina phaseolina* Revealed by RAPD. *Fitopatol. Bras* **2003,** *28,* 279–285.

Almeida, A. M. R.; Torres, E.; Farias, J. R. B.; Benato, L. C.; Pinto, M. C.; Martin, S. S. R. *Macrophomina phaseolinno,* in Soybean: Effect of Tillage System, Survival on Crop Residues and Genetic Diversity. *Londrina PR Embrapa Soja Circular Tecnica no,* **2001,** *34,* 47.

Ammon, V.; Wyllie, T. D.; Brown, M. F. Ultrastructural Investigation of Pathological Alterations Induced by *Macrophomina phaseolina* (Tassi) Goid in Seedlings of Soybean, *Glycine max* (L) Merrill. *Physiol. Plant Pathol.* **1974,** *4,* 1–10.

Ansari, M. M. Integrated Management of Charcoal Rot of Soybean Caused by *Macrophomina phaseolina* (Tassi) Goid. *Soybean Res.* **2010,** *8,* 39–47.

Arias, R. S.; Ray, J. D.; Mengistu, A.; Scheffler, B. E. Discriminating Microsatellites from *Macrophomina phaseolina* and Their Potential Association to Biological Functions. *Pl Pathol.* **2011,** *60,* 709–718.

Arya, V. K.; Vishunavat, K.; Negi, H. Detection, Location and Transmission of Seed Borne Inoculum of *Macrophomina phaseolina* in Soybean. *J. Mycol. Pl Pathol.* **2004,** *34* (2), 233–235.

Babu, B. K.; Reddy, S. S.; Yadav, M. K.; Sukumar, M.; Mishra, V.; Saxena, A. K.; Arora, D. K. Genetic Diversity of *Macrophomina phaseolina* Isolates From Certain Agro-climatic Regions of India by Using RAPD Markers. *Indian J. Microbiol.* **2010,** *50,* 199–204.

Baino, O.; Salazar, S. M.; Ramallo, A. C.; Kirschbaum, D. S. First Report of *Macrophomina phaseolina* Causing Strawberry Crown and Root Rot in Northwestern Argentina. *Plant Dis.* **2011,** *95* (11), 1477.

Baird, R. E.; Wadl, P. A.; Allen, T.; McNeill, D.; Wang, X.; Moulton, J. K.; Rinehart, T. A.; Abbas, H. K.; Shier, T.; Trigiano, R. N. Variability of United States isolates of *Macrophomina phaseolina* Based on Simple Sequence Repeats and Cross Genus Transferability to Related Genera Within Botryosphaeriaceae. *Mycopathologia* **2010,** *170,* 169–80.

Baird, R. E.; Wadl, P. A.; Wang, X.; Johnson, D. H.; Rinehart, T. A.; Abba, H. K.; Shier, T.; Trigiano, R. N. Microsatellites From the Charcoal Rot Fungus *Macrophomina phaseolina. Mol. Ecol. Reso* **2009,** *9,* 946–948.

Baird, R. E.; Watson, C. E.; Scruggs, M. Relative Longevity of *Macrophomina phaseolina* and Associated Mycobiota on Residual Soybean Roots in Soil. *Plant Dis.* **2003,** *87,* 563–566.

Banks, E.; Edgington, L. V. Effect of Integrated Control Practices on the Onion White rot Pathogen in Organic Soil. *Can. J. Plant Pathol.-Rev. Can. Phytopathol.* **1989,** *11,* 268–272.

Basandrai, A. K.; Pandey, A. K.; Somta, P.; Basandrai, D. *Macrophomina phaseolina*–host Interface: Insights Into an Emerging dry Root Rot Pathogen of Mungbean and Urdbean, and its Mitigation Strategies. *Plant Pathol.* **2021,** *70,* 1263–1275.

Bellaloui, N. Soybean Seed Phenol, Lignin, and Isoflavones and Sugars Composition Altered by Foliar Boron Application in Soybean Under Water Stress. *Food Nutr. Sci.* **2012,** *3,* 579–590.

Berestetskiy, A. O. A Review of Fungal Phytotoxins: From Basic Studies to Practical Use. *Appl. Biochem. Microbiol.* **2008,** *44* (5), 453–465.

Bhattacharya, D.; Dhar, T. K.; Siddiqui, K. A. I.; Ali, E. Inhibition of Seed Germination by *Macrophomina phaseolina* is Related to Phaseolinone Production. *J. Appl. Bacteriol.* **1994,** *77* (2), 129–133.

Bhattacharya, D.; Siddiqui, K. A. I.; Ali, E. Phytotoxic Metabolites of *Macrophomina phaseolina*. *Ind. J. Mycol. Plant Pathol.* **1992,** *22,* 54–57.

Bhattacharya, G.; Dhar, T. K.; Bhattacharyya, F. K.; Siddiqui, K. A. Mutagenic Action of Phaseolinone, a Mycotoxin Isolated From *Macrophomina phaseolina*. *Aust. J. Biol. Sci.* **1987,** *40,* 349–353.

Bilgrami, K. S.; Jamaluddin, Rizvi, M. A. *The Fungi of India. Part I. New Delhi, India*; Today and Tomorrow's Printers and Publishers, New Delhi, 1979; pp 467.

Bradely, C. A.; Rio, L. E. First Report of Charcoal Rot on Soybean Caused by *Macrophomina phaseolina* in North Dakota. *Plant Dis.* **2003,** *87* (5), 601.

Bradley, C. A. Effect of Fungicide Seed Treatments on Stand Establishment, Seedling Disease, and Yield of Soybean in North Dakota. *Plant Dis.* **2008,** *92,* 120–125.

Bressano, M.; Giachero, M. L.; Luna, C. M.; Ducasse, D. A. An in Vitro Method for Examining Infection of Soyabean Roots by *Macrophomina phaseolina*. *Physiol. Mol. Plant Pathol.* **2010,** *74,* 201–204.

Bristow, P. R.; Wyllie, T. D. The Effect of Crop Rotation and Date of Planting on Charcoal Rot of Soybean (Abstr). *Proc. Am. Phytopathol. Soc.* **1975,** *2,* 83.

Choudhary, D. K. Plant Growth-promotion (PGP) Activities and Molecular Characterization of Rhizobacterial Strains Isolated From Soybean (*Glycine max* L. Merr) Plants Against Charcoal Rot Pathogen, *Macrophomina Phaseolina*. *Biotechnol. Lett.* **2011,** *33,* 2287–2295.

Collins, D. J.; Wyllie, T. D.; Anderson, S. H. Biological Activity of *Macrophomina phaseolina* in Soil. *Soil Biol. Biochem.* **1991,** *23,* 495–496.

Cook, G. E.; Boosalis, M. G.; Dunkle, L. D.; Odvody, G. N. Survival of *Macrophomina Phaseoli* in Corn and Sorgum Stalk Residue. *Plant Dis. Rptr.* **1973,** *57,* 873–875.

Coser, S. M.; Chowda Reddy, R. V.; Zhang, J.; Mueller, D. S.; Mengistu, A.; Wise, K. A.; Allen, T. W.; Singh, A.; Singh, A. K. Genetic Architecture of Charcoal Rot (*Macrophomina phaseolina*) Resistance in Soybean Revealed Using a Diverse Panel. *Front. Plant Sci.* **2017,** *8,* 1626.

Crous, P. W.; Slippers, B.; Wingfield, M. J.; Rheeder, J.; Marasas, W. F. O.; Philips, A. J. L.; Alves, A.; Burgess, T.; Barber, P.; Groenewald, J. Phylogenetic Lineages in the Botryosphaeriaceae. *Stud. Mycol.* **2006,** *55,* 235–253.

Csondes, I. Mycelial Compatibility of Hungarian *Macrophomina phaseolina* Isolates. *Acta Agronomica Hungarica* **2011,** *59* (4), 371–377.

da Silva, M. P.; Klepadlo, M.; Gbur, E. E.; Pereira, A.; Mason, R. E.; Rupe, J. C.; Bluhm, B. H.; Wood, L.; Mozzoni, L. A.; Chen, P. QTL Mapping of Charcoal rot Resistance in PI 567562A Soybean Accession. *Crop. Sci.* **2019,** *59,* 474–479.

de los Santos, B.; Aguado, A.; Borrero, C.; Viejobueno, J.; Avilés, M. First Report of Charcoal Rot, Caused by *Macrophomina phaseolina*, on Blueberry in Southwestern Spain. *Plant Dis.* **2019,** *103,* 2677.

Deo, S.; James, A.; Singh, S. K.; Singh, C. B.; Rana, M. K. Molecular Diversity Analysis and Cultivar Identification Using Simple Sequence Repeat (SSR) Markers in Soybean [*Glycine max* (L.) Merrill]. *Int. J. Curr. Microbiol. App. Sci.* **2020,** *9* (4), 449–465.

Dhar, T. K.; Siddiqui, K. A. I.; Ali, E. Structure of Phaseolinone, a Novel Phytotoxin From *Macrophomina phaseolina*. *Tetrahedron Lett.* **1982,** *23* (5), 5459–5462.

Dhingra, O. D.; Sinclair, J. B. *An Annotated Bibliography of M. phaseolina 1905–1975*; Universididae Federal de Vicosa: Minas Gerais, Brazil, 1977; pp 166.

Dhingra, O. D.; Sinclair, J. B. *Biology and Pathology of Macrophomina phaseolina*. Universidade Federal De Vicosa, Brazil, 1978; pp 166.

Diourte, M.; Starr, J. L.; Stack, J. P.; Rosenow, D. T. Charcoal Rot (*Macrophomina phaseolina)* Resistance and the Effects of Stress on Disease Development in Sorghum. *Plant Pathol.* **1995,** *44,* 196–202.

Dubey, R. C.; Harish, K.; Pandey, R. R. Combined Effect of Soil Solarization and Neem Amendment of Survival of *Macrophomina phaseolina* Sclerotiaa and Growth of Soybean. *Nat. Sci.* **2009,** *7* (11), 52–57.

Dubey, S. C. Integrated Management of Web Blight of Urd/Mung Bean by Bio-seed Treatment. *Indian Phytopath* **2003,** *56,* 34–38.

Dwivedi, S. K. Guava Wilt Incited by *Macrophomina phaseolina*. *Natl. Acad. Sci. Lett.* **1990,** *13* (8), 301–303.

Egel, D. S.; Guan, W.; Creswell, T.; Bonkowski, J. First Report of *Macrophomina phaseolina* Causing Charcoal Rot of Cucumber in Indiana. *Plant Dis.* **2020,** *104* (7), 2030.

Elaigwu, M.; Oluma, H. O. A.; Onekutu, A. Phytochemical and Antifungal Activity of Leaf Extracts of *Prosopis africana* and *Anacardium occidentale* Against *Macrophomina* Root Rot of *Sesamum indicum* L. in Benue State, Central Nigeria. *J. Geosci. Environ. Prot.* **2018,** *6,* 66–76.

El-Araby, M. E.; Kurle, J. E.; Stetina, S. R. First Report of Charcoal Rot *(Macrophomina phaseolina*) on Soybean in Minnesota. *Plant Dis.* **2003,** *87* (2), 202.

El-Barougy, E.; Awad, N. M.; Turky, A. S.; Hamed, H. A. Antagonistic Activity of Selected Strains of Rhizobacteria Against *Macrophomina phaseolina* of Soybean Plants. *Am. Eurasian J. Agric. Environ. Sci.* **2009,** *5* (3), 337–347.

El-Baz Sahar, M. Inductioon of Resistance in Some Soybean Varieties Against Root rot Diseases by Some Chemical Inducers. *Egypt. J. Appl. Sci.* **2007,** *22* (1), 68–80.

Fehr, W. R.; Caviness, C. E.; Burmood, D. T.; Pennington, J. S. Stage of Development Descriptions for Soybeans, *Glycine max* (L.) Merr. *Crop. Sci.* **1971,** *11,* 929–931.

Franci, L. J.; Wylie, T. D.; Rosenbrock, S. M. Influence of Crop Rotation on Populatio Density of *Macrophomina phaseolina* in Soil Infested with *Heterodera gycines*. *Plant Dis.* **1988,** *72,* 760–764.

Fuhlbohm, M. In *Genotypic Diversity Among Australian Isolates of Macrophomina phaseolina*, XX Biennial Australian Plant Pathology Society Conference, Lincoln University, New Zealand, 1997; 52.

Gacitua, S.; Valiente, C.; Diaz, K.; Hernandez, J.; Uribe, M.; Sanfuentes E. Identification and Biological Characterization of Isolates with Activity Inhibitive Against *Macrophomina phaseolina* (Tassi) Goid. *Chil. J. Agric. Res.* **2009,** *69,* 526–533.

Gangopadhyay, S.; Wyllie, T. D. Melanin-like Compound in the Sclerotia of *Macrophomina phaseolina. Indian Phytopathol.* **1974,** *27,* 661–663.

Gerin, D.; Dongiovanni, C.; De Miccolis Angelini, R. M.; Pollastro, S.; Faretra, F. First Report of *Macrophomina phaseolina* Causing Crown and Root Rot on Strawberry in Italy. *Plant Dis.* **2018,** *102* (9), 1857.

Ghosh, T.; Biswas, M. K.; Soumyajit, D.; Ghosh C. Mechanism Involved in Jute Resistance to *Macrophomina* for Developing Jute Resistant Varieties. *Plant Cell Biotech. Mol. Biol.* **2018,** *19* (3&4), 91–106.

Ghosh, T.; Mukherji, N.; Basak, M. On the Occurrence of a New Species of *Orbilia*. Fr. *Jute Bull.* **1964,** *27,* 134–141.

Gillen, A. M.; Mengistu, A.; Smith, J. R.; Paris, R. L. Registration of DT99-16864 Soybean Germplasm Line with Moderate Resistance to Charcoal Rot [*Macrophomina phaseolina* (Tassi) Goid.]. *J. Plant Registrat.* **2016,** *10,* 309–315.

Goidànish, G. Revisione del Genere *Macrophomina* Petrak. Species Tipica: *Macrophomina phaseolina* (Tassi) G. Goid. n. comb. nov. *M. phaseoli* (Maubl.). *Ashby. Annali Della Sperimentazione Agrarian* **1947,** *1,* 449–461.

Goudarzi, A.; Banihashemi, Z.; Maftoun, M. Effect of Water Potential on Sclerotial Germination and Mycelial Growth of *Macrophomina phaseolina. Phytopathol. Mediterr.* **2008,** *47,* 107–114.

Grau, C. R.; Dorrance, A. E.; Bond, J.; Russin, J. S. Fungal Diseases. In *Soybeans: Improvement, Production, and Uses, Agron*; Boerma, H. R., Specht, J. E., Eds.; Monogr. 16. ASA, CSSA, and SSSA: Madison, WI, 2004; pp 679–763.

Gupta, G. K. Soybean Diseases and their Management. In *Soybean Production and Improvement in India*; Singh, N. B., Chauhan, G. S., Vyas, A. K., Joshi, O. P., Eds.; NRC for Soybean: Indore, India, 2004; pp 145–168.

Gupta, G. K.; Chauhan, G. S. Symptoms, *Identification and Management of Soybean Diseases. Technical Bulletin 10*; National Research Centre for Soybean: Indore, India, 2005.

Gupta, G. K.; Sharma, S. K.; Ramteke, R. Biology, Epidemiology and Management of the Pathogenic Fungus *Macrophomina phaseolina* (Tassi) Goid with Special Reference to Charcoal Rot of Soybean (*Glycine max* (L.) Merr.). *J. Phytopathol.* **2012,** *160,* 167–180.

Hajlaoui, M. R.; Mnari-Hattab, M.; Sayeh, M.; Zarrouk, I.; Jemmali, A.; Koike, S. T. First Report of *Macrophomina phaseolina* Causing Charcoal Rot of Strawberry in Tuniisia. *New Disease Rep.* **2015,** *32,* 14.

Halsted, B. D. Some Fungous Disease of the Sweet Potato. *N. J. Agric. Coll. Exp. Stat. Bulle.* **1890,** *76,* 1–32.

Hartman, G. L.; Sinclair, J. B.; Rupe, J. C. *Compendium of Soybean Diseases*, 4th ed.; St. Paul, Minnesota, 1999.

Hemmati, P.; Zafari, D.; Mahmoodi, S. B.; Hashemi, M.; Gholamhoseini, M.; Dolatabadian, A.; Ataei, R. Histopathology of Charcoal Rot Disease (*Macrophomina phaseolina*) in Resistant and Susceptible Cultivars of Soybean. *Rhizosphere* **2018,** *7,* 27–34.

Hewidy, M. A.; Ismail, I. A.; Morsy, K. M.; Mahmoud Nagwa, M. A.; El-Galay Ola, A. M. Effect of Seed Treatment with Some Fungicides, Biocides and Saponin in Controlling Damping Off on Soybean Disease. *J. Agric. Sci. Mansoura Univ.* **2003,** *28* (4), 2733–2746.

Huda-Shakirah, A. R.; Kee, Y. J.; Mohd Hafifi, A. B.; Mohamad Azni, N. N.; Zakaria, L.; Hawa Mohd M. Identification and Characterization of *Macrophomina phaseolina* Causing Leaf Blight on White Spider Lilies (*Crinum asiaticum* and *Hymenocallis littoralis*) in Malaysia. *Mycobiol* **2019,** *47* (4), 408–414.

Hussain, T.; Khan, A. A. Determining the Antifungal Activity and Characterization of *Bacillus siamensis* AMU03 Against *Macrophomina phaseolina (*Tassi) Goid. *Indian Phytopathol.* **2020,** *73,* 507–516.

Hyder, S.; Gondal, A. S.; Ahmed, R.; Sahi, S. T.; Rehman, A.; Hannan, A. First Report of Charcoal Rot in Tomato Caused by *Macrophomina phaseolina* (Tassi) Goid. From Pakistan. *Plant Dis.* **2018,** *102* (7), 1459.

Ibrahim, M. E.; Abdel-Azeem, A. M. Management of Sesame (*Sesamum Indicum* L.) Charcoal Rot caused by *Macrophomina phaseolina* (Tassi) Goid. through the Application of Different Control Measure. *J. Pure Appl. Microbiol.* **2015,** *9* (1), 1–9.

Ilyas, M. B.; Ellis, M. A.; Sinclair, J. B. Effect of Soil Fungicides on *Macrophomina phaseolina* Sclerotium Viability in Soil and in Soybean Stem Pieces. *Phytopathology* **1976,** *66,* 355–359.

Iriarte, F.; Rosskopf, E.; Hilf, M.; McCollum, G.; Albano, J.; Adkins, S. First Report of *Macrophomina phaseolina* Causing Leaf and Stem Blight of Tropical Soda Apple in Florida. Online Plant Health Prog. 2007; doi: 10.1094/PHP-2007-1115-01-BR.

Jana, T. K.; Singh, N. K.; Koundal, K. R.; Sharma, T. R. Genetic Differentiation of Charcoal Rot Pathogen, *Macrophomina phaseolina,* into Specific Groups Using URP-PCR. *Can. J. Microbiol.* **2005,** *51,* 159–164.

Jimenez, D. R. C. *Influence of Soils, Nutrition, and Water Relations upon Charcoal Rot Disease Processes in Kansas.* M.Sc. Thesis, Kansas State University, Kansas, 2011.

Jordaan, E.; van der Waals, J. E.; McLaren, N. W. Effect of Irrigation on Charcoal Rot Severity, Yield Loss and Colonization of Soybean and Sunflower. *Crop. Prot* **2019,** *122,* 63–69.

Kaisar, S. A. K. M.; Das, S. N. Physical Factors that Influence the Growth ad Spread of Charcoal Rot Pathogen (*Macrophomina phaseolina*) Infecting Maize. *J. Phytopathol.* **1988,** *123* (1), 47–51.

Kamalakannan, A.; Mohan, L.; Valluvaparidasan, V.; Mareeswari, P.; Karuppiah, R. First Report of *Macrophomina* Root Rot (*Macrophomina phaseolina*) on Medicinal Coleus (*Coleus forskohlii*) in India. *Plant Path* **2006,** *55,* 302.

Karadimos, D. A.; Karaoglanidis, G. S.; Klonari, K. First Report of Charcoal Rot of Sugarbeet Caused by Macrophomina Phaseolina in Greece. *Plant Dis.* **2002,** *86* (9), 1051.

Kaur, S.; Dhillon, G. S.; Brar, S. K.; Vallad, G. E.; Chand, R.; Chauhan, V. B. Emerging Phytopathogen *Macrophomina phaseolina:* Biology, Economic Impact, and Current Diagnostic Trends. *Crit. Rev. Microbiol.* **2012,** *38,* 136–151.

Kaushik, C. D.; Chand, J. N.; Satyavir. Seed Born Nature of *Rhizoctonia bataticola* Causing Leaf Blight of Mungbean. *Indian J. Mycol. Pl. Pathol.* **1987,** *17,* 154–157.

Kendig, S. R.; Rupe, J. C.; Scott, H. D. Effect of Irrigation and Soil Water Stress on Densities of *Macrophomina phaseolina* in Soil and Roots of two Soybean Cultivars. *Plant Dis.* **2000,** *84,* 895–900.

Khan, A. N.; Shair, F.; Malik, K.; Hayat, Z.; Khan, M. A.; Hafeez, F. Y.; Hassan, M. N. Molecular Identification and Genetic Characterization of *Macrophomina phaseolina* Strains Causing Pathogenicity on Sunflower and Chickpea. *Front. Microbiol.* **2017,** *8,* 1309.

Khan, S. N. *Macrophomina phaseolina* as Causal Agent for Charcoal Rot of Sunflower. *Mycopathologia* **2007,** *5,* 11–118.

Khangura, R.; Aberra, M. First Report of Charcoal Rot on Canola Caused by *Macrophomina phaseolina* in Western Australia. *Plant Dis* **2009,** *93* (6), 666.

Kitahara, T.; Kiyota, H.; Kurata, H.; Mori, K. Synthesis of Oxygenated Eremophilanes, Gigantenone, Phomenone and Phaseolinone, Phytotoxins from Pathogenic Fungi. *Tetrahedron* **1991,** *47* (9), 649–654.

Kumar, K.; Singh, J. Location, Survival, Transmission and Management of Seed-borne *Macrophomina phaseolina,* Causing Charcoal rot in Soybean. *Ann. Plant Protect. Sci.* **2000,** *8,* 44–46.

Lakhran, L.; Ahir, R. R.; Choudhary, M.; Choudhary, S. Isolation, Purification, Identification and Pathogenicity of *Macrophomina phaseolina* (Tassi) Goid Caused dry Root rot of Chickpea. *J. Pharmacogn. Phytochem.* **2018,** *7,* 3314–3317.

Lodha, S.; Gupta, G. K.; Singh, S. Crop Disease Situation and Some new Records in Indian Arid Zone. *Ann. Arid Zone* **1986,** *25,* 311–320.

Lodha, S.; Sharma, S. K.; Aggarwal, R. K. Solarization and Natural Heating of Irrigated Soil Amended with Cruciferous Residues for Improved Control of *Macrophomina phaseolina. Plant Pathol.* **1997,** *46,* 186–190.

Lorenzetti, E.; Stangarlin, J. R.; Kuhn, O. J.; Portz, R. L. Induction of Resistance to *Macrophomina phaseolina* in Soyben Treated with Rosemary Extract Indução de resistência à *Macrophomina phaseolina* em Soja Tratada com Extrato de Alecrim. *Summa Phytopathol.* **2018,** *44,* 45–50.

Luna, M. P. R.; Mueller, D.; Mengistu, A.; Singh, A. K.; Hartman, G. L.; Wise, K. A. Advancing Our Understanding of Charcoal Rot in Soybeans. *J. Integ. Pest Manag.* **2017,** *8* (1), 1–8.

Lygin, A. V.; Zernova, O. V.; Hill, C. B.; Kholina, N. A.; Widholm, J. M.; Hartman, G. L.; Lozovaya, V. V. Glyceollin is an Important Component of Soybean Plant Defense Against *Phytophthora sojae* and *Macrophomina phaseolina. Phytopathology* **2013,** *103,* 984–994.

Ma, J.; Hill, C. B.; Hartman, G. L. Production of *Macrophomina phaseolina* Conidia by Multiple Soybean Isolates in Culture. *Plant Dis.* **2010,** *94,* 1088–1092.

Mahato, S. B.; Siddiqui, K. A. I.; Bhattacharya, G.; Ghosa, T.; Miyahara, K.; Sholichin, M.; Kawasaki T. Structure and Stereochemistry of Phaseolinic Acid: A new Acid From *Macrophomina phaseolina. J. Nat. Prod.* **1987,** *50,* 245–247.

Mahdizadeh, V.; Safaie, N.; Goltapeh, E. M. Diversity of *Macrophomina phaseolina* Based on Morphological and Genotypic Characteristics in Iran. *Plant Pathol. J.* **2011,** *27,* 128–137.

Mahmoud, A. F. A.; Budak H. First Report of Charcoal Rot Caused by *Macrophomina phaseolina* in Sunflower in Turkey. *Plant Dis.* **2011,** *95* (2), 223–223.

Marquez, N.; Giachero, M. L.; Declerck, S.; Ducasse, D. A. Macrophomina Phaseolina: General Characteristics of Pathogenicity and Methods of Control. *Front. Plant Sci.* **2021,** *12,* 634397.

Marquez, N.; Giachero, M. L.; Gallou, A.; Debat, H. J.; Cranenbrouck, S.; Di Rienzo, J. A.; Pozo, M. J.; Ducasse, D. A.; Declerck, S. Transcriptional Changes in Mycorrhizal and Nonmycorrhizal Soybean Plants Upon Infection with the Fungal Pathogen *Macrophomina phaseolina. Mol. Plant Microbe Interact* **2018,** *31,* 842–855.

Martínez-Salgado, S. J.; Romero-Arenas, O.; Morales-Mora, L. A.; Luna-Cruz, A.; Rivera-Tapia, J. A.; Silva-Rojas, H. V.; Andrade-Hoyos, P. First Report of *Macrophomina phaseolina* Causing Charcoal Rot of Peanut (*Arachis hypogaea* L.) in Mexico. *Plant Dis.* 2021; doi: 10.1094/PDIS-02-21-0337-PDN. Epub ahead of print. PMID: 33754851.

Masafumi, F.; Tokiya, S.; Hidenori, O.; Hidetake, S.; Masanobu, K.; Toyozo, S. Charcoal Rot of Watermelon Newly Found in Nagano and Kanagawa Prefectures and Pathogenicity of Causal Fungus Macrophomina phaseolina. *Ann. Phytopathol. Soc. Jpn.* **2002,** *68* (2), 148–152.

Mayék-Pérez, N.; López-Castañeda, C.; González-Chavira, M.; Garcia-Espinosa, R.; Acosta-Gallegos, J.; De La Vega, O. M.; Simpson, J. Variability of Mexican Isolates of *Macrophomina phaseolina* Based on Pathogenesis and AFLP Genotype. *Physiol. Mol. Plant Pathol.* **2001,** *59,* 257–264.

Mayek-Pérez, N.; López-Castañeda, C.; López-Salinas, E.; Cumpián-Gutiérrez, J.; Joaquín-Torres, I. C.; Padilla-Ramírez, J. S.; Acosta-Gallegos, J. A. Effect of Macrophomina phaseolina (Tassi) Goid. on Grain Yield of Common Beans (*Phaseolus vulgaris* L.) and its Relationship with Yield Stability Parameters. *Rev. Mexicana Fitopatol.* **2003,** *21* (2), 168–175.

Mbaye, N.; Mame, P. S.; Ndiaga, C.; Ibrahima, N. Is the Recently Described *Macrophomina pseudophaseolina* Pathogenically Different From *Macrophomina phaseolina? Afr. J. Microbiol. Res.* **2015,** *9,* 2232–2238.

McCain, A. H.; Scharf, R. F. Effect of Inoculum Density of *Macrophomina phaseolina* on Seedling Suceptibility of Six Conifer Sppecies. *Eur. J. For Pathol.* **1989,** *19,* 119-123.

Mengistu, A.; Arelli, P. A.; Bond, J. P.; Shannon, G. J.; Wrather, A. J. Evaluation of Soybean Genotypes for Resistance to Charcoal Rot. *Plant Health Progress* 2011; doi:10.1094/PHP-2010-0926-01-RS.

Mengistu, A.; Ray, J. D.; Smith, J. R.; Arelli, P. R.; Bellaloui, N.; Chen, P.; Shannon, G.; Boykin, D. Effect of Charcoal Rot on Selected Putative Drought Tolerant Soybean Genotypes and Yield. *Crop. Prot.* **2018,** (105), 90–101.

Mengistu, A.; Ray, J. D.; Smith, J. R.; Paris, R. L. Charcoal rot Disease Assesstment of Soybean Genotypes Using a Colony Forming Unit Index. *Crop. Sci.* **2007,** *47* (6), 2453–2461.

Mengistu, A.; Reddy, K. N.; Zablotowicz, R. M.; Wrather, A. J. Propagule Densities of *Macrophomina phaseolina* in Soybean Tissue and Soil as Affected by Tillage, Cover Crop, and Herbicide. *Plant Health Prog.* **2009,** *10* (1), https://doi.org/10.1094/PHP-2009-0130-01-RS.

Mengistu, A.; Wrather, A.; Rupe, J. C. Charcoal Rot. In *Compendium of Soybean Diseases*, 5th ed.; Hartman, G. L., Rupe, J. C., Sikora, E. J., Domier, L. L., Steffey, J. A., Eds.; APS Press: St. Paul, MN, 2015; pp 67–69.

Meyer, W. A.; Sinclair, J. B.; Khare, M. N. Factors Affecting Charcoal Rot of Soybean Seedlings. *Phytopathology* **1974,** *64,* 845–849.

Mihail, J. D. Macrophomina. In *Methods for Research on Soilborne Phytopathogenic Fungi*; American Phytopathological Society: St. Paul, MN, 1992; pp 134–136.

Mirza, M. S. Occurance of Sunflower Diseasesin Pakistan, *Inproceedings of the National Sunflower Workshop PAR*, 1984; 31–32.

Msikita, W.; Yaninek, J. S.; Ahounou, M.; Baimey, H.; Fagbemissi, R. First Report of Nattrassia Mangiferae Root and Stem Rot of Cassava in West Africa. *Plant Dis.* **1997,** *81* (11), 1332.

Nouri, M. T.; Lawrence, D. P.; Kallsen, C. E.; Trouillas, F. P. *Macrophomina* Crown and Root Rot of Pistachio in California. *Plants* **2020,** *9* (2), 134.

Nouri, M. T.; Zhuang, G.; Culumber, C. M.; Trouillas, F. P. First Report of *Macrophomina phaseolina* Causing Trunk and Cordon Canker Disease of Grapevine in the United States. *Plant Dis.* **2018,** *103* (3), 1–4.

Olaya, G.; Abawi, G. S.; Barnard, J. Influence of Water Potential on Survival of Sclerotia in Soil and on Colonization of Bean Stem Segments by *Macrophomina phaseolina. Plant Dis.* **1996,** *80,* 1351–1354.

Pandey, A. K.; Burlakoti, R. R.; Rathore, A.; Nair, R. M. Morphological and Molecular Characterization of *Macrophomina phaseolina* Isolated From Three Legume Crops and Evaluation of Mungbean Genotypes for Resistance to Dry Root Rot. *Crop. Prot.* **2020,** *127,* 104962.

Paris, R. L.; Mengistu, A.; Tyler, J. M.; Smith, J. R. Registration of Soybean Germplasm Line DT97-4290 with Moderate Resistance to Charcoal Rot. *Crop. Sci.* **2006,** *46,* 2324–2325.

Pawlowski, M. L.; Hill, C. B.; Hartman, G. L. Resistance to Charcoal Rot Identified in Ancestral Soybean Germplasm. *Crop. Sci.* **2015,** *55,* 1230–1235.

Pearson, C. A. S.; Scwenk, F. W.; Crowe, F. J.; Kelly, K. Colonization of Soybean Roots by *Macrophomina phaseolina. Plant Dis.* **1984,** *68* (12), 1086–1088.

Pratt, R. G. A Direct Observation Technique for Evaluating Sclerotium Germination by *Macrophomina phaseolina* and Effects of Biocontrol Materials on Survival of Sclerotia in Soil. *Mycopathologia* **2006,** *162,* 121–131.

Pratt, R. G.; McLaughlin, M. R.; Pederson, G. A.; Rowe, D. E. Pathogenicity of *Macrophomina phaseolina* to Mature Plant Tissues of Alfalfa and White Clover. *Plant Dis.* **1998,** *82* (9), 1033–1038.

Purushotham, N.; Jones, A.; Poudel, B.; Nasim, J.; Adorada, D.; Sparks, A.; Schwessinger, B.; Vaghefi, N. Draft Genome resource for *Macrophomina phaseolina* Associated with Charcoal Rot in Sorghum. *Mol. Plant-Microbe Interac.* **2020,** *33,* 724–726.

Qamar, M. I.; Ghanzafar, M. U.; Abbas, M. F.; Zainab, R.; Andleeb, S.; Shafiq, M.; Mushtaq, S. First Report of Macrophomina Phaseolina Causing Crown and Root Rot on Strawberry in Sargodha, Pakistan. *Plant Dis.* **2019,** *103* (6), 1420.

Ramezani, M.; Shier, W. T.; Abbas, H. K.; Tonos, J. L.; Baird, R. E.; Sciumbato, G. L. Soybean Charcoal Rot Disease Fungus *Macrophomina phaseolina* Mississippi Produces the Phytotoxin Botryodiplodin but no Detectable Phaseolinone. *J. Natural Products* **2007,** *70,* 128–129.

Ramos, A. M.; Gally, M.; Szapiro, G.; Itzcovich, T.; Carabajal, M.; Levin, L. In vitro Growth and Cell Wall Degrading Enzyme Production by Argentinean Isolates of *Macrophomina phaseolina,* the Causative Agent of Charcoal Rot in Corn. *Rev. Argent Microbiol.* **2016,** *48,* 267–273.

Rathore, B. S. Effect of Organic Amendments of Incidence of Seed Rot and Seedling Blight of Mothbean. *Plant Dis. Res.* **2000,** *15,* 217–219.

Raut, J. G. Transmission of Seed Borne *Macrophomina phaseolina* in Seed. *Sci. Technol.* **1983,** *11,* 807–817.

Razavi, S. E.; Pahlavani, M. H. Isolation of the Causal of Charcoal Rot Disease of Safflower and Resistance of Some Cultivars to the Disease. 16th Iranian Plant Protec Cong, Tabriz, Iran; 2004.

Reichert, I.; Hellinger, E. On the Occurrence, Morphology and Parasitism of *Sclerotium bataticola. Palestine J. Bot.* **1947,** *6,* 107–147.

Reznikov, S.; Chiesa, M. A.; Pardo, E. M.; De Lisi, V.; Bogado, N.; González, V.; Ledesma, F.; Morandi, E. N.; Ploper, L. D.; Castagnaro, A. P. Soybean-*Macrophomina phaseolina*-Specific Interactions and Identification of a Novel Source of Resistance. *Phytopathology* **2019,** *109,* 63–73.

Reznikov, S.; Vellicce, G. R.; Mengistu, A.; Arias, R. S.; Gonzalez, V.; De Lisi, V.; Gabriela, G. M.; Lourdes, R. C. M.; Pardo, E. M.; Castagnaro, A. P.; Ploper, L. D. Disease Incidence of Charcoal rot (*Macrophomina phaseolina*) on Soybean in North-western Argentina and Genetic Characteristics of the Pathogen. *Can. J. Plant Pathol.* **2018,** *40,* 423–433.

Saleh, A. A.; Ahmed, H. U.; Todd, T. C.; Travers, S. E.; Zeller, K. A.; Leslie, J. F.; Garrett, K. A. Relatedness of *Macrophomina phaseolina* Isolates from Tallgrass Prairie, Maize, Soybean and Sorghum. *Mol. Ecol.* **2010,** *19,* 79–91.

Santhakumari, P.; Kavitha, K.; Nisha, M. S. Occurrence of Collar Rot in Coral Hibiscus: A New Record. *J. Mycol. Plant Pathol.* **2002,** *32* (2), 258.

Sarr, M. P.; Ndiaye, M.; Groenewald, J. Z.; Crous, P. W. Genetic Diversity in *Macrophomina phaseolina*, the Causal Agent of Charcoal Rot. *Phytopathol. Mediterranea* **2014,** *53,* 250–268.

Schulz, T. J.; Thelen, K. D. Soybean Seed Inoculant and Fungicidal Seed Treatment Effects on Soybean. *Crop. Sci.* **2008,** *48,* 1975–1983.

Senthilkumar, M.; Swarnalakshmi, K.; Govindasamy, V.; Lee, Y.; Annapurna, K. Biocontrol Potential of Soybean Bacterial Endophytes Against Charcoal Rot Fungus, *Rhizoctonia bataticola*. *Curr. Microbiol.* **2009,** *58,* 288–293.

Sett, S.; Mishra, S. K.; Siddiqui, K. A. Avirulent Mutants of *Macrophomina phaseolina* and *Aspergillus fumigatus* Initiate Infection in *Phaseolus mungo* in the Presence of Phaseolinone; Levamisole Gives Protection. *J. Biosci.* **2000,** *25,* 73–80.

Sharda, R.; Shetty, H. S. Location and Transmisssion of *Macrophomina phaseolinain* Black Gram Seeds. *Indian Phytopath* **1987,** *40* (2), 194–196.

Sharifi Tehrani, A.; Shakiba, M.; Okhovat, M.; Zakeri, Z. Biological Control of *Tiarosporella phaseolina* the Causal Agent of Charcoal Rot of Soybean. *Commum. Agric. Appl. Biol. Sci.* **2005,** *70,* 189–192.

Sharma, S.; Kumar, K. Root Rot of *Jatropha curcas* Incited by *Rhizoctonia bataticola* in India. *Indian Forester* **2009,** *135* (3), 433–434.

Shier, W.; Thomas, A.; Hamed, K.; Baird, R. E.; Ramezani, M.; Sciumbato, G. L. (-)-Botryodiplodin, a Unique Ribose-analog Toxin. *Toxin Rev.* **2007,** *26* (4), 343–386.

Short, G. E.; Wyllie, T. D. Inoculum Potential of *Macrophomina phaseolina*. *Phytopathology* **1978,** *68,* 742–746.

Short, G. E.; Wyllie, T. D.; Bristow, P. R. Survival of *Macrophomina phaseolina* in Soil and in Residue of Soybean. *Phytopathology* **1980,** *70,* 13–17.

Siddiqui, K. A.; Gupta, A. K.; Paul, A. K.; Banerjee, A. K. Purification and Properties of Heat-Resistant Exotoxin Produced by *Macrophomina phaseolina* (Tassi) Goid in Culture. *Experientia* **1979,** *35,* 1222–1223.

Simonetti, E.; Viso, N. P.; Montecchia, M.; Zilli, C.; Balestrasse, K.; Carmona, M. Evaluation of Native Bacteria and Manganese Phosphite for Alternative Control of Charcoal Root Rot of Soybean. *Microbiol. Res.* **2015,** *180,* 40–48.

Singh, S. K.; Nene, Y. L.; Reddy, M. V. Influence of Cropping System of *Macrophomina phaseolin* in Soil. *Plant Dis.* **1990,** *74,* 812–814.

Sinha, O. K.; Khare, M. N. Control of Seed Borne *Macrophomina phaseolina* and *Fusarium equisiti* of Cowpea Seeds. *Seeds Res.* **1977,** *5,* 20–22.

Small, W. Notes on Species of *Fusarium* and Sclerotia in Uganda Kew. *Bull. Misc. Inf.* **1925,** *3,* 118–126.

Smith, D.; Chilvers, M.; Dorrance, A.; Hughes, T.; Mueller, D.; Niblack, T.; Wise, K. *Charcoal Rot Management in the North Central Region*; University of Wisconsin Extension Bulletin, 2014; A4037.

Smith, W. H. Germination of *Macrophomina phaseolina* as Affected by Pinus Lambertina Root Exudes. *Can. J. Micrbiol.* **1969,** *15,* 1387–131.

Songa, W.; Hollicks, R. J. Legume Hosts of *Macrophomina phaseolina* in Kenya and Effect of Crop Species on Soil Inoculum Level. *J. Phytopathol.* **1996,** *144,* 387–391.

Spagnoletti, F.; Carmona, M.; Gómez, N. E. T.; Chiocchio, V.; Lavado, R. S. Arbuscular Mycorrhiza Reduces the Negative Effects of *Macrophomina phaseolina* on Soybean Plants in Arsenic-Contaminated Soils. *Appl. Soil Ecol.* **2017,** *121,* 41–47.

Spagnoletti, F. N.; Cornero, M.; Chiocchio, V.; Lavado, R. S.; Roberts, I. N. Arbuscular Mycorrhiza Protects Soybean Plants Against *Macrophomina phaseolina* Even Under Nitrogen Fertilization. *Eur. J. Plant Pathol.* **2020,** *156,* 839–849.

Spagnoletti, F. N.; Leiva, M.; Chiocchio, V.; Lavado, R. S. Phosphorus Fertilization Reduces the Severity of Charcoal Rot (*Macrophomina phaseolina)* and the Arbuscular Mycorrhizal Protection. *J. Plant Nut. Soil Sci.* **2018,** *181* (6), 855–860.

Statista. 2021. https://www.statista.com/statistics/267271/worldwide-oilseed-production-since-2008/

Su, G.; Suh, S. O.; Schneider, R. W.; Russin, J. S. Host Specialization in the Charcoal Rot Fungus, *Macrophomina phaseolina. Phytopathology* **2001,** *91,* 120–126.

Sun, S. L.; Zhu, Z. D.; Duan, C. X.; Zhao, P.; Sun, F.; Deng, D.; He, Y. H. First Report of Charcoal Rot Caused by *Macrophomina phaseolina* on Faba Bean in China. *Plant Dis.* **2019,** *103,* 1415.

Swamy, C.; Naik, M. K.; Amaresh, Y. S.; Chennappa, G.; Jayalakshmi, S. K. Morphological Diversity and Molecular Characterization of *Macrophomina phaseolina* Causing Stem Canker of Pigeonpea in North Eastern Karnataka. *J. Plant Dis. Sci.* **2018,** *13,* 91–98.

Takooree, S. D.; Neetoo, H.; Ranghoo-Sanmukhiya, V. M.; van der Waals, J. E.; Vojvodić, M.; Bulajić, A. First Report of Charcoal Rot Caused by Macrophomina Phaseolina on Potato Tubers in Mauritius. *Plant Dis.* 2021; [Online first] *https://doi.org/10.1094/PDIS-02-21-0258-PDN.*

Tassi, A. Novae Micromycetum Species. *Ser. III Bull. Lab. Orto. Bot. Univ. Siena'* **1901,** *4,* 7–12.

Todd, T. C. Soybean Planting Date and Maturity Effects on *Heterodera gycines* and *Macrophomina phaseolina* in South Eastern. *Kansas J. Nematol.* **1993,** *25,* 731–73.

Tok, F. M. Relationship Between Morphologic, Phenotypic and Pathogenic Characteristics in *Macrophomina phaselina* Isolates From Cucumber Plants. *Int. J. Innov. Approaches Agric. Res.* **2019,** *3,* 651–660.

Torres, M. J.; Brandan, C. P.; Petroselli, G.; Erra-Balsells, R.; Audisio, M. C. Antagonistic Effects of Bacillus Subtilis Subsp. Subtilis and B. Amyloliquefaciens Against *Macrophomina phaseolina:* SEM Study of Fungal Changes and UV-MALDI-TOF MS Analysis of Their Bioactive Compounds. *Microbiol. Res.* **2016,** *182,* 31–39.

Ugulino, A. L. N.; de Mendonça Júnior, A. F.; dos Santos Rodrigues, A. P. M.; Santos, A. B.; da Silva França, K. R.; Cardoso, T. A. L.; do Prado Júnior, L. S. Inhibition Effect of Vegetable Oils on the Mycelial Growth of *Macrophomina phaseolina* (Tassi.). *Goid. J. Agric. Sci.* **2018,** *10* (6), 49–56.

UK, CAB International; Holliday, P.; Punithalingam, E. *Macrophomina phaseolina.* Descriptions of Fungi and Bacteria; CAB International: Wallingford UK, 1970.

Uppal, B. N. *Rhizoctonia bataticola* on Srghum in Bombay Presidency. *Int. Bull. Prot.* **1931,** *5,* 163.

Vasebi, Y.; Safaie, N.; Alizadeh A. Biological Control of Soybean Charcoal Root Rot Disease Using Bacterial and Fungal Antagonists in Vitro and Greenhouse Condition. *J. Crop. Prot.* **2013,** *2,* 139–150.

Veena, A.; Reddy, N. P. E.; Reddy, B. V. B.; Prasanthi, L. Pathogenecity Test and Evaluation of Efficacy of Fungicides Against *Rhizoctonia bataticola*, the Causal Agent of Dry Root Rot of Chickpea. *Inter. J. Appl. Biol. Pharma. Technol.* **2014,** *5,* 283–287.

Wrather, A.; Shannon, G.; Balardin, R.; Carregal, L.; Escobar, R.; Gupta, G.; Ma, Z.; Morel, W.; Ploper, D.; Tenuta, A. Effect of Diseases on Soybean Yield in the Top Eight Producing Countries in 2006. *Plant Health Prog.* **2010,** *11* (1), *doi.org/10.1094/PHP-2010-0102-01-RS.*

Wrather, J. A.; Anderson, T. R.; Arsyad, D. M.; Tan, Y.; Ploper, L. D.; Porta-Puglia, A.; Ram, H. H.; Yorinori, J. T. Soybean Disease Loss Estimates for the top 10 Soybean Producing Countries in 1994. *Plant Dis.* **1997,** *81,* 107–110.

Wrather, J. A.; Koenning, S. R. Soybean Disease Loss Estimates for the United States, 1996–2009. Delta Research Center: Agriculture Experiment Station. University of Missouri, College of Agriculture, Food and Natural Resources, USA, 2010.

Wrather, J. A.; Shannon, J. G.; Carter, T. E.; Bond, J. P.; Rupe, J. C.; Almeida, A. M. R. Reaction of Drought-tolerant Soybean Genotypes to *Macrophomina phaseolina. Plant Health Prog.* **2008,** *9* (1), doi:10.1094/PHP-2008-0618-01-RS.

Wyllie, T.; Calvert, O. Effect of Flower Removal and Pod Set on Formation of Scerotia and Infection of *Glycine max* by *Macrophomina phaseolina. Phytopathology* **1969,** *59,* 1243.

Wyllie, T. D. Charcoal Rot ED. J.B. Sinclair and P.A. Backman, 1989; pp 30–33.

Wyllie, T. D. *Compendium of Soybean Diseases*; 3rd ed.; The American Phytopathological Societ, St. Paul, MN, 1993.

Wyllie, T. D.; Brown, M. F. Ultrastructural Formation of Sclerotia of *Macrophomina phaseoli. Phytopathology* **1970,** *60,* 524–528.

Wyllie, T. D.; Gangopadhyay, S.; Teague, W. R.; Blanchar, R. W. Germination and Production of *Macrophomina phaseolina* Microsclerotia as Affected by Oxygen and Carbon Dioxide Concentration. *Plant Soil* **1984,** *81,* 195–201.

Wyllie, T. D.; Scott, D. H. Charcol Rot of Soyabeans—Current Status. In *Soyabean Diseases of North Central Region*; APS Press Indianapolis: IN, 1988; pp 106–113.

Yang, S. B.; Navi, S. S. First Report of Charcoal Rot Epidemics Caused by *Macrophomina phaseolina* in Soybean in Lowa. *Plant Dis.* **2005,** *89* (5), 526.

Yang, S. M.; Owen, D. F. Symptomatology and Detection of *Macrophomina phaseolina* in Sunflower Plants Parasitized by *Cylindrocopturus adspersus* Larvae. *Phytopathology* **1982,** *72,* 819–821.

Young, P. A. Epidemic of Charcoal Rot of Cornand Other Crops in East Taxes. *Plant Dis. Reptr.* **1944,** *28,* 898–899.

Young, P. A. Charcoal Rot of Plants in East Texas. Bulletin Texas Agricultural Experimental Station No. 33, 1949.

Zhang, J. Q.; Zhu, Z. D.; Duan, C. X.; Wang, X. M.; Li, H. J. First Report of Charcoal Rot Caused by *Macrophomina phaseolina* on Mungbean in China. *Dis Notes* **2011,** *95* (7), 872.

Živanov, D.; Živanov, S. T.; Nagl, N.; Savić, A.; Katanski, S.; Milić, D. First Report of Macrophomina Phaseolina Causing Dry Root Rot of Chickpea (*Cicer arietinum* L.) in Serbia. *Plant Dis.* **2019,** *103,* 2685.

CHAPTER 17

Postharvest Loss of Indigenous Fruits Cops and Its Management Through Interventions of Advanced Technologies

UMME ABIHA[1,2] and SUSMITA SHUKLA[1]

[1]*Applied Plant Biotechnology Research Lab, Centre for Plant and Environment Biotechnology, Amity Institute of Biotechnology, Amity University, Noida, Uttar Pradesh, India*

[2]*Indian Institute of Technology, IDRP- Smart Healthcare, Karwar, Jodhpur, Rajasthan, India*

ABSTRACT

India is one of the world's leading producers of fruit and vegetable products. According to the National Horticulture Database (NHD) published by the National Horticulture Board (NHB) 2019–2020, India accounted for producing 99.07 million tons of fruit and 191.77 million metric tons of vegetables. Fruits such as mangoes, papayas, bananas, and grapes are widely exported to India in the international markets of the UAE, Nepal, Bangladesh, UK, Malaysia, Sri Lanka, the Netherlands, Oman, and Qatar. Indian provinces such as Andhra Pradesh, Uttar Pradesh, Tamil Nadu, Gujarat, Karnataka, Madhya Pradesh, and Kerala are the largest producers of mango, banana, and papaya in the world. Despite such immense production, the share in the global market is very minimal. This chapter is an expatriate documentation on indigenous fruits grown in the lands of an incredible nation–India with diverse landscape and climatic conditions owing to the world's largest producer of fruits and vegetables. Several issues in dealing with fruits after their harvest are discussed and possible solutions are suggested. The

Plant Diseases and Their Management: A Sustainable Approach.
Deepa Srivastava, Rajarshi Kumar Gaur, and Ajay K. Tiwari (Eds.)

plausible reasons for post-harvest losses could be low shelf life, mechanical injury during transportation, the practice of traditional cultivation, use of pesticides, lack of marketing strategies, and poor supply chain management. The transforming era of plant biotechnology has significantly contributed to finding better management tools and techniques of curbing post-harvest losses and strengthened Indian economy's returns from the global market.

17.1 INTRODUCTION

India is the second-biggest Asian country to hit the gigantic 283 million tonnes of its horticulture produce, undoubtedly way beyond the production of food grains in its mainland (Birthare, 2018). The production owing to the vast, irrigated geographic landscape and diverse climatic conditions in India has captured 70 international markets across the world, including Bangladesh, UAE, The Netherland, Nepal, Malaysia, UK, Sri Lanka, Oman, and Qatar (Horticulture, 2020; Crop Diversification in India - C.R. Hazra, n.d.). In the scientific reports of 2018, World Trade Organization ranked India as the 8th major producer of agricultural commodities (WtoOrg. n.d.). Among the total agricultural export by India, around 35% is constituted by horticultural goods. Paradoxically, the 2017 report of the World Bank reveals Indian share in the global market to a bare minimum of 1.7% in vegetables and 0.5% in fruits (Jha, 2007). In the reports of 2019–2020, the Department of Agriculture, Cooperation and Farmers Welfare estimated the total horticulture production to be 320.48 million tonnes (GovIn, n.d.a). The enormous produce of fruits, primarily including banana, papaya, mangoes, and guavas is estimated to be 99.07 MMT (*against 97.97 MT in 2018–19*), while vegetable production is estimated to be around 191.77 MT in 2019–20 (*compared to 183.17 MT in 2018-19*) (TPCI, 2020; Welfare, n.d.). The sector of horticulture has grown immensely with a special focus on growth strategy with the introduction of "Horticulture Mission for North East & Himalayan States" (HMNEH) and the "National Horticulture Mission" (NHM) (Pal et al., n.d.; MIDH, n.d.). Other factors of special importance included increasing productivity through high-density plantations, implementing modern cultivation practices along with micro-irrigation, focus on graded quality on planting, rejuvenating sentinel orchards, and excessive thrust on developing post-harvest management systems to prevent produce-loss and catering to highly competitive markets (Giri Dhari et al., 2020; Uzen and Cetin, 2013; Hoaes, 2010). Refer to Table 17.1 for India's hotspot targets in the international fruit market. This paper discusses the post-harvest losses faced by India in the context to its major produce.

TABLE 17.1 India's Hotspot Targets in International Fruit Market in 2018.

Market	Untapped demand for fruit produce
UK	Grapes (*US$ 46.1 mn*); guavas, mangoes, mangosteens (*US$ 19.8 mn*); bananas (*US$ 17.3 mn*); coconuts (*US$ 1.5 mn*); peaches, pears, papayas, etc. (*US$ 1.9 mn*)
Netherlands	Guavas, mangoes, and mangosteens (*US$ 36 mn*); bananas (*US$ 6.3 mn*); grapes (*US$ 4.2 mn*); watermelons (*US$ 1.5 mn*); oranges (*US$ 1.6 mn*
USA	Grapes (*US$ 74 mn*); guavas, mangoes, and mangosteens (*US$ 13.8 mn*); coconuts (*US$ 8.5 mn*); bananas (*US$ 6.5 mn*); peaches, pears, papayas, etc. (*US$ 5.1 mn*)
Germany	Grapes (*US$ 17.1 mn*); guavas, mangoes, and mangosteens (*US$ 33.9 mn*); bananas (*US$ 10.4 mn*); grapes (*US$ 5.9 mn*); peaches, pears, papayas, etc. (*US$ 4.6 mn*)
Thailand	Grapes (*US$ 32.6 mn*); guavas, mangoes, and mangosteens (*US$ 3.7 mn*); apples (*US$ 976.2 k*); melons (*US$ 884.0 k*); grapes (*US$ 178.8 k*)

Source: *ITC Trade Map.*

17.2 STATUS QUO—CURRENT SCENARIO

The Global Hunger Index (GHI) 2018 ranked India and Nigeria on 103 among 119 countries, calling it to be a serious concern of inadequate food supply (BusinessinsiderIn, 2018). The stench of reality in terms of statistical evidence on grounds is dense. Despite being one of the major producers of the world, India does fight with serious concern of hunger (Gandhi, 2018). The state of worry is such that a considerable portion of its horticultural produce, especially fruits, is wasted every single day (Press Trust of India, 2010). There are many reasons associated with it including cold-chain infrastructure, transportation problems during export, storage facility, processing facilities, and even marketing with 16% of fruits and vegetables being wasted every year (Negi and Anand, 2018; Goedhals-Gerber and Khumalo, 2020; Saurav and Potti, 2016). There are grounded pieces of evidence that account for such losses, one of which was even presented in the meetings of Lok Sabha, regarding it to the study of "Assessment of Quantitative Harvest and Post-Harvest Losses of Major Crops and Commodities in India" primarily conducted by the Central Institute of Post-Harvest Engineering and Technology (CIPHET) (Shyam et al., 2016).

A project called "Coordinated Programme on Horticulture Assessment and Management using geoinformatics" (CHAMAN) was launched from 2014 to 2017 by the Department of Agriculture and Cooperation, Ministry of Agriculture, and Farmers Welfare (GovIn., n.d.a; Press Information Bureau, n.d.; Welfare, n.d.). The main objective was to integrate the development and upgradation of scientific methods for estimating the area and production of Horticulture crops (Horticulture Sector in India, n.d.). This project included two main key components of remote sensing and sample survey (SS), the former was controlled by the Indian Agriculture Statistics Research Institute (IASRI) of ICAR and the latter by Mahalanobis National Cop Forecast Centre (MNCFC) (Home, n.d.). This scheme covered all major states that produced banana, mango, citrus, potato, onion, tomato, and chilli. Another initiative by the government was the "Area Production Information System" (HAPIS), which was a web-enabled information system that facilitated crop data in terms of area production, quarterly and weekly area sown (Dac Gov I, n.d.). The idea was to ensure the availability of relevant information on crops at the district, block, and state levels. All these policies have helped to maintain a common availability of crop registries across the nation.

Andhra Pradesh is one of the biggest fruit producers of India with a production volume of 257.75 lakh MT of fruits and vegetables in the reports of 2019–2020 (Ameen, 2021). The major production capacity comprises banana, mango, orange, papaya, lemon, muskmelon, watermelon, and guava. Much attention has been on improving the markets and networks across the country

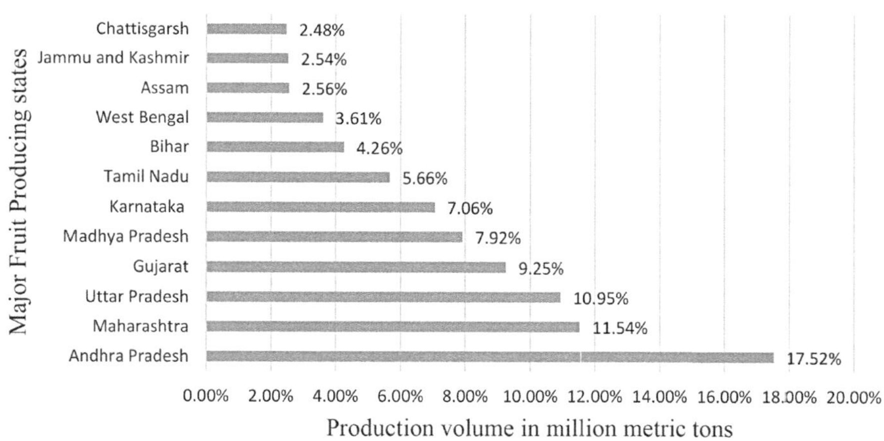

GRAPH 1 Major fruit-producing states of India in 2020 (StatistaCom, n.d.).

to distribute the benefits equally among farmers and consumers; however, there is a need to pay much attention to improving healthy and sustainable food production by upgrading existing technology and also devising solutions to reduce fruit loss. Refer to Graph 1 for major fruit-producing states of India with a production volume of million metric tons.

Let us delve into the next section to understand the trend and current scenario in major fruit-producing states of India.

17.3 FACTORS ATTRIBUTING TO POSTHARVEST LOSSES

Food losses refer to a measurable reduction in foodstuff which affects either food quantity or quality while postharvest food loss is associated reduction in the quality of food after harvesting it (Gardas et al., 2018). There is a clear distinction between damage and loss. A product that gets damaged has a restricted use while a loss makes its use impossible. There are a lot of postharvest losses of fruits that are witnessed in India varying in area and seasons and critical factors are associated with its cause (Gardas et al., 2018; Nath et al., 2018). The critical factors are divided into the basis of two broad categories—external and internal factors.

17.3.1 INTERNAL FACTORS

17.3.1.1 HARVESTING

Fruits are harvested when its color changes and transforms into ripe fruit. The harvesting time of fruit is dependent on the weather, season, and also the crop life cycle. The persistence of traditional pricing and indications of the distance to market fruit and bad weather conditions affect the operation of harvesting machines (Afzal et al., 2020). Other aspects of harvesting include mismanagement, improper harvesting, and poor machinery.

17.3.1.2 PRE-COOLING

Post-harvest losses are due to lack of pre-cooling facilities, inadequate training of pre-cooling technology on commercial scale, and lack of information on the cost benefits of pre-cooling technology (Cherono et al., 2018).

17.3.1.3 TRANSPORTATION

Transportation remains one of the biggest concerns of fruit loss. Major reasons include poor infrastructure (roads, bridges, etc.), lack of proper transport systems, and lack of refrigerated transportation. Many developing countries are experiencing such losses as a result of insufficient access to proper road and transport vehicles, both local and international marketing (Acta Horticulturae, n.d.).

17.3.1.4 FRUIT STORAGE

Fruits are picked during harvesting and collectively stored in closed structures including granaries, warehouses, hermetic bins, and silos. Any change in temperature, humidity, and attack by insects and pests would deteriorate the quality of fruit and lead to losses. Hence, facilities for long-term or short-term storage should be ensured for preventing such fruit loss (Fajola, 1979; Singh and Reddy, 2006).

17.3.1.5 GRADING AND SORTING

There is a lack of a unified standard for grading fruits and a lack of skill and awareness during packaging and fruit sorting. Negligent behavior and aggressive fruit handling lead to mechanical injuries and attribute to enormous losses (Arun Kumar et al., 2020).

17.3.1.6 FRUIT PACKAGING AND LABELLING

Once the fruits are harvested, they are often transported to a local farm or a packing house for packaging. The fruit is either sold at the retailer level or else sent to a distributor market. This process decides the packaging and labeling of the product and impacts the shelf life.

17.3.1.7 BIOLOGICAL

Biological causes include respiratory rate, ethylene production, and action, naming change rates (associated with color, texture, taste, nutritional value), mechanical damage, water stress, root formation, degeneration, and disease progression. Several environmental factors also contribute to degradation, including temperature, relative humidity, air velocity, and atmospheric composition (concentration of carbon, carbon dioxide, and ethylene), as well as sanitation processes (Golob, 2002; Wilson et al., 1991).

17.3.1.8 MICROBIOLOGICAL

Pests and microbes can damage the fruits by attacking the fruit body. Although the attack is a considerably small amount, it becomes unacceptable for marketing (At et al., 2011). Several toxins called mycotoxins also lead to a loss in the nutritional value of the same.

17.3.1.9 CHEMICAL

Protecting fruit crops from pests using pesticides creates a harmful cycle of obnoxious chemicals such as lubricating oil that leads to loss of color, flavor, texture, and nutritional value (At et al., 2011).

17.3.2 EXTERNAL FACTORS

Other factors outside the food chain supply attribute to fruit loss and are discussed below:

17.3.2.1 TEMPERATURE

The shelf life of fruits is greatly reduced with increasing temperature. Evidence on hampered biochemical pathways in fruits due to variable changes in temperature is quoted in many research (At et al., 2011; Bourne, 1977).

17.3.2.2 HUMIDITY

Fresh fruit produce is rich in moisture content and conditions that ensure reduced drying and dehydration should be monitored. Excessive levels of moisture might leverage the quality of fruits and lead to over-ripening (At et al., 2011).

17.3.2.3 ALTITUDE

Fruit storage at higher altitudes tends to prolong shelf life and reduce food loss as long as it is stored in direct sunlight.

17.3.2.4 TIME

Storage time is the most critical factor in loss of fruits. Most of the agricultural production is a sector of rural area and while people have urbanized in cities, this has created to develop more efficient supply chains to minimize the fruit losses incurred during distribution (At et al., 2011; Bourne, 1977; Parra-Coronado et al., 2017).

FIGURE 17.1 Primary, biological, and environmental causes of post-harvest losses of fruits.

17.4 DEMYSTIFYING REASONS FOR LOSSES

Eight major fruits namely apple, banana, citrus, grapes, guava, mango, papaya, and sapota contribute to 84% of the total fruit production of India (Raj et al., 2019). A survey report estimates the post-harvest loss of apples that are grown in the Western Himalayan region of Jammu and Kashmir, Himachal Pradesh, and Uttarakhand. It accounts for farm operation losses to be 9.08% owing to harvesting, sorting/grading, and transportation operations (GovIn, n.d.b; MountaininitiativeIn, n.d.). The farmers of Jammu and Kashmir harvest the fruit by checking for a change in fruit color which is significant to identify its transformation from a raw form to an edible one. These are then collected and stored in temporary racks in an orchard, where they are sorted, graded, and packaged (MountaininitiativeIn, n.d.). On the arrival of the transport, they are loaded and then sent for export and distribution. The prolonged time gap between harvesting and transportation takes a usual gap of 3–10 days and therefore causes a higher loss in sorting operations. However, the farmers of Himachal Pradesh and Uttarakhand harvest the apple after the availability of transport. Cases where unripe fruit is plucked and packaged also account for such losses (GovIn, n.d.b). The present scenario on the post-harvest loss of apples due to transportation is improving due to the availability of better roads and transports. However, the loss at retailer level storage still needs improvement.

Another assessment on losses of bananas on five major agro-climatic zones including Kerala, Karnataka, Tamil Nadu, Andhra Pradesh, and Maharashtra accounts for an overall loss of 7.76%. Out of which 6.04% is accounted by farm operations (NicIn, n.d.; Venkatesh, 2019). The major citrus-producing states like Andhra Pradesh, Punjab, and Maharashtra shared contrasting losses at the regional level. Punjab accounted for a loss of 6.10% and the highest 12.97% in Maharashtra. Kinnow is one such example of a citrus fruit that is harvested in two different seasons in Punjab and Maharashtra (Yogi et al., n.d.; PTI, 2015). Punjab harvests in winter (December–January) while Maharashtra harvests in summer (February–March). Since 2005, many waxing, grading, and packaging plants have been installed in Punjab and have resulted in reduced loss of this fruit. While the production of citrus fruits has increased immensely, the infrastructure facilities to handle the product in the market remain unmet. Hence, the total loss at a national level is 9.69%, with concerns of climatic conditions, mechanization, and storage losses.

Guava is one similar fruit that is harvested in two seasons viz monsoon and winter. The reasons attributed to its post-harvest loss are farm operations, poor management, fall of overripe fruits, poor sorting, an infestation of insects, poor packaging in gunny bags, and overloading. Grapes are majorly grown in states of three agro-climatic zones including Maharashtra, Karnataka, and Andhra Pradesh. The loss of grapes is estimated to be 8.63% at a national level while that of mango and papaya is 9.16% and 6.70% (SfacindiaCom., n.d.). Table 17.2 and Table 17.4 enumerate the reasons for post-harvest losses in banana and guava during handling operations and marketing respectively.

Refer to Table 17.3 to study various diseases that are caused due to various biological agents in oranges.

TABLE 17.2 Causes of Post-harvest Losses in banana During Handling Operations (Davara and Patel, 2016; Cardoen et al., 2015).

Handling operations	Causes of losses
Harvesting and preparation for market	1. Maturity at harvest. 2. Mechanical damage. 3. Failure to protect the commodity for market from sun. 4. Virtual absence of packaging practice.
Transport	1. Inadequacy and delay in transport. 2. Transportation by head-load and bullock-cart before loaded into the truck. 3. No ventilation during transportation. 4. Rough handling and improper stacking in the transport vehicle cause increased mechanical injury. 5. Bulk transport without container.
Handling at destination	1. Rough handling during loading and unloading. 2. Improper ripening and storage methods.
Retailing	1. Absence of grading 2. Exposure to sun and other unfavorable environments. 3. Absence of consumer package system

TABLE 17.3 Diseases Caused by Various Agents in Oranges (Allafi, 2011; Ismail and Zhang, 2004).

Diseases	Causative agent
Stem end rot	*Collectotrichumgloeosporioides, Botryodiploidia theobromae and Phomopsis citri*
Anthracnose	*Geotrichum sp.*
Sour rot	*Coll2ectotrichum gloeosporioides*
Penicillium rot	*Penicillium digitatum*
Aspergillus rot	*Aspergillus niger*

TABLE 17.4 Post-harvest Losses during marketing system of Guava (Gajanana et al., 2015; CAB Direct. CabdirectOrg., n.d.).

Levels	Sound fruits	Damage intensity			Total damaged	Aggregate loss
Farm	100.00	2.12	0.99	0.18	3.29	0.88
Trader	96.71	3.01	-	-	3.01	0.30
Retailer	93.70	2.29	4.27	1.62	8.18	3.99
Total	85.52	7.42	5.26	1.80	14.48	5.17

TABLE 17.5 Stages of Post-Harvest Losses in Grapes—A Case of Maharashtra (GovIn., n.d.b; Ladaniya et al., 2005).

Stage of handling/	Causes of loss		Loss (%) (Pimpalgaon: Nashik)	Loss (%) (Tasgaon: Sangli)	
Fam level	1.Small and mummified berries	0.80	-	1.01	-
	2. Rotting	0.20	-	0.24	-
	3.Bruises/Splitting	0.00	-	0.00	
	Total	1.00	-	1.25	-
Wholesale level		Mumbai	Pune	Nagpur	
	1. Rotting	0.00	0.00	-	0.00
	2.Shattered berries	5.50	6.00	-	8.65
	3. Any Other	0.00	0.00	-	0.00
	Total	5.50	6.00	-	8.65
Retail	1. Rotting	2.50	3.00	-	3.15
	2. Softened berries	2.50	2.50	-	3.60
	3. Shattered berries	7.25	7.50	-	9.25
	4.Any other	0.00	0.00	-	0.00
	Total	12.50	13.00	-	16.00

TABLE 17.6 Layer-wise Post-harvest Losses in Green Mangoes During Ripening (ResIn., n.d.).

Layers of Stack	Perfect fruits	10% damage	10–50% damage	Discards	Aggregate
Top	40.60	39.08	11.16	2.53	12.02
Middle	13.44	63.44	21.50	1.61	18.70
Bottom	16.19	59.09	18.18	4.81	19.82
Overall	23.41	53.87	16.95	2.98	16.85

17.5 CURBING THE PROBLEMS

17.5.1 INFLUENTIAL SECTORS IN HARVEST CHAIN

Three main influential sectors play key roles in agribusiness and develop strategies and policies for infrastructure and capacity building, namely Government, CSO, and private sector (Gardner et al., 2017; IntellecapCom., n.d.). Agriculture is a state study in India; however, the central government plays a major role in implementing national and international initiatives

TABLE 17.7 Pest and Disease Management in Papaya (TFNet News Compilation, 2016).

Disease	Symptom	Effect	Control
Anthracnose	small water-soaked spots that enlarge.	causes small round, dark areas on ripe fruit which later enlarge into sunken lesions.	practice a prophylactic program.
	turn brown or black and become darker.	fruit will rot	
Root rot	rapid browning and wilting of the plants, followed by total collapse within days.	causing root rot, collar rot, and fruit rot and also has a foul odor	spray with metalaxyl
	young leaves wilt and matured leaves turn yellow and drop	if not controlled, the leaves will drop and plant will die.	
Powdery mildew	a superficial white fungus growth on leaf surface	some leaf drop can occur if powdery mildew is not a severe problem	removing infested leaves and removing them from near the plants
	it leads to small, light yellow spots on the lower surfaces of leaves		
Papaya Fruit Fly	may show yellow areas and drop from the tree prematurely.	Lays eggs through the papaya fruit peel into the fruit cavity where the larvae feed. Eventually emerge from the ruined fruit.	place a paper bag over individual fruit when they are small.
			leave the bag on until harvesting.
Oriental Fruit flies		attacks ripe fruit causing fruit damage.	Harvesting done before fruit turn yellow spray with dimethoate 0.1% or acephate
			Mechanical Control
Red Spider Mites	induce premature leaf drop, reduce tree vigor, and cause external blemishes on the fruits		spray with dicofol or propagite 0.1%

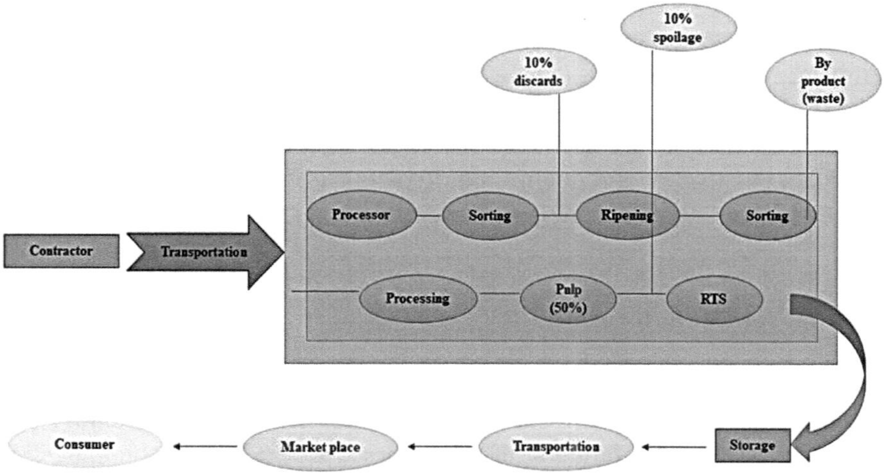

FIGURE 17.2 Post-harvest losses during transportation of Mango (Totapari).

Source: Adapted from Sab et al., 2017

through exemplary actions, funding programs, and regulations to drive infrastructure development. Short-term intergovernmental measures such as the Goods and Services Tax (GST) Act promise to expand smallholder markets leading to better prices, lower price volatility, and market overcrowding, as well as lower PHL (Sharma, 2019). The consolidation of national tax rates under the GST will reduce shipping time on perishable goods and improve market efficiency by promoting the establishment of post-harvest infrastructure based on value chain requirements. Other appropriate measures include the Contract Farming Act, the relaxation of the agricultural sector's Foreign Direct Investment (FDI) restrictions, the restructuring of the Agricultural Commodity Market Committee (APMC) Act, the National Agricultural Market (eNAM), and the funding of store receipts.

On the other hand, CSOs (NGOs, cooperatives, farmers "producers" organizations (FPOs), and self-help groups (SHGs)) support SHFs by providing information dissemination, training, and capacity-building resources before and after harvest stages (Chengappa, 2018; Singh et al., 2020). Agriculture is the backbone of the local business and CSOs have played a key role in building skills before harvest, which has led to in-depth consultations with farming communities. They also help improve communication between farmers and the private sector in order to market the product and get equipment leasing.

The role of the private sector is important at all levels of value chain in order to improve productivity and market demand and reduce PHL by investing in infrastructure, technology, and new SHF collaborative approaches. They are heavily involved in buying, storing, transporting, repairing, and selling the product. Companies are now starting to work with farmers directly to ensure that market capacity and quality requirements are aligned with the production of farmers, which could contribute to the reduction of PHL.

17.5.2 FARMER AGGREGATION

Improving market connectivity can be achieved through effective farmer engagement. Consolidating farmers, the need is to provide training and extension services to farmers and familiarity with new technologies (Jha, 2016). Problems, such as land redistribution, low production rates, limited skills and awareness of modern farming techniques, and the high costs incurred by farmers in transporting and distributing produce, can be solved by mass producers (Dethier and Effenberger, 2011). This model helps to establish a great balance between demand and supply, thus providing greater visibility to farmers in terms of product value and quality. It serves as a reliable model that connects the private sector and farmers, promising to promote greater efficiency in the post-harvest value chain. This is run by farmers or the private sector, FPOs, SHPs, and farmers' interest groups to achieve the size of the group. In addition to these integration models, stakeholders such as government, CSOs, and the private sector also use key mechanisms to expand their interventions, namely Basic Collection Centers, Contract Agreements, and Digital Platforms.

17.5.3 GENE EDITING

The development of genetic engineering and other related technologies has created a buzz in the scientific community, looking for biological solutions to prevent post-harvest fruit loss (Gordon et al., 2021; Brinegar et al., 2017). Since the beginning of genome editing, plant genomes have been overused but the development of a genetic engineering tool, such as Crispr-Cas9, has changed the whole process. Scientific research has successfully calculated the ability to express important traits in many plant species and CRISPR has reportedly reduced the burden of sterility, inconsistency, high heterozygosity, and low frequency recurrence of desired alleles and features and long life cycles that extend or stop normal reproduction (Wang et al., 2012). The CRISPR formulation or RNA synthesis pathway to the nearest proto-spacer motif (PAM) is of interest sequence in which the sequence of the Cas protein is inserted into the cell and processed using a gene expression system (Mekler et al., 2020).

Ethylene synthesis, regulation, and perception lead to the recording of mature genetically regulated genes in climacteric fruits (Payasi and Sanwal, 2010). In silencing the expression of such genes, this expression is strongly suppressed as seen in many examples of melon papaya and kiwi fruit where ethylene production is significantly reduced and shelf life is increased. The quality and health of the fruit shelf increases the cycle that covers the fruit cuticle with triterpenoids and waxes which are a factor in the flowering of grums and blueberries (Chu et al., 2018). The stages involved in the fruit development cycle include cell wall degradation and fruit softening, which is often accompanied by overgrowth and senescence leading to fruit rot (Brummell et al., 1999; Brummell, 2006). The genetic modification of plant endo native enzymes that direct parts of the fungal cell wall and interactions can improve resistance to fungal infections and protect the fruit from loss in this process. Starch and sugar are the main carbohydrates and their conversion can affect the quality of the fruit after harvest. Cracked starch in sugar is not good for potatoes but during ripening, it is important for several varieties, such as apples, bananas, and kiwi fruit. Sweetness is one of the essential qualities of a fruit that is determined by the sugar content of the fruit and the texture of the fruit. Using genetic engineering technology can reduce the loss of tissue sugar that occurs as a result of post-harvest management and ensure future consumer satisfaction. Improving the flavor profiles of the fruit is another gray area surrounding the development of post-harvest losses. The sugar content of korma citrate and volatiles may be used in genetic engineering to enhance the fruit of the appropriate flavor through the novel system of genetic alleles. Finally, reduced brownness due to polyphenol oxidase, peroxidase, and phenylalanine ammonia-lyase has paved the way for reducing consumer waste production of the brown product (Nguyen, 2003; Martínez-Téllez and Lafuente, 1997).

Numerous efforts have been made to prevent post-harvest losses caused by cold damage and the discovery of the new gene SlGRAS4, which is highly depleted in transgender tomatoes by RNAi, promotes cold-tolerant fruit (Liu et al., 2020). Although the future of plant biotechnology has the answers to preventing fruit and vegetable losses, there are many barriers to genetic engineering and plant biotechnology for the following reasons: first, many horticultural plants have been studied traditionally. According to its life cycle, genomic formation is due to its being in a favorable condition for use in active genomics. Second, there is still the need to identify the genes that play a major role in this process. Thus, genetic deception cannot solve the problem of such losses and wastes but the whole issue has many branches that require disciplinary progress with integrated biological engineering and social and cultural solutions.

17.6 IoT AND DIGITAL TWINS—PREDICTIVE FUTURE

Internet of Things (IoT) and Digital Twins are the future of a shelf-shaped life and reduced food loss (Onwude et al., 2020a). Today, IoT has gained momentum in the scientific community in the fields of management, e-commerce, manufacturing, education, medicine, and health care. All credits go to big data banking and internet access. There is no doubt how IoT allows people to connect with objects at any time and from anywhere in the intelligent environment using smart links and links within social and economic contexts. Internal and external interactions enable a sensory system, designed with embedded technology formats that include RFID, printed sensors, web service, machine-to-machine communication, WSN, multi sensors, cloud, and blockchain (Onwude et al., 2020b). There is established evidence of the use of IoT in the field of agricultural production such as organic farming, open field agriculture, and livestock applications. More recently, it has been used as an intelligent packaging technology that uses biosensors and gas sensors to measure time temperatures, hygrothermal burns, and chemical changes in fruit. In addition, this can also be used to inform participants in the supply chain. The growing interest in its use in the food industry, especially product tracking, traceability, and environmental monitoring (e.g., temperature, humidity), weight loss, and overall quality loss in the post-harvest supply chain has opened up new ones, gateways to cold processes, real-time data collection of new computational methods such as artificial intelligence and data analysis (Alfian et al., 2017). This in return helps participants in supply chain management to develop cold chain technology to reduce quality loss and make informed decisions about food safety. However, the use of IoT in controlling cold chain technologies to reduce food loss in the fruit and vegetable supply chain is not enough. Another exciting application is in the fruit delivery sector which uses multiple sensors to control and monitor temperature, humidity, light exposure, and global positioning system (GPS) for tracking (Salam, 2020). The sensors based on the various materials are placed in fruit containers to monitor changes in cold natural environments and are connected to a wireless network and computers to communicate with control channels, manufacturers, or other stakeholders in the supply chain. Collected data may serve as input data for analyzing changes in dietary attributes (e.g., weight loss, shelf health, nutrition, or sensory properties), using a model based on mechanistic physics or digital twins. The new era of AI has revolutionized the agricultural landscape to look for new solutions other than traditional ones.

Digital twins are another way to link mathematical models with computer simulations based on a visual model that contains all the geometric features and physical properties of a product life cycle (Dahmen and Rossmann, 2018). While various methods are provided by their nature, however, only digital-based digital twins can test processes that cause quality loss in a new product that incorporates standardized sensory data (e.g., temperature of the surrounding air). The future of Digital Twins has diagnostic and predictive problems that are likely to be present in a supply chain that will reduce fruit loss (Moshood et al., 2021). These problems can be caused by physiological (e.g., cold injuries), hygrothermal (e.g., severe loss), biotic (e.g., phytosanitary insects, bacteria), and equipment (e.g., puncture injury, bruising) effects. Output data collected from the analysis can be used as an effective preventive measure to minimize quality loss.

17.7 EXISTING CHALLENGES

ITC Trade maps India exports fruits worth US$784.7 million and an untapped export potential worth US$1.3 billion. A major challenge in compensating for fruit loss is the declining results of meeting quality in international standards in overseas markets such as the United States and the European Union. There are EU reports blocking the importation of Indian mangoes from May 2014 to December 2015 due to overcrowding (Mishra, 2014). This creates a need for specialized equipment and monitoring of such issues in order to stabilize international market standards, thus preventing significant fruit losses. Recently, Australia granted market access to the Indian pomegranate and also authorized the surveillance of an additional export center for Indian mangoes. This is clear evidence that India will still fully evaluate its export capacity. In addition, there is intense competition in the international market for additional rotting products, such as the huge demand for a variety of strong grape colors. Weak processing infrastructure, poor on-farm production, unskilled labor, poor farming and management, and a lack of stable communication of fruit production and fruit distribution chain are major barriers to post-harvest harvesting loss.

17.8 CONCLUSION

The key to sustaining tough competition in the international market and curb losses is market intelligence which works on identifying the products in demand and India's potential in fulfilling it. Farmer aggregation models and farm-to-fork traceability are promising approaches that guarantee trust and credibility in the production and distribution process. A homogenized system of global food safety standards to ensure minimal rejection of fruit production in the international market would ensure a reduced loss. Imperatively educating farmers about modern cultivation and management methods to save consignment wastage are also a few approaches to deal with problems. Lastly, with the growing trend in the field of artificial intelligence and IoT to answer agricultural concerns, there is an equal need for smart workforce with upgraded farming technology to address challenges like paucity of land and water.

ACKNOWLEDGMENT

The authors are thankful to Amity University Uttar Pradesh (AUUP) and IIT Jodhpur, Rajasthan for providing necessary platform, infrastructure, and support rendered during the formulation of the chapter.

KEYWORDS

- **national horticulture database**
- **post-harvest loss**
- **indigenous fruits**
- **shelf life**
- **mechanical injury**
- **cultivation, pesticides**
- **supply chain management**
- **plant biotechnology**
- **Indian economy**
- **global market.**

REFERENCES

2nd Advance Estimates of 2019–20 of horticulture crops. GovIn n.d.a https://pib.gov.in/PressReleasePage.aspx?PRID=1628695 (accessed Aug 30, 2021).

Acta Horticulturae. ActahortOrg n.d. [Online]. https://www.actahort.org/members/showpdf?booknrarnr=49_7 (accessed Aug 30, 2021).

Afzal, I.; Kamran, M.; Ahmed Basra, S. M.; Ullah Khan, S. H.; Mahmood, A.; Farooq, M.; et al. Harvesting and Post-harvest Management Approaches for Preserving Cottonseed Quality. *Ind. Crops. Prod.* **2020,** *155,* 112842.

Alfian, G.; Rhee, J.; Ahn, H.; Lee, J.; Farooq, U.; Ijaz, M. F., et al. Integration of RFID, Wireless Sensor Networks, and Data Mining in an e-pedigree Food Traceability System. *J. Food Eng.* **2017,** *212,* 65–75.

Allafi, A. B. Microbiology of Fresh Oranges After Storage at Room and Refrigeration Temperature Conditions. *Coimbatore* **2011,** *3,* 1–3.

Ameen, M. Andhra Pradesh Plans Big Push to Horticulture [Only]. The Hans India 2021. https://www.thehansindia.com/andhra-pradesh/andhra-pradesh-plans-big-push-to-horticulture-crops-in-2021-664889 (accessed Aug 30, 2021).

Arun Kumar, R.; Rajpurohit, V. S.; Kautish, S. A Study on Technology-LED Solutions for Fruit Grading to Address Post-harvest Handling Issues of Horticultural Crops. In *Modern Techniques for Agricultural Disease Management and Crop Yield Prediction*; IGI Global: Hershey, PA, 2020; pp 203–221.

At, A. S. A.; Pessu, P. O.; Agoda, S.; Isong, I. U.; Ikotun, I. The Concepts and Problems of Postharvest Food Losses in Perishable Crops. *Afr. J. Food Sci.* **2011,** *5,* 603–613.

Birthare, S. Scope, Status and Importance of Fruits and Vegetables Industries in INDIA & Role of Fruits and Vegetables in Indian Economy [Online], 2018. https://doi.org/10.13140/RG.2.2.10140.39044.

Bourne, M. *Post Harvest Food Losses – the Neglected Dimension in Increasing the World Food Supply*, 1977.

Brinegar, K.; Yetisen, A.; Choi, S.; Vallillo, E.; Ruiz-Esparza, G. U.; Prabhakar, A. M.; et al. The Commercialization of Genome-editing Technologies. *Crit. Rev. Biotechnol.* **2017,** *37,* 924–932.

Brummell, D. A. Cell Wall Disassembly in Ripening Fruit. *Funct. Plant Biol.* **2006,** *33,* 103–119.

Brummell, D. A.; Harpster, M. H.; Civello, P. M.; Palys, J. M.; Bennett, A. B.; Dunsmuir, P. Modification of Expansin Protein Abundance in Tomato Fruit Alters Softening and Cell Wall Polymer Metabolism During Ripening. *Plant Cell* **1999,** *11,* 2203–2216.

CAB Direct. CabdirectOrg [Online]. n.d. https://www.cabdirect.org/cabdirect/abstract/20103304675 (accessed Aug 30, 2021).

Cardoen, D.; Joshi, P.; Diels, L.; Sarma, P. M.; Pant, D. Agriculture Biomass in India: Part 2. Post-harvest Losses, Cost and Environmental Impacts. *Resour. Conserv. Recycl.* **2015,** *101,* 143–153.

Chengappa, P. G. Development of Agriculture Value Chains as a Strategy for Enhancing Farmers' Income. *Agric. Econ. Res. Rev.* **2018,** *31,* 1.

Cherono, K.; Sibomana, M.; Workneh, T. S. Effect of Infield Handling Conditions and Time to Pre-cooling on the Shelf-life and Quality of Tomatoes. *Braz. J. Food Technol.* **2018,** *21.* https://doi.org/10.1590/1981-6723.01617

Chu, W.; Gao, H.; Chen, H.; Wu, W.; Fang X. Changes in Cuticular Wax Composition of Two Blueberry Cultivars During Fruit Ripening and Postharvest Cold Storage. *J. Agric. Food Chem.* **2018,** *66,* 2870–2876.

Crop Diversification in India - C.R. Hazra [Online]. FaoOrg n.d. http://www.fao.org/3/x6906e/x6906e06.htm (accessed Aug 30, 2021).

Dahmen, U.; Rossmann, J. *Experimentable Digital Twins for a Modeling and Simulation-based Engineering Approach,* In 2018 IEEE International Systems Engineering Symposium (ISSE); IEEE; 2018, pp 1–8.

Davara, P. R.; Patel, N. C. Assessment of Post Harvest Losses in Banana Grown in Gujarat. *J. Horticult. Sci.* **2016,** *4,* 187–190.

Dethier, J.-J.; Effenberger, A. *Agriculture and Development: A Brief Review of the Literature*; The World Bank, 2011.

El-Ramady, H. R.; Domokos-Szabolcsy, É.; Abdalla, N. A.; Taha, H. S.; Fári, M. *Postharvest Management of Fruits and Vegetables Storage. Sustainable Agriculture Reviews*; Springer International Publishing: Cham, 2015; pp 65–152.

Estimated fruit production volume in India FY 2020, by leading state [Online]. StatistaCom n.d. https://www.statista.com/statistics/1036880/india-leading-fruit-producing-state/) (accessed Aug 30, 2021).

Fajola, A. O. The Post-harvest Fruit Rots of Tomato (Lycopersicum esculentum) in Nigeria. *Nahrung* **1979,** *23,* 105–109.

Gajanana, T. M.; Sreenivasa Murthy, D.; Saxena, A. K.; Sudhakar Rao, D. V.; Sudha, M.; Dakshinamoorthy, V. Economic Analysis of Post-Harvest Loss and Marketing Efficiency in Guava (cv. Allahabad safeda) in Karnataka. *J. Horticult. Sci.* **2015,** *10,* 70–73.

Gandhi, V. *Fighting Hunger is India's Greatest Challenge*; Hindustan Times 2018.

Gardas, B. B.; Raut, R. D.; Narkhede, B. Evaluating Critical Causal Factors for Post-Harvest Losses (PHL) in the Fruit and Vegetables Supply Chain in India Using the DEMATEL Approach. *J. Clean Prod.* **2018,** *199,* 47–61.

Gardner, C. J.; Rocliffe, S.; Gough, C.; Levrel, A.; Singleton, R. L.; Vincke, X.; et al. *Value Chain Challenges in Two Community-managed Fisheries in Western Madagascar: Insights for the Small-scale Fisheries Guidelines. MARE Publication Series*; Springer International Publishing: Cham, 2017; pp 335–354.

Subedi, G. D.; Atreya, P. N.; Gurung, C. R.; Giri, R. K.; Gurung, Y. R. In*High Density Cultivation of Major Fruit Crops: Opportunities and Challenges in Nepal*, Proceedings of the 11th National Conference on Horticulture Seminar; Kirtipur, Kathmandu, Nepal, 2020.

Goedhals-Gerber, L. L.; Khumalo, G. Identifying Temperature Breaks in the Export Cold Chain of Navel Oranges: A Western Cape Case. *Food Control* **2020,** *110,* 107013.

Golob, P. *Crop Post-harvest: Science and Technology, Volume 1: Principles and Practice*; Blackwell Science: Philadelphia, PA, 2002.

Gordon, D. R.; Jaffe, G.; Doane, M.; Glaser, A.; Gremillion, T. M.; Ho, M. D. Responsible Governance of Gene Editing in Agriculture and the Environment. *Nat. Biotechnol.* **2021,** 1–3.

GovIn n.d.b [Online] https://midh.gov.in/VCS%20Reports/Uttarakhand%20Value%20Chain%20Analysis%20and%20Market%20Assessment%20Report%20on%20Apple.pdf (accessed Aug 30, 2021).

GovIn. n.d.a [Online] http://ncfc.gov.in/po/chaman/srsc/2018-19/mp.pdf (accessed Aug 30, 2021).

GovIn. n.d.b [Online] https://www.midh.gov.in/VCS%20Reports/7-Value%20chain%20 for%20Grapes%20crop%20in%20Nasik%20district%20of%20Maharashtra.pdf (accessed Aug 30, 2021).

GovIn. n.d.c [Online] https://darpg.gov.in/sites/default/files/enam_0.pdf (accessed Aug 30, 2021). GovIn. n.d.c [Online] https://midh.gov.in/VCS%20Reports/Value%20Chain%20 Study%20on%20Kinnow%20in%20Punjab.pdf (accessed Aug 30, 2021).

Hoaes, I. E. *Does Post-settlement Support Play a Role in the Success of Land Reform?: The Case of Resettlement Beneficiaries in Namibia*; University of Stellenbosch: Stellenbosch, 2010.

Home. GovIn n.d. [Online] https://midh.gov.in/FAQ.html (accessed Aug 30, 2021).

Horticulture 4.0: Time to play to India's strengths – India Business & Trade, an initiative of Trade Promotion Council of India [Online], TpciIn 2020. https://www.tpci.in/indiabusinesstrade/blogs/horticulture-4-0-time-to-play-to-indias-strengths/ (accessed Aug 30, 2021).

Horticulture Sector in India [Online]. DrishtiiasCom n.d. https://www.drishtiias.com/to-the-points/paper3/horticulture-sector-in-india (accessed Aug 30, 2021).

Dac Gov I. Horticulture Area Production Information System. GovIn n.d. https://aps.dac.gov.in/HIS_UserManual.pdf (accessed Aug 30, 2021).

India suffers serious level of hunger, ranks 103 on global hunger index 2018. BusinessinsiderIn 2018. https://www.businessinsider.in/india-suffers-serious-level-of-hunger-ranks-103-on-global-hunger-index-2018/articleshow/66235076.cms (accessed Aug 30, 2021).

IntellecapCom. n.d. http://intellecap.com/wp-content/themes/intellecap/pdf/Public-Facing-Report.pdf (accessed Aug 30, 2021).

Ismail, M.; Zhang, J. Post-harvest Citrus Diseases and Their Control. *Outlooks Pest Manag.* **2004,** *15,* 29–35.

Jha, S. K. Indian Institute of Management Bangalore, Pinsonneault A, Dubé L, McGill University, McGill University. The Evolution of an ICT Platform-enabled Ecosystem for Poverty Alleviation: The Case of eKutir. *MIS Q.* **2016,** *40,* 431–445.

Jha, S. *World Bank Report: India's Share in Global Horticulture*; Hindustan Times, 2007.

Ladaniya, M. S.; Wanjari, V.; Mahalle, B. Marketing of Grapes and Raisins and Post-Harvest Losses of Fresh Grapes in Maharashtra. *Indian J. Agric. Res.* **2005,** *39,* 167–176.

Liu, Y.; Shi, Y.; Zhu, N.; Zhong, S.; Bouzayen, M.; Li, Z. SlGRAS4 Mediates a Novel Regulatory Pathway Promoting Chilling Tolerance in Tomato. *Plant Biotechnol. J.* **2020,** *18,* 1620–1633.

Martínez-Téllez, M. A.; Lafuente, M. T. Effect of High Temperature Conditioning on Ethylene, Phenylalanine Ammonia-lyase, Peroxidase and Polyphenol Oxidase Activities in Flavedo of Chilled ‹Fortune› Mandarin Fruit. *J. Plant Physiol.* **1997,** *150,* 674–678.

Mekler, V.; Kuznedelov, K.; Severinov, K. Quantification of the Affinities of CRISPR-Cas9 Nucleases for Cognate Protospacer Adjacent Motif (PAM) Sequences. *J. Biol. Chem.* **2020,** *295,* 6509–6517.

Mishra, A. R.; PTI. EU Bans Import of Alphonso Mangoes, Vegetables From India. Mint 2014. https://www.livemint.com/Politics/dhIDf4UArH2czVOf3Y7EVJ/EU-bans-Indian-Alphonso-mangoes-vegetables-from-1-May.html (accessed Aug 30, 2021).

Moshood, T. D.; Nawanir, G.; Sorooshian, S.; Okfalisa, O. Digital Twins Driven Supply Chain Visibility Within Logistics: A New Paradigm for Future Logistics. *Appl. Syst. Innov.* **2021,** *4,* 29.

MountaininitiativeIn [Online] n.d. https://www.mountaininitiative.in/images/IMI_FAO_Report_on_SOMA.pdf (accessed Aug 30, 2021).

Nabi, S. U.; Raja, W. H.; Kumawat, K. L.; Mir, J. I.; Sharma, O. C.; Singh, D. B.; et al. Post Harvest Diseases of Temperate Fruits and Their Management Strategies-A Review. *Int. J. Pure Appl. Biosci.* **2017,** *5,* 885–98.

Nath, A.; Meena, L. R.; Kumar, V.; Panwar, A. S. Postharvest Management of Horticultural Crops for Doubling Farmer's Income; 2018.

Negi, S.; Anand, N. *Wastage and Cold Chain Infrastructure Relationship in Indian Food Supply Chain: A Study From Farm to Retail. Supply Chain Management Strategies and Risk Assessment in Retail Environments*; Hershey, PA: IGI Global, 2018, pp 247–266.

Uzen, N.; Cetin, O. In *Micro Irrigation for Modern Agriculture*, Proceedings of the 1st Central Asia Congress on Modern Agricultural Techniques and Plant Nutrition, Oct 01–03, 2013; Bishkek, Kyrgyzstan; 2131–2138.

Nguyen, T. Relationship Between Browning and the Activities of Polyphenoloxidase and Phenylalanine Ammonia Lyase in Banana Peel During Low Temperature Storage. *Postharvest. Biol. Technol.* **2003,** *30,* 187–193.

NicIn [Online] n.d. https://agricoop.nic.in/sites/default/files/Service%204%20SRS%201.1%20NEW.pdf (accessed Aug 30, 2021).

Onwude, D. I.; Chen, G.; Eke-emezie, N.; Kabutey, A.; Khaled, A. Y.; Sturm, B. Recent Advances in Reducing Food Losses in the Supply Chain of Fresh Agricultural Produce. *Processes (Basel)* **2020a,** *8,* 1431.

Onwude, D. I.; Chen, G.; Eke-emezie, N.; Kabutey, A.; Khaled, A. Y.; Sturm, B. Recent Advances in Reducing Food Losses in the Supply Chain of Fresh Agricultural Produce. *Processes (Basel)* **2020b,** *8,* 1431.

Operational Guidelines. Mission for integrated development of horticulture [Online]. GovIn n.d. http://midh.gov.in/PDF/MIDH_Guidelines (final).pdf (accessed Aug 30, 2021).

Pal, G.; Dipak, K. S., Indusnettechnologies, S. W. D. About HMNEH: About HMNEH [Online]. GovIn. n.d. https://shm.uk.gov.in/pages/display/149-about-hmneh (accessed Aug 30, 2021).

Parra-Coronado, A.; Fischer, G.; Camacho-Tamayo, J. H. Post-harvest Quality Model of Pineapple Guava Fruit According to Storage and Weather Conditions of Cultivation. *Rev. Bras Eng. Agric. Ambient/Braz J. Agric. Environ. Eng.* **2017,** *21,* 634–639.

Payasi, A.; Sanwal, G. G. Ripening of Climacteric Fruits and Their Control: Control of Ripening in Fruits. *J. Food Biochem.* **2010,** *34,* 679–710.

Press Information Bureau [Online]. GovIn n.d. https://pib.gov.in/PressReleseDetailm.aspx?PRID=1741940 (accessed Aug 30, 2021).

Press Trust of India. India wastes 59 MT fruits, vegetables every year [Online]. Business-StandardCom 2010. https://www.business-standard.com/article/markets/india-wastes-59-mt-fruits-vegetables-every-year-110122100215_1.html (accessed Aug 30, 2021).

PTI. Kinnow Growers in Punjab see 50% drop in income. Hindu Bus Line; 2015.

Raj, A.; Jhariya, M. K.; Yadav, D. K.; Banerjee, A. Agroforestry with Horticulture: A New Strategy Toward a Climate-resilient Forestry Approach. In *Agroforestry and Climate Change, Includes Bibliographical References and Index*; Apple Academic Press, 2019; pp 67–96.

ResIn [Online]. n.d. https://iihr.res.in/sites/default/files/post%20harvestfinalcolour_0.pdf (accessed Aug 30, 2021).

Sab, M.; Ashok, M. B.; Sudhakara, S. N. Estimation of Post-harvest Losses of Mangoes at Different Stages From Harvesting to Consumption. *Int. J. Curr. Microbiol. Appl. Sci.* **2017,** *6,* 310–318.

Salam, A. Internet of Things for Environmental Sustainability and Climate Change. In *Internet of Things*; Cham: Springer International Publishing; 2020, pp 33–69.

Saurav, S.; Potti, R. Cold Chain Logistics in India: A Study. In *Innovative Solutions for Implementing Global Supply Chains in Emerging Markets*; Hershey, PA: IGI Global, 2016, pp 159–172.

SfacindiaCom [Online]. n.d. http://sfacindia.com/PDFs/SFAC_Value-Chain-Analysis.pdf (accessed Aug 30, 2021).

Sharma, S. D. Modinomics in India: The Promise and the Reality. *Asian Surv.* **2019,** *59,* 548–572.

Jha, S. N.; Vishwakarma, R. K.; Ahmed, T.; et al. Assessment of Quantitative Harvest and Post-Harvest Losses of Major Crops/Commodities in India. 2016.

Singh, A. K.; Pinto, A. R.; Singh, P.; De, A. Transforming Rural Farm Livelihoods; 2020. https://doi.org/10.1596/34726.

Singh, K. K.; Reddy, B. S. Post-harvest Physico-mechanical Properties of Orange Peel and Fruit. *J. Food Eng.* **2006,** *73,* 112–120.

TFNet News Compilation. PAPAYA – Pest & Disease Management [Online]. ItfnetOrg 2016. https://www.itfnet.org/v1/2016/05/papaya-pest-disease-management/ (accessed Aug 30, 2021).

Venkatesh. Agro-Climatic zones of India [Online]. AgrivergeIn 2019. https://agriverge.in/agro-climatic-zones-of-india/ (accessed Aug 30, 2021).

Wang, T. L.; Uauy, C.; Robson, F.; Till, B. Tilling in Extremis. *Plant Biotechnol. J.* **2012,** *10,* 761–772.

Welfare, F. Horticultural Statistics at a Glance 2018 [Online]. NicIn n.d. https://agricoop.nic.in/sites/default/files/Horticulture%20Statistics%20at%20a%20Glance-2018.pdf (accessed Aug 30, 2021).

Welfare, F. Horticultural Statistics at a Glance 2018 [Online]. NicIn n.d. https://agricoop.nic.in/sites/default/files/Horticulture%20Statistics%20at%20a%20Glance-2018.pdf (accessed Aug 30, 2021).

Wilson, C. L.; Wisniewski, M. E.; Biles, C. L.; McLaughlin, R.; Chalutz, E.; Droby, S. Biological Control of Post-harvest Diseases of Fruits and Vegetables: Alternatives to Synthetic Fungicides. *Crop. Prot.* **1991,** *10,* 172–177.

WtoOrg. n.d. [Online]. https://www.wto.org/english/res_e/statis_e/wts2019_e/wts2019_e.pdf (accessed Aug 30, 2021).

Yogi, V.; Kumar, P.; Kumar, M. किन्नू फल उत्पादन का अर्थशास्त्र – Krishisewa [Online]. KrishisewaCom n.d. https://www.krishisewa.com/miscellaneous-articles/1025-economics-of-kinnow-fruit-production.html (accessed Aug 30, 2021).

Index

C

D

E

F

K

L

M

N

O

P

R

S

T

U

V

W

Y

Z